151305867 3

WITHDRAWN

AF616346

Small DNA Tumour Viruses

Edited by

Kevin Gaston

School of Biochemistry
University of Bristol
Bristol
UK

Caister Academic Press

Caister Academic Press
Norfolk, UK

www.caister.com

British Library Cataloguing-in-Publication Data
A catalogue record for this book is available from the British Library

ISBN: 978-1-904455-99-8

Description or mention of instrumentation, software, or other products in this book does not imply endorsement by the author or publisher. The author and publisher do not assume responsibility for the validity of any products or procedures mentioned or described in this book or for the consequences of their use.

Cover design adapted from Figure 1.3

Printed and bound in Great Britain

Contents

Contributors v

Preface ix

1 Human Papillomavirus Infection and its Association with Neoplasia: From Molecular Biology to Prevention and Treatment 1
Richard Oparka and C. Simon Herrington

2 The Art and Science of Obtaining Virion Stocks for Experimental Human Papillomavirus Infections 19
Michelle A. Ozbun and Michael P. Kivitz

3 The Regulation of Human Papillomavirus Gene Expression by the E2 Protein: Keeping a Finger in Every Pie 37
Sheila V. Graham and Kevin Gaston

4 HPV E5: An Enigmatic Oncoprotein 55
Laura F. Wetherill, Rebecca Ross and Andrew Macdonald

5 E6 Oncoproteins: Structure and Associations 71
Scott B. Vande Pol

6 Biochemical and Structure–Function Analyses of the HPV E7 Oncoprotein 99
Leonardo G. Alonso, Lucía B. Chemes, María L. Cerutti, Karina I. Dantur and Gonzalo de Prat-Gay

7 Replication and Maintenance of Viral Genomes by Association with Host Chromatin 125
Koenraad Van Doorslaer, Vandana Sekhar, Jameela Khan and Alison A. McBride

8 Alterations in Cellular miRNAs Induced by Human Papillomaviruses 151
Amy S. Gardiner, Abigail I. Wald and Saleem A. Khan

9 Viral Deregulation of DNA Damage Responses 175
Sergei Boichuk and Ole Gjoerup

10 Structural 'Snap-shots' of the Initiation of SV40 Replication 195
Gretchen Meinke and Peter A. Bullock

11 Human Papillomavirus DNA Replication: Insights into the Structure and Regulation of a Eukaryotic DNA Replisome 217
Claudia M. D'Abramo, Amélie Fradet-Turcotte and Jacques Archambault

12 Induction of Genomic Instability by Human Papillomavirus Oncoproteins 239
Karl Münger and Stefan Duensing

13 Targeting of Promyelocytic Leukaemia Proteins and Promyelocytic Leukaemia Nuclear Bodies by DNA Tumour Viruses 255
Keith N. Leppard and Jordan Wright

14 Adenoviruses and Gene Therapy: The Role of the Immune System 281
Laura White and G. Eric Blair

Index 321

Contributors

Leonardo G. Alonso
XBio Inc.
Buenos Aires
Argentina

lalonso@leloir.org.ar

Jacques Archambault
Molecular Virology Laboratory
Institut de Recherches Cliniques de Montréal (IRCM) and Department of Biochemistry
Université de Montréal
Montreal, QC
Canada

jacques.archambault@ircm.qc.ca

G. Eric Blair
Institute of Molecular and Cellular Biology
University of Leeds
Leeds
UK

g.e.blair@leeds.ac.uk

Sergei Boichuk
University of Pittsburgh Cancer Institute
Pittsburgh, PA
USA

boichuksergei@mail.ru

Peter A. Bullock
Department of Biochemistry
Tufts University School of Medicine
Boston, MA
USA

peter.bullock@tufts.edu

María L. Cerutti
XBio Inc.
Buenos Aires
Argentina

mcerutti@leloir.org.ar

Lucía B. Chemes
Protein Structure, Function and Engineering Laboratory
Fundación Instituto Leloir and Instituto de Investigaciones Bioquímicas, CONICET
Buenos Aires
Argentina

lchemes@leloir.org.ar

Claudia M. D'Abramo
Molecular Virology Laboratory
Institut de Recherches Cliniques de Montréal (IRCM) and Department of Biochemistry
Université de Montréal
Montreal, QC
Canada

claudia.dabramo@ircm.qc.ca

Karina I. Dantur
Estación Experimental Agroindustrial Obispo Colombres
Sección Biotecnología
Las Talitas
Argentina

karinaines.dantur@gmail.com

Stefan Duensing
Section of Molecular Urooncology
Department of Urology
University of Heidelberg School of Medicine
Heidelberg
Germany;
Cancer Virology Program
University of Pittsburgh Cancer Institute
Pittsburgh, PA
USA;
Department of Microbiology and Molecular Genetics
University of Pittsburgh School of Medicine
Pittsburgh, PA
USA

duensing@pitt.edu

Amélie Fradet-Turcotte
Molecular Virology Laboratory
Institut de Recherches Cliniques de Montréal (IRCM) and Department of Biochemistry
Université de Montréal
Montreal, QC
Canada

amelie.fradet-turcotte@ircm.qc.ca

Amy S. Gardiner
Department of Microbiology and Molecular Genetics
University of Pittsburgh School of Medicine
Pittsburgh, PA
USA

agardiner@salud.unm.edu

Kevin Gaston
School of Biochemistry
University of Bristol
Bristol
UK

kevin.gaston@bristol.ac.uk

Ole Gjoerup
Tufts Medical Center/MORI
Boston, MA
USA

ogjoerup@tuftsmedicalcenter.org

Sheila V. Graham
MRC-University of Glasgow Centre for Virus Research
College of Medical, Veterinary and Life Sciences
University of Glasgow
Glasgow
UK

s.v.graham@bio.gla.ac.uk

C. Simon Herrington
Department of Pathology
Division of Medical Sciences
University of Dundee
Ninewells Hospital and Medical School
Dundee
UK

s.herrington@dundee.ac.uk

Jameela Khan
Laboratory of Viral Diseases
National Institute of Allergy and Infectious Diseases
National Institutes of Health
Bethesda, MD
USA

khanj2@mail.nih.gov

Saleem A. Khan
Department of Microbiology and Molecular Genetics
University of Pittsburgh School of Medicine
Pittsburgh, PA
USA

khan@pitt.edu

Michael P. Kivitz
Departments of Molecular Genetics & Microbiology and Obstetrics & Gynecology
The UNM Cancer Center
The UNM Interdisciplinary HPV Prevention Center
The University of New Mexico School of Medicine and Cancer Center
Albuquerque, NM
USA

mpkivitz@salud.unm.edu

Keith N. Leppard
School of Life Sciences
University of Warwick
Coventry
UK

keith.leppard@warwick.ac.uk

Alison A. McBride
Laboratory of Viral Diseases
National Institute of Allergy and Infectious Diseases
National Institutes of Health
Bethesda, MD
USA

amcbride@niaid.nih.gov

Andrew Macdonald
Institute of Molecular and Cellular Biology
Faculty of Biological Sciences
University of Leeds
Leeds
UK

a.macdonald@leeds.ac.uk

Gretchen Meinke
Department of Biochemistry
Tufts University School of Medicine
Boston, MA
USA

gretchen.meinke@tufts.edu

Karl Münger
The Channing Laboratory
Brigham & Women's Hospital
Harvard Medical School
Boston, MA
USA

kmunger@rics.bwh.harvard.edu

Richard Oparka
Department of Pathology
Ninewells Hospital and Medical Schools
Dundee
UK

roparka@nhs.net

Michelle A. Ozbun
Departments of Molecular Genetics & Microbiology and Obstetrics & Gynecology
The UNM Cancer Center
The UNM Interdisciplinary HPV Prevention Center
The University of New Mexico School of Medicine and Cancer Center
Albuquerque, NM
USA

mozbun@salud.unm.edu

Gonzalo de Prat-Gay
Protein Structure, Function and Engineering Laboratory
Fundación Instituto Leloir and Instituto de Investigaciones Bioquímicas, CONICET
Buenos Aires
Argentina

gpg@leloir.org.ar

Rebecca Ross
Institute of Molecular and Cellular Biology
Faculty of Biological Sciences
University of Leeds
Leeds
UK

r.l.ross@leeds.ac.uk

Vandana Sekhar
Laboratory of Viral Diseases
National Institute of Allergy and Infectious Diseases
National Institutes of Health
Bethesda, MD
USA

sekharv@mail.nih.gov

Koenraad Van Doorslaer
Laboratory of Viral Diseases
National Institute of Allergy and Infectious Diseases
National Institutes of Health
Bethesda, MD
USA

vandoorslaerkm@mail.nih.gov

Scott B. Vande Pol
Department of Pathology
University of Virginia
Charlottesville, VA
USA

vandepol@virginia.edu

Abigail I. Wald
Department of Microbiology and Molecular Genetics
University of Pittsburgh School of Medicine
Pittsburgh, PA
USA

aib7@pitt.edu

Laura F. Wetherill
Institute of Molecular and Cellular Biology
Faculty of Biological Sciences
University of Leeds
Leeds
UK

bslfw@leeds.ac.uk

Laura White
Institute of Molecular and Cellular Biology
University of Leeds
Leeds
UK

bslw@leeds.ac.uk

Jordan Wright
School of Life Sciences
University of Warwick
Coventry
UK

jordan.wright@warwick.ac.uk

Preface

The small DNA tumour viruses continue to provide new insights into many fundamental aspects of biology, including gene regulation, DNA replication, cell transformation, cell cycle control and tumorigenesis. The causal link between papillomaviruses and some human cancers is well known and has resulted in the development of effective vaccines that are coming into widespread use. A role for polyomavirus in human cancer has recently been established. Adenoviruses do not cause cancer in humans but, as well as providing excellent tools for the study of many host cell processes, these viruses have been exploited as delivery vehicles in gene therapy. A common feature of the small DNA tumour viruses is their heavy reliance on the host for survival and replication. Understanding the virus–host relationship is critical to understanding the tumorigenic process and how these viruses subvert the host's immune system. The aim of this book is to provide an overview of the molecular biology of these viruses and some of their interactions with the host. Many of the chapters focus on human papillomavirus, reflecting the current volume of research in this area. However, in many cases comparisons are made with other viruses. Other chapters go beyond the papillomaviruses and focus on cellular pathways that are targeted by many viruses. I hope that this collection is a useful reference for those currently working in this field and a stimulating introduction for those new to this area.

Kevin Gaston

1 Human Papillomavirus Infection and its Association with Neoplasia: From Molecular Biology to Prevention and Treatment

Richard Oparka and C. Simon Herrington

Abstract

The papillomaviruses are diverse, predominantly epitheliotropic, viruses that are ubiquitous throughout the world. Well over 100 different types are known to infect humans, affecting particularly the squamous epithelia of the anogenital region, the skin and the upper aerodigestive tract. The majority of infections remain subclinical and, in many cases, human papillomavirus (HPV) infection results in benign lesions such as warts that regress with elimination of the infection. However, infection with some HPV types, for example HPV16 and -18, can also lead to malignant transformation. This association with malignant transformation has led to the development of vaccines against HPV and, in some countries, implementation of a vaccination programme. This chapter highlights the natural history of HPV-associated disease as well as the effects of the virus at a molecular level, and addresses the future implications of our knowledge of HPV infection for both prevention and treatment of HPV-associated disease.

Introduction

The human papillomaviruses (HPVs) have been well conserved throughout evolution, are almost ubiquitous across mammalian species and are known to infect humans throughout the world. In most cases infection remains subclinical, but in certain situations persistent or repeated infection occurs resulting in a variety of pathological lesions. The viruses show a strong propensity for infecting squamous cells throughout the body and infection has been implicated most importantly in the development of cervical cancer. HPV infection is also associated with lesions [viral warts, intraepithelial neoplasia and squamous cell carcinoma (SCC)] of the upper aerodigestive tract, skin and other parts of the anogenital region in both men and women.

At present, more than 100 different types of the HPV have been identified (de Villiers *et al.*, 2004). However, the link between these viruses and their associated diseases was not clear until relatively recently. Hippocrates first described cervical SCC as early as the fourth century BC (Gasparini and Panatto, 2009), but an understanding of the aetiology of the disease remained elusive for a long time. It was not until the mid-nineteenth century that Rigoni-Stern, a surgeon practising in Padua, Italy, made the observation that the incidence of cervical cancer amongst nuns was almost zero (Rigoni-Stern, 1842). Over the next century or so this observation was consolidated and the presence of increased disease in sex workers was identified (Mak *et al.*, 2004). Other risk factors have since been well established. All relate to sexual activity and include early age at first coitus, multiple sexual partners, a partner with multiple sexual partners as well as a lack of use of barrier contraception methods. In the late 1970s, zur Hausen and colleagues began to postulate that HPVs may have a role in the development of benign condylomas and cervical SCCs (zur Hausen, 1976, 1977). In 1983, HPV16 was isolated from a cervical carcinoma and in 1985 the presence of transcriptionally active HPV DNA was confirmed in cervical cancer cells (Dürst *et al.*,

1983; Schwarz *et al.*, 1985). In 2008 zur Hausen was awarded the Nobel prize in Physiology or Medicine for his work.

HPVs have a strong propensity for infecting squamous cells. The alpha genus consists of approximately 30 types that have a tendency to infect the mucosa of the anogenital tract, aerodigestive tract as well as other mucosa within the head and neck region. Some viruses belonging to the alpha group are also known to produce variants of benign cutaneous warts (HPV2). Viruses of the beta genus preferentially infect the skin and have been implicated in the development of skin cancers whilst those of the gamma, mu and nu genera are more typically associated with benign cutaneous warts and papillomas (Ghittoni *et al.*, 2010). The remaining genera are not found in humans (Bernard *et al.*, 2010) (Table 1.1). Within the alpha genus it has emerged that not all viruses have the same propensity for causing cancer. Accordingly the viruses have been grouped into low-risk types, responsible for causing benign warts/condyloma and high-risk types responsible for malignant transformation. The low-risk group includes the two most common viruses, HPV6 and -11, as well as types 42, 44, 53, 54, 62 and 66. The most common high-risk types in the UK include HPV16 and -18. Less commonly encountered high-risk types include types 31, 33, 35, 39, 45, 51, 52, 56, 58, 59 and 68 (zur Hausen, 1996).

The pathology of HPV infection-associated neoplasia

HPV infection

HPVs are DNA viruses of the *Papillomaviridae* family. The HPV genome is circular and relatively small, consisting of approximately 8000 bp (Hebner and Laimins, 2006) (Fig. 1.1). The production of virions occurs in tandem with the maturation of keratinocytes through the thickness of the squamous epithelium. The viral DNA contains eight open reading frames (ORFs). These encode proteins that perform a wide range of functions that promote viral replication and survival. Seven of the genes and proteins (E1, E2, E4, E5, E6, E7) are given the prefix 'E' as they are preferentially expressed early in the viral life cycle.

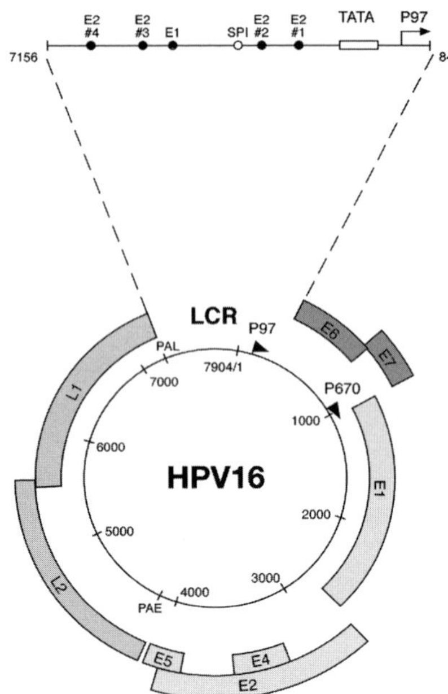

Figure 1.1 Pictorial representation of the human papillomavirus genome. Reproduced with permission from Doorbar (2006).

The remaining two ORFs and the capsid proteins they encode are designated late ('L') genes/proteins and are preferentially expressed as the virus reaches the epithelial surface. The products of the HPV early genes mediate specific functions that control viral replication and in the case of the oncogenic viruses, cellular transformation. The E1 protein is involved in virus replication and genome maintenance. It is a nuclear phosphoprotein that binds specifically to the virus origin of replication, unwinding the DNA template and acting as a helicase (Wilson *et al.*, 2002). The E2 protein is a transcriptional regulator and is also involved, with E1, in viral DNA replication (Longworth and Laimins, 2004). Moreover, E2 acts to trans-repress the expression of E6 and E7. The E4 protein, which is formed as a fusion protein incorporating part of the E1 protein (E1^E4), is the most abundant viral gene product, constituting approximately 90% of viral transcripts in condylomata (Fang *et al.*, 2006). The role of this fusion protein is complex and it appears to exert several effects on the cell cycle. The function of the E5 protein has remained unclear but its interaction with the vacuolar ATPase may be pathogenetically

Table 1.1 Papillomavirus genera and species. Modified from Bernard *et al.* (2010)

Genus	Species (common use)	Species (ICTV)	Disease
Alphapapillomavirus	Alpha 1	HPV32	Associated with mucosal and cutaneous lesions. Subdivided into high- and low-risk types according to propensity for malignant transformation
	Alpha 2	HPV10	
	Alpha 3	HPV61	
	Alpha 4	HPV2	
	Alpha 5	HPV26	
	Alpha 6	HPV53	
	Alpha 7	HPV18	
	Alpha 8	HPV7	
	Alpha 9	HPV16	
	Alpha 10	HPV6	
	Alpha 11	HPV34	
	Alpha 12	*Macca mulata Papillomavirus 1*	
	Alpha 13	HPV54	
	Alpha 14	HPV90	
Betapapillomavirus	Beta 1	HPV5	Associated with benign and malignant cutaneous lesions particularly when associated with *Epidermodysplasia verruciformis*
	Beta 2	HPV9	
	Beta 3	HPV49	
	Beta 4	HPV92	
	Beta 5	HPV96	
	Beta 6	*Macca fascicularis Papillomavirus 2*	
Gammapapillomavirus	Gamma 1	HPV4	Cutaneous lesions (typically benign)
	Gamma 2	HPV48	
	Gamma 3	HPV50	
	Gamma 4	HPV60	
	Gamma 5	HPV88	
	Gamma 6	HPV101	
	Gamma 7	HPV109	
	Gamma 8	HPV112	
	Gamma 9	HPV116	
	Gamma 10	HPV121	
Mupapillomavirus	Mu 1	HPV1	Cutaneous lesions (typically benign)
	Mu 2	HPV63	
Nupapillomavirus	Nu 1	HPV43	Cutaneous lesions (typically benign)

ICTV, International Committee on Taxonomy of Viruses

important (DiMaio and Mattoon, 2001). There is also emerging evidence that the E5 protein down-regulates expression of surface HLA-A2, thereby aiding the evasion of $CD8^+$ T cells (Campo *et al.*, 2010). The E6 and E7 genes encode the main transforming proteins. These are capable of immortalization and neoplastic transformation under appropriate conditions. The L1 and L2 open reading frames (ORFs) encode the major and minor capsid proteins, respectively, that make up the icosahedral capsid coat of the virus (Zhou and Frazer, 1996). A non-coding region referred to as the upstream regulatory region (URR), or long control region (LCR), containing glucocorticoid response elements and the origin of DNA replication is present between the early and late gene regions (Moodley *et al.*, 2003).

The viral life-cycle is dependent on infection of replicative basal cells within squamous epithelia (Chow *et al.*, 2009). Metaplastic epithelium between squamous and glandular epithelia (also known in the cervix as the transformation zone) is a particular target for viral infection as the replicating cell pool is not protected by a thick layer of non-dividing cells. Infection of these basal cells switches on the viral early proteins that begin the process of manipulating the host cell's replicative machinery in order to support vegetative viral production (Doorbar, 2006). The life cycle of stratified squamous epithelium is unique in that it permits the virus to mature and assemble in tandem with the migration of keratinocytes to the superficial layers of the epithelium where they are shed naturally, permitting release of mature virions (Fig. 1.2).

Both L1 and L2 are involved in the initial binding of the virus to basal keratinocytes, which occurs by a number of mechanisms depending on the specific viral type. However, the current consensus is that the majority of viral types bind via an interaction with heparan sulphate proteoglycans expressed on the keratinocyte cell surface (Schiller *et al.*, 2010). Other secondary receptors are, however, probably involved and these may include the integrin family of proteins (McMillan *et al.*, 1999). The virus then enters the cell through a slow process of endocytosis and undergoes disassembly in late endosomes and/or lysosomes. Disassembly is necessary for the release of viral DNA prior to its relocation to the host nucleus.

Once within the basal keratinocytes, the E2 protein facilitates the transport of viral DNA to the host nucleus and, once within the nucleus, promotes binding to host DNA. Only the early promoter (for example p97 in HPV16; see Fig. 1.1) is active in basal cells as it functions independently from differentiation and is active before viral production occurs (Conway and Meyers, 2009). E2, in concert with E1, binds to various host proteins including a subset of DNA polymerases and replication protein A (Loo and Melendy, 2004). In basal epithelial cells, the viral genome is maintained as a stable episome and low-level replication occurs in tandem with the host cell cycle during S phase. This process occurs relatively slowly in comparison to other viruses and transcription products can be detected in cells between 12 and 24 hours after infection (Day *et al.*, 2004). The E6 and E7 proteins are expressed at this stage and promote expansion of the infected

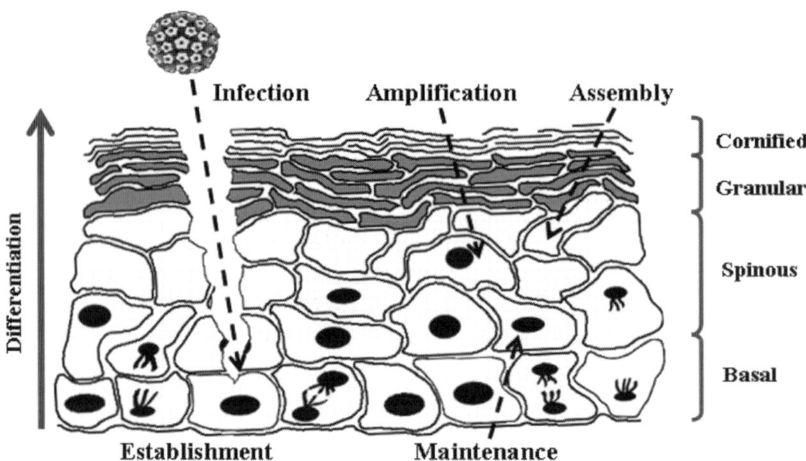

Figure 1.2 The relationship between the HPV life cycle and keratinocyte differentiation. Reproduced with permission from Conway and Meyers (2009).

cell population, allowing suprabasal cells to continue to express S phase-associated proteins by uncoupling cell cycle exit from squamous epithelial differentiation.

The differentiation-dependent promoter, which is contained within the E7 ORF in many HPV types (for example p670 in HPV16; see Fig. 1.1), is only activated within the suprabasal 'differentiated' cells. Activation of this promoter leads to an increase in the levels of E1, E2, E4 and E5, with consequent viral genome amplification; and to the expression of the L1 and L2 proteins. A fusion protein formed from E1 and E4, termed E1^E4, is thought to disrupt cellular scaffolding and permit the release of mature virions, completing the life cycle of the virus (Bryan and Brown, 2000). The L1 and L2 proteins form the viral capsid that is necessary for final assembly before the superficial cells are shed. The most superficial layers of the epithelium, the granular and cornified layers in keratinizing epithelium, are characterized by the production of keratin and the shedding of completely keratinized cells. HPV has therefore evolved to replicate in concert with the maturation of keratinocytes and does not need to initiate premature cell death in order to release mature virions.

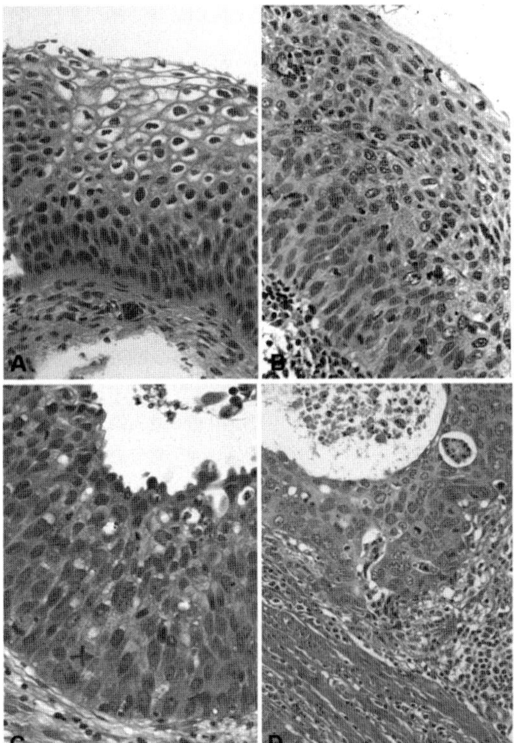

Figure 1.3 Representative histological images of (A) HPV infection; (B) CIN 2; (C) CIN 3; and (D) invasive squamous cell carcinoma. Note the presence of koilocytosis in (A). A tetrapolar mitosis is present in (C).

HPV and intraepithelial neoplasia

From a morphological point of view, productive HPV infection produces a characteristic cytopathic effect, termed koilocytosis, in which nuclear enlargement is accompanied by the formation of a perinuclear halo (Fig. 1.3A). This indicates vegetative viral replication and is seen typically in viral warts and low-grade intraepithelial lesions, in which the viral life cycle proceeds relatively normally. In a small subset of HPV infections, infection persists and leads to the development of pre-cancers and cancers i.e. intraepithelial (pre-invasive) lesions and invasive carcinomas. In these cases the viral DNA usually becomes integrated into the host genome and the normal processes that regulate E6 and E7 expression (discussed below) are lost. In such situations the suprabasal keratinocytes retain the capacity to undergo mitosis and relatively 'undifferentiated' cells expand to occupy the upper layers of the epithelium. These features are easily identified in histological sections and permit grading of lesions based on the extent to which these changes are present throughout the thickness of the epithelium. This is the grading system employed for, and exemplified by, cervical lesions (Fig. 1.3). For example, dysplasia limited to the basal third of the epithelium is designated cervical intraepithelial neoplasia (CIN) 1, and so on to include CIN 2 and CIN 3. In the alternative terminology used for these lesions, CIN 1 is encompassed in low-grade squamous intraepithelial lesions (SILs) and CIN 2 and CIN 3 are broadly equivalent to high-grade SIL. Similar terminology is used in other sites in which HPV infection produces similar lesions e.g. vulvar intraepithelial neoplasia (VIN). Infection of the cervical glandular epithelium is less common as, given the requirement of HPV infection for squamous differentiation, it is non-productive. However, some viral types (particularly HPV18) are associated with the development of cervical glandular intraepithelial neoplasia (CGIN)/adenocarcinoma *in situ* (AIS) and adenocarcinoma (Clifford *et al.*, 2003).

The role of HPV in cervical cancer

The association between HPV infection and cervical cancer is well established and strong, with HPV DNA being identified in 99.7% of invasive cervical carcinomas in the landmark study by Walboomers and co-workers (Walboomers *et al.*, 1999). However, despite our increasing understanding of the natural history of this disease the burden of HPV related cervical pathology remains large. Worldwide, cervical cancer alone accounts for in excess of 270,000 deaths each year, representing 9% of all cancer deaths (http://info.cancerresearchuk.org/cancerstats/types/cervix/mortality/). Rates vary greatly around the world as a result of a number of factors such as sexual practices, the prevalence of HIV infection and the availability of national screening programmes. In South Africa and Central America cervical cancer is the most common cancer diagnosed in women and in certain populations the prevalence of HPV infection approaches 80% (Brown *et al.*, 2005). In the UK, the incidence of cervical cancer is approximately 8.4/100,000 with mortality from the disease 2.4/100,000 (http://info.cancerresearchuk.org/cancerstats/types/cervix). The majority of cases occur in women of a young age but the large majority of fatal cases occur in an older age group.

In the UK a screening system aimed at detecting premalignant lesions was instituted in 1988 by the Department of Health. The programme consists of the cytological analysis of Pap (Papanicolaou) smears taken from the cervix. Smears are performed on a 3- to 5-yearly basis with screening beginning at 20 or 25 years until the age of 60 or 64 years. The smears allow the detection of abnormal cells from the cervical transformation zone. These atypical cells correspond closely with the CIN lesions (also known as SILs) identified in limited cervical resections (for example large loop excision of the transformation zone or LLETZ) which can then be graded as outlined above. These grades correlate with a loosely defined risk of progression to malignancy and can thus inform decisions regarding treatment and prognosis.

HPV in cancers of the head and neck

The role of HPV infection in cancers of the upper aerodigestive tract is now well established. However, whilst HPV is implicated in the aetiology of almost all cervical carcinomas (Walboomers *et al.*, 1999), it is involved much less frequently in head and neck cancers. Estimates vary greatly but, at most, HPV is thought to be associated with up to 25% of SCCs (Gillison *et al.*, 2000; Herrero *et al.*, 2003). The highest rates of infection appear to be within the oropharynx and particularly the tonsil (approximately 35%) (Kreimer *et al.*, 2005). Infection rates in the oral cavity proper and larynx are comparatively lower (approximately 23% and 24%, respectively). In contrast to cervical lesions, where HPV16 is isolated from approximately 55% of SCCs (Walboomers *et al.*, 1999; Clifford *et al.*, 2003), HPV16 is found in excess of 85% of HPV-positive oral lesions. The other viral types appear to have much lower oncogenic potential at this site. For example, HPV18 is found in less than 8% of lesions (Kreimer *et al.*, 2005). Furthermore, and again in contrast to cervical lesions, infection with multiple viral types is extremely rare (<4%). In the absence of HPV, the most significant risk factors are smoking and alcohol intake. Interestingly, in comparison to these lesions, those associated with HPV are associated with a much better prognosis. In some studies HPV infection in SCC of the head and neck was found to be associated with a 60–80% reduction in the risk of death (Psyrri *et al.*, 2009). The exact mechanism behind this remains unclear but may be related to differences in radiosensitivity.

HPV in cancers of the skin

The role of HPV in the development of cutaneous malignancy remains unclear. Viruses of the beta genus are known to preferentially infect the skin but the extent to which they promote the development of SCCs has not been fully elicited. The evidence for a causal role for HPV in cutaneous carcinogenesis is not as well established as it is in other sites. However, the presence of beta papillomavirus has been detected at higher levels in SCC and premalignant actinic keratoses than in healthy skin (Forslund *et al.*, 2007). Recently, persistent HPV infection in the skin has been shown to result in a significantly higher risk of developing UV damage and cutaneous SCC (Bouwes Bavinck *et al.*, 2010).

HPV and its effects on the cell cycle

Many of the pathological lesions associated with HPV infection occur as a result of a series of subtle and elegant alterations to the cell cycle. In order to appreciate the role of HPV in cellular transformation it is therefore important to understand the normal events involved in the cell cycle.

The purpose of cell division is to produce two daughter cells that are identical to the parent cell. The cell cycle is broadly divided into four stages. In G1, a stepwise progression occurs to prepare the cell for DNA synthesis. G1 is followed by S phase – during which DNA synthesis occurs. Following this the cell undergoes a series of checks to ensure it is ready to undergo division. This phase is designated G2. Finally cell division occurs during M (mitosis) phase. The progression of the cell through these phases is mediated and monitored by a variety of proteins, some of the most important of which are the cyclins, the cyclin-dependent kinases (CDKs) and the cyclin-dependent kinase inhibitors (CDKIs).

CDKs are the motors that drive the cell through the various stages of the cell cycle by activating a series of critical target proteins. All of the CDKs are expressed throughout the cell cycle in an inactive form. It is the cyclins that are responsible for the sequential activation of the CDKs. Unlike the CDKs, levels of the different cyclins rise and fall at different stages of the cell cycle ensuring that events take place in an orderly progression. As is often the case in biological systems, these stimulatory elements are in turn counteracted by a number of inhibitory signals, the CDKIs. These molecules fall into two broad categories, the CIP/KIP family (including p21, p27, p57) and the INK family (including p14, p15, p16 and p18).

G1–S phase

The retinoblastoma protein (pRb), and its association with the E2F family of transcription factors, are central to progression at the G1–S phase boundary. In the resting state, hypophosphorylated pRb forms a complex with E2F. As part of this complex, pRb utilizes histone deacetylases to maintain the chromatin in a configuration that is unsuitable for transcription. Early in the cell cycle, in response to a variety of external stimuli, levels of cyclin D1 increase. The purpose of cyclin D1 is to activate CDK4 (and CDK6). CDK4 in turn phosphorylates pRb leading to its dissociation from E2F and permitting activation of transcription. Some of the targets of E2F include cyclins E and A, thereby continuing the progressive and stepwise production of these proteins throughout the cell cycle. E2F also leads to transcription of DNA polymerases and thymidine kinase as well as a host of other molecules required for DNA synthesis. Increased expression of cyclin E therefore lags slightly behind that of cyclin D1 and it is the complex of cyclin E and CDK2 that permits further progression of the cell through S phase.

G2–M phase

The G2-M phase checkpoint represents the next important step in cell cycle control. Progression through these phases is mediated first by the cyclin A/CDK2 complex and then by the cyclin B/CDK1 complex. Cyclin A/CDK2 levels reach their peak during prophase. The cyclin A/CDK2 complex regulates cellular events preparing the cell for progression through to metaphase and beyond. Finally the cyclin B and CDK1 (also known as cdc25) complex promotes breakdown of the nuclear envelope and regulates microtubule stability as well as the separation of centrosomes and chromosome condensation. Completion of mitosis requires the degradation and eventual absence of this final cyclin B/CDK1 complex.

Control of the cell cycle

Brakes can be applied at various stages throughout the cell cycle and, as outlined above, a variety of molecules act to prevent (or pause) progression through the cycle. Of these, p53 is perhaps the most well known. p53 has a role at various points in the cell cycle and is important as its function is altered in a large number of the common visceral malignancies. In G1/S phase, p53 acts to increase levels of p21, which binds to and inactivates a number of cyclin/CDK complexes, including cyclin D1/CDK4 (thus preventing phosphorylation of pRb and release of E2F), cyclin E/CDK2 and cyclin A/CDK2. In turn this leads to G1 arrest, preventing cells from entering S phase. If DNA is repaired, p53 activates MDM2 which degrades p53 through ubiquitination. If DNA cannot be repaired, p53 acts to induce proapoptotic factors such as BAX.

Members of the CIP/KIP family of CDKIs, particularly p21 and p27, act at more than one point in the cell cycle, although the predominant effect of p21, which is a downstream target of p53, is at the G1/S transition and p27 acts predominantly on the cyclin E/CDK2 complex. In general, the INK family of molecules have more specific effects. The downstream effects of p14 are to increase the levels of p53 and therefore p21. p14 disrupts the degradation of p53 as mediated by MDM2. p16 acts in a similar fashion to p21 but is a more specific inhibitor of the cyclin D1/CDK4 complex.

The cell cycle targets of HPV

Fig. 1.4 demonstrates that there are numerous potential cell cycle targets that may be involved in cellular transformation. HPV must exploit several of the 'major players' in order to drive viral DNA replication: neoplastic transformation, when it occurs, is a secondary consequence.

The E6 protein

The E6 protein is one of the first to be expressed in viral infection and interacts with numerous cell cycle regulatory proteins. The E6 ORF encodes an approximately 150 amino acid protein (Ghittoni *et al.*, 2010) which bears a C-terminal region that in high-risk types contains a PSD-95/Dig/Z01 (PDZ) domain allowing interaction with many PDZ domain-containing proteins. Moreover, at the base of a series of zinc fingers, the E6 protein contains several Cys-x-x-Cys motifs responsible

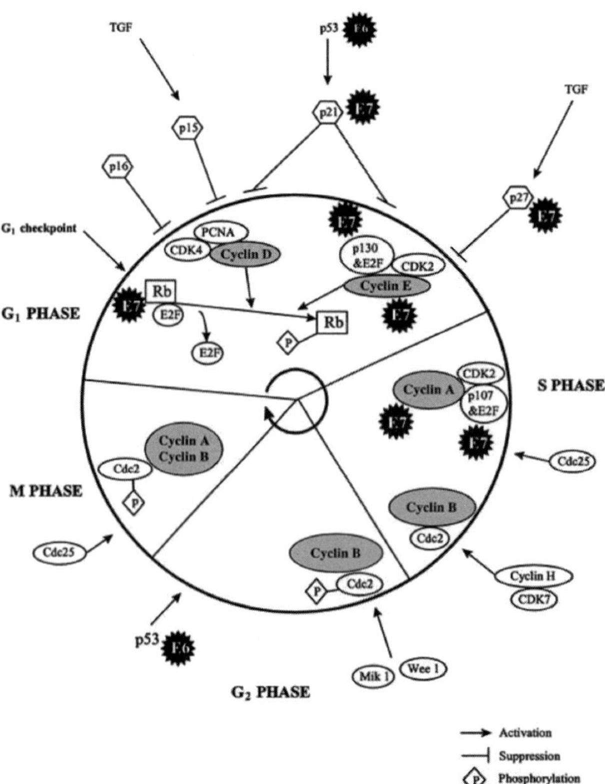

Figure 1.4 The main effects of HPV16 E6 and E7 proteins on the host cell cycle. Reproduced with permission from Southern and Herrington (2000).

for many of the processes that lead to cellular transformation (reviewed by Vande Pol, 2011).

The most important function of the E6 protein is to counteract the effects of p53 (Werness *et al.*, 1990). The E6-associated protein (E6AP), an endogenous 100-kDa ubiquitin ligase, binds to E6 at a conserved LXXLL motif. This complex can then bind to p53 with subsequent recruitment of the ubiquitin enzyme complex leading to ubiquitination of lysine residues on p53 and then proteolysis. E6-driven ubiquitination of p53 can also occur in the absence of E6AP as this phenomenon has been shown to occur in E6AP null mice (Massimi *et al.*, 2008). As detailed in the above discussion of the normal cell cycle, loss of p53 leads to loss of inhibition by p21 and potentiation of the effects of the various cyclin/CDK complexes. Abrogation of p53 function by E6 results in the failure of G1/S arrest and predisposes the cell to accumulating DNA damage and developing a transformed phenotype (Thomas *et al.*, 1998). Several other members of the p53 family have recently been discovered but there are conflicting data about the effect of HPV on these molecules (Marin *et al.*, 1998; Park *et al.*, 2001).

Of further but as yet not fully elucidated importance in the role of malignant transformation is the manipulation of membrane-associated guanylate kinases (MAGUK) by the E6 protein. The MAGUK family is one group of proteins that contain the PDZ domain referred to above. The MAGUK proteins are located in the cytoplasmic membrane and, as well as regulating cell to cell contact at tight junctions and cell polarity (Funke *et al.*, 2005), many are putative tumour suppressors. Interference with the function of MAGUK proteins probably contributes to the well-documented morphology of high-grade CIN/SIL. The observations that (i) the transforming properties of an HPV16 variant that encodes an E6 protein lacking a PDZ binding domain are significantly reduced (Nguyen *et al.*, 2003; Watson *et al.*, 2003) and (ii) wild-type low-risk HPV encodes an E6 protein that lacks a PDZ-binding domain (Jackson *et al.*, 2000) suggest that the E6/MAGUK interaction is important in neoplastic transformation. Moreover, a short peptide within the PDZ binding domain of E6 has been shown to interact with several MAGUK proteins including the tumour suppressors MAGI-1 and SAP97/hDlg (Jackson *et al.*, 2000).

A further function of E6 production is the transcriptional activation of telomerase reverse transcriptase (hTERT). In senescent cells telomerase activity is absent and the length of telomeres shortens with each progressive cell division providing a ticking clock before eventual cell death. By activating hTERT, the E6 protein prevents progressive shortening of telomeres: this occurs through a number of mechanisms. Echoing the degradation of p53 and Bak (see below), E6 in combination with E6AP leads to the degradation of NFX1-91, a transcriptional repressor of hTERT. However, hTERT is also up-regulated in E6AP binding deficient viruses (Sekaric *et al.*, 2008), highlighting that other mechanisms are involved. E6 may also directly bind to Myc and this complex may directly regulate hTERT levels (Veldman *et al.*, 2003). Furthermore, E7 appears to act synergistically with E6 in activating hTERT but is unable to do so in isolation (Liu *et al.*, 2008).

The E7 protein

The E7 gene encodes a low molecular weight protein approximately 100 amino acids in length (Ghittoni *et al.*, 2010). Several other human DNA viruses, including adenoviruses and simian virus 40 produce structurally and functionally similar proteins, the E1A and large T antigen respectively (Munger *et al.*, 2001). The shared domains within these proteins are referred to as conserved regions (CRs) of which there are three. The first CR is the N-terminus, which has a crucial role in pRb degradation. The second CR contains a LxCxE core sequence and a casein kinase II phosphorylation site (CKII). This domain allows E7 to bind to a variety of endogenous 'pocket proteins', the most important of which is pRb. The function of pRb in preventing progression of G1/S phase has been described above, particularly its role in inhibiting the E2F family of proteins. By binding to pRb, E7 interrupts the inhibitory effect of hypophosphorylated pRb on E2F. This therefore leads to unchecked transcription of a number of cell cycle mediators including cyclins E and A. The second CR of E7 also binds to the pocket proteins p107 and p130 (Helt and Galloway, 2003). These proteins exhibit similar effects to pRb on other members of the E2F family (E2F4 and E2F5 respectively). p130 is expressed in differentiated cells within higher levels of the epithelium, whereas p107 levels are increased when quiescent cells begin to divide (Classon and Dyson, 2001). The E7 protein therefore has multiple target binding sites throughout the epithelium as cells progress through the maturation process. p130 exerts its regulatory function at the G0/G1 checkpoint whereas p107 functions both at the G1/S checkpoint as well as in G2 (Cobrinik, 2005; Dimova and Dyson, 2005). The third CR, in the C-terminal domain, consists of two C-x-x-C motifs separated by a 29/30 amino acid spacer. Similar to the second CR, this third region is crucial for binding to pRb.

The ability of the early proteins to interact with a number of molecules and different pathways to produce the same net result is further highlighted by the observation that E7 can activate E2F related transcription without binding to pRb (Hwang *et al.*, 2002). E7 has also been shown to interact with other cellular complexes, including the TATA box binding proteins (TBP), that may have wider-ranging effects on transcription (Massimi *et al.*, 1997).

Overlapping with the function of E6, E7 also blocks the effects of a variety of CDKIs. E7 has been shown to abrogate the inhibitory role of CIP/KIP proteins, in particular p21 and p27 (Zerfass-Thome *et al.*, 1996; Jones *et al.*, 1997). Interestingly, pRb is known to reduce the expression of p16 (Li *et al.*, 1994). The absence of pRb induced in HPV infected cells by E7 expression therefore leads to an increase in p16 levels. At first this effect seems counterintuitive and of little evolutionary sense. However, closer consideration reveals that E7 effectively substitutes for the phosphorylation of pRb by CDK4. As CDK4 is the target of p16 inhibition, increased levels of p16 are of little consequence, the virus having bypassed the checkpoints at this point in the cell cycle. In clinical practice the increase in p16 expression is of some relevance as it may be used as a surrogate marker for E7-expressing high-risk HPV infection (Klaes *et al.*, 2001, 2002).

Another of the many effects of E7 is to promote chromosomal instability by inducing abnormal numbers of centrosomes. E6 shares a similar function, acting in synergy to promote further abnormalities in host chromosome and DNA structure (Duensing *et al.*, 2000).

Apoptosis

All viruses tread a tightrope between preserving host cells for viral replication and initiating cell death to allow subsequent infection of other cells. Viruses that kill cells too readily may be viewed as inefficient as fewer numbers of the virus are produced whereas the converse is also true, as viruses that remain within the host cell will replicate but have little means of infecting other cells. Thus, many viruses, including HPV, have developed mechanisms for inhibiting apoptosis as well as hijacking DNA synthesis to effect viral DNA replication.

In the normal cell apoptosis is mediated either through an extrinsic or an intrinsic pathway. The extrinsic pathway involves interaction with a membrane bound receptor, most often the tumour necrosis factor (TNF) receptor and the Fas (CD95) receptor. The Fas receptors have cytoplasmic death domains. When bound by FasL (expressed on many host cell including peripheral T cells), several receptors combine their death domains and form a binding site for the Fas-associated death domain (FADD). FADD in turn binds an inactive form of caspase 8. The caspases (cysteine aspartyl-specific proteases) are a group of proteins that degrade normal intracellular structures during apoptosis. They act in a cascade-like fashion and activation of caspase 8 initiates that cascade.

The intrinsic pathway does not involve death receptors and instead relies on changes in mitochondrial permeability that result in leakage of pro-apoptotic molecules. In the healthy state, signals for survival maintain the production of a family of anti-apoptotic molecules, including Bcl-2 and Bcl-x. When subjected to stress, or removal of survival signals, these molecules are replaced by several pro-apoptotic molecules including Bax, Bak, Bim and Bad. These proteins act to increase the permeability of the mitochondrial membrane. Cytochrome c is released from the mitochondria and binds to apoptosis activating factor 1 (Apaf1) which activates caspase 9. Apoptosis inducing factor (AIF) is also released from mitochondria and binds and inhibits many of the anti-apoptotic proteins.

The effects of HPV on apoptosis

HPV has evolved many mechanisms to manipulate apoptotic pathways. Many of these effects are pro-apoptotic, and linked to virion release, whilst those that lead to cellular transformation are inevitably anti-apoptotic.

E6 has several pro-apoptotic effects in certain circumstances. The presumed function of this effect is the release of virions after sufficient viral replication has occurred. E6 undergoes post-translational splicing to form a truncated protein termed E6*I, which is approximately half the size of the normal protein (Filippova *et al.*, 2004). E6*I interferes with the interaction of E6 and E6AP and therefore obstructs their effects on p53. It is also postulated that E6*I can directly induce apoptosis (Pim and Banks, 1999).

As well as trans-repressing the function of the main oncogenic viral proteins, E2 has strong pro-apoptotic effects. The N-terminal of the protein has been shown to be responsible for this effect and induces the caspase cascade directly by activating caspase 8. E2 itself is then cleaved as part of the cascade, functioning as both an inducer and target of the cascade's activity (Demeret *et al.*, 2003).

HPV E1^E4 also has potent pro-apoptotic potential. This protein has a crucial role in degradation of the host cell to enable release of the virus after replication. It appears to do this by first binding and then degrading the intracellular network of intermediate filaments. Subsequently E1^E4 binds to mitochondria altering the mitochondrial membrane potential and pushing the cell towards apoptosis (Raj *et al.*, 2004).

In contrast to these functions, in order for immortalization of the cell to occur, the high-risk HPV types must produce strong anti-apoptotic signals. Unsurprisingly, the E6 and E7 proteins play an important role in this function. One of the many roles of p53 is, in response to cellular damage, to stimulate pro-apoptotic molecules. Clearly, with E6 mediated ubiquitination of p53 these pro-apoptotic factors are lost. Moreover, E6 also enhances the cell's ability to evade apoptosis by directly causing the ubiquitination of Bak (Jackson *et al.*, 2000). This finding may be of particular significance in cutaneous lesions where additional DNA damage occurs through UV damage. It has been shown that, in the healthy epidermis, Bak is expressed at high levels preventing damaged cells from replicating. Clearly, in HPV-infected cells this protective mechanism is lost.

Additionally, E6 can prevent apoptosis by p53-independent mechanisms. E6 can bind to the cytoplasmic death domain of the TNF receptor and prevent binding to FADD and the downstream activation of caspase 8. By binding to both Fas and the TNF receptor, E6 also provides a mechanism for evading host immunity and promoting viral survival (Filippova *et al.*, 2004).

E7 has similar effects, many of which counteract the functions of E2. E7 decreases the activation of the caspase cascade initiator pro-caspase-8 (Duensing and Münger, 2004). In experimental models, E7 immortalized cells appear almost completely resistant to cell death via the extrinsic pathway.

High-risk versus low-risk HPV types

Whilst all HPV types express E6 and E7 proteins, what is not clear from the above discussion is why certain HPV types have a far greater propensity for malignant transformation than others. Importantly, the DNA from the low-risk HPV types remains in an extrachromosomal episomal form, whereas the DNA from the high-risk types often integrates into the host DNA during transformation (Pett and Coleman, 2007; Winder *et al.*, 2007). Furthermore, whilst the viral DNA is integrated randomly into host DNA, the viral breakpoint during integration is relatively constant, occurring downstream of the E6 and E7 coding regions in the E1/E2 region. As outlined previously, the role of E2 is to trans-repress the expression of the E6 and E7 proteins. Therefore, when integration occurs at this site, the negative feedback exerted by E2 is lost resulting in higher levels of the E6 and E7 proteins in comparison to lower risk HPV types (Woodman *et al.*, 2007). Furthermore, as outlined above, E2 has strong pro-apoptotic effects by stimulating the caspase cascade (Demeret *et al.*, 2003): with integration into host DNA this potential to induce apoptosis is lost. The E2 protein may also induce the caspase cascade via a p53-dependent mechanism: by causing p53 levels to fall, E6 protein expression may therefore block E2 induced apoptosis (Webster *et al.*, 2000). Interestingly, in in-vivo studies, herpes simplex virus VP22-HPV E2 fusion proteins were shown to induce apoptosis in HPV transformed cervical squamous cell carcinoma cell lines (Roeder *et al.*, 2004), a finding of potential therapeutic benefit. Finally, loss of E2 function also leads to disruption of the E1/E2 hetero-oligomer, which is required for viral DNA replication. Hence, vegetative viral replication cannot occur.

There are also subtle differences in the function and efficacy of the E6 and E7 proteins between high- and low-risk groups. In the previous discussion of the transforming properties of E6, it was noted that PDZ binding domains allow the virus to interact with many host proteins (e.g. MAGUK) promoting transformation. Low-risk types lack these PDZ binding domains (Jackson *et al.*, 2000). In addition, high-risk HPV E6 proteins degrade p53 much more effectively than low-risk HPV E6 proteins (Hiller *et al.*, 2006). Similarly high-risk HPV E7 proteins have been shown to bind pRb with greater affinity than low-risk HPV E7 (Münger *et al.*, 1989). This appears to be a direct result of a change in one amino acid, aspartic acid 21 in high-risk types and glycine 22 in low-risk types (Sang and Barbosa, 1992). As always, the story may not be quite so simple as not all E7 proteins that bind pRb with high affinity (HPV1 and HPV32 for example) carry the same intrinsic risk of carcinogenesis (Dong *et al.*, 2001). Whether this is because of additional interactions or a difference in the inherent function of the protein remains unclear.

These differences between low- and high-risk HPVs go some way to explaining the differences in the pathology associated with these HPV types. Low-risk HPVs, which do not integrate into host chromosomes, interfere with cell cycle control to effect viral DNA replication in post-mitotic differentiated squamous epithelial cells, primarily through the action of E7. Co-ordinated expression of viral proteins occurs and intact virions are produced after desquamation of surface epithelial cells. High-risk HPVs can produce similar effects but, when co-ordinated expression of HPV genes is disrupted, for example by HPV integration, inappropriate high-level expression of particularly E6 and E7 in dividing cells leads to genetic instability with accumulation of secondary genetic changes, and inhibition of apoptosis, resulting eventually in neoplastic transformation.

Prevention, surveillance and treatment of HPV infection and HPV-associated disease

Despite our increasing understanding of the role of HPV in carcinogenesis, the virus still exerts a large burden on health services around the world. At present, treatment of HPV-associated lesions involves destruction or surgical excision for pre-invasive lesions such as CIN/SIL or CGIN/AIS; surgical resection with adjuvant radiotherapy and/or chemotherapy is employed for invasive lesions, as appropriate. Owing to the young age of those affected, the treatment of cervical cancer in particular raises several issues that are not typically encountered when considering treatment of other malignancies. Many patients may have yet to complete their families and therefore the consequences of irreversible surgical procedures are very significant. Continued exploration of other potential therapeutic avenues is therefore important: these include the use of apoptosis-inducing E2 homologues, therapeutic vaccination and the use of silencing RNAs against the E6 and E7 transcripts (Chang *et al.*, 2010). Unfortunately, many of these options are either impractical or remain at an early experimental stage and therefore a great deal of attention has been centred on disease prevention.

The cervical screening programme has been running in the UK for more than 20 years and has resulted in a striking reduction in mortality from cervical carcinoma. The screening programme aims to detect and treat premalignant (intraepithelial) lesions thereby preventing cancer, but does not aid in preventing initial infection. Furthermore, many resource-poor countries lack a screening programme. Therefore, once it was established that these diseases were in part caused by an infectious agent, the development of a successful vaccine became a major research focus. Development of a successful vaccine has however proved difficult. HPV is extremely difficult to culture and the fact that it contains several potent oncogenes means that an inactivated viral vaccine was not possible (Roden and Wu, 2006). Furthermore, HPV replicates in the genital mucosa and avoids many of the constituents of the host's humoral immunity. In addition, levels of antibodies of the IgG class have been shown to fluctuate throughout the menstrual cycle (Nardelli-Haefliger *et al.*, 2003).

Hopes were first raised after a successful hepatitis B vaccine was made using viral capsid proteins. Trials with denatured papillomavirus capsid proteins were, however, unsuccessful and protection appeared to be achieved only with the native full-length product (Lin *et al.*, 1993). Experiments with recombinant L1 were more successful and led to the generation of virus-like particles (VLPs) that share homology with the native virus and were found to induce specific antibody production in animal models (Marais *et al.*, 1999). Further problems were encountered when it emerged that antibodies were only raised to specific HPV types and that a single vaccine would not offer protection against the whole spectrum of HPV genera (Jarrett *et al.*, 1990). In order to prevent 90% of HPV-related cervical cancers it has been estimated that vaccination against eight different subtypes is required (Muñoz *et al.*, 2004).

At present two vaccines have emerged that share similar efficacy and safety profiles. 'Gardasil' is manufactured by Merck, offers protection against HPV types 6, 11, 16 and 18, and comprises L1 VLPs from recombinant yeast. GlaxoSmithKline produced 'Cervarix' which offers protection against HPV16 and 18. Initial phase 1 trials showed a 40-fold increase in IgG titres against the VLPs in comparison to levels observed in natural infection (Harro *et al.*, 2001). In one study, vaccination three times with HPV16 L1 VLPs prevented infection with HPV16 for approximately 18 months in all individuals receiving the vaccine (Koutsky *et al.*, 2002). Subsequent randomized controlled trials (Dillner *et al.*, 2010; Muñoz *et al.*, 2010) have found that quadrivalent vaccines are over 95% successful in preventing HPV16 and 18 related high grade lesions when compared with a placebo-treated group. Similar figures have been obtained for HPV related lesions outwith the cervix (Dillner *et al.*, 2010). Both of these trials extended over a 4 year period illustrating to a limited extent that the vaccine confers at least some long-term protection.

In the UK a nationwide HPV vaccination programme was rolled out in 2008. This programme offers a series of three vaccinations to girls aged between 12 and 13. As with all vaccines, adequate uptake is required in an attempt to have a significant impact on prevalence and to produce a degree of herd immunity. In the UK, early estimates of vaccination uptake were encouraging at around 70% (Brabin *et al.*, 2008) but this figure will require continued monitoring in the future if vaccination is to be a successful public health measure. The implementation of the vaccination programme raises several interesting questions. As the programme is in its infancy, few data are available regarding the longevity of protection offered by the vaccine. Studies have, however, estimated that antibody levels should remain above those produced by natural infection for at least 12 years and that antibodies will persist for life at detectable levels in 99% of patients (Fraser *et al.*, 2007). Clearly, it is important that future studies confirm these estimates as the first cohort of vaccinated patients grows older.

Another important issue raised by the vaccination programme is the effect it will have on the current screening programme. The incidence of premalignant lesions should drop dramatically in the coming years making any screening programme increasingly less cost effective. Should screening stop? Clearly, the current vaccine only offers protection against HPV types 16 and 18 and pre-malignant and malignant lesions caused by the many other high-risk viruses will still occur. It is important therefore that screening continues and that the public remain aware of this. A related concern is whether vaccination against types 16 and 18 will lead to an evolutionary advantage for the other high-risk types. It is possible that the incidence of HPV-related malignancy may not decrease as much as predicted as other types inhabit the niche vacated by types 16 and 18. However, the numbers of non-16/18 related lesions based on early follow-up studies does not seem to be increasing (Harper *et al.*, 2006).

Many of the traditional treatments for

HPV-associated lesions employ corrosive chemical agents such as salicylic acid or toxic compounds that are used only in specific clinical situations, such as podophyllotoxin. Although the use of the immunomodulatory compound imiquimod is gaining increasing clinical acceptance, for example for the treatment of genital warts (Komericki *et al.*, 2011) and vulval intraepithelial neoplasia (Daayana *et al.*, 2010) there has been little development of specific agents designed to modulate HPV infection itself. One potential approach to this is exemplified by the investigation of agents that activate the p53 pathway, which is abrogated by HPV infection (Grey *et al.*, 2007; Jolly *et al.*, 2009). This has the logic that the induction of apoptosis in HPV-infected cells can be achieved selectively.

Many of the questions regarding the prevention and treatment of HPV related disease in the future remain unanswered. Time will show whether the vaccination programme is as successful as is hoped or whether other viral types come to the fore. The next 50 years may see marked changes in the demographic of the screened population and indeed the frequency of screening. In parts of the world where no screening programme exists, and particularly in resource-poor countries, the challenge will be different, centring initially on costs and the practicalities of implementation. Finally, the need to develop HPV-specific therapies is likely to remain.

References

Bernard, H. -U., Burk, R.D., Chen, Z., van Doorslaer, K., Hausen, H. z., and de Villiers, E. -M. (2010). Classification of papillomaviruses (PVs) based on 189 PV types and proposal of taxonomic amendments. Virology *401*, 70–79.

Bouwes Bavinck, J.N., Neale, R.E., Abeni, D., Euvrard, S., Green, A.C., Harwood, C.A., de Koning, M.N.C., Naldi, L., Nindl, I., Pawlita, M., *et al.* (2010). Multicenter study of the association between betapapillomavirus infection and cutaneous squamous cell carcinoma. Cancer Res. *70*, 9777–9786.

Brabin, L., Roberts, S.A., Stretch, R., Baxter, D., Chambers, G., Kitchener, H., and McCann, R. (2008). Uptake of first two doses of human papillomavirus vaccine by adolescent schoolgirls in Manchester: prospective cohort study. BMJ *336*, 1056–1058.

Brown, D.R., Shew, M.L., Qadadri, B., Neptune, N., Vargas, M., Tu, W., Juliar, B.E., Breen, T.E., and Fortenberry, J.D. (2005). A longitudinal study of genital human papillomavirus infection in a cohort of closely followed adolescent women. J. Infect. Dis. *191*, 182–192.

Bryan, J.T., and Brown, D.R. (2000). Association of the human papillomavirus type 11 E1() E4 protein with cornified cell envelopes derived from infected genital epithelium. Virology *277*, 262–269.

Campo, M.S., Graham, S.V., Cortese, M.S., Ashrafi, G.H., Araibi, E.H., Dornan, E.S., Miners, K., Nunes, C., and Man, S. (2010). HPV16 E5 down-regulates expression of surface HLA class I and reduces recognition by CD8 T cells. Virology *407*, 137–142.

Chang, J.T. -C., Kuo, T. -F., Chen, Y. -J., Chiu, C. -C., Lu, Y. -C., Li, H. -F., Shen, C. -R., and Cheng, A. -J. (2010). Highly potent and specific siRNAs against E6 or E7 genes of HPV16- or HPV18-infected cervical cancers. Cancer Gene Ther. *17*, 827–836.

Chow, L.T., Duffy, A.A., Wang, H. -K., and Broker, T.R. (2009). A highly efficient system to produce infectious human papillomavirus: Elucidation of natural virus–host interactions. Cell Cycle *8*, 1319–1323.

Classon, M., and Dyson, N. (2001). p107 and p130: versatile proteins with interesting pockets. Exp. Cell Res. *264*, 135–147.

Clifford, G.M., Smith, J.S., Plummer, M., Muñoz, N., and Franceschi, S. (2003). Human papillomavirus types in invasive cervical cancer worldwide: a meta-analysis. Br. J. Cancer *88*, 63–73.

Cobrinik, D. (2005). Pocket proteins and cell cycle control. Oncogene *24*, 2796–2809.

Conway, M.J., and Meyers, C. (2009). Replication and assembly of human papillomaviruses. J. Dent. Res. *88*, 307–317.

Daayana, S., Elkord, E., Winters, U., Pawlita, M., Roden, R., Stern, P.L., and Kitchener, H.C. (2010). Phase II trial of imiquimod and HPV therapeutic vaccination in patients with vulval intraepithelial neoplasia. Br. J. Cancer *102*, 1129–1136.

Day, P.M., Baker, C.C., Lowy, D.R., and Schiller, J.T. (2004). Establishment of papillomavirus infection is enhanced by promyelocytic leukemia protein (PML) expression. Proc. Natl. Acad. Sci. U.S.A. *101*, 14252–14257.

Demeret, C., Garcia-Carranca, A., and Thierry, F. (2003). Transcription-independent triggering of the extrinsic pathway of apoptosis by human papillomavirus 18 E2 protein. Oncogene *22*, 168–175.

Dillner, J., Kjaer, S.K., Wheeler, C.M., Sigurdsson, K., Iversen, O. -E., Hernandez-Avila, M., Perez, G., Brown, D.R., Koutsky, L.A., Tay, E.H., *et al.* (2010). Four year efficacy of prophylactic human papillomavirus quadrivalent vaccine against low grade cervical, vulvar, and vaginal intraepithelial neoplasia and anogenital warts: randomised controlled trial. BMJ *341*, c3493.

DiMaio, D., and Mattoon, D. (2001). Mechanisms of cell transformation by papillomavirus E5 proteins. Oncogene *20*, 7866–7873.

Dimova, D.K., and Dyson, N.J. (2005). The E2F transcriptional network: old acquaintances with new faces. Oncogene *24*, 2810–2826.

Dong, W.L., Caldeira, S., Sehr, P., Pawlita, M., and Tommasino, M. (2001). Determination of the binding affinity of different human papillomavirus E7 proteins for the tumour suppressor pRb by a plate-binding assay. J. Virol. Methods *98*, 91–98.

Doorbar, J. (2006). Molecular biology of human papillomavirus infection and cervical cancer. Clin. Sci. *110*, 525–541.

Duensing, S., Lee, L.Y., Duensing, A., Basile, J., Piboonniyom, S., Gonzalez, S., Crum, C.P., and Munger, K. (2000). The human papillomavirus type 16 E6 and E7 oncoproteins cooperate to induce mitotic defects and genomic instability by uncoupling centrosome duplication from the cell division cycle. Proc. Natl. Acad. Sci. U.S.A. *97*, 10002–10007.

Duensing, S., and Münger, K. (2004). Mechanisms of genomic instability in human cancer: insights from studies with human papillomavirus oncoproteins. Int. J. Cancer *109*, 157–162.

Dürst, M., Gissmann, L., Ikenberg, H., and zur Hausen, H. (1983). A papillomavirus DNA from a cervical carcinoma and its prevalence in cancer biopsy samples from different geographic regions. Proc. Natl. Acad. Sci. U.S.A. *80*, 3812–3815.

Fang, L., Budgeon, L.R., Doorbar, J., Briggs, E.R., and Howett, M.K. (2006). The human papillomavirus type 11 E1/\E4 protein is not essential for viral genome amplification. Virology *351*, 271–279.

Filippova, M., Parkhurst, L., and Duerksen-Hughes, P.J. (2004). The human papillomavirus 16 E6 protein binds to Fas-associated death domain and protects cells from Fas-triggered apoptosis. J. Biol. Chem. *279*, 25729–25744.

Forslund, O., Iftner, T., Andersson, K., Lindelof, B., Hradil, E., Nordin, P., Stenquist, B., Kirnbauer, R., Dillner, J., de Villiers, E. -M., *et al.* (2007). Cutaneous human papillomaviruses found in sun-exposed skin: B-papillomavirus species 2 predominates in squamous cell carcinoma. J. Infect. Dis. *196*, 876–883.

Fraser, C., Tomassini, J.E., Xi, L., Golm, G., Watson, M., Giuliano, A.R., Barr, E., and Ault, K.A. (2007). Modeling the long-term antibody response of a human papillomavirus (HPV) virus-like particle (VLP) type 16 prophylactic vaccine. Vaccine *25*, 4324–4333.

Funke, L., Dakoji, S., and Bredt, D.S. (2005). Membrane-associated guanylate kinases regulate adhesion and plasticity at cell junctions. Annu. Rev. Biochem. *74*, 219–245.

Gasparini, R., and Panatto, D. (2009). Cervical cancer: from Hippocrates through Rigoni-Stern to zur Hausen. Vaccine *27 (Suppl 1)*, A4–5.

Ghittoni, R., Accardi, R., Hasan, U., Gheit, T., Sylla, B., and Tommasino, M. (2010). The biological properties of E6 and E7 oncoproteins from human papillomaviruses. Virus Genes *40*, 1–13.

Gillison, M.L., Koch, W.M., Capone, R.B., Spafford, M., Westra, W.H., Wu, L., Zahurak, M.L., Daniel, R.W., Viglione, M., Symer, D.E., *et al.* (2000). Evidence for a causal association between human papillomavirus and a subset of head and neck cancers. J. Natl. Cancer Inst. *92*, 709–720.

Gray, L., Bjelogrlic, P., Appleyard, V., Thompson, A., Jolly, C., Lain, S., and Herrington, C.S. (2007). Selective induction of apoptosis by leptomycin B in keratinocytes expressing HPV oncogenes. Int. J. Cancer *120*, 2317–2324.

Harper, D., Franco, E., Wheeler, C., Moscicki, A., Romanowski, B., Roteli-Martins, C., Jenkins, D., Schuind, A., Costa Clemens, S., and Dubin, G. (2006). Sustained efficacy up to 4.5 years of a bivalent L1 virus-like particle vaccine against human papillomavirus types 16 and 18: follow-up from a randomised control trial. Lancet *367*, 1247–1255.

Harro, C.D., Pang, Y.Y., Roden, R.B., Hildesheim, A., Wang, Z., Reynolds, M.J., Mast, T.C., Robinson, R., Murphy, B.R., Karron, R.A., *et al.* (2001). Safety and immunogenicity trial in adult volunteers of a human papillomavirus 16 L1 virus-like particle vaccine. J. Natl. Cancer Inst. *93*, 284–292.

zur Hausen, H. (1976). Condylomata acuminata and human genital cancer. Cancer Res. *36*, 794.

zur Hausen, H. (1977). Human papillomaviruses and their possible role in squamous cell carcinomas. Curr. Top. Microbiol. Immunol. *78*, 1–30.

zur Hausen, H. (1996). Papillomavirus infections – a major cause of human cancers. Biochim. Biophys. Acta *1288*, F55–78.

Hebner, C.M., and Laimins, L.A. (2006). Human papillomaviruses: basic mechanisms of pathogenesis and oncogenicity. Rev. Med. Virol. *16*, 83–97.

Helt, A. -M., and Galloway, D.A. (2003). Mechanisms by which DNA tumour virus oncoproteins target the Rb family of pocket proteins. Carcinogenesis *24*, 159–169.

Herrero, R., Castellsagué, X., Pawlita, M., Lissowska, J., Kee, F., Balaram, P., Rajkumar, T., Sridhar, H., Rose, B., Pintos, J., *et al.* (2003). Human papillomavirus and oral cancer: the International Agency for Research on Cancer multicenter study. J. Natl. Cancer Inst. *95*, 1772–1783.

Hiller, T., Poppelreuther, S., Stubenrauch, F., and Iftner, T. (2006). Comparative analysis of 19 genital human papillomavirus types with regard to p53 degradation, immortalization, phylogeny, and epidemiologic risk classification. Cancer Epidemiol. Biomarkers Prev. *15*, 1262–1267.

Hwang, S.G., Lee, D., Kim, J., Seo, T., and Choe, J. (2002). Human papillomavirus type 16 E7 binds to E2F1 and activates E2F1-driven transcription in a retinoblastoma protein-independent manner. J. Biol. Chem. *277*, 2923–2930.

Jackson, S., Harwood, C., Thomas, M., Banks, L., and Storey, A. (2000). Role of Bak in UV-induced apoptosis in skin cancer and abrogation by HPV E6 proteins. Genes Dev. *14*, 3065–3073.

Jarrett, W.F., O'Neil, B.W., Gaukroger, J.M., Smith, K.T., Laird, H.M., and Campo, M.S. (1990). Studies on vaccination against papillomaviruses: the immunity after infection and vaccination with bovine papillomaviruses of different types. Vet. Rec. *126*, 473–475.

Jolly, C.E., Gray, L.J., Parish, J.L., Lain, S., and Herrington, C.S. (2009). Leptomycin B induces apoptosis in cells containing the whole HPV16 genome. Int. J. Oncol. *35*, 649–656.

Jones, D.L., Alani, R.M., and Münger, K. (1997). The human papillomavirus E7 oncoprotein can uncouple cellular differentiation and proliferation in human keratinocytes by abrogating p21Cip1-mediated inhibition of cdk2. Genes Dev. *11*, 2101–2111.

Klaes, R., Friedrich, T., Spitkovsky, D., Ridder, R., Rudy, W., Petry, U., Dallenbach-Hellweg, G., Schmidt, D., and von Knebel Doeberitz, M. (2001). Overexpression of p16(INK4A) as a specific marker for dysplastic and neoplastic epithelial cells of the cervix uteri. Int. J. Cancer *92*, 276–284.

Klaes, R., Benner, A., Friedrich, T., Ridder, R., Herrington, C.S., Jenkins, D., Kurman, R., Schmidt, D., Stoler, M., and von Knebel Doeberitz, M. (2002). p16INK4a immunohistochemistry improves interobserver agreement in the diagnosis of cervical intraepithelial neoplasia. Am. J. Surg. Pathol. *26*, 1389–1399.

Komericki, P., Akkilic-Materna, M., Strimitzer, T., and Aberer, W. (2011). Efficacy and safety of imiquimod versus podophyllotoxin in the treatment of anogenital warts. Sex Transm. Dis. *38*, 216–218.

Koutsky, L.A., Ault, K.A., Wheeler, C.M., Brown, D.R., Barr, E., Alvarez, F.B., Chiacchierini, L.M., Jansen, K.U., and Investigators, P. o. P. S. (2002). A controlled trial of a human papillomavirus type 16 vaccine. N. Engl. J. Med. *347*, 1645–1651.

Kreimer, A.R., Clifford, G.M., Boyle, P., and Franceschi, S. (2005). Human papillomavirus types in head and neck squamous cell carcinomas worldwide: a systematic review. Cancer Epidemiol. Biomarkers Prev. *14*, 467–475.

Li, Y., Nichols, M.A., Shay, J.W., and Xiong, Y. (1994). Transcriptional repression of the D-type cyclin-dependent kinase inhibitor p16 by the retinoblastoma susceptibility gene product pRb. Cancer Res. *54*, 6078–6082.

Lin, Y.L., Borenstein, L.A., Ahmed, R., and Wettstein, F.O. (1993). Cottontail rabbit papillomavirus L1 protein-based vaccines: protection is achieved only with a full-length, nondenatured product. J. Virol. *67*, 4154–4162.

Liu, X., Roberts, J., Dakic, A., Zhang, Y., and Schlegel, R. (2008). HPV E7 contributes to the telomerase activity of immortalized and tumorigenic cells and augments E6-induced hTERT promoter function. Virology *375*, 611–623.

Longworth, M.S., and Laimins, L.A. (2004). Pathogenesis of human papillomaviruses in differentiating epithelia. Microbiol. Mol. Biol. Rev. *68*, 362–372.

Loo, Y. -M., and Melendy, T. (2004). Recruitment of replication protein A by the papillomavirus E1 protein and modulation by single-stranded DNA. J. Virol. *78*, 1605–1615.

McMillan, N.A., Payne, E., Frazer, I.H., and Evander, M. (1999). Expression of the alpha6 integrin confers papillomavirus binding upon receptor-negative B-cells. Virology *261*, 271–279.

Mak, R., Van Renterghem, L., and Cuvelier, C. (2004). Cervical smears and human papillomavirus typing in sex workers. Sex Transm. Infect. *80*, 118–120.

Marais, D., Passmore, J.A., Maclean, J., Rose, R., and Williamson, A.L. (1999). A recombinant human papillomavirus (HPV) type 16 L1-vaccinia virus murine challenge model demonstrates cell-mediated immunity against HPV virus-like particles. J. Gen. Virol. *80 (Pt 9)*, 2471–2475.

Marin, M.C., Jost, C.A., Irwin, M.S., DeCaprio, J.A., Caput, D., and Kaelin, W.G. (1998). Viral oncoproteins discriminate between p53 and the p53 homolog p73. Mol. Cell Biol. *18*, 6316–6324.

Massimi, P., Pim, D., and Banks, L. (1997). Human papillomavirus type 16 E7 binds to the conserved carboxy-terminal region of the TATA box binding protein and this contributes to E7 transforming activity. J. Gen. Virol. *78* 2607–2613.

Massimi, P., Shai, A., Lambert, P., and Banks, L. (2008). HPV E6 degradation of p53 and PDZ containing substrates in an E6AP null background. Oncogene *27*, 1800–1804.

Moodley, M., Moodley, J., Chetty, R., and Herrington, C.S. (2003). The role of steroid contraceptive hormones in the pathogenesis of invasive cervical cancer: a review. Int. J. Gynecol. Cancer *13*, 103–110.

Münger, K., Werness, B.A., Dyson, N., Phelps, W.C., Harlow, E., and Howley, P.M. (1989). Complex formation of human papillomavirus E7 proteins with the retinoblastoma tumour suppressor gene product. EMBO J. *8*, 4099–4105.

Münger, K., Basile, J., Duensing, S., Eichten, A., Gonzalez, S., Grace, M., and Zacny, V. (2001). Biological activities and molecular targets of the human papillomavirus E7 oncoprotein. Oncogene *20*, 7888–7898.

Muñoz, N., Bosch, F.X., Castellsagué, X., Díaz, M., de Sanjose, S., Hammouda, D., Shah, K.V., and Meijer, C.J. L. M. (2004). Against which human papillomavirus types shall we vaccinate and screen? The international perspective. Int. J. Cancer *111*, 278–285.

Muñoz, N., Kjaer, S.K., Sigurdsson, K., Iversen, O. -E., Hernandez-Avila, M., Wheeler, C.M., Perez, G., Brown, D.R., Koutsky, L.A., Tay, E.H., *et al.* (2010). Impact of human papillomavirus (HPV)-6/11/16/18 vaccine on all HPV-associated genital diseases in young women. J. Natl. Cancer Inst. *102*, 325–339.

Nardelli-Haefliger, D., Wirthner, D., Schiller, J.T., Lowy, D.R., Hildesheim, A., Ponci, F., and De Grandi, P. (2003). Specific antibody levels at the cervix during the menstrual cycle of women vaccinated with human papillomavirus 16 virus-like particles. J. Natl. Cancer Inst. *95*, 1128–1137.

Nguyen, M.L., Nguyen, M.M., Lee, D., Griep, A.E., and Lambert, P.F. (2003). The PDZ ligand domain of the human papillomavirus type 16 E6 protein is required for E6's induction of epithelial hyperplasia *in vivo*. J. Virol. *77*, 6957–6964.

Park, J.S., Kim, E.J., Lee, J.Y., Sin, H.S., Namkoong, S.E., and Um, S.J. (2001). Functional inactivation of p73, a homolog of p53 tumour suppressor protein, by human papillomavirus E6 proteins. Int. J. Cancer *91*, 822–827.

Pett, M., and Coleman, N. (2007). Integration of high-risk human papillomavirus: a key event in cervical carcinogenesis? J. Pathol. *212*, 356–367.

Pim, D., and Banks, L. (1999). HPV18 E6*I protein modulates the E6-directed degradation of p53 by binding to full-length HPV18 E6. Oncogene *18*, 7403–7408.

Psyrri, A., Gouveris, P., and Vermorken, J.B. (2009). Human papillomavirus-related head and neck tumours: clinical and research implication. Curr. Opin. Oncol. *21*, 201–205.

Raj, K., Berguerand, S., Southern, S., Doorbar, J., and Beard, P. (2004). E1^E4 protein of human papillomavirus type 16 associates with mitochondria. J. Virol. *78*, 7199–7207.

Rigoni-Stern, D. (1842). Fatti statistica relativi alle malattie cancerose. Giornale per servire ai progressi della Paologica e della Terapoa 2, 507–517.

Roden, R., and Wu, T. -C. (2006). How will HPV vaccines affect cervical cancer? Nat. Rev. Cancer *6*, 753–763.

Roeder, G.E., Parish, J.L., Stern, P.L., and Gaston, K. (2004). Herpes simplex virus VP22-human papillomavirus E2 fusion proteins produced in mammalian or bacterial cells enter mammalian cells and induce apoptotic cell death. Biotechnol. Appl. Biochem. *40*, 157–165.

Sang, B.C., and Barbosa, M.S. (1992). Single amino acid substitutions in 'low-risk' human papillomavirus (HPV) type 6 E7 protein enhance features characteristic of the 'high-risk' HPV E7 oncoproteins. Proc. Natl. Acad. Sci. U.S.A. *89*, 8063–8067.

Schiller, J.T., Day, P.M., and Kines, R.C. (2010). Current understanding of the mechanism of HPV infection. Gynecol. Oncol. *118*, S12–17.

Schwarz, E., Freese, U.K., Gissmann, L., Mayer, W., Roggenbuck, B., Stremlau, A., and zur Hausen, H. (1985). Structure and transcription of human papillomavirus sequences in cervical carcinoma cells. Nature *314*, 111–114.

Sekaric, P., Cherry, J.J., and Androphy, E.J. (2008). Binding of human papillomavirus type 16 E6 to E6AP is not required for activation of hTERT. J. Virol. *82*, 71–76.

Southern, S., and Herrington, C.S. (2000). Disruption of cell cycle control by human papillomaviruses with special reference to cervical carcinoma. Int. J. Gynecol. Cancer *10*, 263–274.

Thomas, J.T., Laimins, L.A., and Ruesch, M.N. (1998). Perturbation of cell cycle control by E6 and E7 oncoproteins of human papillomaviruses. Papillomavirus Rep. *9*, 59–64.

Vande Pol, S.B. (2011) E6 oncoproteins: Structure and associations. In Small DNA Tumour Viruses, Gaston, K., ed. (Horizon Scientific Press, Norfolk, UK), pp. 77–97.

Veldman, T., Liu, X., Yuan, H., and Schlegel, R. (2003). Human papillomavirus E6 and Myc proteins associate *in vivo* and bind to and cooperatively activate the telomerase reverse transcriptase promoter. Proc. Natl. Acad. Sci. U.S.A. *100*, 8211–8216.

de Villiers, E. -M., Fauquet, C., Broker, T.R., Bernard, H. -U., and zur Hausen, H. (2004). Classification of papillomaviruses. Virology *324*, 17–27.

Walboomers, J., Jacobs, M., Manos, M., Bosch, F., Kummer, J., Shah, K., Snijders, P., Peto, J., Meijer, C., and Munoz, N. (1999). Human papillomavirus is a necessary cause of invasive cervical cancer worldwide. J. Pathol. *189*, 12–19.

Watson, R.A., Thomas, M., Banks, L., and Roberts, S. (2003). Activity of the human papillomavirus E6 PDZ-binding motif correlates with an enhanced morphological transformation of immortalized human keratinocytes. J. Cell Sci. *116*, 4925–4934.

Webster, K., Parish, J., Pandya, M., Stern, P.L., Clarke, A.R., and Gaston, K. (2000). The human papillomavirus (HPV) 16 E2 protein induces apoptosis in the absence of other HPV proteins and via a p53-dependent pathway. J. Biol. Chem. *275*, 87–94.

Werness, B.A., Levine, A.J., and Howley, P.M. (1990). Association of human papillomavirus types 16 and 18 E6 proteins with p53. Science *248*, 76–79.

Wilson, V.G., West, M., Woytek, K., and Rangasamy, D. (2002). Papillomavirus E1 proteins: form, function, and features. Virus Genes *24*, 275–290.

Winder, D., Pett, M., Foster, N., Shivji, M., Herdman, M., Stanley, M., Venkitaraman, A., and Coleman, N. (2007). An increase in DNA double-strand breaks, induced by Ku70 depletion, is associated with human papillomavirus 16 episome loss and *de novo* viral integration events. J. Pathol. *213*, 27–34.

Woodman, C.B.J., Collins, S.I., and Young, L.S. (2007). The natural history of cervical HPV infection: unresolved issues. Nat. Rev. Cancer *7*, 11–22.

Zerfass-Thome, K., Zwerschke, W., Mannhardt, B., Tindle, R., Botz, J.W., and Jansen-Dürr, P. (1996). Inactivation of the cdk inhibitor p27KIP1 by the human papillomavirus type 16 E7 oncoprotein. Oncogene *13*, 2323–2330.

Zhou, J., and Frazer, I.H. (1996). Papovaviridae: capsid structure and capsid protein function. In: Papillomavirus Reviews: Current Research on Papillomaviruses, Lacey, C., ed. (Leeds University Press, Leeds, UK), pp. 93–100.

The Art and Science of Obtaining Virion Stocks for Experimental Human Papillomavirus Infections

2

Michelle A. Ozbun and Michael P. Kivitz

Abstract

Some human papillomaviruses (HPVs) have been propagated in the laboratory for ≈20 years; currently, there are multiple means for obtaining infectious virion stocks for experimental infections. Processes dependent upon epithelial differentiation for achieving virion production include natural warts, rodent xenografts, and cultured organotypic epithelial tissues. However, these options are not amenable for every HPV genotype, and virion yields are variable and viral genotype-dependent. Differentiation-dependent laboratory approaches are limited to virus production from replication-competent viral genomes and typically only viral genomes capable of conferring a significant cellular growth advantage can be maintained long-term. Cutaneous HPV and animal PV lesions yield the highest virion levels and pure stocks can be obtained for many cutaneous PVs. Only crude, low-purity virion stocks have been reported for carcinogenic HPVs (e.g. types 16, 18, 31) typically via organotypic tissue propagation. Advances in virion production methods independent of epithelial differentiation permit the isolation of high-titre, high-purity viral stocks from virtually any cloned PV genotype. Based on the self assembly of transiently expressed capsid proteins to package DNA molecules of ≈5 to 8 kb in size, this approach provides the ability to produce infectious particles encapsidating any viral or reporter genome regardless of its replicative ability. These particles share many structural and functional similarities with differentiation-induced virions; however, questions remain about the absolute physiological likeness among PV stocks derived by different means. Herein we discuss advantages and disadvantages among the varied means of virion production, detail known similarities, and make note of potential disparities that have yet to be tested.

Introduction

The family *Papillomaviridae* comprises a group of small, DNA-containing viruses long recognized to infect mucosal and cutaneous epithelium to cause benign and malignant tumours in their natural hosts (Hurst, 1933; Shope, 1935; Strauss *et al.*, 1949, 1950; zur Hausen, 1976; Lancaster and Olson, 1982; Pfister, 1984; Shope Howley, 1996). Papillomaviruses (PVs) demonstrate remarkable host restrictions, and productive PV infections are not observed to cross species *in vivo*. Thus, human PVs (HPVs) infect only humans (Rowson and Mahy, 1967; Koller and Olson, 1972). In addition to a narrow host range, productive HPV infection *in vivo* is limited to differentiating squamous epithelial cells and human keratinocytes. Papillomaviruses have co-evolved with their hosts for millions of years, leading to tightly synchronized viral gene expression with the differentiation state of infected cells in stratifying epithelium. Early viral gene products are both poorly immunogenic and expressed at very low levels, and only cells committed to terminal differentiation are observed to express late gene products at high levels (reviewed by Doorbar, 2005). Following assembly, there is scant physical evidence that PV virions (see Table 2.1 for PV particle terminology) are ever overtly released from cells; rather, it

Table 2.1 Papillomavirus particle terminology

Virion: a complete virus particle composed of L1, L2 and viral genome; capable of initiating an infection and expressing viral genes
Virus-like particle (VLP): a non-infectious particle composed of L1 with or without L2, but encapsidating no specific nucleic acid
Pseudovirion (PsV): a virus particle composed of L1 and L2 capable of transducing a reporter gene expression plasmid
'Quasi-virion': coined by Culp and Christensen to refer to infectious particles assembled *in vitro* (293T cells) 'until their virion-like nature is more fully evaluated' (Culp *et al.*, 2006c)
'Native' virion: coined by Meyers *et al.* to refer to HPV virions obtained from differentiated tissues including warts, xenografts and the *in vitro* organotypic culture system (Conway *et al.*, 2009b)

is probable that they are shed with desquamating cells for transmission (Bryan and Brown, 2000). Additionally, over 30% of women testing positive for the presence of HPV DNA fail to show signs of seroconversion, further emphasizing the virus's ability to stay below the immunological radar (Carter *et al.*, 1996).

PV virions are made up of 360 copies of the L1 major capsid protein, 12–72 copies of L2, which is mostly internal to the particle, and the circular viral genome condensed by cellular histones (Howley and Lowy, 2007). According to current models, HPVs gain access to the mitotically active basal cell layer through a microabrasion or wound in the epithelium to set up a persistent infection (White *et al.*, 1963; Broker and Botchan, 1986). Although wounding is an obvious barrier to pathogen access and has been shown to potentiate infection (Friedewald, 1942; Roberts *et al.*, 2007; Cladel *et al.*, 2008), abrasion has never been tested as an *essential* component of epithelial infection. Nevertheless, once past the barrier at the outer surface of the skin, HPV particles bind with high affinity to laminin 5 and heparan-sulfonated proteoglycan ectodomains, which are secreted by human keratinocytes onto the extracellular matrix (Culp *et al.*, 2006a,b; Smith *et al.*, 2008b; Kines *et al.*, 2009). These moieties appear to hold virions at the wound site for eventual transfer to the plasma membrane of adjacent susceptible keratinocytes. Filopodia, which can be induced by wounding and by HPV virion exposure, take up particles from the environment and the extracellular matrix (Schelhaas *et al.*, 2008; Smith *et al.*, 2008b), and migrating cells in a wound may move over matrix-bound virions. Studies in monolayer cell cultures identify α6-integrin and heparan-sulfonated proteoglycans, specifically syndecan-1, as cellular attachment factors (Evander *et al.*, 1997; Joyce *et al.*, 1999; Giroglou *et al.*, 2001; Drobni *et al.*, 2003; Shafti-Keramat *et al.*, 2003). New evidence from our lab suggests that HPV virions become associated with heparan sulfate and growth factors, which then can use cognate growth factor receptors to activate signalling for infectious entry (Surviladze *et al.*, 2011). The literature contains much disagreement over the requirement for specific attachment moieties, as well as the routes of internalization and cellular organelles involved in authentic HPV infections. The purity of virus particles, varied experimental designs and cell lines used in each study probably contribute to the disparate conclusions: the difficulty in obtaining high-titre infectious stocks for most PVs has necessitated the use of virus-like particles (VLPs) and pseudovirions (PsVs) where bona fide infection (i.e. nuclear delivery of and expression from the encapsidated nucleic acid) often cannot be verified in host epithelial cells. For example, although VLPs facilitate cell-surface binding studies, they cannot be used to assess functional infection, and working with PsVs often has predominantly relied on SV40 T-antigen expression in non-host cells to facilitate sensitive reporter gene detection. In the past few years, it has become possible to efficiently assemble high-titre infectious virions encapsidating viral genomes as well as PsVs carrying reporter plasmids (pseudo-genomes) in the absence of differentiating tissues. Typically, transformed human embryonic kidney cells expressing SV40 large T-antigen (HEK293T, aka 293T) are used for ectopic expression of the capsid proteins L1 and L2 and packaging of reporter plasmids or viral genomes (Buck *et al.*, 2004; Pyeon *et al.*, 2005).

As discussed below, the 293T-derived particles share many structural and functional features of virions obtained from tissues *in vivo*. However, a number of arguments have arisen whether the 293T-derived particles function like infectious virions derived from warts; various terminology has been coined (see Table 2.1) to differentiate the particles obtained from the spectrum of sources: warts, xenografts, organotypic tissues, 293T cells, insect cells, yeast, etc. Our laboratory has been studying HPVs for nearly fifteen years and has extensive experience generating virions from both organotypic raft tissue cultures as well as the 293T system; we have also studied virion stocks from warts, xenografts, rafts and 293T cells. With this expertise, the goal of this chapter is to review the benefits and limitations among the varied means of virion isolation, discuss the known likenesses, and note disparities. We also address remaining issues and propose guidelines we feel will be beneficial to interpreting the many, varied studies of laboratory PV production in an attempt to standardize future work for the benefit of all – and, ultimately, for those whose lives we seek to improve by studying these important pathogens.

Overview of the papillomavirus life cycle as it pertains to virion production

The ability of PVs to undergo a complete replication cycle culminating in the production of infectious progeny (i.e. virions) is tightly linked to the differentiation state of the infected cells (zur Hausen and Schneider, 1987; Howley and Lowy, 2007). This occurs via strict regulation of gene expression from the major early and late cluster of promoters as discussed in detail in other chapters of this issue (see Chapter 3; Oparka and Herrington, 2011). Although basal epithelial cells cannot initially fulfil the requirements for the HPV life cycle on their own, they become virally permissive after HPV genome establishment via detachment from the basement membrane, at which point they begin differentiating and superficially migrate to replace older cells and maintain full thickness epithelium (reviewed by Doorbar, 2005). Thus, HPVs cannot be replicatively propagated in traditional undifferentiated monolayer cell cultures. In anogenital epithelia *in vivo*, productive HPV infection appears to occur predominantly in benign or lower grade lesions referred to as condylomata acuminata, cervical or anal low-grade squamous intraepithelial lesions (LSIL), which are slightly altered in their differentiation scheme compared with normal epithelium (Koss, 1987). In higher-grade lesions (e.g. HSIL, carcinoma *in situ*, or invasive carcinoma), the cells remain undifferentiated, HPV genomes may be present at high levels and/or may be integrated into the host genome, E6 and E7 oncogene expression is maintained and/or increased, and virion production is not typically observed (Matsukura *et al.*, 1989; Morrison *et al.*, 1991; Howley, 1996; Forslund *et al.*, 1997; Swan *et al.*, 1999).

Obtaining papillomavirus virion stocks

Naturally occurring warts

Viral stocks can be readily purified from cutaneous skin lesions caused by the BPV types, CRPV, HPVs, and from the mucosotropic canine oral PV (COPV) (Crawford and Crawford, 1963). However, virions from most HPV types are produced in relatively smaller amounts *in vivo* (Pfister, 1984) and obtaining sufficient virion quantities necessary for infectivity studies has historically been severely limited, inhibiting many studies of HPV biology, especially for oncogenic genotypes (Table 2.2). As a rule, the more carcinogenic the HPV genotype, the fewer virus particles produced in warts/lesions *in vivo*. For example, plantar and common warts, which do not progress to malignancies, can yield up to 7×10^9 HPV particles per mg of tissue, whereas laryngeal papillomas caused by HPV6 or HPV11 rarely lead to cancer and produce $<10^3$ HPV particles per mg of papilloma (Barrera-Oro *et al.*, 1962; Butel, 1972; Boyle *et al.*, 1973). In the mid-1970s, Favre *et al.* used electron microscopy (EM), SDS-PAGE and Coomassie staining to show the high purity and morphology of virion stocks isolated from rabbit, cow, and human warts (Favre *et al.*, 1975); they also demonstrated histone-sized proteins in pure virus preparations via SDS-PAGE and Coomassie staining as well as chromatin-like structures by

Table 2.2 Yield of purified papillomavirus virions from various sources

Source	Yield (per mg of starting tissue)	Reference
Plantar and common warts (probably HPV1, HPV2)	7×10^9 particles	Barrera-Oro (1962), Butel (1972)
Laryngeal papillomas (HPV6, HPV11)	$<10^3$ particles	Boyle *et al.* (1973)
HPV11-infected xenografts	4.5×10^8 particles	Bonnez (personal communication), Smith *et al.* (1995)
HPV31-infected raft tissues	4×10^5 vge[1]	Ozbun (2002a,b)
HPV18-infected raft tissue	4×10^5 vge	Meyers *et al.* (1997), Ozbun (unpublished)
HPV18-infected raft tissue	4×10^6 vge	Wang *et al.* (2009)
HPV16-infected raft tissue	$\approx10^4$ vge[2]	Ozbun (unpublished)
HPV16 from 293T cells	2.5×10^9 vge	Campos and Ozbun (2009)
HPV31 from 293T cells	5×10^8 vge	Smith *et al.* (2007)

[1]Based on estimating rafts as 10^7 cells, and 62 mg of tissue per raft.

[2]Before DNase treatment.

EM after disrupting the PV particles (Favre *et al.*, 1975, 1977). There have been no reports of viral particle isolation from the typically small lesions from anogenital warts or those that can progress to malignancies arising from carcinogenic HPV types. Although some of the studies described above and elsewhere attempted experimental infections with human wart-derived virions in various cells and in mouse xenografts (see below), such work has not been undertaken in monolayer cells or within the era of RT-qPCR detection of viral transcripts. Therefore, little is known about the cell and molecular biology of early infection by cutaneous HPV types.

The development and implementation of assays for HPV infection has been a challenge. Infectivity methods based on late life cycle activities are not feasible (e.g. a plaque assay or an assay for viral particles) because PVs are differentiation dependent and the viruses are not lytic. To date, the only true, cell-based quantitative assay for PV infectivity involves the use of non-host cells whereby BPV1 and BPV2 can cause focal transformation in mouse cells; however, no other PVs have been found to possess this ability (Dvoretzky *et al.*, 1980). These methods indicate a particle-to-infectious-unit ratio of 10^4:1 for the BPVs in non-host murine C127 cells (Roden *et al.*, 1996b). Although there is no such assay for HPVs, available data suggest infectivity is similarly inefficient using raft tissue-derived viruses (see below) in current experimental systems (Ozbun, 2002a,b).

Murine xenografts

HPV11 has been the type most readily propagated and studied in the xenograft system. Although details of the actual purification and quantification of particles are limited, HPV11 infected xenograft tissues can yield $\approx4.5\times10^8$ particles per mg of starting material, with purity estimated at 90% (Bonnez, personal communication; (Smith *et al.*, 1995)). One report cites the production of infectious HPV16 from xenografted SCID mice, but the details of virus purification were not published (White *et al.*, 1998). Although there is no documentation of the quantities of virions obtained, HPV16 virion production in the xenograft model appears to be lower than that of HPV11 (Bonnez, personal communication). Despite the fact that few HPV types have been cultivated effectively in the xenograft system, this model permitted examination of differentiation-regulated viral gene and protein expression (Brown *et al.*, 1995; Stoler *et al.*, 1990), examination of tissue tropism (Kreider *et al.*, 1987), the first experimental HPV infectivity studies in cultured cells (Angell *et al.*, 1992; Smith *et al.*, 1993, 1995), and importantly, evaluation of antibody-mediated infection neutralization (Christensen and Kreider, 1990; Christensen *et al.*, 1992, 1995).

Organotypic (raft) epithelial tissues

The culture of organotypic epithelial tissues has greatly benefited HPV research as the differentiation achieved gives rise to an environment permissive for the complete viral life cycle (McCance *et al.*, 1988; Meyers, 1996). This is the only *in vitro* system that can support full epithelial differentiation and, thus, provides a permissive environment for the complete viral replicative life cycle including virion production (Fig. 2.1A). Meyers and coworkers were the first to reproduce the HPV life cycle *in vitro* using a continuous cell line, CIN 612 9E, established from a low-grade cervical biopsy and persistently infected with HPV31 subtype b (Bedell *et al.*, 1991; Meyers *et al.*, 1992). Over the years, there have been an increasing number of publications describing HPV virion production in the raft tissue culture system from naturally infected cells (Dollard *et al.*, 1992; Meyers *et al.*, 1992; Ozbun, 2002a,b) and from human keratinocytes stably transfected with and maintaining episomal HPV genomes

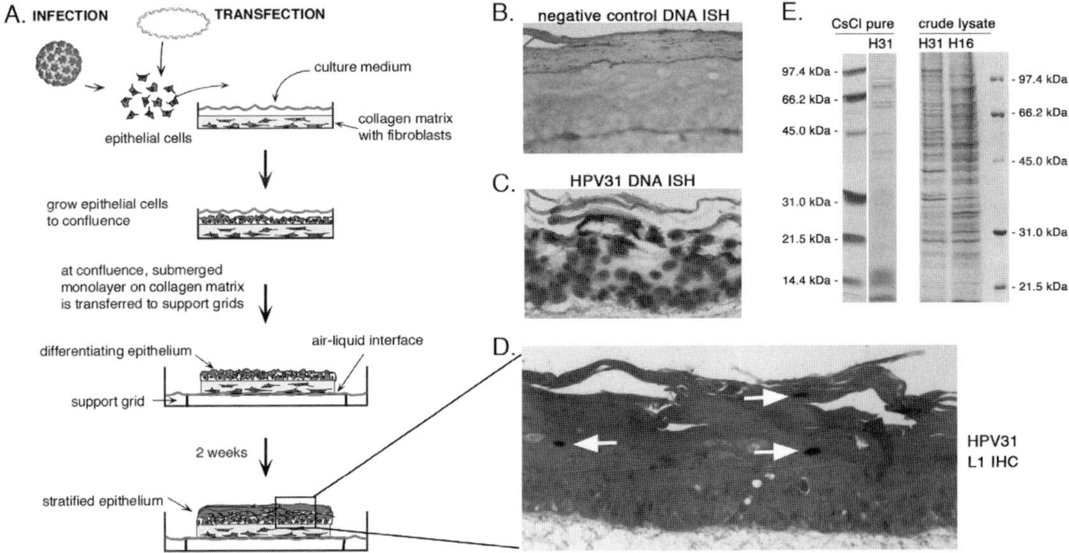

Figure 2.1 HPV virion production from the organotypic (raft) tissue culture system. (A) Epithelial cells are seeded onto 3.5 cm submerged type I collagen matrices containing fibroblasts. When the epithelial cells reach confluence, the collagen matrices are lifted onto stainless steel support grids. The collagen matrix on the support grid suspends the epithelial cells as a 'raft' at the air–liquid interface to promote differentiation. The cells are fed by diffusion from below; the collagen-fibroblast matrix acts as a dermal equivalent. Epithelial tissues differentiate over a 10–14 day period (Meyers, 1996; Ozbun and Meyers, 1998a; Steele *et al.*, 2002). (B, C) *In situ* hybridization for (B) lambda phage DNA and (C) HPV31 DNA in CIN-612 9E raft tissue sections. (D) Immunohistochemistry for HPV31 L1 where arrows point to cells positive for L1 protein. (E) Raft tissues (100–300) were extracted as previously described (Favre *et al.*, 1975). After the low speed clarifications, the viral DNA-containing particles were pelleted based on a sedimentation coefficient of 296S–300S for 'full' viral DNA-containing particles (Crawford and Crawford, 1963) for 1 h at 4°C in a swinging bucket rotor at 130,000×g. The supernatants were discarded and the virus pellets suspended in 2 ml of buffer B (0.05 M NaCl, 0.01 M EDTA, 0.05 M sodium phosphate buffer, pH 7.4) using a Dounce homogenizer. The debris was pelleted at 8000×g for 10 min at 4°C. The supernatants were kept on ice while the pellet was re-extracted with 2 ml of buffer B and then pelleted again under the same conditions ('crude lysate'). Alternatively, caesium chloride was added to the final pooled 'crude' supernatants and the refractive index was used to verify the density at 1.3 g/ml. A gradient was formed by centrifugation at 135,000×g for 24 h at 4°C. The tubes were bottom-punctured and 0.5 ml fractions were collected. Each fraction was placed in SpectraPor tubing (12,000–14,000 MW cut-off) and the fractions were dialysed against four to five changes of PBS. An aliquot from the gradient fraction containing viral DNA was subjected 10% SDS-PAGE and Coomassie Blue staining for proteins (CsCl pure) (Ozbun, 2002a).

(Meyers *et al.*, 1997; McLaughlin-Drubin *et al.*, 2003; Holmgren *et al.*, 2005; Conway *et al.*, 2009b). Adenovirus transduction and *Cre-LoxP* recombination has also been used to deliver replicating viral genomes to human keratinocytes for virion biosynthesis (Lee *et al.*, 2004; Wang *et al.*, 2009). In our work we have striven to obtain the purest possible HPV virion stocks of HPV16, HPV18 and HPV31 from raft tissues for studies of early infection events (Ozbun, 2002a,b; Lee *et al.*, 2004; Holmgren *et al.*, 2005; Patterson *et al.*, 2005; Smith *et al.*, 2007; Song *et al.*, 2010). Our procedures for virion isolation from tissues were adapted from those of Favre and co-workers (Favre *et al.*, 1975; Meyers *et al.*, 1992) wherein differentiated epithelial raft tissues were ground in sea sand or in a sealed BeadBeater™ homogenizer (see below), virion suspensions clarified by low-speed centrifugation, pelleted based on the sedimentation coefficient for DNA-containing particles, and resuspended in a small volume of buffer. In the majority of cases, the preparations were centrifuged to equilibrium in a CsCl or Opti-Prep™ iodixanol gradient and fractionated (Ozbun, 2002a,b). It must be noted that the carcinogenic HPV types, especially HPV16 and HPV18, are thought to cause a subset of head-and-neck cancers (reviewed by Chung and Gillison, 2009). Owing to the carcinogenic nature of these HPVs, we use BSL2$^+$ biosafety procedures and HEPA-filtered masks to reduce possible exposure to aerosols that might be generated during the purification of HPV stocks from organotypic tissues. We find the BeadBeater™ (BioSpec Products), which utilizes glass beads in a sealed mechanical Teflon® homogenization device, to reduce aerosol generation and yield virus preps of reproducibly high quality (Ozbun, 2002a,b).

Although most of the cells in the raft tissue culture exhibit high levels of viral DNA, only a few cells in the epithelial granular layer express detectable capsid proteins (Fig. 2.1B–D) (McLaughlin-Drubin *et al.*, 2004; Holmgren *et al.*, 2005). Thus, the few cells permissive for viral morphogenesis in a background predominantly made up of non-productive cells makes it quite challenging to isolate high-titre or pure virion stocks from raft tissues. Unlike the HPV virion stocks purified under the same conditions from cutaneous warts, which show discrete major and minor capsid bands by SDS-PAGE and Coomassie staining (Favre *et al.*, 1975), raft-derived HPV stocks are quite contaminated with other proteins (Fig. 2.1E) and difficult to image clearly by EM owing to low concentrations (Meyers *et al.*, 1992, 1997). As indicated above, there is currently no quantitative or cell-based infectivity assay for HPV virions. HPV virion stocks can be enumerated by protein quantity of L1 and/or by quantifying viral genomes from virion preparations to determine the number of viral DNA-containing particles. In the latter scenario, the physical (as opposed to infectious) 'titre' units are referred to as viral genome equivalents (vge) (Roden *et al.*, 1996b; Meyers *et al.*, 1997; Ozbun, 2002a,b; Conway *et al.*, 2009b).

Raft virus yields and specific infectivity reports vary

Based on vge, we have found that we can consistently obtain HPV31 $\approx 3 \times 10^7$ vge per raft, which turns out to be ≤ 1 vge/cell or 4×10^5 vge per mg tissue. This average is 3–4 orders of magnitude lower than yields based on tissue weight from cutaneous HPVs, consistent with virion yields decreasing as oncogenicity increases. We have achieved similar yields of HPV18 from rafts; however, the amounts of HPV16 based on vge quantification of CsCl gradient-banded particles acquired from W12-E, variant 114/B, or NIKS-SG3 [genome from W12 cells (Genther *et al.*, 2003)] clonal cell lines are lower by at least another order of magnitude. For particles carrying viral genomes, the detection and quantification of newly synthesized, spliced viral RNAs is currently the most useful and common means of assaying early infection; targeting spliced E1^E4 is the most sensitive readout (Angell *et al.*, 1992; Smith *et al.*, 1995; White *et al.*, 1998; Ozbun, 2002a). Using 40–50 cycles of endpoint RT-PCR, we are consistently able to detect HPV16 and HPV31 infection as measured by E1^E4 presence by incubating human keratinocyte cell lines such as HaCaT, NIKS, and N/TERT-1 with a viral dose between 1 and 10 vge/cell and assaying at 48 hours post infection (Fig. 2.2E) (Ozbun, 2002a,b; Lee *et al.*, 2004; Holmgren *et al.*, 2005; Patterson *et al.*, 2005). The low levels of viral transcripts

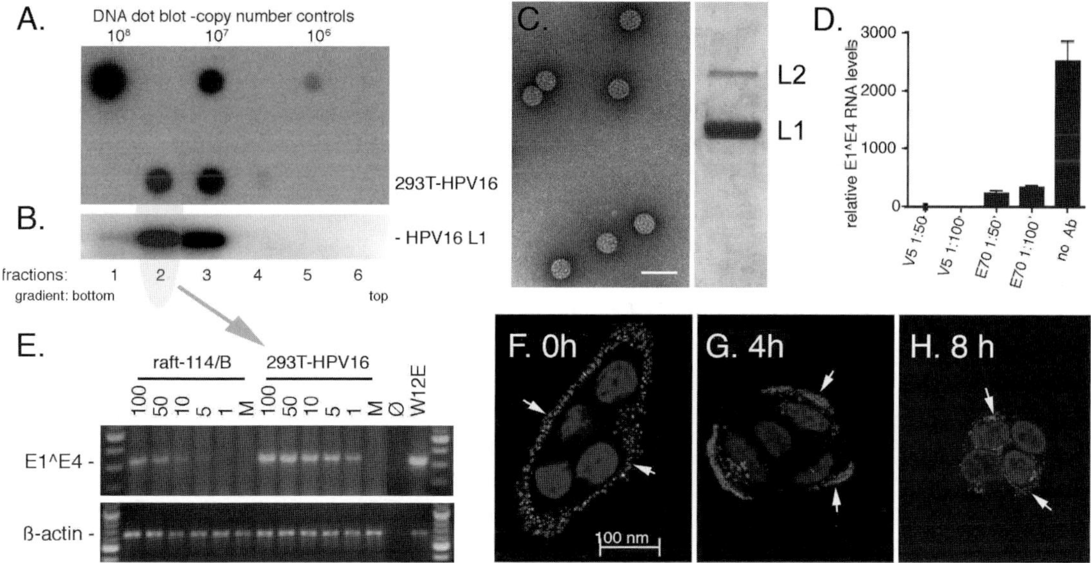

Figure 2.2 Infectious 293T-virion generation. HEK293T cells were transfected with a plasmid expressing HPV16 L1 and L2 (Buck *et al.*, 2004; Pyeon *et al.*, 2005) and with circular HPV16 genome. After 48 h, the cells were lysed and treated with Benzonase to digest unpackaged DNA; capsids were allowed to mature (37°C, 16–24 h) (Buck *et al.*, 2005). (A and B) Fractions were collected following CsCl density gradient centrifugation and analysed by dot blot hybridization for HPV16 genome quantification (A) and L1 capsid protein by Western blot (B) (fraction collected from the top of the gradient are shown aligned in columns in A and B). (C) Pure 293T-HPV16 virion preparation. 293T-HPV16 particles encapsidating HPV16 genomes were CsCl gradient purified and analysed by transmission EM (left) or 10% SDS-PAGE and Coomassie gel staining revealing major and minor capsid proteins (right). (D) Antibody-mediated neutralization of 293T-HPV16. Virions were incubated with dilutions of HPV16 NAb H16.V5 and H16.E70 (1 h, 37°C) immediately prior to infection. E, HPV virion infectivity comparing raft tissue-derived HPV16 virions (114/B) to 293T-HPV16 virions. Fraction 2 of 293T-HPV16 and a raft-derived HPV16 (114/B) virion stock were tested for infectivity by incubating HaCaT cells with doses of 100, 50, 10, 5 or 1 vge/cell. Total RNAs were purified 48 h pi and analysed for viral E1^E4 transcripts or cellular β-actin by endpoint RT-PCR as we reported (Lee *et al.*, 2004). Mock infections (M), persistently HPV16-infected cells (W12E), and no RNA input (Ø) served as controls. F-H, HaCaT cell binding and entry using Alexa Fluor-594 labelled 293T-HPV16 particles (arrows point to virion spots). Cells were incubated with ≤10,000 vge/cell; photos were taken of cells at 0, 4, 8 h pi to show attachment and entry. DAPI was used to visualize nuclei; Bar = 10 μm.

detected following experimental infection *in vitro* suggest a very high particle-to-infectious unit ratio as reported for BPV1 by Roden *et al.* (Roden *et al.*, 1996b; Ozbun, 2002a). Although Conway *et al.* reported obtaining up to 2×10^9 vge of HPV16 per raft, 30–1500 vge/cell of crude cell lysates were required in order to detect infection by E1^E4 amplification by RT and nested PCR (Conway *et al.*, 2009b). The Broker-Chow lab reported $3–5 \times 10^8$ vge per HPV18 raft culture following the *Cre-LoxP* transduction approach and density gradient purification of virions (Wang *et al.*, 2009). Following infection with decreasing virus doses, E1^E4 detection by 30 cycles of endpoint RT-PCR required >40 vge/cell. Thus, the infectious 'titres' of virions produced in the raft system are within an order of magnitude of one another, making no single approach significantly more robust than any other.

An essential aspect of establishing the specificity of infection by a viral stock is demonstrating antibody-mediated neutralization. The fact that VLP-based vaccines are effective at eliciting protection against PV infections and pathology in animal models and humans (Christensen *et al.*, 1994; Rose *et al.*, 1994; Fligge *et al.*, 2001a; Suzich *et al.*, 1995; Roden *et al.*, 1996a; Lu *et al.*, 2011) underscores the relevance of antibody-mediated neutralization in verifying the specificity of infection and quality of HPV virion or PsV stocks. VLP-based vaccines are effective at provoking functional, robust antibody protection against natural virion transmission *in vivo*, lending support to the proposed structural similarity amongst VLPs and naturally occurring virions. Although we readily observed specific antibody-mediated neutralization of CsCl-banded HPV31 virion stocks from raft tissues *in vitro* (Ozbun, 2002a), we found some raft-derived HPV16 stocks to be resistant to antibody-mediated neutralization as measured by RT-PCR detection of spliced viral RNAs. Furthermore, DNase I treatment of these HPV16 virion stocks very significantly reduced the vge titres and diminished the readout of infection as measured by RT-PCR, whereas DNase I had no effect on infections with HPV31 virion stocks (Ozbun, 2002a). The issue with high levels of viral genomes and low L1 levels in the raft tissues, which appears more pronounced in HPV16 raft tissues, indicates that virion production is below the 10^4 vge per mg of tissue suggested by viral DNA analysis described above. Similarly, the Meyers lab showed that only 2% of viral DNA present in raft tissues was resistant to Benzonase (a promiscuous nucleic acid endonuclease), more quantitatively underscoring the low level of viral genome encapsidation and high levels of potentially contaminating unpackaged viral DNA in raft tissues (Conway *et al.*, 2009b). Since some HPV16 raft-derived virion preparations are not susceptible to neutralizing antibodies, and DNase I treatment of the virus reduces the readout of viral genome quantification, we interpret these findings as evidence of unpackaged viral nucleic acids in the virion preparations that could be transduced into, or otherwise associated with, cells exposed to the virus solutions giving a false-positive result for infection. There is precedent for transduction of unpackaged DNA as the Coursaget group linked a reporter gene to purified capsids for PsV production and gene transfer experiments (Combita *et al.*, 2001). There are also reports indicating that cells, including human keratinocytes, can take up and express free DNA via macropinocytosis (Basner-Tschakarjan *et al.*, 2004; Wittrup *et al.*, 2007). This emphasizes the need to show the specificity of capsid-mediated transfer of viral genomes in infection studies. There have been a number of recent reports involving experimental infections using crude lysates of raft tissues (Alam *et al.*, 2008; Conway *et al.*, 2009a,b, 2011; Conway and Meyers, 2009; Chen *et al.*, 2011). Because of our experience working with the most pure raft tissue-derived virion stocks, which we show still have many contaminating proteins and DNA (Fig. 2.1E), it seems that caution should be used in interpreting infection data stemming from crude cell lysates. Certainly, controls for infectivity and to rule out transduction, including antibody-mediated neutralization, should be demonstrated before conclusions can be made from experiments that employ crude preparations.

Qualitative versus quantitative studies

Quantitative comparisons have been made among various chimeric and mutant HPV genomes in the organotypic epithelial raft system (Meyers *et al.*, 2002; Conway *et al.*, 2009a,b; Chen *et al.*, 2011). However, there have been no reports that have included a quantitative investigation of the reproducibility of growth, differentiation, and – most importantly for the topic at hand – virion production from this model to lay the foundation for such studies. As the raft tissue culture system is notoriously fastidious in growth requirements, the level of differentiation, and hence virion production, can vary widely (Meyers, 1996). Experimental details impacting differentiation and late gene expression include the initial seeding density of epithelial cells, the species and passage number of the fibroblasts that are embedded in the collage matrix, the collagen lot and acidity, the number of days rafts are permitted to differentiate (Ozbun and Meyers, 1997, 1998b; Steele *et al.*, 2002; Conway *et al.*, 2009b), number of rafts per grid and dish, volume of media used to feed, presence of various cytokines and growth factors (Ozbun and Meyers, 1996), and type of protein kinase C inducer used to promote differentiation (Meyers *et al.*, 1992; Meyers, 1996). Although viral gene mutants studied in raft tissues have greatly facilitated our understanding of the qualitative functions of HPV gene products during the viral replicative cycles (Stubenrauch *et al.*, 1992, 1998, 2000; Fehrmann *et al.*, 2003; Genther *et al.*, 2003; Holmgren *et al.*, 2005), there have also been disparate results. For example, E7 is required for genome maintenance in low passage human foreskin keratinocytes, but not in immortalized human foreskin keratinocytes (Thomas *et al.*, 1999; Flores *et al.*, 2000). The ability to study differentiation-dependent HPV activities or aspects of long-term genome maintenance have been limited thus far to carcinogenic HPV types, which is probably due to the inability of non-oncogenic types to provide cells with the growth advantage necessary for viral genome maintenance *in vitro*.

Finally, the low purity of raft-derived virions prevents their use in a number of important experimental situations, including fluorescent labelling and particle trafficking by microscopy, virion-initiated cell signalling studies, mass spectroscopy analyses, structural analysis, and binding affinity assays. As discussed in greater detail below, raft tissue-derived virions have never been compared with either xenograft-derived particles or virions obtained from warts. Thus, it is unclear if raft cultivated virions are representative of PV particles produced in natural infections.

In vitro assembly of PV virions and PsVs

High-titre production of HPV virions independent of viral replication was first reported by Pyeon *et al.* (2005). The method builds upon work by Buck and co-workers whereby small DNA pseudogenomes, like reporter plasmids, are packaged in self-assembling L1 plus L2 capsids to make PsV in transiently transfected 293T mammalian cells (Buck *et al.*, 2004). A key element of the system is the use of L1 and L2 genes that carry an extensive series of silent mutations designed to eradicate the inhibitory RNA sequences that prevent the expression of L1 and L2 in basal keratinocytes and in cultured monolayer cells (Zhou *et al.*, 1999; Leder *et al.*, 2001; Hindmarsh and Laimins, 2007). Fig. 2.2 shows the characterization of HPV16 virion stocks made via transfection of 293T cells. Transfection included a plasmid expressing optimized HPV16 L1 and L2 open reading frames, which has a >9 kb size that precludes efficient packaging into the 55 nm icosahedral viral particles (Buck *et al.*, 2004; Pyeon *et al.*, 2005) plus viral genomes that had been re-circularized following release from a plasmid backbone by restriction endonuclease digestion. The density gradient-purified 293T virion fractions that contain L1 also contain HPV16 genomes (Fig. 2.2A,B fractions 2–3). We typically use a reduced amount of transfected L1/L2 capsid expression vector to minimize the production of empty particles, which could interfere with our infection studies. With this approach, we generally produce 2.5×10^9 vge (5×10^9 vge/ml) from $\approx 10^7$ cells, >100× more than we can produce from the same cell numbers in the raft tissue culture system (Meyers *et al.*, 1997; Smith *et al.*, 2007, 2008a,b; Campos and Ozbun, 2009). The 293T-derived DNA and L1 positive fractions confer infection at a dose as low as 1.0 vge/cell, and the levels of infectivity are at least as high as those of raft-derived HPV16 (Fig. 2.2E; 'raft-114/B'). Importantly, the infectivity of these particles can be neutralized with HPV16-specific monoclonal antibodies (Fig. 2.2D) (Christensen *et al.*, 1996). As shown by the TEM and Coomassie staining (Fig. 2.2C), the virions from the 293T system are quite pure and thus can be externally labelled with a fluorophore (e.g. Alexa Fluor dye 'AF594', Invitrogen-Molecular Probes). This permits multiparameter microscopic tracking of particle entry and trafficking (Fig. 2.2F–H). Such techniques are not possible with the low yields and poor purification of raft-derived virus particles. Similar 293T transfections have been repeated hundreds of times in our lab with highly reproducible yields and titres. In addition to the high yields, another advantage of this system over the raft system includes the ability to assemble particles from virtually any PV type if the capsid genes have been cloned. A major benefit to this method is that it is not limited to encapsidating replication-competent viral genomes: virtually any type of chimera or mutant genome can be created from cloned genomes. As discussed below, rabbit infections were virtually indistinguishable when 293T-derived CRPV virion inoculations were compared with tissue-derived CRPV inoculations (Culp *et al.*, 2006c).

Comparisons: transfection-based particles versus virions from tissues

Similarities among transfection-based virions and those from tissues

Structure

EM analyses revealed that VLPs are structurally indistinguishable from virions isolated from naturally occurring lesions (Hagensee *et al.*, 1994). In addition, VLPs compete with virions for binding to the cellular receptor (Roden *et al.*, 1994) and L1-only containing VLPs are capable of normal internalization in cells (Kämper *et al.*, 2006). VLPs are antigenically similar to natural virions, they induce neutralizing antibodies in many systems (Christensen *et al.*, 1994; Rose *et al.*, 1994; Suzich *et al.*, 1995; Roden *et al.*, 1996a; Fligge *et al.*, 2001a), and VLP-based vaccines are highly efficacious in preventing persistent infections and cervical diseases associated with vaccine-specific HPV types among young women (Lu *et al.*, 2011).

The redox environment is important for the formation of disulfide bonds between and amongst individual capsid proteins that leads to stable, mature particles (Buck and Trus, 2011). In normal skin, the redox gradient never seems to reach a fully oxidized state and the skin's outermost surface continues to be rich in free thiol groups, which probably have a role in barrier maintenance (Ogawa *et al.*, 1979; Pickard *et al.*, 2009). It has been argued that differences in the redox states of monolayer versus organotypic cultures or *in vivo* tissues may play a role in the structural discrepancies noted between immature synthetic papillomavirus particles and tissue-derived virions (Conway and Meyers, 2009; Conway *et al.*, 2009b). However, it must be emphasized that there have been no structural assessments of raft- or natural infection-derived anogenital HPV virions owing to very low yield and low purity. As anogenital HPVs have never been purified from the typically small lesions *in vivo* or in sufficient quantities or purities from raft tissue, Buck and Trus (2011) purified HPV1 and HPV2 in milligram amounts from human skin wart specimens for structural investigations. These analyses of HPV1 and HPV2 virions revealed a slightly increased maturation of particles compared with HPV16 PsV from 293T cells. However, more specific control of PsV maturation in a neutral-buffer of pH 7.5 lead to more fully mature HPV16 capsids that were dramatically more regular, permitting a map of ~12 Å resolution (Buck and Trus, 2011).

Function

One of the most thorough comparisons was performed by Christensen's group, in which New Zealand White rabbits were inoculated with either xenograft tissue-derived CRPV virions or 293T-produced CRPV virions (Culp *et al.*, 2006c). They carefully established that the virion stocks were essentially indistinguishable as assayed by susceptibility to antibody-mediated neutralization, papilloma induction, and gene expression within lesions in rabbits indicating that 293T-derived virions are biologically indistinguishable from tissue-derived virions (Culp *et al.*, 2006c). They also found no major difference if OptiPrep gradient-purified virions were used directly or dialysed first against saline. We showed the kinetics of entry are similar for raft- and 293T-derived HPV31 virions (Smith *et al.*, 2007). Although HPV16 L1 C175S mutant PsVs were found to be physically fragile, they remained capable of transducing reporter plasmids into cells at levels that approach wild-type mature PsV (Buck and Trus, 2011). This suggests that virion maturation is not crucial for engagement of the various steps of infectious entry, but primarily contributes to the environmental stability of the virus. Thus far, no clear differences have been identified in comparisons between the infectious entry pathways used by tissue-derived PV virions and the pathways used by reporter PsV *in vitro*.

Documented differences among transfection-based virions and those from tissues

Structure

Sapp *et al.* found that only half of the L1 molecules in VLPs produced in insect cells were disulfide bonded into trimers, whereas all of the L1 molecules in virions isolated from warts were disulfide bonded (Sapp *et al.*, 1998). However, this difference was remedied by DNA encapsidation into VLPs, which promotes increased cross-linking between L1 molecules comparable to that seen in virions (Fligge *et al.*, 2001b). The neutralization reported for a panel of monoclonal antibodies (Christensen *et al.*, 1996) varied dependent upon the system from which the particles were obtained: HPV16 PsV are neutralized by monoclonals H16.V5, H16.U4, H16.E70, H16.7E (as in Fig. 2.2) (Roden *et al.*, 1997), whereas xenograft-derived HPV16 virions are not neutralized by H16.U4 (White *et al.*, 1998). There is anecdotal evidence that these antibodies do not effectively neutralize raft-derived HPV16 preparations, but this must be verified (C. Meyers, personal communication). It is worth reiterating that the currently available HPV VLP vaccines appear to elicit functional neutralizing antibodies *in vivo* against natural HPV infections (Lu *et al.*, 2011), suggesting that conformational differences between VLPs and natural virions do not impact proper immune recognition *in vivo*.

Function

When considering functional differences that have been reported and those yet to be tested, we stress the importance of considering the source and purity of virion stocks being compared. We feel that the majority of discrepancies in function reported between tissue-derived and 293T-derived virus stocks are due to the fact that rafts yield impure preps even when density-gradient purified, whereas remarkably pure particles can be obtained from the 293T system (Fig. 2.1E vs. Fig. 2.2C). We found that raft-derived HPV31 can infect HaCaT cells in an HSPG-independent manner (Patterson *et al.*, 2005), whereas more pure HPV31 from the 293T system is HSPG-dependent in the same cells (Johnson *et al.*, 2009). We recently demonstrated that PV particles must maintain their association with HSPG during the infection process as opposed to actual dissociation from HSPG and transfer to an internalization receptor (Surviladze *et al.*, 2011). We suggest that the less pure raft-derived virions may become associated with HSPG during the isolation process and that this explains the different results with raft versus 293T HPV31 in HSPG dependence. Virion modifications that occur during virus isolation from the raft system might also explain why raft-derived HPV virions can readily infect human foreskin keratinocytes (Ozbun, 2002b), whereas virus preparations from 293T cells require furin pretreatment (Day *et al.*, 2008). Lastly, the time and expense of virion production differ significantly between raft and 293T systems normalized to equivalent virion outputs. Based on our HPV31 biosynthesis via raft tissues, cultivation and purification takes 2.5–3 weeks with media changes every other day, and an average supply cost estimated at \$6800 for ≈300 rafts yielding 10^{10} vge. For the same number of vge from the 293T system, cultivation and purification takes ≈5 days with an average materials expense of \$800 for 40 × 100 mm dishes. Thus, virion production in the raft system takes three times longer and costs at least eight times more in supplies.

Comparisons that have yet to be initiated

Studying the HPV life cycle in the raft epithelial culture system has undoubtedly provided significant insights into the differentiation-dependent nature of the replication of these significant pathogens. During complex tissue specific processes, organotypic cultures surely rely on regulated differentiation and redox gradients that are similar to epithelial tissues *in vivo*. However, whether these and other crucial conditions are fully supported in raft tissue is not known. It is noteworthy that virions from rafts have never been compared directly to those obtained from naturally occurring warts – high-risk HPVs have never been purified from lesions; low-risk and non-oncogenic HPVs have yet to be propagated in raft tissues. Thus, it is not prudent to claim or assume that raft-derived virions completely represent virions from warts arising in a whole organism.

It is realistic to consider that virus particles assembled in non-keratinocyte monolayer cells in the absence of the full complement of viral gene expression could have functional differences from wart-derived virions. There is no evidence that naturally occurring virions carry structural proteins, lipids, or nucleic acids other than L1, L2, cellular histones, and viral genomes, although they probably package some free cellular DNA fragments non-specifically. However, until quantitative assessment of virion constituents is carried out, one must consider this possibility and that wart-derived particles could differ from synthetic virions from any *in vitro* system. The 293T system may fail to provide important keratinocyte- or tissue-specific protein or nucleic acid modifications. Bacterially derived genomes may be differentially methylated from transformation into bacteria or from low-level 293T cell replication compared with viral genomes that tend to be hypomethylated in differentiated epithelial tissues where assembly takes place naturally (Kim *et al.*, 2003). The Laimins lab has shown that there is a change in chromatin structure, including differential histone modifications, on HPV genomes in differentiated cells (del Mar Peña and Laimins, 2001; Wooldridge and Laimins, 2008). Although the histone modifications have not yet been investigated in virions from any model or *in vivo*, chromatin structure and viral genome methylation states could impact early gene expression upon new infections. Differences in capsid conformations due to the incorrect redox state of the cells or L1:L2 ratios should be considered, but the efficacy of the HPV vaccines suggest this is not an issue, at least from an immunological standpoint.

The take-home message from this chapter is that we simply do not know whether raft tissue-derived virions are more like virions from warts, more like monolayer-assembled virions, or a hybrid between the two. As VLP-based vaccines are effective in eliciting strong antibody protection against natural virion transmission *in vivo*, there must be a strong structural similarity amongst VLPs and naturally occurring virions, at least from the immune system's perspective. It is most important to emphasize that each method for obtaining virions has advantages and disadvantages, which must be acknowledged and considered so that data are not overinterpreted.

Recommendations for future studies

Use of gradient-purified virions

Many cell proteins co-purify with raft-derived virus particles following tissue disruption and laboratory purification techniques (see Fig. 2.1E). The biological relevance of these co-purifying proteins from cell lysates to the *in vivo* situation where virions are shed in the context of desquamating cells is not entirely clear. Our recent work has found that some cellular proteins bound to virus particles are needed for infection (Surviladze *et al.*, 2011), and it would seem that these could associate with the virus during the release of shed virions or during interaction of the virions with uninfected host cells. Whenever possible, we recommend that virion preparations be subjected at minimum to density gradient purification, and if not, that the purity of the preparations be specifically discussed as a possible confounding factor. With these virion production parameters in place, many confounding variables may be eliminated, allowing for greater reproducibility and reduction of inconsistencies amongst labs.

Repetition of studies

If groups of raft tissues are to be used for a quantitative assessment of viral functions, the reproducibility of obtaining pure virus particles for infectivity studies must be demonstrated among raft tissues grown at separate times. Based on our lab's experience with variable raft growth, we find the reports of the quantitative comparisons of 'infection' among crude cell lysates from a small number of rafts to be questionable without (1) a rigorous demonstration of the batch-to-batch reproducibility of this approach, and/or (2) some form of standardized neutralization assay to verify the means of nucleic acids delivery to the cell.

Avoid the use of inaccurate terminologies

Adjectives such as native, authentic, and quasi- should not be used to characterize virus particles that have been obtained by laboratory propagation until these particles have been rigorously compared with wart-derived virions. These qualifiers are inaccurate and unnecessary, whether referring to lab-derived PsV or infectious, genome-containing virions. A more accurate way to refer to particles is to indicate their source: 293T-derived viral particles, raft-derived viral particles, yeast derived VLPs, etc.

Agree to a standard for proving infection

To ensure that encapsidated nucleic acids are conferring what is scored as infection (RT-PCR, reporter gene expression), we recommend providing a control neutralization of infection. Some viruses can clearly enter via pathways independent of receptor engagement, and thus may be less affected by some types of antibody-mediated neutralization (Mercer and Helenius, 2009). The *in vivo* efficacy of the VLP vaccine suggests this is not the case for HPVs. As the *in vivo* function of the VLP-based vaccines is well documented, using polyclonal antisera to neutralize infection should be the most appropriate control to ensure HPV particles are entering via the expected route.

Standardized extraction/purification techniques

Standardization would also help reduce inter-lab variation. High salt buffer may alter virion structure, prevent DNase digestion of unpackaged DNA, reduce otherwise neutralizing antibody interactions, and diminish microbicide neutralization. Development of a tractable animal model would be useful to monitor biological activity of virion stocks. Whereas infection models in rodents and monkeys are advantageous for some work, they are not practical or affordable for many types of experiments.

Acknowledgements

We are grateful to members of our laboratory for stimulating discussions, to Neil Christensen for monoclonal antibodies, to Neil Christensen, Craig Meyers, Robert Bonnez, Christopher Buck, Patricia Day, and Samuel Campos for sharing unpublished data. The US National Institutes of Health funded this work (CA85747, CA132136, AI084081).

References

Alam, S., Conway, M.J., Chen, H. -S., and Meyers, C. (2008). The Cigarette smoke carcinogen benzo[a] pyrene enhances human Papillomavirus synthesis. J. Virol. *82*, 1053–1058.

Angell, M.G., Christensen, N.D., and Kreider, J.W. (1992). An *in vitro* system for studying the initial stages of cottontail rabbit papillomavirus infection. J. Virol. Meth. *39*, 207–216.

Barrera-Oro, J.G., Smith, K.O., and Melnick, J.L. (1962). Quantitation of papova virus particles in human warts. J. Natl. Cancer Inst. *29*, 583–595.

Basner-Tschakarjan, E., Mirmohammadsadegh, A., Baer, A., and Hengge, U.R. (2004). Uptake and trafficking of DNA in keratinocytes: evidence for DNA-binding proteins. Gene Ther. *11*, 765–774.

Bedell, M.A., Hudson, J.B., Golub, T.R., Turyk, M.E., Hosken, M., Wilbanks, G.D., and Laimins, L.A. (1991). Amplification of human papillomavirus genomes *in vitro* is dependent on epithelial differentiation. J. Virol. *65*, 2254–2260.

Boyle, W.F., Riggs, J.L., Oshiro, J.S., and Lennette, E.H. (1973). Electron microscopic identification of papovavirus in laryngeal papilloma. Laryngoscope *83*, 1102–1108.

Broker, T.R., and Botchan, M. (1986). Papillomaviruses: retrospectives and prospectives. Cancer Cells *4*, 17–36.

Brown, D.R., Bryan, J.T., Pratt, L., Handy, V., Fife, K.H., and Stoler, M.H. (1995). Human papillomavirus type 11 E1^E4 and L1 proteins colocalize in the mouse xenograft system at multiple time points. Virology *214*, 259–263.

Bryan, J.T., and Brown, D.R. (2000). Transmission of human papillomavirus type 11 infection by desquamated cornified cells. Virology *281*, 35–42.

Buck, C.B., and Trus, B.L., eds. (2011). The Papillomavirus Virion: A Machine Built to Hide Molecular Achilles' Heels (Springer, New York), vol. 726, p. 722.

Buck, C.B., Pastrana, D.V., Lowy, D.R., and Schiller, J.T. (2004). Efficient intracellular assembly of papillomaviral vectors. J. Virol. *78*, 751–757.

Buck, C.B., Thompson, C.D., Pang, Y. -Y. S., Lowy, D.R., and Schiller, J.T. (2005). Maturation of Papillomavirus Capsids. J. Virol. *79*, 2839–2846.

Butel, J.S. (1972). Studies with human papilloma virus modeled after known papovavirus systems. J. Natl. Cancer Inst. *48*, 285–299.

Campos, S.K., and Ozbun, M.A. (2009). Two highly conserved cysteine residues in HPV16 L2 form an intramolecular disulfide bond and are critical for infectivity in human keratinocytes. PLoS ONE *4*, e4463.

Carter, J.J., Koutsky, L.A., Wipf, G.C., Christensen, N.D., Lee, S.K., Kuypers, J., Kiviat, N., and Galloway, D.A. (1996). The natural history of human papillomavirus type 16 capsid antibodies among a cohort of university women. J. Infect. Dis. *174(5)*, 927–936.

Chen, H. -S., Conway, M.J., Christensen, N.D., Alam, S., and Meyers, C. (2011). Papillomavirus capsid proteins mutually impact structure. Virology *412*, 378–383.

Christensen, N.D., and Kreider, J.W. (1990). Antibody-mediated neutralization *in vivo* of infectious papillomaviruses. J. Virol. *64*, 3151–3156.

Christensen, N.D., Kreider, J.W., Shah, K.V., and Randon, R.F. (1992). Dectection of human serum antibodies that neutralize infectious human papillomavirus type 11 virions. J. Gen. Virol. *73*, 1261–1267.

Christensen, N.D., Hopfl, R., DiAngelo, S.L., Cladel, N.M., Patrick, S.D., Welsh, P.A., Budgeon, L.R., Reed, C.A., and Kreider, J.W. (1994). Assembled baculovirus-expressed human papillomavirus type 11 L1 capsid protein virus-like particles are recognized by neutralizing monoclonal antibodies and induce high titres of neutralizing antibodies. J. Gen. Virol. *75*, 2271–2276.

Christensen, N.D., Cladel, N.M., and Reed, C.A. (1995). Postattachment neutralization of papillomaviruses by monoclonal and polyclonal antibodies. Virology *207*, 136–142.

Christensen, N.D., Dillner, J., Eklund, C., Carter, J.J., Wipf, G.C., Reed, C.A., Cladel, N.M., and Galloway, D.A. (1996). Surface conformational and linear epitopes on HPV16 and HPV18 L1 virus-like particles as defined by monoclonal antibodies. Virology *223*, 174–184.

Chung, C.H., and Gillison, M.L. (2009). Human papillomavirus in head and neck cancer: its role in pathogenesis and clinical implications. Clin. Cancer Res. *15*, 6758–6762.

Cladel, N.M., Hu, J., Balogh, K., Mejia, A., and Christensen, N.D. (2008). Wounding prior to challenge substantially improves infectivity of cottontail rabbit papillomavirus and allows for standardization of infection. J. Virol. Meth. *148*, 34–39.

Combita, A.L., Touze, A., Bousarghin, L., Sizaret, P. -Y., Munoz, N., and Coursaget, P. (2001). Gene transfer using human papillomavirus pseudovirions varies according to virus genotype and requires cell surface heparan sulfate. FEMS Microbiol. Lett. *204*, 183–188.

Conway, M.J., and Meyers, C. (2009). Replication and assembly of human Papillomaviruses. J. Dent. Res. *88*, 307–317.

Conway, M.J., Alam, S., Christensen, N.D., and Meyers, C. (2009a). Overlapping and independent structural roles for human papillomavirus type 16 L2 conserved cysteines. Virology *393*, 295–303.

Conway, M.J., Alam, S., Ryndock, E.J., Cruz, L., Christensen, N.D., Roden, R.B.S., and Meyers, C. (2009b). Tissue-spanning redox gradient-dependent assembly of native human Papillomavirus type 16 virions. J. Virol. *83*, 10515–10526.

Conway, M.J., Cruz, L., Alam, S., Christensen, N.D., and Meyers, C. (2011). Cross-neutralization potential of native human papillomavirus N-terminal L2 epitopes. PLoS ONE *6*, e16405.

Crawford, L.V., and Crawford, E.M. (1963). A comparative study of polyoma and papilloma viruses. Virology *21*, 258–263.

Culp, T.D., Budgeon, L.R., and Christensen, N.D. (2006a). Human papillomaviruses bind a basal extracellular matrix component secreted by keratinocytes which is distinct from a membrane-associated receptor. Virology *347*, 147–159.

Culp, T.D., Budgeon, L.R., Marinkovich, M.P., Meneguzzi, G., and Christensen, N.D. (2006b). Keratinocyte-secreted laminin 5 can function as a transient receptor for human papillomaviruses by binding virions and transferring them to adjacent cells. J. Virol. *80*, 8940–8950.

Culp, T.D., Cladel, N.M., Balogh, K.K., Budgeon, L.R., Mejia, A.F., and Christensen, N.D. (2006c). Papillomavirus particles assembled in 293TT cells are infectious *in vivo*. J. Virol. *80*, 11381–11384.

Day, P.M., Lowy, D.R., and Schiller, J.T. (2008). Heparan sulfate-independent cell binding and infection with furin-precleaved Papillomavirus capsids. J. Virol. *82*, 12565–12568.

Dollard, S.C., Wilson, J.L., Demeter, L.M., Bonnez, W., Reichman, R.C., Broker, T.R., and Chow, L.T. (1992). Production of human papillomavirus and modulation of the infectious program in epithelial raft cultures. Genes Dev. *6*, 1131–1142.

Doorbar, J. (2005). The papillomavirus life cycle. J. Clin. Virol. *32S*, S7-S15.

Drobni, P., Mistry, N., McMillan, N., and Evander, M. (2003). Carboxy-fluorescein diacetate, succinimidyl ester labeled papillomavirus virus-like particles fluoresce after internalization and interact with heparan sulfate for binding and entry. Virology *310*, 163–172.

Dvoretzky, I., Shober, R., Chattopadhyay, S.K., and Lowy, D.R. (1980). A quantitative *in vitro* focus assay for bovine papilloma virus. Virology *103*, 369–375.

Evander, M., Frazer, I.H., Payne, E., Qi, Y.M., Hengst, K., and McMillan, N.A. J. (1997). Identification of the a_6 integrin as a candidate receptor for papillomaviruses. J. Virol. *71*, 2449–2456.

Favre, M., Breitburd, F., Croissant, O., and Orth, G. (1975). Structural polypeptides of rabbit, bovine, and human papillomaviruses. J. Virol. *15*, 1239–1247.

Favre, M., Breitburd, F., Croissant, O., and Orth, G. (1977). Chromatin-like structures obtained after alkaline disruption of bovine and human papillomaviruses. J. Virol. *21*, 1205–1209.

Fehrmann, F., Klumpp, D.J., and Laimins, L.A. (2003). Human papillomavirus type 31 E5 protein supports cell cycle progression and activates late viral functions upon epithelial differentiation. J. Virol. *77*, 2819–2831.

Fligge, C., Giroglou, T., Streeck, R.E., and Sapp, M. (2001a). Induction of type-specific neutralizing antibodies by capsomeres of human Papillomavirus type 33. Virology *283*, 353–357.

Fligge, C., Schäfer, F., Selinka, H. -C., Sapp, C., and Sapp, M. (2001b). DNA-induced structural changes in the papillomavirus capsid. J. Virol. *75*, 7727–7731.

Flores, E.R., Allen, H.B.L., Lee, D., and Lambert, P.F. (2000). The human papillomavirus type 16 E7 oncogene is required for the productive stage of the viral life cycle. J. Virol. *74*, 6622–6631.

Forslund, O., Lindqvist, P., Haadem, K., Czegledy, J., and Hansson, B.G. (1997). HPV16 DNA and mRNA in cervical brush samples quantified by PCR and microwell hybridization. J. Virol. Meth. *69*, 209–222.

Friedewald, W.F. (1942). Cell state as affecting susceptibility to a virus: enhanced effectiveness of the rabbit papilloma virus on hyperplastic epidermis J. Exp. Med. *75*, 197–220.

Genther, S.M., Sterling, S., Duensing, S., Münger, K., Sattler, C., and Lambert, P.F. (2003). Quantitative role of the human papillomavirus type 16 E5 gene during the productive stage of the viral life cycle. J. Virol. *77*, 2832–2842.

Giroglou, T., Florin, L., Schafer, F., Streeck, R.E., and Sapp, M. (2001). Human papillomavirus infection requires cell surface heparan sulfate. J. Virol. *75*, 1565–1570.

Hagensee, M.E., Olson, N.H., Baker, T.S., and Galloway, D.A. (1994). Three dimensional structure of vaccinia-virus produced human papillomavirus type 1 capsid. J. Virol. *68*, 4503–4505.

Hindmarsh, P.L., and Laimins, L.A. (2007). Mechanisms regulating expression of the HPV31 L1 and L2 capsid proteins and pseudovirion entry. Virol. J. *4*, 19.

Holmgren, S.C., Patterson, N.A., Ozbun, M.A., and Lambert, P.F. (2005). The minor capsid protein L2 contributes to two steps in the human papillomavirus type 31 life cycle. J. Virol. *79*, 3938–3948.

Howley, P.M. (1996). *Papillomavirinae*: The viruses and their replication. In Fields Virology, Third Edition, Fields, B.N., and Knipe, D.M., eds. (Raven Press, New York), pp. 2045–2076.

Howley, P.M., and Lowy, D.R. (2007). Papillomaviruses. In Fields Virology, Knipe, D.M., and Howley, P.M., eds. (Lippincott Williams & Wilkins, Philadelphia), pp. 2299–2354.

Johnson, K.M., Kines, R.C., Roberts, J.N., Lowy, D.R., Schiller, J.T., and Day, P.M. (2009). Role of heparan sulfate in attachment to and infection of the murine female genital tract by human Papillomavirus. J. Virol. *83*, 2067–2074.

Joyce, J.G., Tung, J.-S., Przysiecki, C.T., Cook, J.C., Lehman, E.D., Sands, J.A., Jansen, K.U., and Keller, P.M. (1999). The L1 major capsid protein of human papillomavirus type 11 recombinant virus-like particles interacts with heparin and cell-surface glycosaminoglycans on human keratinocytes. J. Biol. Chem. *274*, 5810–5822.

Kämper, N., Day, P.M., Nowak, T., Selinka, H.-C., Florin, L., Bolscher, J., Hilbig, L., Schiller, J.T., and Sapp, M. (2006). A membrane-destabilizing peptide in capsid protein L2 is required for egress of papillomavirus genomes from endosomes. J. Virol. *80*, 759–768.

Kim, K., Garner-Hamrick, P.A., Fisher, C., Lee, D., and Lambert, P.F. (2003). Methylation patterns of Papillomavirus DNA, its influence on E2 function, and implications in viral infection. J. Virol. *77*, 12450–12459.

Kines, R.C., Thompson, C.D., Lowy, D.R., Schiller, J.T., and Day, P.M. (2009). The initial steps leading to papillomavirus infection occur on the basement membrane prior to cell surface binding. Proc. Natl. Acad. Sci. U.S.A. *106*, 20458–20463.

Koller, L.D., and Olson, C. (1972). Attempted transmission of warts from man, cattle, and horses and of deer fibroma, to selected hosts. J. Invest. Dermatol. *58*, 366–368.

Koss, L.G. (1987). Cytologic and histologic manifestations of human papillomavirus infection of the female genital tract and their clinical significance. Cancer *60*, 1942–1950.

Kreider, J.W., Howett, M.K., Stoler, M.H., Zaino, R.J., and Welsh, P. (1987). Susceptibility of various human tissues to transformation *in vivo* with human papillomavirus type 11. Int. J. Cancer. *39*, 459–465.

Lancaster, W.D., and Olson, C. (1982). Animal papillomaviruses. Microbiol. Rev. *46*, 191–207.

Leder, C., Kleinschmidt, J.A., Wiethe, C., and Muller, M. (2001). Enhancement of Capsid gene expression: preparing the human Papillomavirus type 16 major structural gene L1 for DNA vaccination purposes. J. Virol. *75*, 9201–9209.

Lee, J.H., Yi, S.M.P., Anderson, M.E., Berger, K.L., Welsh, M.J., Klingelhutz, A.J., and Ozbun, M.A. (2004). Propagation of infectious human papillomavirus type 16 by using an adenovirus and Cre/LoxP mechanism. Proc. Natl. Acad. Sci. U.S.A. *101*, 2094–2099.

Lu, B., Kumar, A., Castellsagué, X., and Giuliano, A.R. (2011). Efficacy and safety of prophylactic vaccines against cervical HPV infection and diseases among women: a systematic review & meta-analysis. BMC Infect. Dis. *11*, 13.

McCance, D.J., Kopan, R., Fuchs, E., and Laimins, L.A. (1988). Human papillomavirus type 16 alters human epithelial cell differentiation *in vitro*. Proc. Natl. Acad. Sci. U.S.A. *85*, 7169–7173.

McLaughlin-Drubin, M.E., Wilson, S., Mullikin, B., Suzich, J., and Meyers, C. (2003). Human papillomavirus type 45 propagation, infection, and neutralization. Virology *312*, 1–7.

McLaughlin-Drubin, M.E., Christensen, N.D., and Meyers, C. (2004). Propagation, infection, and neutralization of authentic HPV16 virus. Virology *322*, 213–219.

del Mar Peña, L., and Laimins, L.A. (2001). Differentiation-dependent chromatin rearrangement coincides with activation of human papillomavirus type 31 late gene expression. J. Virol. *75*, 10005–10013.

Matsukura, T., Koi, S., and Sugase, M. (1989). Both episomal and integrated forms of human papillomavirus type 16 are involved in invasive cervical cancers. Virology *172*, 63–72.

Mercer, J., and Helenius, A. (2009). Virus entry by macropinocytosis. Nat. Cell Biol. *11*, 510–520.

Meyers, C. (1996). Organotypic (raft) epithelial tissue culture system for the differentiation-dependent replication of papillomavirus. Methods Cell Sci *18*, 201–210.

Meyers, C., Frattini, M.G., Hudson, J.B., and Laimins, L.A. (1992). Biosynthesis of human papillomavirus from a continuous cell line upon epithelial differentiation. Science *257*, 971–973.

Meyers, C., Mayer, T.J., and Ozbun, M.A. (1997). Synthesis of infectious human papillomavirus type 18 in differentiating epithelium transfected with viral DNA. J. Virol. *71*, 7381–7386.

Meyers, C., Bromberg-White, J.L., Zhang, J., Kaupas, M.E., Bryan, J.T., Lowe, R.S., and Jansen, K.U. (2002). Infectious virions produced from a human papillomavirus type 18/16 genomic DNA chimera. J. Virol. *76*, 4723–4733.

Morrison, E.A., Ho, G.Y., Vermund, S.H., Goldberg, G.L., Kadish, A.S., Kelley, K.F., and Burk, R.D. (1991). Human papillomavirus infection and other risk factors for cervical neoplasia: a case–control study. Int. J. Cancer. *49*, 6–13.

Ogawa, H., Taneda, A., Kanaoka, Y., and Sekine, T. (1979). The histochemical distribution of protein bound sulfhydryl groups in human epidermis by the new staining method. J. Histochem. Cytochem. *27*, 942–946.

Oparka R., and Herrington, C.S. (2011) Human papillomavirus infection and its association with neoplasia: From molecular biology to prevention and treatment. In Small DNA Tumour Viruses, Gaston, K., ed. (Horizon Scientific Press, Norfolk, UK), pp. 1–17.

Ozbun, M.A. (2002a). Infectious human papillomavirus type 31b: purification and infection of an immortalized human keratinocyte cell line. J. Gen. Virol. *83*, 2753–2763.

Ozbun, M.A. (2002b). Human papillomavirus type 31b infection of human keratinocytes and the onset of early transcription. J. Virol. *76*, 11291–11300.

Ozbun, M.A., and Meyers, C. (1996). Transforming growth factor ß1 induces differentiation in human papillomavirus-positive keratinocytes. J. Virol. *70*, 5437–5446.

Ozbun, M.A., and Meyers, C. (1997). Characterization of late gene transcripts expressed during vegetative replication of human papillomavirus type 31b. J. Virol. *71*, 5161–5172.

Ozbun, M.A., and Meyers, C. (1998a). Human papillomavirus type 31b E1 and E2 transcript expression correlates with vegetative viral genome amplification. Virology *248*, 218–230.

Ozbun, M.A., and Meyers, C. (1998b). Temporal usage of multiple promoters during the life cycle of human papillomavirus type 31b. J. Virol. *72*, 2715–2722.

Patterson, N.A., Smith, J.L., and Ozbun, M.A. (2005). Human papillomavirus type 31b infection of human keratinocytes does not require heparan sulfate. J. Virol. *79*, 6838–6847.

Pfister, H. (1984). Biology and biochemistry of papillomaviruses. Rev. Physiol. Biochem. Pharmacol. *99*, 111–181.

Pickard, C., Louafi, F., McGuire, C., Lowings, K., Kumar, P., Cooper, H., Dearman, R.J., Cumberbatch, M., Kimber, I., Healy, E., *et al.* (2009). The cutaneous biochemical redox barrier: a component of the innate immune defenses against sensitization by highly reactive environmental xenobiotics. J. Immunol. *183*, 7576–7584.

Pyeon, D., Lambert, P.F., and Ahlquist, P. (2005). Production of infectious human papillomavirus independently of viral replication and epithelial cell differentiation. Proc. Natl. Acad. Sci. U.S.A. *102*, 9311–9316.

Roberts, J.N., Buck, C.B., Thompson, C.D., Kines, R., Bernardo, M., Choyke, P.L., Lowy, D.R., and Schiller, J.T. (2007). Genital transmission of HPV in a mouse model is potentiated by nonoxynol-9 and inhibited by carrageenan. Nat. Med. *13*, 857–861.

Roden, R.B.S., Kirnbauer, R., Jenson, A.B., Lowy, D.R., and Schiller, J.T. (1994). Interaction of papillomaviruses with the cell surface. J. Virol. *68*, 7260–7266.

Roden, R.B., Greenstone, H.L., Kirnbauer, R., Booy, F.P., Jessie, J., Lowy, D.R., and Schiller, J.T. (1996a). *In vitro* generation and type-specific neutralization of a human papillomavirus type 16 virion pseudotype. J. Virol. *70*, 5875–5883.

Roden, R.B.S., Greenstone, H.L., Kirnbauer, R., Booy, F.P., Jessie, J., Lowy, D.R., and Schiller, J.T. (1996b). *In vitro* generation and type-specific neutralization of a human papillomavirus type 16 virion pseudotype. J. Virol. *70*, 5875–5883.

Roden, R.B.S., Armstrong, A., Haderer, P., Christensen, N.D., Hubbert, N.L., Lowy, D.R., Schiller, J.T., and Kirnbauer, R. (1997). Characterization of human papillomavirus type 16 variant-dependent neutralizing epitope. J. Virol. *71*, 6247–6252.

Rose, R.C., Reichman, R.C., and Bonnez, W. (1994). Human papillomavirus (HPV) type 11 recombinant virus-like particles induce the formation of neutralizing antibodies and detect HPV-specific antibodies in human sera. J. Gen. Virol. *75*, 2075–2079.

Rowson, K.E.K., and Mahy, B.W. J. (1967). Human papova (wart) virus. Bacteriol. Rev. *31*, 110–131.

Sapp, M., Fligge, C., Petzak, I., Harris, J.R., and Streeck, R.E. (1998). Papillomavirus assembly requires trimerization of the major capsid protein by disulfides between two highly conserved cysteines. J. Virol. *72*, 6186–6189.

Schelhaas, M., Ewers, H., Rajamäki, M. -L., Day, P.M., Schiller, J.T., and Helenius, A. (2008). Human papillomavirus type 16 entry: retrograde cell surface transport along actin-rich protrusions. PLoS Pathogens *4*, e1000148.

Shafti-Keramat, S., Handisurya, A., Kriehuber, E., Meneguzzi, G., Slupetzky, K., and Kirnbauer, R. (2003). Different heparan sulfate proteoglycans serve as cellular receptors for human papillomaviruses. J. Virol. *77*, 13125–13135.

Shope, R.E. (1935). Serial transmission of virus if infectious papillomatasis in domestic rabbits. Proc. Soc. Exp. Biol. Med. *32*, 830–832.

Shope, R.E., and Hurst, E.W. (1933). Infectious papillomatosis of rabbits: with a note on the histopathology. J. Exp. Med. *58*, 607 – 624.

Smith, J.L., Campos, S.K., and Ozbun, M.A. (2007). Human papillomavirus type 31 uses a caveolin 1- and dynamin 2-mediated entry pathway for infection of human keratinocytes. J. Virol. *81*, 9922–9931.

Smith, J.L., Campos, S.K., Wandinger-Ness, A., and Ozbun, M.A. (2008a). Caveolin-1 dependent infectious entry of human papillomavirus type 31 in human keratinocytes proceeds to the endosomal pathway for pH-dependent uncoating. J. Virol. *82*, 9505–9512.

Smith, J.L., Lidke, D.S., and Ozbun, M.A. (2008b). Virus activated filopodia promote human papillomavirus type 31 uptake from the extracellular matrix. Virology *381*, 16–21.

Smith, L.H., Foster, C., Hitchcock, M.E., and Isseroff, R. (1993). *In vitro* HPV-11 infection of human foreskin. J. Invest. Dermatol. *101*, 292–295.

Smith, L.H., Foster, C., Hitchcock, M.E., Leiserowitz, G.S., Hall, K., Isseroff, R., Christensen, N.D., and Kreider, J.W. (1995). Titration of HPV-11 infectivity and antibody neutralization can be measured *in vitro*. J. Invest. Dermatol. *105*, 438–444.

Song, H., Moseley, P., Lowe, S.L., and Ozbun, M.A. (2010). Inducible heat shock protein 70 enhances HPV-31 viral genome replication and virion production during the differentiation-dependent life cycle in human keratinocytes. Virus Res. *147*, 113–122.

Steele, B.K., Meyers, C., and Ozbun, M.A. (2002). Variable expression of some 'housekeeping' genes during human keratinocyte differentiation. Anal. Biochem. *307*, 341–347.

Stoler, M.H., Whitbeck, A., Wolinsky, S.M., Broker, T.R., Chow, L.T., Howett, M.K., and Kreider, J.W. (1990). Infectious cycle of human papillomavirus type 11 in human foreskin xenografts in nude mice. J. Virol. *64*, 3310–3318.

Strauss, M.J., Shaw, E.W., Bunting, H., and Melnick, J.L. (1949). 'Crystalline' virus-like particles from skin papillomas characterized by intranuclear inclusion bodies. Proc. Soc. Exp. Biol. Med. *72*, 46–50.

Strauss, M.J., Bunting, H., and Melnick, J.L. (1950). Virus-like particles and inclusion bodies in skin papillomas. J. Invest. Dermatol. *15*.

Stubenrauch, F., Malejczyk, J., Fuchs, P.G., and Pfister, H. (1992). Late promoter of human papillomavirus type 8 and its regulation. J. Virol. *66*, 3485–3493.

Stubenrauch, F., Colbert, A.M.E., and Laimins, L.A. (1998). Transactivation by the E2 protein of oncogenic human papillomavirus type 31 is not essential for early and late viral functions. J. Virol. *72*, 8115–8123.

Stubenrauch, F., Hummel, M., Iftner, T., and Laimins, L.A. (2000). The E8^E2C protein, a negative regulator of viral transcription and replication, is required for extrachromosomal maintenance of human papillomavirus type 31 in keratinocytes. J. Virol. *74*, 1178–1186.

Surviladze, Z., Dziduszko, A., and Ozbun, M.A. (2011). Essential Roles for Soluble Virion-associated Heparin Sulfonated Proteoglycans and Growth Factors in Human Papillomavirus Infections. Submitted.

Suzich, J.A., Ghim, S., Palmer-Hill, F.J., White, W.I., Tamura, J.K., Bell, J.A., Newsome, J.A., Jenson, A.B., and Schlegel, R. (1995). Systemic Immunization with Papillomavirus L1 Protein Completely Prevents the Development of Viral Mucosal Papillomas. Proc. Natl. Acad. Sci. U.S.A. *92*, 11553–11557.

Swan, D.C., Tucker, R.A., Tortolero-Luna, G., Mitchell, M.F., Wideroff, L., Unger, E.R., Nisenbaum, R.A., Reeves, W.C., and Icenogle, J.P. (1999). Human papillomavirus (HPV) DNA copy number is dependent on grade of cervical disease and HPV type. J. Clin. Microbiol. *37*, 1030–1034.

Thomas, J.T., Hubert, W.G., Ruesch, M.N., and Laimins, L.A. (1999). Human papillomavirus type 31 oncoproteins E6 and E7 are required for the maintenance of episomes during the viral life cycle in normal keratinocytes. Proc. Natl. Acad. Sci. U.S.A. *96*, 8449–8454.

Wang, H.K., Duffy, A.A., Broker, T.R., and Chow, L.T. (2009). Robust production and passaging of infectious HPV in squamous epithelium of primary human keratinocytes. Genes Dev. *23*, 181–194.

White, D.O., Huebener, R.J., Rowe, W.P., and Traub, R. (1963). Studies on the virus of rabbit papilloma. I. Methods of assay. Aust. J. Exp. Biol. *41*, 41–50.

White, W.I., Wilson, S.D., Bonnez, W., Rose, R.C., Koenig, S., and Suzich, J.A. (1998). *In vitro* infection and type-restricted antibody-mediated neutralization of authentic human papillomavirus type 16. J. Virol. *72*, 959–964.

Wittrup, A., Sandgren, S., Lilja, J., Bratt, C., Gustavsson, N., M√∂rgelin, M., and Belting, M. (2007). Identification of proteins released by mammalian cells that mediate DNA internalization through proteoglycan-dependent macropinocytosis. J. Biol. Chem. *282*, 27897–27904.

Wooldridge, T.R., and Laimins, L.A. (2008). Regulation of human papillomavirus type 31 gene expression during the differentiation-dependent life cycle through histone modifications and transcription factor binding. Virology *374*, 371–380.

Zhou, J., Liu, W.J., Peng, S.W., Sun, X.Y., and Frazer, I. (1999). Papillomavirus capsid protein expression level depends on the match between codon usage and tRNA availability. J. Virol. *73*, 4972–4982.

zur Hausen, H. (1976). Condylomata acuminata and human genital cancer. Cancer Res. *36*, 794.

zur Hausen, H., and Schneider, A. (1987). The role of papillomaviruses in human anogenital cancer. In The Papovaviridae, Vol. 2. The Papillomaviruses, Salzman, N.P., and Howley, P.M., eds. (Plenum Press, New York, NY), pp. 245–263.

The Regulation of Human Papillomavirus Gene Expression by the E2 Protein: Keeping a Finger in Every Pie

3

Sheila V. Graham and Kevin Gaston

Abstract

The human papillomavirus (HPV) genome is around 8000 base pairs in length and it encodes only eight proteins, a limited number of protein isoforms and no known microRNAs. Despite this relative paucity of genes and gene products these viruses are highly successful. Over 120 HPV types have been identified and they are the causative agents of a wide range of endemic prevalent diseases such as genital warts and common warts as well as rarer but much more serious diseases such as cervical cancer and penile cancer. In order to complete its life cycle HPV must harness the activities of the host cell to transcribe, translate and replicate the viral genome while simultaneously evading a battery of host defences. To facilitate these processes virally encoded non-structural proteins interact with a plethora of host cell proteins and manipulate many host cell regulatory pathways. This review focuses on the regulation of viral gene expression and in particular the diverse roles played by the viral E2 protein and its multiple cellular and viral binding proteins. E2 appears to function as a 'master regulator', controlling viral gene expression at many levels while also enabling viral DNA replication and ensuring equal viral genome segregation during cell division. This impressive feat of multitasking is achieved via a network of E2-interacting proteins that includes almost all of the other viral proteins and a wide range of cellular partners.

Introduction

Human papillomaviruses (HPVs) are non-enveloped DNA tumour viruses that infect cutaneous and mucosal epithelia and cause numerous diseases ranging from harmless warts to potentially fatal carcinomas (reviewed by Chow *et al.*, 2010; Stanley, 2010a; Oparka and Herrington, 2011). Over 120 HPV types have been identified and these can be grouped into genera on the basis of sequence homology (de Villiers *et al.*, 2004; Bernard *et al.*, 2010). The α-papillomaviruses are associated with both cutaneous and mucosal lesions whereas the remaining genera (principally β and γ) are associated with cutaneous lesions. The α- and β-papillomaviruses can be further subdivided into 'high-risk' (HR-HPV) or 'low-risk' (LR-HPV) types depending on their ability to bring about malignant transformation of their host cells whilst the γ-papillomaviruses are typically LR-HPV types. HPV6 and HPV11, for example, are LR α -HPV that cause genital warts, a common viral sexually transmitted disease that can have a significant negative impact on quality of life (Mortensen and Larsen, 2010). In contrast, HR α-HPV types such as HPV16 and HPV18, which also infect the genital tract, are causative agents of cervical cancer, the second most common cancer in women (Munoz *et al.*, 2003; Parkin *et al.*, 2005). The causative role of HR-HPV in this disease has been established through the finding that almost all cervical cancers contain DNA from either HPV16, HPV18 or one of the other HR-HPV types (Walboomers *et al.*, 1999). Moreover, the viral oncogenes E6 and E7 are expressed in cervical cancer cells and their inactivation reverses the transformed phenotype (Goodwin *et al.*, 1998; Jiang and Milner, 2002; Hall and Alexander, 2003). HR-HPV are also involved in many vulvar, anal and penile cancers as well as in a significant proportion of head and neck squamous cell carcinomas (SCC) (Gillison *et al.*, 2000; Jones, 2001). HR β-HPV types such as HPV5 and 8, are associated with SCC of the skin although a causative role for these viruses has not yet been established (reviewed by Oparka and Herrington, 2011).

The HPV genome

The HPV genome is a circular double-stranded DNA molecule with around 8000 base pairs (Fig. 3.1). It exists as multiple plasmids (20–100 copies), sometimes referred to as episomes, replicating independently of the host genome in infected cells. The HPV genome can be divided into an 'early' and a 'late' coding region and a long control region (LCR) also known as the upstream regulatory region (URR). The LCR is largely responsible for determining the tissue specificity of each HPV type and for regulating viral gene expression and replication (reviewed by Thierry, 2009). It contains multiple binding sites for cellular transcription factors as well as the viral origin of replication. The genome encodes 8 major proteins that are relatively well characterized and listed in Table 3.1 as well as a number of less well understood isoforms some of which are also listed in the table. The HPV genome is not thought to produce microRNAs although this has not yet been conclusively ruled out (Gardiner *et al.*, 2011).

The viral proteins can be loosely classified as early or late depending on their time of expression after infection. Thus E1, E2, E6 and E7 are early proteins involved in viral DNA replication and manipulation of the environment within the host cell, E4 and E5 are expressed in early and late infection and they also manipulate the host cell to facilitate viral replication while L1 and L2 are structural proteins expressed late in the life cycle that form the virus capsid. E1 is a helicase that is required for replication of the viral DNA (Hughes and Romanos, 1993; Yang *et al.*, 1993). E1 is recruited to the viral origin of replication by E2 (Mohr *et al.*, 1990), a sequence-specific DNA-binding protein with four binding sites in the LCR and multiple functions in viral replication and the control of viral gene expression. E4 is required for genome replication and it is thought to enable the virus to disrupt the keratin network in late infected cells to facilitate virus release (Doorbar *et al.*, 1991; Roberts *et al.*, 1993). E5, E6 and E7 encode viral oncoproteins that promote replication of the infected cells while also inhibiting cell differentiation and apoptosis. The functions of E1, E5, E6 and E7 are described in other chapters in this book and will only be discussed in detail here in terms of their relationship to E2 and the control of viral gene expression.

L2 is the minor virus capsid protein but it also plays a role in the encapsidation of the viral genome (Holmgren *et al.*, 2005). L1 is the major protein in the viral capsid and it is important for the binding of virions to host cells and therefore for infection. The L1 protein self-assembles into virus-like particles (VLP) which resemble native papillomavirus virions (reviewed by Ozbun and Kivitz, 2011). These VLP can produce high numbers of neutralizing antibodies and two VLP-based vaccines are currently available. A bivalent HPV16/18 vaccine is produced by GSK and a quadrivalent HPV6/11/16/18 vaccine by

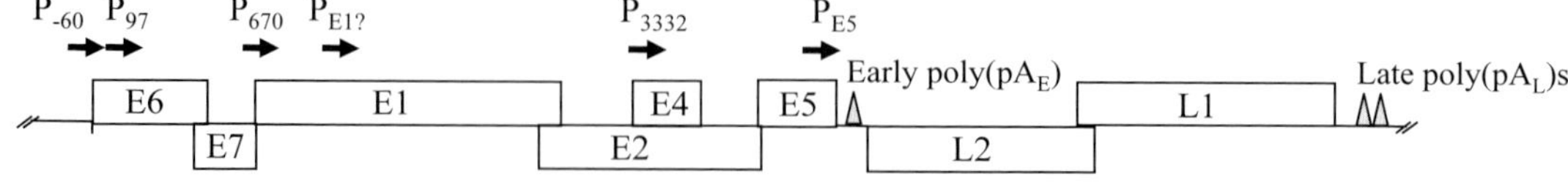

Figure 3.1 The HPV16 genome. A schematic representation of the HPV16 genome linearized at the first nucleotide. The open boxes represent the open reading frames. Transcription start points are indicated by arrows and the starting nucleotide. The early (pA$_E$) and the late (pA$_L$) polyadenylation sites are indicated by the triangular shapes.

Table 3.1 HPV proteins and their functions

Protein	Function or target
E1	Genome replication, ATP-dependent helicase
E2	Genome replication, segregation, encapsidation
E4	Remodelling of cytokeratin, cell growth arrest, virion assembly
E5	Modulation of cell growth, immune evasion
E6	Degradation of p53 and other cellular proteins, inhibition of apoptosis, regulation of cell morphology and cell signalling
E7	Cell cycle regulation via Rb and related proteins, centrosome binding
L1	Major capsid protein
L2	Minor capsid protein
Isoforms	
E6*	Antagonist of E6
E8^E2	Transcriptional regulation
E1^E2	Transcriptional regulation

Merck. Both should reduce the incidence of HPV-associated diseases in the long term (reviewed by Stanley, 2010b). However, the other HPV types will remain a problem for the foreseeable future and vaccine uptake, especially in developing countries and in certain religious and socio-economic groups in developed countries, will probably be low or at best highly variable.

The HPV life cycle

The HPV life cycle begins with the infection of cutaneous or mucosal epithelial cells depending on the viral type (reviewed by Doorbar, 2005). In normal human epithelia, cell proliferation takes place in the basal layer of keratinocytes attached to the basement membrane. Following cell division most basal epithelial cells remain in the basal compartment. However, in some cases a daughter cell migrates towards the surface of the epithelium and undergoes terminal differentiation. Normally, these transiting cells exit the cell cycle and start to synthesize high molecular weight keratins that accumulate in the stratum granulosum and stratum corneum, the highest strata of the epithelium. In mucosal epithelia the uppermost layers of the squamous mucosa have a moist surface and do not develop a stratum corneum (reviewed by Chow *et al.*, 2010).

HPV infection is thought to require a break or microlesion in the stratified epithelium that exposes the basal layer of cells to infection. Post-infection, the virus genome is maintained at a low copy number in the nuclei of the basal layer cells and this requires the E1 and E2 proteins. During this replication phase the E2 protein ensures equal segregation of the HPV genome between daughter cells by simultaneously binding to the viral genome and to host chromosomes via an adapter molecule (reviewed by Van Doorslaer *et al.*, 2011). Productive viral replication requires the differentiation and transit of the infected cell through the various epithelial layers. However, host cell division is promoted in these differentiating cells by the actions of the E5, E6 and E7 viral oncoproteins resulting in hyperproliferation and a slowing of cell differentiation. Presumably this occurs in order to allow time for the host cells to renter S-phase and for viral genome amplification and packaging in the upper layers of the infected epithelium. Amplification of the HPV genomes occurs in the upper cell layers and requires E1 and E2 and the E4 and E5 proteins. The viral capsid proteins are also produced in the upper layers of the epithelium allowing the assembly of viral particles that are released at the surface of the epithelium resulting in transmission of the virus. Virus-infected cells are also sloughed off as the cells die and these probably also play an important role in viral transmission possibly by allowing newly formed virions to exist for longer in the environment.

HPV gene expression

The HPV genome is transcribed by host RNA polymerase II which initiates transcription from at least two viral promoters and produces multiple polycistronic mRNAs that can undergo alternative splicing (Fig. 3.2). Since these mRNAs lack internal ribosome entry sites each mRNA species is presumed to produce predominantly the protein encoded by the first open reading frame. However, E6 splice isoform mRNAs exhibit splicing to at least three different downstream splice acceptor sites indicating that subsequent open reading frames may be translated perhaps owing to changes in codon usage (Ding *et al.*, 2010). In the α-HPV, transcription is initiated at an early promoter (or promoters) at the 3′ end of the LCR at or near the start of the E6 open reading frame; p97 for HPV16 and HPV31, p105 for HPV18 and p90 for HPV6 (Smotkin and Wettstein, 1986; Ozbun and Meyers, 1998; Chow *et al.*, 1987). A cell differentiation-dependent late promoter initiates transcription within the E7 open reading frame; p670 for HPV16 and p742 for HPV31 (Grassmann *et al.*, 1996). In addition to the early and late promoters, several other promoters have been reported to occur in different HPV types. For example, in HPV16 and HPV6 a promoter in E5 produces mRNAs that could encode the L2 protein (Karlen *et al.*, 1996; Milligan *et al.*, 2007).

Polyadenylation of HPV transcripts can occur at an early polyadenylation site (pA_E) that follows the E5 open reading frame or at two late

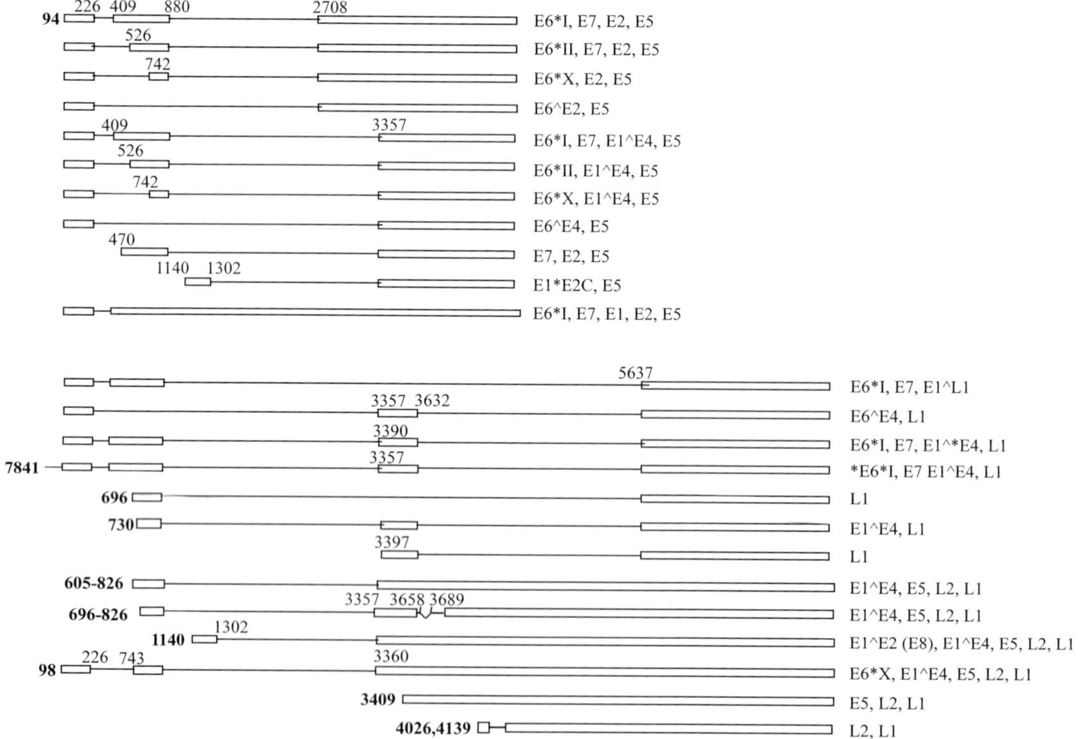

Figure 3.2 HPV16 transcripts in W12 cells. A summary of the HPV16 transcripts identified in W12 cells. The numbers in bold indicate transcription start sites. Non-bold numbers indicate splice donor and acceptor sites. Adapted from Graham (2010).

polyadenylation (pA_L) sites that follow the L1 open reading frame (Milligan *et al.*, 2007). The early polyadenylation site is used to produce transcripts that encode the early proteins as well as several early proteins isoforms produced by alternative splicing (see Fig. 3.2). Cleavage and polyadenylation specificity factor (CPSF) binds to the early polyadenylation signal. However, cleavage stimulatory factor (CstF) binds only weakly to three sites downstream of the early cleavage site and possibly as a consequence of this, 3′ end formation at this site is inefficient. Various cis-acting sequence elements upstream and downstream of the pA_E that bind cellular proteins appear to regulate its efficiency. This might be important to regulate read through of the early polyadenylation site to allow the production of late mRNAs that use the late polyadenylation sites (Terhune *et al.*, 1999, 2001; Zhao *et al.*, 2005).

The late polyadenylation sites produce transcripts that encode early proteins and in addition transcripts that encode the late proteins. The downstream pA_L is used much more efficiently than the upstream site, at least in the W12 model of virus replication and this is probably due to the presence of a strong downstream CstF binding region (Milligan *et al.*, 2007). Alternative splicing of late transcripts also results in the production of multiple RNA species.

Transformation by HPV

Integration of the viral genome into the host genome is not part of the HPV life cycle. However, in many HPV-transformed cells, viral genomes are found integrated into the host genome (Fig. 3.3) or both free plasmid HPV genomes and integrated genomes are present (Baker *et al.*, 1987; Schneider-Maunoury *et al.*, 1987). Integration mostly occurs at random positions in the host genome that probably correspond to chromosomal fragile sites (Smith *et al.*, 1992; Thorland *et al.*, 2000, 2003). However, upon integration, the viral genome is usually disrupted in the E2 or E1 open reading frame. This leaves the E6 and E7 open reading frames intact and under the control of the HPV early promoter as well as neighbouring cellular regulatory elements that will differ at each integration site. The E6 and E7 oncoproteins are key factors in cell transformation and they continue to be expressed post-integration, often at an elevated level (Jeon *et al.*, 1995; Jeon and Lambert, 1995). Presumably integration events that disrupt E6 and E7 also occur but these do not give the host cell a selective advantage and they do not result in tumorigenesis. As will be discussed in more detail below, viral integration results in the misregulation of E6 and E7 expression due in large part to the loss of E2 protein. However, integration also results in misregulation of the early

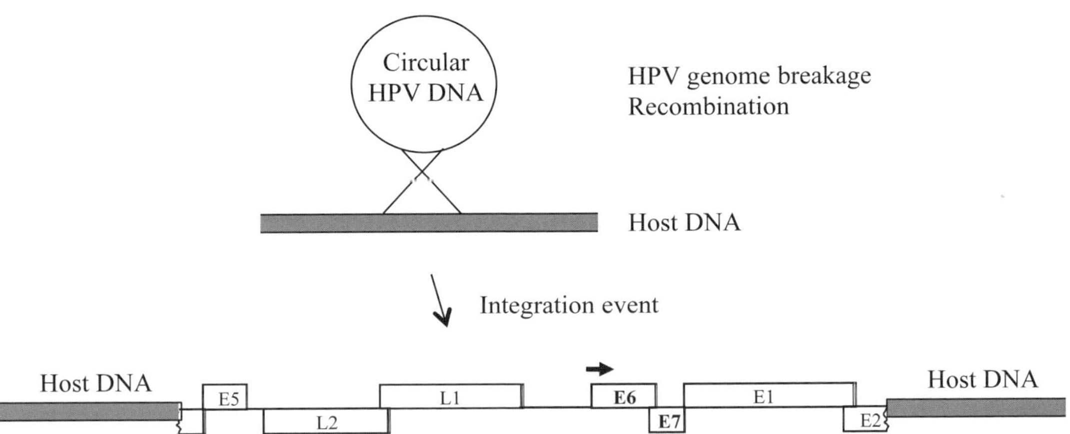

Figure 3.3 The integration of HPV DNA during cell transformation. Viral integration during tumorigenesis often disrupts the E2 ORF as shown here or disrupts the E1 ORF. In most cases expression of E6 and E7 is maintained and expression of E2 and the other HPV genes is lost. Circular and integrated genomes coexist in some cancer cells.

promoter by cellular factors and in changes in E6 and E7 mRNA expression patterns and stability.

The regulation of HPV transcription

The HPV early promoter is constitutively active and it is controlled by the viral E2 protein and by a large number cellular transcription factors many of which have multiple binding sites within the LCR. The late promoter is initially silent or active at a low level but it becomes more active as the host cell differentiates (Doorbar, 2005; Spink and Laimins, 2005). The near overwhelming complexity of the transcription factor binding site array found in the early promoter and the differentiation-specific activity of the late promoter mean that both of these promoters are surprisingly poorly understood. This has been reviewed previously (Hamid *et al.*, 2009; Thierry, 2009) and we will focus here on the role of E2 protein, E2 isoforms and E2-interacting proteins in the regulation of HPV gene expression.

The E2 proteins are around 360 amino acids in length and contain an amino terminal transcription activation domain (TAD) and a carboxyl terminal DNA-binding domain (DBD) separated by a region that probably functions as an intrinsically disordered flexible linker or hinge (Fig. 3.4) (Giri and Yaniv, 1988). The DBD is an obligate dimer, meaning that it must dimerize in order to fold correctly and bind DNA (Hegde *et al.*, 1992; de Prat Gay *et al.*, 2008). The LCR has four highly conserved E2 binding sites and the relative positions of these sites are very well conserved in the α-HPV (Sanchez *et al.*, 2008). The TAD also forms dimers although this domain is not an obligate dimer and monomeric species probably occur or at least there could be rapid exchange of TADs between nearby E2 dimers (Antson *et al.*, 2000). The interaction of TADs from different E2 dimers is thought to facilitate the interaction of E2 proteins bound at distant sites to form multimeric species and DNA loops (Knight *et al.*, 1991; Hernandez-Ramon *et al.*, 2008). This and the conservation of the distances between E2 binding sites in the LCR, suggests that E2 oligomers condense the LCR into a specific 3D protein-DNA structure. However, the nature of this higher-order complex has not yet been determined.

Truncated E2 isoforms have been described in animal and human papillomaviruses. In Bovine papillomavirus (BPV) so-called E2C proteins are produced that lack the TAD (Lambert *et al.*, 1987; Choe *et al.*, 1989). In several HPV types including HPV16 and HPV31, an E8^E2C protein is produced which contains 12 amino acids encoded by a small E8 open reading frame fused upstream of the E2 hinge region and DBD (Stubenrauch *et al.*, 2000, 2001; Lace *et al.*, 2008). These truncated E2 isoforms bind to the same DNA sequences as E2 but because they lack the TAD they would be

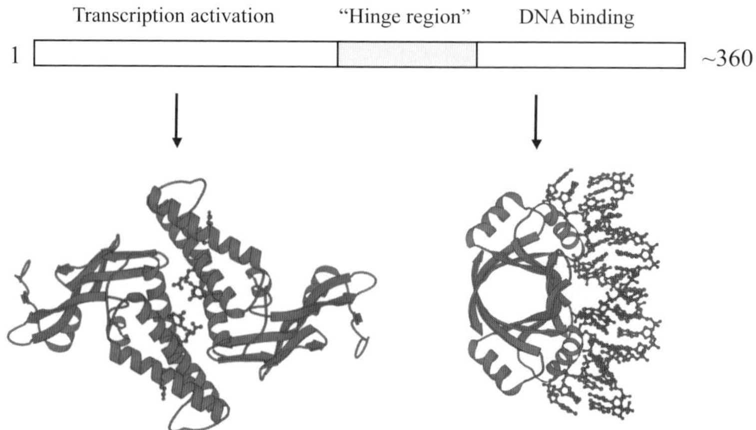

Figure 3.4 The E2 protein. A schematic representation of the HPV16 E2 protein (top). Bottom right, a model of the HPV16 E2 transcription activation domain based on 1DTO. Bottom left, a model of the HPV16 E2 DNA-binding domain bound to DNA based on 2BOP.

unable to form DNA loops. They could however compete with E2 for its binding sites and disrupt any higher–order complexes formed by interactions between E2 proteins bound at different sites. They could also form heterodimeric species with full-length E2 linked by a shared DBD and in this case DNA looping might still occur.

The E2 proteins can activate transcription when they bind upstream of a heterologous promoter. However, when the E2 proteins from a number of HPV types and the related BPV-1 E2 protein bind to an HPV LCR they generally appear to repress promoter activity (Hou *et al.*, 2002; Ottinger *et al.*, 2009) or to produce relatively modest transcription activation at low E2 expression levels and repression at higher E2 expression levels (Bouvard *et al.*, 1994; Hubert and Laimins, 2002; Schweiger *et al.*, 2007). If the major role of the TAD is to activate transcription it is surprising that this is the case. Moreover, mutations in the TAD that prevent HPV31 E2 from activating transcription do not prevent HPV replication in organotypic cell cultures that faithfully mimic the HPV life cycle (Stubenrauch *et al.*, 1998). These data suggest that the ability to activate transcription is not the primary purpose of this protein. On the other hand, HPV16 E2 does coactivate the early promoter when it is co-expressed with p300 (Kruppel *et al.*, 2008) so a biological role for transcription activation by E2 cannot be easily ruled out. In contrast, transcriptional repression by E2 is well established. The binding of E2 to the LCR can sterically hinder the binding of cellular Sp1 and TBP to their sites and thereby repress transcription (Dostatni *et al.*, 1991; Tan *et al.*, 1992, 1994; Dong *et al.*, 1994). Moreover, E2 can recruit cellular proteins and complexes that repress transcription (Wu *et al.*, 2006; Smith *et al.*, 2010). The E8^E2C isoforms do not activate transcription and in fact they recruit protein complexes that actively repress early promoter activity (Ammermann *et al.*, 2008; Fertey *et al.*, 2010; Powell *et al.*, 2010). They could also act as dominant negative proteins for the full-length E2 proteins through the mechanisms described above.

The precise roles of E2 and its isoforms and of any heterodimeric species that may form *in vivo* are still very poorly understood. However, one point that has been established beyond all doubt is that when HPV genomes integrate during tumorigenesis expression of the E2 protein and its isoforms is lost. When E2 or E8^E2C are reintroduced into HPV-transformed cells the early promoter is repressed resulting in decreased E6 and E7 expression and the reestablishment of cellular growth control (Goodwin *et al.*, 1998; Goodwin and DiMaio, 2000; Stubenrauch *et al.*, 2007; Fertey *et al.*, 2011). This in turn can increase cell death via apoptosis and induce cell senescence (Desaintes *et al.*, 1997; Sanchez-Perez *et al.*, 1997; Webster *et al.*, 2000; Parish *et al.*, 2006b; Burns *et al.*, 2010). However, this ability of E2 to repress E6 and E7 transcription may or may not be of great importance in the HPV life cycle. When transcription from the early promoter was examined in isogenic cell lines containing either integrated or plasmid HPV genomes, HPV16 E2 was found to repress transcription from the integrated genomes but it failed to repress transcription from the plasmid genomes (Bechtold *et al.*, 2003). Although these data and the data described above suggest that the ability of E2 to regulate transcription on its own is very probably secondary to its role in viral DNA replication, it is important to point out that many cellular and viral proteins interact with E2 and that this can modulate E2 activity and in turn modulate the activity of the interacting partners. These interactions often have multiple roles in the regulation of HPV gene expression and in other aspects of the HPV life cycle.

Cellular E2-interacting proteins

E2 interacts with many cellular proteins that are important in gene regulation including: the basal transcription factors TFIIB and TBP (Rank and Lambert, 1995), gene-specific transcription factors such as p53 and Sp1 (Massimi *et al.*, 1999; Steger *et al.*, 2002), transcription cofactors such as p300/CBP, TaxBP1 and NuA4/TIP60 (Lee *et al.*, 2000; Wang *et al.*, 2009; Smith *et al.*, 2010), proteins involved in chromatin remodelling such as AMF-1/Gps2, Brd4, CHD6 and NAP-1 (Breiding *et al.*, 1997; Rehtanz *et al.*, 2004; You *et al.*, 2004; Wu *et al.*, 2006; Fertey *et al.*, 2010), as well as proteins that may or may not have roles in transcription such as TopBP1 and ChlR1 (Boner *et al.*, 2002; Parish *et al.*, 2006a). Given the sheer number of E2-interacting proteins that regulate transcription it would seem probable that these interactions are important in the regulation of HPV gene expression. However, if the ability of E2 to regulate transcription is of relatively little importance in HPV biology, these interacting partners may in fact be playing more significant roles in other aspects of the viral life cycle. In keeping with this view, non-transcriptional roles are now well established for some of these interactions or at least functions additional to transcriptional regulation have been defined. The E2–ChlR1 and E2–TopBP1 interactions for example, are important in the tethering of HPV E2 proteins to cellular chromatin to ensure equal segregation of E2-bound viral genomes during cell division and the multiplicity of interacting partners might simply reflect a high level of degeneracy in this function (reviewed by Van Doorslaer *et al.*, 2011). Many of the other E2-cellular protein interactions are less well understood. The E2–p53 interaction for instance, allows p53 to inhibit HPV DNA replication (Lepik *et al.*, 1998; Lepik and Ustav, 2000; Brown *et al.*, 2008). However, the importance of this is unclear since p53 is down-regulated by the E6 protein in HPV-infected cells; it could be that it is advantageous to HPV to recruit p53 to the origin for example to facilitate the rescue of stalled replication forks even at the cost of repressing viral replication. Of course it is always possible that the E2 proteins act as adaptor proteins that allow crosstalk between various cellular proteins at the HPV LCR and that this is not directly linked to transcriptional regulation in many cases (see below).

Another aspect of transcriptional regulation by E2 that is very poorly understood is its role in the control of cellular genes. The regulation of cellular genes by E2 might require some or all of the interactions described above. However, very few cellular targets for E2 have been identified and the co-regulator requirements at these promoters have not yet been established. E2 up-regulates the expression of several serine-arginine (SR)-rich protein family members involved in the control of mRNA splicing. While for some of these genes this up-regulation could be indirect, the activation of SRSF1 transcription by E2 is direct involving the binding of E2 to the SRSF1 promoter (Mole *et al.*, 2009). Moreover, the activation of SRSF1 transcription by E2 is required for the expression of L1 and L2 (SG unpublished observations). The HPV8 E2 protein binds to the β4-integrin promoter in normal human keratinocytes resulting in the displacement of cellular AP-1 (JunB/Fra-1) and the repression of transcription (Oldak *et al.*, 2004, 2010). The repression of this promoter by HPV8 E2 would be expected to alter β4-integrin levels and the adhesion of infected keratinocytes and this could play an important role in the viral life cycle (Oldak *et al.*, 2004). A thorough analysis of other cellular genes that are regulated by E2 either directly, via DNA binding or indirectly, via protein–protein interactions, could reveal other important targets for E2 in the host genome.

The E8^E2C isoform also interacts with several cellular proteins. In this case the role of the E8^E2C-interacting partners appears to be more clear cut. The HPV31 E8^E2C protein binds to nuclear receptor corepressor (NCoR1) and transducin beta-like protein 1 (TBLR1), components of complexes known to facilitate transcriptional repression by nuclear hormone receptors (Powell *et al.*, 2010). This complex binds to the E8 amino acid sequence in E8^E2C and contributes to repression of the early promoter (Powell *et al.*, 2010). In addition, the HPV31 E8^E2C protein has been shown to bind to complexes containing tripartite motif-containing protein 28 (TRIM28) and histone deacetylases (Ammermann *et al.*, 2008). However, it is not clear how important histone deacetylation is in transcriptional repression by E8^E2C. The E8^E2C proteins from HR α-HPV also bind to chromodomain helicase DNA-binding domain 6 protein (CHD6), another protein that functions as part of repressive complexes (Fertey *et al.*, 2010). CHD6 binds to the DBD of E8^E2C and to the full-length E2 DBD and acts as a co-repressor in both cases. The HPV31 E8^E2C proteins also repress HPV DNA replication and some of the E8^E2C interacting proteins are important in this activity (Zobel *et al.*, 2003; Ammermann *et al.*, 2008). It is also possible that cellular repressive complexes bound to E8^E2C crosstalk with other cellular E2-binding proteins that contact the DBD or hinge region or via heterodimerization of E8^E2C and E2.

Viral E2-interacting proteins

It is surprising how many HPV proteins interact with E2. The importance of the E1–E2 interaction is well known; in contrast to the SV40 T helicase, which does not require a loading factor to bind and unwind the SV40 origin of replication, E1 does not bind to the HPV origin of replication with high affinity. Instead the E2 protein binds to its sites in the origin and recruits E1 via protein–protein interactions (reviewed by D'Abramo *et al.*, 2011). The other E2-viral protein interactions are much less well understood.

The HR-αHPV E6 proteins bind to their cognate E2 proteins across the E2 DBD (Grm *et al.*, 2005). The binding of E6 to E2 alters the localization of E2 and both proteins appear to accumulate in nuclear bodies known as splicing factor compartments (Grm *et al.*, 2005). The binding of E6 to E2 does not alter the DNA binding activity of E2 although it does alter the ability of E2 to regulate transcription and facilitate HPV DNA replication. The binding of E2 to E6 also impacts on E6 function, inhibiting the degradation of PDZ proteins by E6 whilst having little or no effect on the degradation of p53 (Grm *et al.*, 2005). The situation is complicated by the fact that these proteins can also interact indirectly to regulate each other's activity. The E6 and E2 proteins from HPV8 and HPV16 for instance all bind to p300 directly or indirectly (Zimmermann *et al.*, 1999; Muller *et al.*, 2002; Thomas and Chiang, 2005; Muller-Schiffmann *et al.*, 2006) and this could influence transcriptional regulation by E2 (and E6). Whatever the precise mechanism for the control of E2 by E6, any change in E2 activity brought about by E6 could allow the autoregulation of E6 levels in infected cells.

More recently a function in splicing regulation has been postulated for the E2/E6 interaction. The N-terminal domain and the hinge region of E2 and the first half of the E6 protein were required to inhibit splicing of a model E6 mRNA splicing construct. Both proteins from HPV16, were able to bind the intronic RNA produced from this construct and to interact with SR proteins (Bodaghi *et al.*, 2009). However, a role for this interaction in the virus life cycle has not been elucidated.

The HPV16 E7 protein also binds to E2 (Gammoh *et al.*, 2006). E7 binds to the E2 DBD (Smal *et al.*, 2009) although the hinge region may either bind to E7 independently or bind E7 in concert with the DBD (Gammoh *et al.*, 2006). The binding of E7 to E2 does not appear to influence the regulation of transcription by E2 (Gammoh *et al.*, 2006) and *in vitro* the E7–E2 complex can be disrupted by DNA carrying an E2 binding site (Smal *et al.*, 2009). However, the binding of E2 to E7 certainly modulates E7 activity, inhibiting transformation by E7 while increasing E7 stability (Gammoh *et al.*, 2009). Therefore, although the importance of E2–E7 interaction in the HPV life cycle is still unclear, it appears to be more significant in the control of E7 function than in the control of E2 function. Loss of the E2–E7 interaction following HPV integration would be expected to result in increased transformation by E7. Loss of the E2–E6 interaction described above would be expected to result in increased degradation of PDZ proteins. Both of these consequences of viral integration might contribute to tumorigenesis.

The E1^E4 protein is expressed at low levels in early infection and at higher levels during late infection. The functions of this protein are poorly understood; although it is important for HPV31 and HPV16 early replication and genome amplification (Nakahara *et al.*, 2005; Wilson *et al.*, 2005) it is not essential for HPV11 DNA amplification (Fang *et al.*, 2006). E1^E4 binds to the E2 TAD (Davy *et al.*, 2009). E1^E4 stabilizes E2 and can relocalize E2 to the cytoplasm but surprisingly, this it does not seem to alter the ability of E2 to regulate transcription (Davy *et al.*, 2009). It would therefore seem unlikely that the role of the E1^E4–E2 interaction is to modulate E1^E4 expression levels. More probably it is involved in other aspects of HPV replication.

PML nuclear bodies are nuclear domains involved in a variety of cellular activities and they are thought to play a central role in the antiviral response (reviewed by Leppard and Wright, 2011). PML bodies are formed by the self-association of promyelocytic leukaemia (PML) proteins although they also contain other proteins. There are several PML isoforms and they can act as co-repressors or co-activators for DNA binding proteins and chromatin modifiers. Importantly, PML proteins are involved in the activation of p53 and in the control of E2F activity by Rb and the disruption of PML bodies can block p53-induced cell senescence (Vernier *et al.*, 2011). The BPV and HPV16 L2 proteins co-localize with PML proteins in PML bodies (Day *et al.*, 1998). This appears to cause partial disruption of PML body function (Florin *et al.*, 2002) consistent with a description of these domains as PML body-like structures. The L2 proteins bring about the colocalization of E2 to these PML body-like structures. Moreover, in the absence of L2, the HPV11 E2 and E1 proteins both colocalize with PML to an extent and they colocalize more prominently with PML when they are co-expressed in the presence of an HPV11 origin containing plasmid (Swindle *et al.*, 1999). This suggests that HPV DNA replication occurs in PML body-like compartments although viral DNA replication can still occur in PML-null cells indicating that PML itself is not essential for the formation of these structures (Nakahara and Lambert, 2007).

Interestingly, E1^E4 also interacts with PML bodies (Roberts *et al.*, 2003). Furthermore, E6 interacts with some PML isoforms directly, colocalizing with these proteins and inducing their degradation (Guccione *et al.*, 2004). This can inhibit PML-induced cell senescence (Guccione *et al.*, 2004). E7 also modulates PML function and again blocks PML-induced cell senescence (Bischof *et al.*, 2005). However, the interactions that target E7 to PML bodies are not well understood. In short, it would appear probable that HPV proteins collectively target PML bodies in order to block the antiviral response and possibly to usurp some of their functions to enable viral replication. In agreement with this view, HPV replication in raft cultures initially causes an increase in the number of PML bodies but at later stages of the life cycle the number of these bodies appears to be reduced (Nakahara and Lambert, 2007). There is clearly much still to learn about the role of PML bodies in HPV replication during normal infection. However, the reintroduction of E2 into HPV-transformed cells would be expected to re-establish PML-induced cell senescence and this indeed appears to be the case (DeFilippis *et al.*, 2003; Horner *et al.*, 2004).

The oligomeric nature of E2, especially when it is bound to multiple DNA sites, suggests that this protein could form a scaffold that allows its many viral interacting proteins to contact each other or to contact E2-binding cellular proteins (Fig. 3.5). It is becoming increasing clear that many cellular transcription factors function in exactly this manner, enabling the formation of large dynamic multi–protein complexes that can perform multiple functions and integrate diverse signals. The intracellular localization of E2 is important in determining which proteins can bind to E2 and whether E2 can mediate interactions between its binding partners. E1^E4 and E7 for example both appear to bind to a population of E2 proteins in the cytoplasm (Davy *et al.*, 2006, 2009). This suggests that there could be complexes in the cytoplasm containing these three proteins as well as some or all of their respective cellular binding partners. Cytoplamic localization also appears to be important in E2-induced apoptosis probably via the formation of specific protein–protein interactions in this compartment (Blachon *et al.*, 2005). Nuclear complexes containing E2 and some or all of its viral and cellular proteins might regulate HPV gene expression and/or HPV DNA replication. In this regard it is important to point out that E2 proteins can be localized in PML body-like structures (Day *et al.*, 1998), associated with chromatin and bound to the nuclear matrix (Zou *et al.*, 2000; Donaldson *et al.*, 2007), the network of proteins and RNA molecules that remains after the removal of chromatin. In all of these different subcellular domains E2 could have specific sets of interacting partners. For example, a subpopulation of both the E2 and the E7 proteins localize to the nuclear matrix and these proteins might interact in this compartment (Greenfield *et al.*, 1991; Zou *et al.*, 2000). E2 post-translation modifications may also play a role in E2 function. E2 can be phosphorylated on multiple sites and this controls its DNA binding activity and stability during S-phase (Penrose *et al.*, 2004; Johansson *et al.*, 2009; Chang *et al.*, 2011). However, E2 may undergo other modifications, such as sumoylation and ubiquitination, that would also be predicted to alter its functions and localization.

Post-transcriptional regulation

Polyadenylation

The choice of early or late polyadenylation site is obviously important in determining the mRNAs that are produced at different stages of infection. However, the mechanisms that regulate this choice are poorly understood. Regulatory

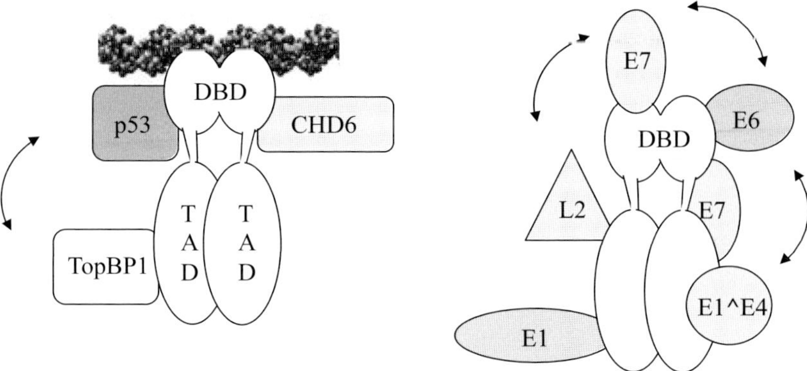

Figure 3.5 E2 and some of the E2-interacting proteins. The diagrams show the HPV E2 protein as a dimer either bound to DNA (right) or unbound (left). DBD, DNA binding domain; TAD, transcription activation domain. Some but not all of the E2-interacting proteins described in the text are shown here. For simplicity, cellular and viral proteins are shown in the left and right panels respectively, but it is important to note that complexes containing mixtures of cellular and viral partner proteins are possible and indeed probable. Binding to the E7 protein and binding to DNA are competitive. Some of the other interactions shown here may also be competitive. However, binding to E2 could allow further interactions between different E2 partner proteins, indicated for some partners by double-headed arrows.

elements that increase usage of the early polyadenylation site have been identified. These include a splicing enhancer within the E4 open reading frame that binds the E2-regulated protein SRSF1 (Rush *et al.*, 2005; Somberg and Schwartz, 2010), an element at the start of L2 that recruits CstF and one upstream that binds PTB, hFIP1, CstF-64 subunit and hnRNP C1/C2 (Zhao *et al.*, 2005; Somberg *et al.*, 2008). Conversely, sequences that repress the early polyadenylation site are present within the L2 open reading frame but the proteins that bind to these sites are not known. The change in polyadenylation site usage during host cell differentiation could be brought about by changes in the levels or activity of proteins that bind to these regulatory sites. However, changes in the transcription rate of the genome brought about by changes in RNA polymerase loading or changes in chromatin structure, perhaps in association with binding of various E2 protein complexes, could also influence the choice of polyadenylation site. The switch from usage of the early promoter to the late promoter might also play a role in polyadenylation site choice. Finally, the switch from use of the early to late polyadenylation site could be dictated by differentiation stage-specific splicing events that splice out the early polyadenylation site in late mRNAs.

Splicing

The regulation of HPV splicing is very poorly understood. Splicing is controlled by cis-acting RNA binding proteins. In general, serine-arginine (SR)-rich proteins bind to exonic and intronic sequence enhancers (ESE and ISE, respectively) and promote splicing while hnRNP proteins bind to exonic and intronic sequence silencers (ESS and ISS, respectively) and inhibit splicing although SR-related inhibitory and hnRNP-related enhancement activity have been noted. As mentioned earlier, E2 binds to the SRSF1 promoter and activates the transcription of this SR gene (McPhillips *et al.*, 2004; Mole *et al.*, 2009). Activation of SRSF1 transcription by E2 is required for HPV31 late gene expression, most probably owing to the regulation of splice site selection or mRNA stability regulation (Stubenrauch *et al.*, 1998; Mole *et al.*, 2009). Several E2-interacting proteins might influence the regulation of SRSF1 and other SR genes by E2. The possibility of a role for E8^E2C in the regulation of these genes has not been investigated.

Various members of the SR and hnRNP protein families are required for viral alternative splicing events and for the early to late switch in virus gene expression that is dependent upon epithelial differentiation. Cycles of phosphorylation/dephosphorylation of SR proteins are required for their diverse functions in splicing. Although E2 may regulate some of these factors directly, E2 recruitment of TopBP1 and through this, Topo1 a known SR protein kinase may regulate splicing indirectly (Soret and Tazi, 2003).

mRNA stability, transport and translation

The expression of L1 and L2 is regulated by changes in mRNA stability. A regulatory element at the 3′ end of the L1 open reading frame induces mRNA instability during early infection (reviewed in Graham, 2005). The cellular regulator of mRNA stability and transport HuR binds to the L1 regulatory region and increases L1 expression (Cumming *et al.*, 2009). The mechanism that results in increased HuR activity at this regulatory region in late infection is currently unknown. Elements that inhibit translation are present in the HPV16 L2 open reading frame and the HPV1 3′ UTR. How these elements are controlled and how they contribute to the viral life cycle is not known.

As well as controlling viral gene expression during HPV infection, mRNA stability is also important in tumorigenesis. Viral integration results in increased stability of E6 and E7 mRNAs and this probably contributes to the continued and increased expression of these oncoproteins in cancer cells (Jeon and Lambert, 1995). At present there is no evidence to implicate E2 directly or indirectly in the regulation of mRNA stability.

Conclusions

The regulation of HPV gene expression during the viral life cycle is poorly understood. Many cellular proteins have been shown to bind to the HPV genome and to regulate viral gene expression. However, we are far from an understanding of how the virus integrates the inputs from these proteins in order to fine-tune gene expression at the levels of transcription, mRNA processing, mRNA stability and translation.

We have focussed here on what might be expected to be a simpler puzzle; unravelling the role of the E2 protein and its isoforms in the control of HPV gene expression. However, the unexpectedly large number of cellular and viral proteins that interact with these apparently simple proteins again produces a complex web of relationships. A detailed biochemical analysis of each of these interactions is required and ideally E2 proteins defective for single interactions must be studied in the context of the HPV life cycle. So far however, very few of these interactions have been studied in this depth. Unravelling the role of these numerous interactions and the crosstalk between E2-interaction proteins in HPV biology will require a multifaceted approach unlike most current studies which focus on individual E2-interacting proteins. Working out which interactions are sequential in the viral life cycle, which are simultaneous and competitive or non-competitive, and which are localized to specific subcellular compartments is a daunting task. However, each interaction provides new insights into virus and host biology as well as a new potential targets for drug discovery.

References

Ammermann, I., Bruckner, M., Matthes, F., Iftner, T., and Stubenrauch, F. (2008). Inhibition of transcription and DNA replication by the papillomavirus E8-E2C protein is mediated by interaction with corepressor molecules. J. Virol. *82*, 5127–5136.

Antson, A.A., Burns, J.E., Moroz, O.V., Scott, D.J., Sanders, C.M., Bronstein, I.B., Dodson, G.G., Wilson, K.S., and Maitland, N.J. (2000). Structure of the intact transactivation domain of the human papillomavirus E2 protein. Nature *403*, 805–809.

Baker, C.C., Phelps, W.C., Lindgren, V., Braun, M.J., Gonda, M.A., and Howley, P.M. (1987). Structural and transcriptional analysis of human papillomavirus type 16 sequences in cervical carcinoma cell lines. J. Virol. *61*, 962–971.

Bechtold, V., Beard, P., and Raj, K. (2003). Human papillomavirus type 16 E2 protein has no effect on transcription from episomal viral DNA. J. Virol. *77*, 2021–2028.

Bernard, H.U., Burk, R.D., Chen, Z., Van, D.K., Hausen, H., and de Villiers, E.M. (2010). Classification of papillomaviruses (PVs) based on 189 PV types and proposal of taxonomic amendments. Virology *401*, 70–79.

Bischof, O., Nacerddine, K., and Dejean, A. (2005). Human papillomavirus oncoprotein E7 targets the promyelocytic leukemia protein and circumvents cellular senescence via the Rb and p53 tumor suppressor pathways. Mol. Cell Biol. *25*, 1013–1024.

Blachon, S., Bellanger, S., Demeret, C., and Thierry, F. (2005). Nucleo-cytoplasmic shuttling of high risk human Papillomavirus E2 proteins induces apoptosis. J. Biol. Chem. *280*, 36088–36098.

Bodaghi, S., Jia, R., and Zheng, Z.M. (2009). Human papillomavirus type 16 E2 and E6 are RNA-binding proteins and inhibit *in vitro* splicing of pre-mRNAs with suboptimal splice sites. Virology *386*, 32–43.

Boner, W., Taylor, E.R., Tsirimonaki, E., Yamane, K., Campo, M.S., and Morgan, I.M. (2002). A Functional interaction between the human papillomavirus 16 transcription/replication factor E2 and the DNA damage response protein TopBP1. J. Biol. Chem. *277*, 22297–22303.

Bouvard, V., Storey, A., Pim, D., and Banks, L. (1994). Characterization of the human papillomavirus E2 protein: evidence of trans-activation and trans-repression in cervical keratinocytes. EMBO J. *13*, 5451–5459.

Breiding, D.E., Sverdrup, F., Grossel, M.J., Moscufo, N., Boonchai, W., and Androphy, E.J. (1997). Functional interaction of a novel cellular protein with the papillomavirus E2 transactivation domain. Mol. Cell Biol. *17*, 7208–7219.

Brown, C., Kowalczyk, A.M., Taylor, E.R., Morgan, I.M., and Gaston, K. (2008). p53 represses human papillomavirus type 16 DNA replication via the viral E2 protein. Virol. J. *5*.

Burns, J.E., Walker, H.F., Schmitz, C., and Maitland, N.J. (2010). Phenotypic effects of HPV16 E2 protein expression in human keratinocytes. Virology *401*, 314–321.

Chang, S.W., Tsao, Y.P., Lin, C.Y., and Chen, S.L. (2011). NRIP, a novel calmodulin binding protein, activates calcineurin to dephosphorylate HPV E2 protein. J. Virol. *85*, 6750–6763.

Choe, J., Vaillancourt, P., Stenlund, A., and Botchan, M. (1989). Bovine papillomavirus type 1 encodes two forms of a transcriptional repressor: structural and functional analysis of new viral cDNAs. J. Virol. *63*, 1743–1755.

Chow, L.T., Nasseri, M., Wolinsky, S.M., and Broker, T.R. (1987). Human papillomavirus types 6 and 11 mRNAs from genital condylomata acuminata. J. Virol. *61*, 2581–2588.

Chow, L.T., Broker, T.R., and Steinberg, B.M. (2010). The natural history of human papillomavirus infections of the mucosal epithelia. APMIS *118*, 422–449.

Cumming, S.A., Chuen-Im, T., Zhang, J., and Graham, S.V. (2009). The RNA stability regulator HuR regulates L1 protein expression *in vivo* in differentiating cervical epithelial cells. Virology *383*, 142–149.

D'Abramo, C.M., Fradet-Turcotte, A., and Archambault, J. (2011). Human papillomavirus DNA replication: Insights into the structure and function of a eukaryotic DNA replisome. In Small DNA Tumour Viruses, Gaston, K., ed. (Horizon Scientific Press, Norfolk, UK), pp. 217–240.

Davy, C., McIntosh, P., Jackson, D.J., Sorathia, R., Miell, M., Wang, Q., Khan, J., Soneji, Y., and Doorbar, J. (2009). A novel interaction between the human papillomavirus type 16 E2 and E1–E4 proteins leads to stabilization of E2. Virology *394*, 266–275.

Davy, C.E., Ayub, M., Jackson, D.J., Das, P., McIntosh, P., and Doorbar, J. (2006). HPV16 E1–E4 protein is phosphorylated by Cdk2/cyclin A and relocalizes this complex to the cytoplasm. Virology *349*, 230–244.

Day, P.M., Roden, R.B., Lowy, D.R., and Schiller, J.T. (1998). The papillomavirus minor capsid protein, L2, induces localization of the major capsid protein, L1, and the viral transcription/replication protein, E2, to PML oncogenic domains. J. Virol. *72*, 142–150.

DeFilippis, R.A., Goodwin, E.C., Wu, L., and DiMaio, D. (2003). Endogenous human papillomavirus E6 and E7 proteins differentially regulate proliferation, senescence, and apoptosis in HeLa cervical carcinoma cells. J. Virol. *77*, 1551–1563.

Desaintes, C., Demeret, C., Goyat, S., Yaniv, M., and Thierry, F. (1997). Expression of the papillomavirus E2 protein in HeLa cells leads to apoptosis. EMBO J. *16*, 504–514.

Ding, J., Doorbar, J., Li, B., Zhou, F., Gu, W., Zhao, L., Saunders, N.A., Frazer, I.H., and Zhao, K.N. (2010). Expression of papillomavirus L1 proteins regulated by authentic gene codon usage is favoured in G2/M-like cells in differentiating keratinocytes. Virology *399*, 46–58.

Donaldson, M.M., Boner, W., and Morgan, I.M. (2007). TopBP1 regulates human papillomavirus type 16 E2 interaction with chromatin. J. Virol. *81*, 4338–4342.

Dong, G., Broker, T.R., and Chow, L.T. (1994). Human papillomavirus type 11 E2 proteins repress the homologous E6 promoter by interfering with the binding of host transcription factors to adjacent elements. J. Virol. *68*, 1115–1127.

Doorbar, J. (2005). The papillomavirus life cycle. J. Clin. Virol. *32* (Suppl. 1), S7–15.

Doorbar, J., Ely, S., Sterling, J., McLean, C., and Crawford, L. (1991). Specific interaction between HPV16 E1-E4 and cytokeratins results in collapse of the epithelial cell intermediate filament network. Nature *352*, 824–827.

Dostatni, N., Lambert, P.F., Sousa, R., Ham, J., Howley, P.M., and Yaniv, M. (1991). The functional BPV-1 E2 trans-activating protein can act as a repressor by preventing formation of the initiation complex. Genes Dev. *5*, 1657–1671.

Fang, L., Budgeon, L.R., Doorbar, J., Briggs, E.R., and Howett, M.K. (2006). The human papillomavirus type 11 E1/\E4 protein is not essential for viral genome amplification. Virology *351*, 271–279.

Fertey, J., Ammermann, I., Winkler, M., Stoger, R., Iftner, T., and Stubenrauch, F. (2010). Interaction of the papillomavirus E8–E2C protein with the cellular CHD6 protein contributes to transcriptional repression. J. Virol. *84*, 9505–9515.

Fertey, J., Hurst, J., Straub, E., Schenker, A., Iftner, T., and Stubenrauch, F. (2011). Growth inhibition of

HeLa cells is a conserved feature of high-risk human papillomavirus E8^E2C proteins and can also be achieved by an artificial repressor protein. J. Virol. *85*, 2918–2926.

Florin, L., Schafer, F., Sotlar, K., Streeck, R.E., and Sapp, M. (2002). Reorganization of nuclear domain 10 induced by papillomavirus capsid protein l2. Virology *295*, 97–107.

Gammoh, N., Grm, H.S., Massimi, P., and Banks, L. (2006). Regulation of human papillomavirus type 16 E7 activity through direct protein interaction with the E2 transcriptional activator. J. Virol. *80*, 1787–1797.

Gammoh, N., Isaacson, E., Tomaic, V., Jackson, D.J., Doorbar, J., and Banks, L. (2009). Inhibition of HPV16 E7 oncogenic activity by HPV16 E2. Oncogene *28*, 2299–2304.

Gardiner, A.S., Wald, A.I., and Khan, S.A. (2011). Alterations in cellular miRNAs induced by human papillomaviruses. In Small DNA Tumour Viruses, Gaston, K., ed. (Horizon Scientific Press, Norfolk, UK), pp. 151–174.

Gillison, M.L., Koch, W.M., Capone, R.B., Spafford, M., Westra, W.H., Wu, L., Zahurak, M.L., Daniel, R.W., Viglione, M., Symer, D.E., *et al.* (2000). Evidence for a causal association between human papillomavirus and a subset of head and neck cancers. J. Natl. Cancer Inst. *92*, 709–720.

Giri, I., and Yaniv, M. (1988). Structural and mutational analysis of E2 trans-activating proteins of papillomaviruses reveals three distinct functional domains. EMBO J. *7*, 2823–2829.

Goodwin, E.C., and DiMaio, D. (2000). Repression of human papillomavirus oncogenes in HeLa cervical carcinoma cells causes the orderly reactivation of dormant tumour suppressor pathways. Proc. Natl. Acad. Sci. U.S.A. *97*, 12513–12518.

Goodwin, E.C., Naeger, L.K., Breiding, D.E., Androphy, E.J., and DiMaio, D. (1998). Transactivation-competent bovine papillomavirus E2 protein is specifically required for efficient repression of human papillomavirus oncogene expression and for acute growth inhibition of cervical carcinoma cell lines. J. Virol. *72*, 3925–3934.

Graham, S.V. (2005). Late events in the life cycle of human papillomaviruses. In Papillomavirus Research: From Natural History to Vaccines and Beyond, Campo, M.S., ed. (Caister Academic Press, UK), pp. 193–212.

Graham, S.V. (2010). Human papillomavirus: gene expression, regulation and prospects for novel diagnostic methods and antiviral therapies. Future Microbiol. *5*, 1493–1506.

Grassmann, K., Rapp, B., Maschek, H., Petry, K.U., and Iftner, T. (1996). Identification of a differentiation-inducible promoter in the E7 open reading frame of human papillomavirus type 16 (HPV16) in raft cultures of a new cell line containing high copy numbers of episomal HPV16 DNA. J. Virol. *70*, 2339–2349.

Greenfield, I., Nickerson, J., Penman, S., and Stanley, M. (1991). Human papillomavirus 16 E7 protein is associated with the nuclear matrix. Proc. Natl. Acad. Sci. U.S.A. *88*, 11217–11221.

Grm, H.S., Massimi, P., Gammoh, N., and Banks, L. (2005). Crosstalk between the human papillomavirus E2 transcriptional activator and the E6 oncoprotein. Oncogene *24*, 5149–5164.

Guccione, E., Lethbridge, K.J., Killick, N., Leppard, K.N., and Banks, L. (2004). HPV E6 proteins interact with specific PML isoforms and allow distinctions to be made between different POD structures. Oncogene *23*, 4662–4672.

Hall, A.H., and Alexander, K.A. (2003). RNA interference of human papillomavirus type 18 E6 and E7 induces senescence in HeLa cells. J. Virol. *77*, 6066–6069.

Hamid, N.A., Brown, C., and Gaston, K. (2009). The regulation of cell proliferation by the papillomavirus early proteins. Cell Mol. Life Sci. *66*, 1700–1717.

Hegde, R.S., Grossman, S.R., Laimins, L.A., and Sigler, P.B. (1992). Crystal structure at 1.7 A of the bovine papillomavirus-1 E2 DNA-binding domain bound to its DNA target. Nature *359*, 505–512.

Hernandez-Ramon, E.E., Burns, J.E., Zhang, W., Walker, H.F., Allen, S., Antson, A.A., and Maitland, N.J. (2008). Dimerization of the human papillomavirus type 16 E2 N terminus results in DNA looping within the upstream regulatory region. J. Virol. *82*, 4853–4861.

Holmgren, S.C., Patterson, N.A., Ozbun, M.A., and Lambert, P.F. (2005). The minor capsid protein L2 contributes to two steps in the human papillomavirus type 31 life cycle. J. Virol. *79*, 3938–3948.

Horner, S.M., DeFilippis, R.A., Manuelidis, L., and DiMaio, D. (2004). Repression of the human papillomavirus E6 gene initiates p53-dependent, telomerase-independent senescence and apoptosis in HeLa cervical carcinoma cells. J. Virol. *78*, 4063–4073.

Hou, S.Y., Wu, S.Y., and Chiang, C.M. (2002). Transcriptional activity among high and low risk human papillomavirus E2 proteins correlates with E2 DNA binding. J. Biol. Chem. *277*, 45619–45629.

Hubert, W.G., and Laimins, L.A. (2002). Human papillomavirus type 31 replication modes during the early phases of the viral life cycle depend on transcriptional and posttranscriptional regulation of E1 and E2 expression. J. Virol. *76*, 2263–2273.

Hughes, F.J., and Romanos, M.A. (1993). E1 protein of human papillomavirus is a DNA helicase/ATPase. Nucleic Acids Res. *21*, 5817–5823.

Jeon, S., and Lambert, P.F. (1995). Integration of human papillomavirus type 16 DNA into the human genome leads to increased stability of E6 and E7 mRNAs: implications for cervical carcinogenesis. Proc. Natl. Acad. Sci. U.S.A. *92*, 1654–1658.

Jeon, S., len-Hoffmann, B.L., and Lambert, P.F. (1995). Integration of human papillomavirus type 16 into the human genome correlates with a selective growth advantage of cells. J. Virol. *69*, 2989–2997.

Jiang, M., and Milner, J. (2002). Selective silencing of viral gene expression in HPV-positive human cervical carcinoma cells treated with siRNA, a primer of RNA interference. Oncogene *21*, 6041–6048.

Johansson, C., Graham, S.V., Dornan, E.S., and Morgan, I.M. (2009). The human papillomavirus 16 E2 protein is stabilised in S phase. Virology *394*, 194–199.

Jones, R.W. (2001). Vulval intraepithelial neoplasia: current perspectives. Eur. J. Gynaecol. Oncol. *22*, 393–402.

Karlen, S., Offord, E.A., and Beard, P. (1996). Functional promoters in the genome of human papillomavirus type 6b. J. Gen. Virol. *77 (Pt 1)*, 11–16.

Knight, J.D., Li, R., and Botchan, M. (1991). The activation domain of the bovine papillomavirus E2 protein mediates association of DNA-bound dimers to form DNA loops. Proc. Natl. Acad. Sci. U.S.A. *88*, 3204–3208.

Kruppel, U., Muller-Schiffmann, A., Baldus, S.E., Smola-Hess, S., and Steger, G. (2008). E2 and the co-activator p300 can cooperate in activation of the human papillomavirus type 16 early promoter. Virology *377*, 151–159.

Lace, M.J., Anson, J.R., Thomas, G.S., Turek, L.P., and Haugen, T.H. (2008). The E8^E2 gene product of human papillomavirus type 16 represses early transcription and replication but is dispensable for viral plasmid persistence in keratinocytes. J. Virol. *82*, 10841–10853.

Lambert, P.F., Spalholz, B.A., and Howley, P.M. (1987). A transcriptional repressor encoded by BPV-1 shares a common carboxy-terminal domain with the E2 transactivator. Cell *50*, 69–78.

Lee, D., Lee, B., Kim, J., Kim, D.W., and Choe, J. (2000). cAMP response element-binding protein-binding protein binds to human papillomavirus E2 protein and activates E2-dependent transcription. J Biol. Chem. *275*, 7045–7051.

Lepik, D., Ilves, I., Kristjuhan, A., Maimets, T., and Ustav, M. (1998). p53 protein is a suppressor of papillomavirus DNA amplificational replication. J. Virol. *72*, 6822–6831.

Lepik, D., and Ustav, M. (2000). Cell-specific modulation of papovavirus replication by tumour suppressor protein p53. J. Virol. *74*, 4688–4697.

Leppard, K.N., and Wright, J. (2011). Targeting of PML proteins and PML nuclear bodies by DNA tumour viruses. In Small DNA Tumour Viruses, Gaston, K., ed. (Horizon Scientific Press, Norfolk, UK), pp. 255–280.

McPhillips, M.G., Veerapraditsin, T., Cumming, S.A., Karali, D., Milligan, S.G., Boner, W., Morgan, I.M., and Graham, S.V. (2004). SF2/ASF binds the human papillomavirus type 16 late RNA control element and is regulated during differentiation of virus-infected epithelial cells. J. Virol. *78*, 10598–10605.

Massimi, P., Pim, D., Bertoli, C., Bouvard, V., and Banks, L. (1999). Interaction between the HPV16 E2 transcriptional activator and p53. Oncogene *18*, 7748–7754.

Milligan, S.G., Veerapraditsin, T., Ahamet, B., Mole, S., and Graham, S.V. (2007). Analysis of novel human papillomavirus type 16 late mRNAs in differentiated W12 cervical epithelial cells. Virology *360*, 172–181.

Mohr, I.J., Clark, R., Sun, S., Androphy, E.J., MacPherson, P., and Botchan, M.R. (1990). Targeting the E1 replication protein to the papillomavirus origin of replication by complex formation with the E2 transactivator. Science *250*, 1694–1699.

Mole, S., Milligan, S.G., and Graham, S.V. (2009). Human papillomavirus type 16 E2 protein transcriptionally activates the promoter of a key cellular splicing factor, SF2/ASF. J. Virol. *83*, 357–367.

Mortensen, G.L., and Larsen, H.K. (2010). The quality of life of patients with genital warts: a qualitative study. BMC Pub. Health *10*, 113.

Muller, A., Ritzkowsky, A., and Steger, G. (2002). Cooperative activation of human papillomavirus type 8 gene expression by the E2 protein and the cellular coactivator p300. J. Virol. *76*, 11042–11053.

Muller-Schiffmann, A., Beckmann, J., and Steger, G. (2006). The E6 protein of the cutaneous human papillomavirus type 8 can stimulate the viral early and late promoters by distinct mechanisms. J. Virol. *80*, 8718–8728.

Munoz, N., Bosch, F.X., de Sanjose, S., Herrero, R., Castellsague, X., Shah, K.V., Snijders, P.J., and Meijer, C.J. (2003). Epidemiologic classification of human papillomavirus types associated with cervical cancer. N. Engl. J. Med. *348*, 518–527.

Nakahara, T., and Lambert, P.F. (2007). Induction of promyelocytic leukemia (PML) oncogenic domains (PODs) by papillomavirus. Virology *366*, 316–329.

Nakahara, T., Peh, W.L., Doorbar, J., Lee, D., and Lambert, P.F. (2005). Human papillomavirus type 16 E1^E4 contributes to multiple facets of the papillomavirus life cycle. J. Virol. 79, 13150–13165.

Oldak, M., Smola, H., Aumailley, M., Rivero, F., Pfister, H., and Smola-Hess, S. (2004). The human papillomavirus type 8 E2 protein suppresses beta4-integrin expression in primary human keratinocytes. J. Virol. *78*, 10738–10746.

Oldak, M., Maksym, R.B., Sperling, T., Yaniv, M., Smola, H., Pfister, H.J., Malejczyk, J., and Smola, S. (2010). Human papillomavirus type 8 E2 protein unravels JunB/Fra-1 as an activator of the beta4-integrin gene in human keratinocytes. J. Virol. *84*, 1376–1386.

Oparka, R., and Herrington, C.S. (2011). Human papillomavirus infection and its association with neoplasia: From molecular biology to prevention and treatment. In Small DNA Tumour Viruses, Gaston, K., ed. (Horizon Scientific Press, Norfolk, UK), pp. 1–18.

Ottinger, M., Smith, J.A., Schweiger, M.R., Robbins, D., Powell, M.L., You, J., and Howley, P.M. (2009). Cell-type specific transcriptional activities among different papillomavirus long control regions and their regulation by E2. Virology *395*, 161–171.

Ozbun, M.A., and Kivitz, M.P. (2011). The Art and science of obtaining virion stocks for experimental human papillomavirus infections. In Small DNA Tumour Viruses, Gaston, K., ed. (Horizon Scientific Press, Norfolk, UK), pp. 19–36.

Ozbun, M.A., and Meyers, C. (1998). Human papillomavirus type 31b E1 and E2 transcript expression correlates with vegetative viral genome amplification. Virology *248*, 218–230.

Parish, J., Kowalczyk, A.M., Chen, H., Roeder G. E., Sessions, R., Buckle, M., and Gaston, K. (2006b). The E2 proteins from high- and low-risk HPV types differ in their ability to bind p53 and induce apoptotic cell death. J. Virol. *80*, 4580–4590.

Parish, J.L., Bean, A.M., Park, R.B., and Androphy, E.J. (2006a). ChlR1 is required for loading papillomavirus E2 onto mitotic chromosomes and viral genome maintenance. Mol. Cell *24*, 867–876.

Parkin, D.M., Bray, F., Ferlay, J., and Pisani, P. (2005). Global cancer statistics, 2002. CA Cancer J. Clin. *55*, 74–108.

Penrose, K.J., Garcia-Alai, M., Prat-Gay, G., and McBride, A.A. (2004). Casein kinase II phosphorylation-induced conformational switch triggers degradation of the papillomavirus E2 protein. J. Biol. Chem. *279*, 22430–22439.

de Prat Gay, G., Gaston, K., and Cicero, D.O. (2008). The papillomavirus E2 DNA binding domain. Frontiers Biosci. *13*, 6006–6021.

Powell, M.L., Smith, J.A., Sowa, M.E., Harper, J.W., Iftner, T., Stubenrauch, F., and Howley, P.M. (2010). NCoR1 mediates papillomavirus E8;E2C transcriptional repression. J. Virol. *84*, 4451–4460.

Rank, N.M., and Lambert, P.F. (1995). Bovine papillomavirus type 1 E2 transcriptional regulators directly bind two cellular transcription factors, TFIID and TFIIB. J. Virol. *69*, 6323–6334.

Rehtanz, M., Schmidt, H.M., Warthorst, U., and Steger, G. (2004). Direct interaction between nucleosome assembly protein 1 and the papillomavirus E2 proteins involved in activation of transcription. Mol. Cell Biol. *24*, 2153–2168.

Roberts, S., Ashmole, I., Johnson, G.D., Kreider, J.W., and Gallimore, P.H. (1993). Cutaneous and mucosal human papillomavirus E4 proteins form intermediate filament-like structures in epithelial cells. Virology *197*, 176–187.

Roberts, S., Hillman, M.L., Knight, G.L., and Gallimore, P.H. (2003). The ND10 component promyelocytic leukemia protein relocates to human papillomavirus type 1 E4 intranuclear inclusion bodies in cultured keratinocytes and in warts. J. Virol. *77*, 673–684.

Rush, M., Zhao, X., and Schwartz, S. (2005). A splicing enhancer in the E4 coding region of human papillomavirus type 16 is required for early mRNA splicing and polyadenylation as well as inhibition of premature late gene expression. J. Virol. *79*, 12002–12015.

Sanchez, I.E., Dellarole, M., Gaston, K., and de Prat Gay, G. (2008). Comprehensive comparison of the interaction of the E2 master regulator with its cognate target DNA sites in 73 human papillomavirus types by sequence statistics. Nucleic Acids Res. *36*, 756–769.

Sanchez-Perez, A.M., Soriano, S., Clarke, A.R., and Gaston, K. (1997). Disruption of the human papillomavirus type 16 E2 gene protects cervical carcinoma cells from E2F-induced apoptosis. J. Gen. Virol. *78*, 3009–3018.

Schneider-Maunoury, S., Croissant, O., and Orth, G. (1987). Integration of human papillomavirus type 16 DNA sequences: a possible early event in the progression of genital tumours. J. Virol. *61*, 3295–3298.

Schweiger, M.R., Ottinger, M., You, J., and Howley, P.M. (2007). Brd4-independent transcriptional repression function of the papillomavirus e2 proteins. J. Virol. *81*, 9612–9622.

Smal, C., Wetzler, D.E., Dantur, K.I., Chemes, L.B., Garcia-Alai, M.M., Dellarole, M., Alonso, L.G., Gaston, K., and de Prat-Gay, G. (2009). The human papillomavirus E7–E2 interaction mechanism *in vitro* reveals a finely tuned system for modulating available E7 and E2 proteins. Biochemistry *48*, 11939–11949.

Smith, J.A., White, E.A., Sowa, M.E., Powell, M.L., Ottinger, M., Harper, J.W., and Howley, P.M. (2010). Genome-wide siRNA screen identifies SMCX, EP400, and Brd4 as E2-dependent regulators of human papillomavirus oncogene expression. Proc. Natl. Acad. Sci. U.S.A. *107*, 3752–3757.

Smith, P.P., Friedman, C.L., Bryant, E.M., and McDougall, J.K. (1992). Viral integration and fragile sites in human papillomavirus-immortalized human keratinocyte cell lines. Genes Chromosomes Cancer *5*, 150–157.

Smotkin, D., and Wettstein, F.O. (1986). Transcription of human papillomavirus type 16 early genes in a cervical cancer and a cancer-derived cell line and identification of the E7 protein. Proc. Natl. Acad. Sci. U.S.A. *83*, 4680–4684.

Somberg, M., and Schwartz, S. (2010). Multiple ASF/SF2 sites in the human papillomavirus type 16 (HPV16) E4-coding region promote splicing to the most commonly used 3′-splice site on the HPV16 genome. J. Virol. *84*, 8219–8230.

Somberg, M., Zhao, X., Frohlich, M., Evander, M., and Schwartz, S. (2008). Polypyrimidine tract binding protein induces human papillomavirus type 16 late gene expression by interfering with splicing inhibitory elements at the major late 5′ splice site, SD3632. J. Virol. *82*, 3665–3678.

Soret, J., and Tazi, J. (2003). Phosphorylation-dependent control of the pre-mRNA splicing machinery. Prog. Mol. Subcell. Biol. *31*, 89–126.

Spink, K.M., and Laimins, L.A. (2005). Induction of the human papillomavirus type 31 late promoter requires differentiation but not DNA amplification. J. Virol. *79*, 4918–4926.

Stanley, M. (2010a). Pathology and epidemiology of HPV infection in females. Gynecol. Oncol. *117*, S5–10.

Stanley, M. (2010b). Prophylactic human papillomavirus vaccines: will they do their job? J. Intern. Med. *267*, 251–259.

Steger, G., Schnabel, C., and Schmidt, H.M. (2002). The hinge region of the human papillomavirus type 8 E2 protein activates the human p21(WAF1/CIP1) promoter via interaction with Sp1. J. Gen. Virol. *83*, 503–510.

Stubenrauch, F., Colbert, A.M., and Laimins, L.A. (1998). Transactivation by the E2 protein of oncogenic human papillomavirus type 31 is not essential for early and late viral functions. J. Virol. *72*, 8115–8123.

Stubenrauch, F., Hummel, M., Iftner, T., and Laimins, L.A. (2000). The E8E2C protein, a negative regulator of viral transcription and replication, is required for extrachromosomal maintenance of human papillomavirus type 31 in keratinocytes. J. Virol. *74*, 1178–1186.

Stubenrauch, F., Zobel, T., and Iftner, T. (2001). The E8 domain confers a novel long-distance transcriptional repression activity on the E8E2C protein of high-risk human papillomavirus type 31. J. Virol. *75*, 4139–4149.

Stubenrauch, F., Straub, E., Fertey, J., and Iftner, T. (2007). The E8 repression domain can replace the E2 transactivation domain for growth inhibition of HeLa cells by papillomavirus E2 proteins. Int. J. Cancer *121*, 2284–2292.

Swindle, C.S., Zou, N., Van Tine, B.A., Shaw, G.M., Engler, J.A., and Chow, L.T. (1999). Human papillomavirus DNA replication compartments in a transient DNA replication system. J. Virol. *73*, 1001–1009.

Tan, S.H., Gloss, B., and Bernard, H.U. (1992). During negative regulation of the human papillomavirus-16 E6 promoter, the viral E2 protein can displace Sp1 from a proximal promoter element. Nucleic Acids Res. *20*, 251–256.

Tan, S.H., Leong, L.E., Walker, P.A., and Bernard, H.U. (1994). The human papillomavirus type 16 E2 transcription factor binds with low cooperativity to two flanking sites and represses the E6 promoter through displacement of Sp1 and TFIID. J. Virol. *68*, 6411–6420.

Terhune, S.S., Milcarek, C., and Laimins, L.A. (1999). Regulation of human papillomavirus type 31 polyadenylation during the differentiation-dependent life cycle. J. Virol. *73*, 7185–7192.

Terhune, S.S., Hubert, W.G., Thomas, J.T., and Laimins, L.A. (2001). Early polyadenylation signals of human papillomavirus type 31 negatively regulate capsid gene expression. J. Virol. *75*, 8147–8157.

Thierry, F. (2009). Transcriptional regulation of the papillomavirus oncogenes by cellular and viral transcription factors in cervical carcinoma. Virology *384*, 375–379.

Thomas, M.C., and Chiang, C.M. (2005). E6 oncoprotein represses p53-dependent gene activation via inhibition of protein acetylation independently of inducing p53 degradation. Mol. Cell *17*, 251–264.

Thorland, E.C., Myers, S.L., Persing, D.H., Sarkar, G., McGovern, R.M., Gostout, B.S., and Smith, D.I. (2000). Human papillomavirus type 16 integrations in cervical tumors frequently occur in common fragile sites. Cancer Res. *60*, 5916–5921.

Thorland, E.C., Myers, S.L., Gostout, B.S., and Smith, D.I. (2003). Common fragile sites are preferential targets for HPV16 integrations in cervical tumors. Oncogene *22*, 1225–1237.

Van Doorslaer, K., Sekhar, V., Khan, J., and McBride, A.A. (2011). Replication and maintenance of viral genomes by association with host chromatin. In Small DNA Tumour Viruses, Gaston, K., ed. (Horizon Scientific Press, Norfolk, UK), pp. 125–150.

Vernier, M., Bourdeau, V., Gaumont-Leclerc, M.F., Moiseeva, O., Begin, V., Saad, F., Mes-Masson, A.M., and Ferbeyre, G. (2011). Regulation of E2Fs and senescence by PML nuclear bodies. Genes Dev. *25*, 41–50.

de Villiers, E.M., Fauquet, C., Broker, T.R., Bernard, H.U., and zur Hausen, H. (2004). Classification of papillomaviruses. Virology *324*, 17–27.

Walboomers, J.M., Jacobs, M.V., Manos, M.M., Bosch, F.X., Kummer, J.A., Shah, K.V., Snijders, P.J., Peto, J., Meijer, C.J., and Munoz, N. (1999). Human papillomavirus is a necessary cause of invasive cervical cancer worldwide. J. Pathol. *189*, 12–19.

Wang, X., Naidu, S.R., Sverdrup, F., and Androphy, E.J. (2009). Tax1BP1 interacts with papillomavirus E2 and regulates E2-dependent transcription and stability. J. Virol. *83*, 2274–2284.

Webster, K., Parish, J., Pandya, M., Stern, P.L., Clarke, A.R., and Gaston, K. (2000). The human papillomavirus (HPV) 16 E2 protein induces apoptosis in the absence of other HPV proteins and via a p53-dependent pathway. J. Biol. Chem. *275*, 87–94.

Wilson, R., Fehrmann, F., and Laimins, L.A. (2005). Role of the E1–E4 protein in the differentiation-dependent life cycle of human papillomavirus type 31. J. Virol. *79*, 6732–6740.

Wu, S.Y., Lee, A.Y., Hou, S.Y., Kemper, J.K., Erdjument-Bromage, H., Tempst, P., and Chiang, C.M. (2006). Brd4 links chromatin targeting to HPV transcriptional silencing. Genes Dev. *20*, 2383–2396.

Yang, L., Mohr, I., Fouts, E., Lim, D.A., Nohaile, M., and Botchan, M. (1993). The E1 protein of bovine papilloma virus 1 is an ATP-dependent DNA helicase. Proc. Natl. Acad. Sci. U.S.A. *90*, 5086–5090.

You, J., Croyle, J.L., Nishimura, A., Ozato, K., and Howley, P.M. (2004). Interaction of the bovine papillomavirus E2 protein with Brd4 tethers the viral DNA to host mitotic chromosomes. Cell *117*, 349–360.

Zhao, X., Oberg, D., Rush, M., Fay, J., Lambkin, H., and Schwartz, S. (2005). A 57-nucleotide upstream early polyadenylation element in human papillomavirus type 16 interacts with hFip1, CstF-64, hnRNP C1/C2, and polypyrimidine tract binding protein. J. Virol. *79*, 4270–4288.

Zimmermann, H., Degenkolbe, R., Bernard, H.U., and O'Connor, M.J. (1999). The human papillomavirus type 16 E6 oncoprotein can down-regulate p53 activity by targeting the transcriptional coactivator CBP/p300. J. Virol. *73*, 6209–6219.

Zobel, T., Iftner, T., and Stubenrauch, F. (2003). The papillomavirus E8-E2C protein represses DNA replication from extrachromosomal origins. Mol. Cell Biol. *23*, 8352–8362.

Zou, N., Lin, B.Y., Duan, F., Lee, K.Y., Jin, G., Guan, R., Yao, G., Lefkowitz, E.J., Broker, T.R., and Chow, L.T. (2000). The hinge of the human papillomavirus type 11 E2 protein contains major determinants for nuclear localization and nuclear matrix association. J. Virol. *74*, 3761–3770.

HPV E5: An Enigmatic Oncoprotein

4

Laura F. Wetherill, Rebecca Ross and Andrew Macdonald

Abstract

Mucosal papillomaviruses contain a short open reading frame (ORF) at the 3′ end of the early region, termed E5. The E5 protein is a hydrophobic, membrane-integrated protein that localizes to the endoplasmic reticulum, Golgi apparatus, endosomes and the nuclear envelope. Structural similarity to the bovine papillomavirus (BPV-1) E5 protein prompted an examination of the transforming abilities of human papillomavirus (HPV) E5, which revealed E5 to be a third oncoprotein encoded by HPV. In comparison to E6 and E7, less is understood of the role that E5 plays in host transformation and the HPV life cycle. Although recent findings have increased our knowledge of E5, the field is controversial and there is much work to be accomplished before we fully understand the many functions of this protein. This chapter summarizes our latest understanding of the HPV E5 protein and its interactions with host cells, demonstrating that this protein is indeed of critical importance for HPV and worthy of further investigation.

Introduction

The E5 ORF of mucosal HPVs is located downstream of the E2 ORF at the 3′ end of the early coding region. Sequence analysis predicts the presence of an E5 ORF in mucosal HPVs, ungulate fibropapillomaviruses and animal and cutaneous HPVs, although it is absent in the epidermodysplasia veruciformis (EV) and melanoma-associated HPVs (beta PVs) (Alonso and Reed, 2002). This region is highly divergent between HPV types, often with the inclusion of several ORFs putatively termed E5. Sequence alignments reveal these putative E5 proteins to be poorly conserved. Phylogenetic studies from the Alonso group suggest that there are in fact four subgroups of E5 protein (alpha–delta), with the E5 protein of high-risk mucosal types (HPV16/18/31) belonging to subgroup alpha (Bravo and Alonso, 2004). It is this first subgroup that has received the most attention. Whether E5 proteins from other subgroups have similar characteristics and roles in the HPV life cycle is yet to be determined.

In a subset of cervical cancers, the E5 ORF is deleted after integration of the viral genome into host chromosomes (Pater and Pater, 1985; Schwarz *et al.*, 1985). For this reason it has long been considered that E5 has a function early in the development of malignancy but is dispensable for the maintenance of the malignant phenotype. However, as only 55% of HPV16 positive cervical cancers contain integrated genomes (Vinokurova *et al.*, 2008) and high levels of E5 transcript are detected in cervical cancer biopsies (Schrevel *et al.*, 2011), it is an opportune moment to reconsider the role of E5 in HPV-mediated pathogenesis.

General characteristics of HPV E5

The HPV16 E5 protein is 83 amino acids in length and strongly hydrophobic in nature (Borzacchiello *et al.*, 2010). Hydropathy analysis predicts the presence of three putative transmembrane domains (TM1–3) with short regions at each terminus that extend beyond the lipid bilayers (Fig. 4.1A and B). Membrane permeabilization assays revealed that the amino terminus of E5 is present within the lumen of endomembranes, whilst the hydrophilic carboxyl-terminus resides in the cytosol (Krawczyk *et al.*, 2010). Some human and animal E5 proteins are significantly shorter than HPV16 E5 (e.g. HPV2a E5 is 48 amino acids in length) and are predicted to have only a single transmembrane domain, but the impact of this difference has not been studied in detail. The inherent hydrophobic nature of E5 has precluded the purification of E5 protein in sufficient quantities for structural analysis (Yang *et al.*, 2003). To circumvent this, preliminary biophysical characterization has been performed using peptides generated against each of the predicted transmembrane domains in isolation. Whilst these studies have provided some insight into the possible folding of each domain, they provide little understanding of the overall structure of E5 (Alonso and Reed., 2002). A recent breakthrough in the purification of full-length native HPV16 E5, in a lipid environment should allow a comprehensive structural analysis of E5 (Wetherill *et al.*, unpublished). Research has also been hindered by the lack of effective antibody reagents against E5 proteins. Accordingly, most studies utilize overexpression systems with epitope tagged E5 proteins. Moreover, the extremely low expression levels of this protein in stable cell culture hampers analysis of the functional effects of E5. Sequence analysis reveals that E5 contains a significant quantity of rare codons (40%). In an effort to overcome this, a codon-modified form of E5 has been generated, which exhibits higher levels of expression in cell culture (Disbrow *et al.*, 2003). Analysis of mammalian expressed E5 by SDS PAGE reveals that in addition to a monomeric species, overexpressed E5 migrates as both a dimer and higher order oligomer (Gieswein *et al.*, 2003). Interestingly, BPV1 E5 has been shown to form dimers through

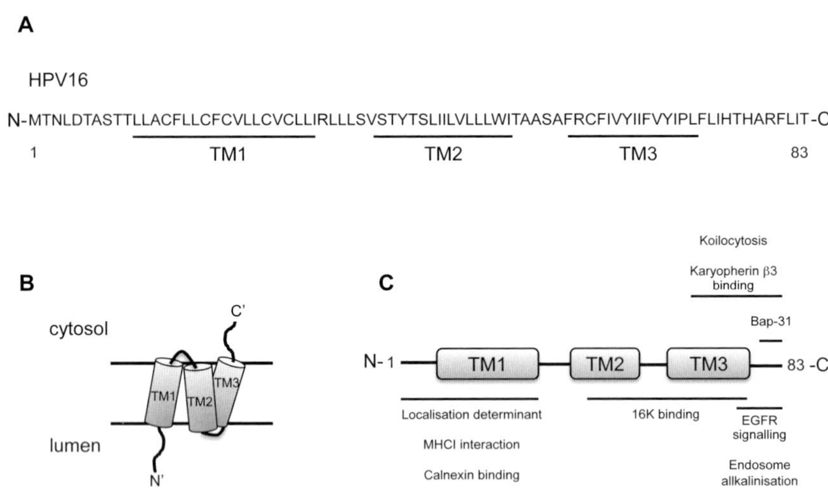

Figure 4.1 Schematic of the HPV16 E5 protein. (A) Primary sequence and predicted transmembrane (TM) regions of the HPV16 E5 protein. (B) *In silico* modelling and membrane permeabilization assays demonstrate that HPV16 E5 encodes a short luminal amino-terminus followed by three transmembrane (TM) domains and a short hydrophilic carboxyl-terminus that residues in the cytosol. (C) The amino terminus and TM1 are responsible both for subcellular localization and an interaction with MHCI, whereas the carboxy-terminal third of E5 appears to mediate the majority of known functions. These include binding to the 16K subunit of the v-ATPase, activating EGFR signalling, inducing endosome alkalinization, increasing koilocytosis and binding to Bap-31.

disulphide linkages (Surti *et al.*, 1998; Mattoon *et al.*, 2001) and HPV E5 has been suggested to oligomerize through hydrophobic regions or cysteine residues present in the first transmembrane domain. Subsequent observations of HPV E5 oligomerization in the presence of reducing agents are suggestive of a mechanism other than disulphide binding (Gieswein *et al.*, 2003). Furthermore, recent analysis of recombinant HPV16 E5 by native PAGE identifies the presence of a single species of approximately 60 kDa, indicative of a hexameric assembly (Wetherill *et al.*, unpublished). Controversy surrounds the purpose of E5 dimerization/oligomerization, with speculation that it is an artefact of strong overexpression conditions. Of note, one of the few studies to analyse E5 expression in samples from HPV16 positive cervical scrapes, highlighted the presence of E5 dimers, supporting the theory that dimerization/oligomerization is required for function (Kell *et al.*, 1994). In line with these findings, recent studies have suggested a role for E5 as a novel viroporin (Wetherill *et al.*, unpublished). These virally encoded channels are small hydrophobic proteins that oligomerize to mediate the permeabilization of cell membranes. The role of E5 channel activity *in vivo* is a focus of current research.

Expression of E5 mRNA during the HPV life cycle

In the absence of appropriate antibody detection reagents, mRNA expression profiles have been used as a surrogate to examine the E5 expression profile during the HPV life cycle. In undifferentiated cultures of CIN-612 (HPV31) or W12 (HPV16) cells, abundant mRNA sequences contain the E5 ORF (Doorbar *et al.*, 1990; Hummel *et al.*, 1992). Although these early transcripts all contain the E5 ORF, it is unlikely that they express high levels of E5 protein owing to the leaky ribosome-scanning mechanism used by HPV to translate proteins from polycistronic mRNA. As a consequence, it is more probable that E5 protein is primarily synthesized in differentiating cells from late transcripts due to a shift of E5 from the fourth to the second open reading frame in the transcripts. E5 sequences have also been detected in human CIN lesions and cervical carcinomas (Stoler and Broker, 1986; Durst *et al.*, 1992) and recent attempts to define E5 mRNA expression in commonly utilized cervical cell lines reveal abundant E5 expression in HPK-1A cells compared with low levels of expression in CaSki cells and an absence of E5 transcripts in SiHa cells (Schmitt and Pawlita, 2011). These changes in E5 abundance are thought to correlate with the physical state of the HPV genome.

Subcellular localization

Localization studies have relied upon the overexpression of epitope tagged E5 proteins. Whilst the addition of epitope tags does not appear to affect known E5 functions, it is unclear whether they have a more subtle phenotypic effect. There is great need, therefore, to develop effective detection reagents against E5. HPV16 E5 shows an endomembrane localization, with predominant staining in the endoplasmic reticulum (ER), early endosomes and nuclear membrane. As expression levels increase, E5 can also be found in the Golgi apparatus (Conrad *et al.*, 1993; Krawczyk *et al.*, 2010). Interestingly, E5 from cutaneous HPV2a and HPV6b types appear to localize predominantly to the Golgi apparatus. Whether these differences in localization reflect alterations in E5 functions remains to be determined, although, preliminary functional data suggest that this is not the case (Cartin and Alonso, 2003). The first hydrophobic transmembrane domain (TM1) is responsible for mediating endomembrane localization, as its deletion results in loss of co-localization with ER and other membranous structures (Lewis *et al.*, 2008). Some studies also suggest that the E5 protein localizes to the plasma membrane and that its carboxyl-terminus faces the extracellular environment, where it induces cell–cell fusion (Hu and Ceresa, 2009). These findings are controversial and have yet to be confirmed.

E5 regulates late functions in the viral life cycle

Development of differentiation-inducible cell culture assays, including the organotypic raft culture system (McCance *et al.*, 1988) and single cell suspensions in methylcellulose (Flores and Lambert, 1997; Ruesch and Laimins, 1998) has greatly facilitated research into the role of individual HPV proteins in the HPV life cycle. Initial predictions, based upon the levels of E5 containing mRNA transcripts, suggested a role for E5 during the early stages of the HPV life cycle. Mayer and Meyers (1998) cast doubt on these assumptions by analysing the expression of HPV31 E5 protein during keratinocyte differentiation in raft cultures (Mayer and Meyers, 1998). For this, they developed one of the few antisera against HPV31 derived peptides. Contrary to earlier speculation, E5 was detected throughout the epidermis with the highest levels recorded in terminally differentiated cells, where virion morphogenesis takes place. Unfortunately, a lack of antibody reagents has precluded confirmation of these data in other high-risk HPV types. However, mutant genomes have been used to interrogate the functions of the HPV31 and HPV16 E5 proteins (Fehrmann *et al.*, 2003; Genther *et al.*, 2003). These studies highlight a critical role for E5 in the later stages of the HPV life cycle rather than at the early stages. HPV31 E5 was shown to play a role in differentiation-induced genome amplification and in viral gene transcription from the late (P742) promoter. In line with these differences expression of the major late protein product E1^E4 was significantly repressed in the absence of E5 (Fehrmann *et al.*, 2003). Importantly, HPV31 E5 was also responsible for maintaining suprabasal cells in mitosis and active cell division following differentiation induced cell cycle arrest. This is a critical requirement in enabling the maintenance of viral replication in the epithelium to ensure efficient virus production (Fehrmann *et al.*, 2003).

A subtler phenotype was observed in cells lacking HPV16 E5 (Genther *et al.*, 2003). Whilst E5 also supported DNA synthesis in the suprabasal cells, its absence did not affect viral genome amplification in differentiating cells. Additionally, viral transcription and E1^E4 protein levels were similar to wild type controls. Interestingly, HPV16 genomes lacking E7 also display a defect in DNA synthesis, and it is plausible that E5 may co-operate with E7 to reprogramme differentiated cells to allow vegetative DNA replication (Bouvard *et al.*, 1994; Flores *et al.*, 2000). The differences in results between the two studies are possibly due to different functional properties attributed to the 30% sequence variation between HPV16 and HPV31 or to the experimental assays used. Further work using specific E5 point mutations or knockout genomes generated from other HPV types may help to explain these differences.

Mechanisms of transformation

Investigations into the mechanisms by which viruses, including HPV, transform cells provide valuable insight into cellular function and disease pathogenesis. In comparison to oncogenic HPV types whose major transforming proteins are E6 and E7, the major transforming protein of BPV1 is E5 (DiMaio *et al.*, 1986). The structural similarities of HPV E5 to BPV1 E5 prompted studies of the potential transforming activities of HPV E5 proteins. The first indication that HPV E5 had transforming potential came from studies with HPV6, where E5 was shown to be capable of stimulating anchorage-independent growth in established murine cells (Chen and Mounts, 1990). Subsequently, high-risk E5 has also been shown to be capable of stimulating established murine cells to grow in an anchorage-independent manner (Pim *et al.*, 1992), and injection of E5-expressing transfected fibroblasts into nude mice induces tumour formation (Leechanachai *et al.*, 1992). E5 has also been shown to enhance the immortalization potential of E6 and E7 and to act synergistically with E7 to stimulate the proliferation of human and mouse primary cells (Bouvard *et al.*, 1994; Valle and Banks, 1995). These results clearly demonstrate that E5 represents the third transforming protein encoded by HPV.

Recent progress has been made towards clarifying the *in vivo* transforming potential of E5. In a series of elegant experiments, the Lambert group generated mice that expressed the HPV16 E5 protein from a keratin-14 promoter, which directs expression to the stratified squamous epithelium. High-levels of E5 expression in the skin induced epithelial hyperproliferation resulting in spontaneous tumour formation (Genther Williams *et al.*, 2005). Interestingly, in this skin model, E5 appeared to act as a more potent oncoprotein than E6 or E7 skin models, and was shown to contribute to both tumour promotion and progression (Maufort *et al.*, 2007). These mice also displayed increased dysplastic disease in the cervical epithelium and when treated with oestrogen for nine months developed malignant cancers (Maufort *et al.*, 2010). Co-expression of E5 with E7 increased the level of suprabasal DNA synthesis in the cervical epithelium similar to what was observed in cell culture studies (Genther *et al.*, 2003) and increased the number of tumours developed (Maufort *et al.*, 2010). Of note, expression of all three oncoproteins provided limited enhancement in tumour development compared with E6 and E7, suggesting that E6 expression obviates the need for E5. These findings hint at the potential interplay between the actions of the three HPV oncoproteins and warrant further investigation.

The epidermal growth factor receptor is a major target of E5

To decipher the molecular mechanisms of E5 transformation, focus centred on the BPV1 E5 protein, which mediates transformation through an interaction with receptor tyrosine kinases (DiMaio and Mattoon, 2001). The epidermal growth factor receptor (EGFR) is expressed in all epithelial cells, including the mucosal keratinocytes that line the cervix. EGFR is one of the most important proto-oncogenes whose expression is increased during progression of cervical cancer (Pfeiffer *et al.*, 1989; Boiko *et al.*, 1998; Zimmermann *et al.*, 2006), and overexpression of EGFR has been associated with poor disease prognosis (Kersemaekers *et al.*, 1999). Epidermal growth factor (EGF) is the EGFR ligand and is an essential mitogen for primary keratinocyte growth *in vitro*. EGF binds to the extracellular domain of the EGFR, initiating receptor dimerization and auto-phosphorylation of the internal tyrosine residues of the receptor. This in turn causes an intracellular cascade of signalling events, which are transmitted to the nucleus to activate a vast array of cellular genes that control cell differentiation, proliferation, mobility and survival. EGF binding also triggers internalization of receptor–ligand complexes and subsequent degradation by the lysosomal pathway (Yarden, 2001). EGFR is recruited to clathrin-coated pits and rapidly internalized as the pits mature into clathrin-coated vesicles before fusing with early endosomes. At this stage, non-EGF bound receptors are recycled back to the cell surface whilst EGF–EGFR complexes continue onto receptor-kinase-dependent sorting in multivesicular bodies (MVBs) (Futter *et al.*, 1996). The mature late endosomes fuse with pre-existing lysosomes, causing dissociation of the EGFR–EGF complex and degradation by lysosomal proteases (Authier and Chauvet, 1999). This process clears the cell surface of activated EGFR and desensitizes cells to extracellular mitogens.

In both fibroblasts and keratinocytes, the expression of E5 leads to EGFR hyper-activation in a ligand-dependent manner (Straight *et al.*, 1993). This might expand the population of basal or stem cell-like keratinocytes and thereby increase the probability that some of these cells would undergo malignant transformation. The *in vivo* significance of EGFR activation for the high-risk E5 phenotype is being dissected with the advent of transgenic mouse models. The overt phenotype of mice overexpressing E5 includes epidermal hyperproliferation, and is strikingly similar to mice overexpressing ligands of the EGFR in the epidermis (Genther Williams *et al.*, 2005). Importantly, the epithelial hyperproliferation induced by E5 in these mice was abrogated when crossed with mice carrying an allele encoding a dominant defective EGFR. This indicates that EGFR signalling is required for alterations in keratinocyte growth and differentiation associated with E5 expression (Genther Williams *et al.*, 2005).

Despite the apparent importance of this pathway for the E5 phenotype, the molecular mechanisms for EGFR activation remain elusive, although several putative mechanisms have been suggested. Similar to BPV1 E5, it is plausible that E5 interacts with EGFR directly to modulate activation, although these results are controversial (Conrad *et al.*, 1994; Hwang *et al.*, 1995). Using radiolabelled ligand binding assays, Straight and colleagues (1993) observed an increase in the levels of EGFR on the cell surface (E5 40% vs. control 7%) (Straight *et al.*, 1993). This finding indicated that E5 was able to enhance the recycling of activated EGFR in response to EGF treatment. Nevertheless, this increase was only observed after prolonged EGF treatment, whereas addition of EGF for shorter periods of time led to increased EGFR activation in the absence of enhanced receptor recycling (Crusius *et al.*, 1998). In spite of this, these data indicated that the presence of E5 was somehow perturbing the intracellular trafficking process, resulting in enhanced EGFR signalling. Ubiquitylation of growth factor receptors is essential for the normal degradation process, and c-Cbl mediates degradation of EGFR in response to EGF treatment (Ettenberg *et al.*, 2001). Interestingly, levels of ubiquitylated EGFR are lower in cells expressing E5 and this correlates with impaired recruitment of c-Cbl to the EGFR receptor (Zhang *et al.*, 2005). How E5 prevents this interaction and whether this impairment is essential for E5 transformation is currently unclear.

Is there a role for the vacuolar ATPase in E5 mediated EGFR activation?

A number of studies also indicate that E5 may interfere with the acidification of early endosomes containing activated EGFR (Straight *et al.*, 1995; Thomsen *et al.*, 2000; Disbrow *et al.*, 2005). Once again, taking a cue from BPV1 E5, it has been shown that HPV16 E5 can bind to the 16kDa subunit of the vacuolar ATPase (16K) (Table 4.1) (Conrad *et al.*, 1993; Rodriguez *et al.*, 2000). The v-ATPase is a membrane-bound proton pump that facilitates the acidification of organelles such as the Golgi apparatus, endosomes and lysosomes to a luminal pH of 4.5–5, utilizing the energy from ATP hydrolysis to generate an influx of protons into the vesicle (Schneider, 1987). On this basis it has been postulated that potential impairment of the proton ATPase would result in incorrect endosomal pH and, consequently, ligand separation from the endocytosed receptor and increased EGFR recycling to the surface. This would explain both the increased cell surface levels of EGFR and enhanced receptor phosphorylation in E5-expressing cells. Subsequent studies suggest that the role of 16K in E5-induced EGFR activation is more ambiguous. E5 binding to 16K can be functionally separated from EGFR activation; whereas 16K binding has been mapped to a region within TM2 of E5, truncation of the last five amino acids impairs the ability of E5 to increase EGFR phosphorylation (Rodriguez *et al.*, 2000). In addition, E5 has been reported to both inhibit (Briggs *et al.*, 2001) and have no effect (Ashby *et al.*, 2001) on vacuolar acidification in studies using yeast. Furthermore, initial observations of impaired endosomal acidification made use of pH-sensitive reagents that require

Table 4.1 Cellular proteins shown to interact with E5

Protein partner	Functional consequence	References
16K v-ATPase	Directly modulates vATPase activity and prevents endosome acidification, potential mechanism for EGFR activation	Andresson *et al.* (1995)
Bap-31	Binding to Bap-31 mediated by extreme carboxyl-terminus of E5. Mutation of this region or depletion of endogenous Bap-31 using siRNA impairs proliferative capacity of HPV-infected cells following differentiation. May also de-regulate cellular trafficking and aid in E5-mediated subversion of MHCI presentation	Regan and Laimins (2008)
MHCI	Direct binding to the heavy chain of MHCI mediated by residues in TM1. May sequester MHCI in the ER and reduce surface expression to prevent detection of HPV-encoded antigens by T cells	Ashrafi *et al.* (2006b)
EGFR	A direct interaction observed in Cos-7 cells overexpressing E5, although not repeated by other groups	Hwang *et al.* (1995)
Karyopherin β3	A nuclear import protein that also has roles in the secretory pathway and as an ER chaperone. Binds to the final ten amino acids of E5 but the functional consequences of this interaction are not known	Krawczyk *et al.* (2008a)
Calnexin	E5 forms a ternary complex with MHCI and calnexin that is dependent on TM1. Calnexin binding appears required for MHCI down-regulation	Gruener *et al.* (2007)

B-cell-associated protein-31 (Bap-31) mediated trafficking from the ER (Annaert *et al.*, 1997). Bap-31 is an E5 binding partner and as such may cause retention in the ER lumen, which is not acidic, and therefore may provide misleading data (Regan and Laimins, 2008) (Table 4.1; Fig. 4.1B and 4.2). Until recently, all studies that have focussed on the E5–16K interaction have relied upon overexpression of epitope tagged proteins *in vitro*, owing to a lack of specific antibody reagents. This problem was recently solved when the Schlegel group generated a specific antibody directed against the 16K subunit (Suprynowicz *et al.*, 2010). This allowed them to interrogate the interaction between E5 and endogenous 16K, with startling results. In primary keratinocytes, only a small portion of the endogenous 16K interacts with E5, precluding the possibility that E5 directly inhibits v-ATPase function in cells (Suprynowicz *et al.*, 2010). Instead, they found that E5 expression prevented the acidification of EGFR-containing early endosomes by impeding their fusion with later endocytic compartments. With these revelations suggesting new functions for E5 within the EGFR signalling pathway, the hunt is now on to discover the specific cellular proteins that regulate endocytic trafficking that may be targeted by E5.

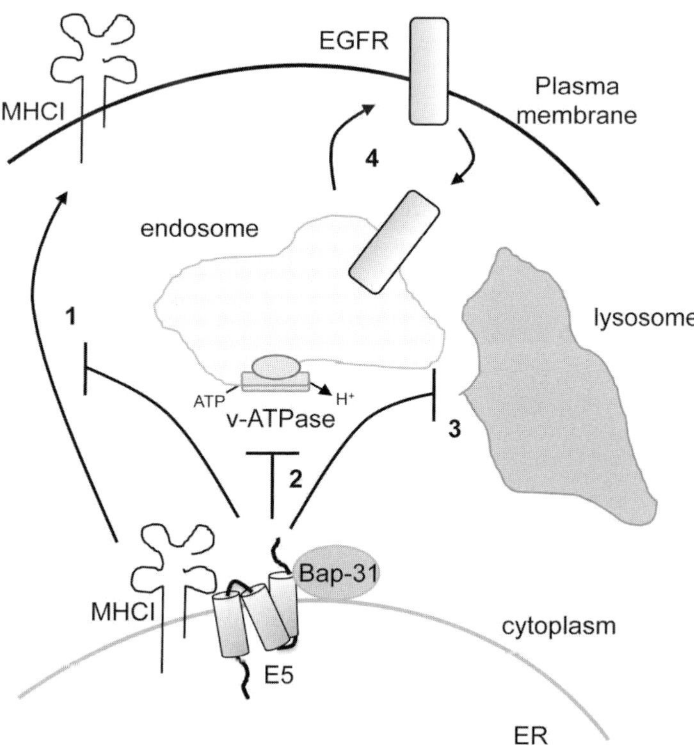

Figure 4.2 Modulation of cellular trafficking subverts EGFR and MHCI signalling. E5 contributes to HPV pathogenesis by modulating the transit of signalling proteins through the ER as well as interacting with cellular factors including MHCI and Bap-31 and the v-ATPase. (1) E5 expression results in decreased cell surface expression of MHCI, which may be mediated by a direct interaction with MHCI or by subverting the ER resident Bap-31, resulting in reduced T-cell surveillance and an inadequate cellular immune response to HPV infection. E5 increases EGFR signalling and activation of the MAPK pathway, resulting in aberrant proliferation. (2) The interaction of E5 with the vacuolar ATPase or alternatively perturbation of endosome fusion to acidic lysosomes (3) may promote EGFR recycling to the cell surface, resulting in constitutive signalling (4).

E5 modulates numerous signalling pathways – more than one pathway to transformation

EGFR activation by E5 initiates a diverse signalling cascade culminating in the activation of several proto-oncogenes. In particular, enhanced activation of the extracellular regulated kinases/mitogenic activated protein kinases (ERK/MAPK) has been observed in E5-expressing cells, coupled to increased transcription of early genes (Bouvard *et al.*, 1994). Levels of the activating protein-1 (AP-1) component c-Fos are increased upon EGF treatment, and this is augmented in the presence of E5. Since the HPV enhancer contains AP-1 binding sites, E5 may potentiate virus gene expression via the activation of c-Fos (Chen *et al.*, 1996). Ultimately, this may increase levels of E6/E7, stimulating E7-mediated transformation (Bouvard *et al.*, 1994). Thus, E5 may also play an additional role in transformation by regulating the expression of other cellular and viral genes. Interestingly, expression of E5 alone is sufficient to induce a small, but detectable, increase in c-Fos expression in the absence of EGF treatment (Crusius *et al.*, 1997). Either this is due to greater sensitivity to lower levels of growth factor or E5 activates an EGFR-independent signalling pathway. One potential candidate is protein kinase C (PKC), which is hyper-activated after phorbol ester treatment in E5 expressing cells and is able to induce the phosphorylation of ERK/MAPK (Crusius *et al.*, 1997). The precise contribution of these EGFR-independent pathways to transformation remains undetermined.

E5 also enhances cell proliferation by inhibiting the key tumour suppressors $p21^{WAF1}$ and $p27^{KIP1}$. These cyclin-dependent kinase inhibitors (CKI) regulate cell cycle progression. E5 from low- and high-risk mucosal types was able to repress $p21^{WAF1}$ transcription in immortalized keratinocyte cells (Tsao *et al.*, 1996). This reduction was correlated with transforming ability and relied upon increased expression of the AP-1 component c-Jun. In addition, HPV16 E5 down-regulated the half-life of $p27^{KIP1}$ in NIH3T3 cells (Pedroza-Saavedra *et al.*, 2010), resulting in cell cycle progression and increased DNA synthesis. Interestingly, the effects of E5 on the cell cycle were enhanced in EGF treated cells, but abrogated by an EGFR inhibitor, suggesting that down-regulation of $p27^{KIP1}$ is EGFR-dependent.

E5 also perturbs gap-junction mediated intercellular communication by targeting the major gap-junction component Connexin-43 (Oelze *et al.*, 1995). Levels of Connexin-43 phosphorylation were reduced in HaCaT cells expressing E5. Whether E5 disables a protein kinase or up-regulates a protein phosphatase is currently unclear, but it is probable that by impairing gap-junction function, E5 is able to separate HPV infected cells from control signals emitted by neighbouring cells. Interestingly, a similar effect was observed for the E5 protein of BPV-4, suggesting that deregulation of gap-junctions is a shared function of disparate E5 proteins (Faccini *et al.*, 1996).

Formation of koilocytes, which are epithelial cells containing a heterochromatic nucleus that is asymmetrically displaced by perinuclear vacuoles, requires co-operation between the E5 and E6 oncoproteins (Krawczyk *et al.*, 2008b). Whilst it is currently unclear why these morphological changes take place, they have been utilized to aid pathologists in the identification of HPV-infected epithelial cells in Pap-smears. Interestingly, they are produced both by low- and high-risk HPV infection and mutation of the final 20 amino acids of E5 abrogates koilocyte formation (Krawczyk *et al.*, 2008b). It is plausible that koilocyte formation is one consequence of the vesicular trafficking properties of E5, since mutants unable to induce koilocytosis are also defective for endosome alkalinization and EGFR activation. This suggests that modulation of vesicular trafficking and membrane modulation may be a common aspect of E5 function.

Subversion of the immune response

The immune system plays a pivotal role in determining the clinical outcome of HPV infection, as evidenced by increased persistence and neoplastic progression in individuals with immune deficiencies (Ozsaran *et al.*, 1999). In particular, clearance of virus infection requires antigen-specific cytotoxic T-lymphocytes that can either directly kill infected cells (CTLs) or B-lymphocytes to secrete antibodies to limit virus infectivity. CTL recognize virally infected cells via ligation to cell surface class I major histocompatibility complex (MHCI) bound to viral peptides on the surface of infected cells. HPV induced tumours often exhibit loss of MHCI molecules (Bontkes *et al.*, 1998). Accordingly, research has focussed on the identification of HPV-mediated mechanisms of immune subversion. In this regard, E5 is capable of perturbing the cell surface expression of several immune activation molecules, including MHCI. Down-regulation of cell surface MHCI has been shown for E5 from multiple HPV types, although the mechanisms of down-regulation are not fully understood. It has been proposed that arrest of MHCI in the Golgi apparatus is due to E5-induced alkalinization of endomembrane compartments (Schapiro *et al.*, 2000), although a direct interaction between the MHCI heavy chain (HC) and the first TM domain of E5 has also been reported (Table 4.1) (Ashrafi *et al.*, 2006a). Other indirect mechanisms of action include an interaction between E5 and the ER-resident proteins calnexin and Bap-31 (Table 4.1). Both of these proteins are necessary for the efficient trafficking of MHCI to the cell surface, theoretically binding of E5 to these host proteins could interfere with MHCI processing and trafficking (Gruener *et al.*, 2007; Regan and Laimins, 2008). Decreased cell surface expression of MHCI correlated with impaired CD8 T-cell responses in cell culture (Campo *et al.*, 2010) and in high-grade neoplasias (Gill *et al.*, 1998). E5 may also contribute directly to CD8 T-cell dysfunction by up-regulating the cell surface expression of ganglioside GM1 on the surface of infected cells (Suprynowicz *et al.*, 2008). Gangliosides are often upregulated on tumour cells and strongly inhibit CD8 T-cell activation by interfering with immune synapse formation between CTL and target cells.

E5 can also arrest the cell surface expression of another MHCI family member, CD1d (Miura *et al.*, 2010). CD1d is an MHCI-like glycoprotein that presents antigens to natural killer T (NKT) cells, and as such represents a bridge between the innate and cellular immune responses. Cell surface levels of CD1dare significantly reduced in HPV positive biopsy samples compared with controls and this down-regulation can be traced to expression of E5 (Miura *et al.*, 2010). E5 promotes the proteasomal degradation of CD1d, mediated by an interaction with calnexin in the ER (Table 4.1). As NKT cells play an important role in the generation of innate and adaptive immune responses in the cervix, decreased CD1d expression would allow HPV to evade the host immune surveillance and establish a persistent infection at the primary transmission site.

E5 may also play a more significant role in subverting intrinsic anti-viral processes within keratinocytes. Immunoprecipitation analysis demonstrated that high-risk E5 interacts with EVER proteins and the zinc transporter ZnT1 (Lazarczyk *et al.*, 2008). EVER proteins, mutated in the rare genetic disorder Epidermodysplasia verruciformis (EV), are thought to regulate levels of free zinc within cells and to act as part of an intrinsic defence mechanism against virus infection. The delicate intracellular zinc balance is subverted in the presence of E5, and this is thought to favour virus transcription. Interestingly the beta HPV, which lack an E5 protein, are harmless in most individuals but associated with significant disease in EV patients. This suggests that keratinocyte intrinsic defences normally counteract beta HPV infection, but in the presence of an E5 protein or mutated EVER proteins, this defence breaks down and HPV replication proceeds unhindered (Orth, 2008). Clearly, more research is required to determine whether there are other host targets and processes modulated by E5 to counteract intrinsic keratinocyte defences.

Together, these findings demonstrate that E5 plays an important role in the establishment and persistence of HPV infection by allowing infected cells to escape host immunosurveillance. Importantly, persistent lesions arising from this immune evasion would be at greater risk of undergoing additional oncogenic events, which would trigger malignant transformation.

Inhibition of apoptosis

Apoptosis is an important process for the elimination of virus-infected cells and consequently many viruses that establish a chronic infection have evolved mechanisms to evade this response to ensure their persistence. Keratinocytes expressing high-risk E5 are protected from the apoptotic effects of the death receptor ligands FasL and tumour necrosis factor-related apoptosis-inducing ligand (TRAIL) and this appears to be mediated by two distinct mechanisms (Kabsch and Alonso, 2002b; Kabsch *et al.*, 2004). Firstly, levels of cell surface FasL receptor (CD95/Fas) are significantly reduced in E5-expressing cells (Kabsch and Alonso, 2002b). Secondly, formation of the death-receptor inducible signalling complex (DISC) is abolished after treatment with TRAIL in E5-expressing cells (Kabsch and Alonso, 2002b). DISC formation is essential for propagation of the apoptotic signal from external sources and disruption of this complex is likely to prevent immune mediated apoptosis. Importantly, raft cultures from E5-expressing cells were protected from FasL and TRAIL-induced apoptosis, emphasizing the importance of E5 in preventing cell death.

E5 also protects tumour cells from a variety of cellular insults and environmental stresses. UV-irradiation induced apoptosis is strongly reduced by E5-mediated PI3K signalling (Zhang *et al.*, 2002), whilst hydrogen peroxide induced cell death is impaired by proteasomal degradation of the pro-apoptotic Bcl-2 protein Bax (Oh *et al.*, 2010). In addition, levels of survival factors including the anti-apoptotic Bcl-2 protein increase in E5 expressing cells (Zhang *et al.*, 2002; Auvinen *et al.*, 2004).

In contrast, E5 may act as a pro-apoptotic factor under osmotic stress. When expressed in human keratinocytes and mouse fibroblasts, E5 sensitized these cells to sorbitol-induced apoptosis (Kabsch and Alonso, 2002a). The molecular details explaining the switch to pro-apoptotic factor, in this instance, are not clear. However, the liability of E5-expressing cells to osmotic stress possibly results from modifications to cellular membranes owing to the presence of the strongly hydrophobic viral protein (Crusius *et al.*, 1999). This link remains to be verified by experimental evidence.

By modulating apoptosis, HPV-infected cells can survive apoptotic stimuli from numerous environmental insults and evade host immunosurveillance and CTL-mediated killing via extrinsic factors such as FasL and TRAIL.

Modulation of stress-associated and inflammatory signalling

In the context of viral pathogenesis during a persistent infection, arguably an important factor in the development of carcinogenesis is modulation of stress and inflammatory signalling. Studies have focussed on stress-associated MAPK pathways and the products of cyclo-oxygenase-2 signalling (Cox-2), often with conflicting findings. Reports highlight that E5 expression is sufficient to activate both stress-activated MAPK signalling modules, which includes the p38 MAPK and the c-Jun N-terminal kinase (JNK) (Crusius *et al.*, 2000; Chen *et al.*, 2007). Activation of p38 MAPK by E5 appears restricted to anisomycin treatment (Crusius *et al.*, 2000), whilst E5 expression alone is sufficient to induce phosphorylation of the JNK substrate c-Jun (Chen *et al.*, 2007). The significance of activated JNK and p38 MAPK is not clear, since these kinases are associated with increased cellular stress and apoptosis. Further investigation is required to confirm and extend these data.

Controversy surrounds the role of Cox-2 during the HPV life cycle. Whilst it is known that Cox-2 plays an important role in carcinogenesis and cancer progression in various malignancies, including cervical cancer (Kulkarni *et al.*, 2001) and Cox-2 levels can be increased by both E6 and E7, it is not yet clear whether Cox-2 can be modulated by E5. Studies in cervical cancer cell lines (C33A/HaCaT) highlight an EGFR pathway-dependent increase in the levels of Cox-2 expression in cells containing E5, which results in increased cell survival (Kim *et al.*, 2009). Contradictory microarray data, obtained from primary foreskin keratinocytes constitutively expressing E5, demonstrates reduced levels of Cox-2 (Sudarshan *et al.*, 2010). This study suggests that Cox-2 down-regulation is part of a general suppression of the ER-stress response in the presence of E5 from high-risk HPV types. How these conflicting observations can be reconciled is not yet clear. It is possible that Cox-2 levels may change depending on the cellular context or that the results reflect different experimental systems; transformed cell lines versus primary keratinocytes. Clearly, the role of Cox-2 signalling in E5-dependent processes requires further investigation.

Is E5 a target for anti-viral therapeutics?

Despite the availability of prophylactic vaccines and interventionist surgery, there is still an unmet need to develop a wider array of therapeutic options to treat the many millions infected with HPV. Targeting HPV proteins directly is an avenue of research, but is the E5 oncoprotein a valid target? Certainly, E5 contributes to carcinogenesis by modulating signalling pathways associated with malignancy. Moreover, as E5 appears to be expressed in abundance in pre-cancerous lesions and is required for efficient HPV replication, then E5 or constituents of the affected pathways are potential targets to prevent viral persistence and lesions from progressing into invasive cancers.

The recent observation that E5 displays channel-forming activity highlights a previously unidentified therapeutic opportunity (Wetherill *et al.*, unpublished). By far the best-characterized virus-encoded ion channel is the M2 proton channel encoded by the influenza A virus, which is the target for the antiviral drugs amantadine and rimantadine (Schnell and Chou, 2008). M2 sets a precedent for viral ion channels as therapeutic targets that could drive research into the development of E5-specific small molecule inhibitors. Alternatively, tumour vaccines could target E5-expressing cells. Vaccination with recombinant adenoviruses expressing E5 or CTL-specific E5 peptides with CpG-oligodeoxynucleotides significantly reduced tumour size in preliminary animal experiments (Chen *et al.*, 2004). Other therapeutic strategies targeting E5 could include the use of oncolytic viruses, gene-silencing using short interfering RNA (siRNA) and peptide/RNA aptamers, which have been originally used to target the E6/E7 oncoproteins (Nicol *et al.*, 2011). Lastly, strategies could instead target specific host-expressed enzymes that HPV requires to replicate in keratinocytes. Importantly, targeting these cellular proteins would provide broad-spectrum protection against multiple HPV types. One potential target is the EGFR-MAPK pathway, since small molecule inhibitors against constituents of this signalling cascade are already in clinical use or trials. Blocking the E5-mediated activation of MAPK may hinder HPV replication by preventing AP-1-mediated activation of the HPV enhancer. If verified as effective and safe in clinical studies, E5-targeted therapies in combination with standard therapeutic modalities could improve the treatment outcomes in HPV-associated malignancies.

Conclusions and future directions

It is now generally accepted that E5 from high-risk HPV functions within the later stages of the HPV life cycle, to enhance genomic DNA replication in suprabasal cells and to maintain terminally differentiated keratinocytes within the cell cycle. However, it is unclear how this is achieved and whether E5 proteins from other HPV types behave in a similar manner. An important task for the future will be to decipher the molecular mechanisms and critical host interactions that allow E5 to achieve this perturbation of host keratinocytes and to determine whether this is restricted to high-risk E5 proteins. A recent finding that E5 is the latest member of the viroporin family holds great interest. In particular, viroporins play roles in virus genome replication and in virus assembly. Since early studies demonstrated that the highest levels of HPV31 E5 expression coincide with virion morphogenesis it will be critical to dissect whether viroporin function is required during these later stages of HPV replication. If so, this would increase the known roles of E5 within the virus life cycle.

Findings from the Lambert and Schlegel groups have also widened interest in E5-induced cellular transformation. Mouse models have definitively demonstrated the transforming functions of E5 and have shown it to be a more potent oncoprotein than originally assumed. Whilst it is believed that the E5 phenotype, most frequently linked to the development of cancer, is enhanced EGFR activation, it is now clear that inhibition of v-ATPase alone cannot account for this phenotype. We must now add perturbation of endosomal maturation to the list of potential mechanisms by which this is achieved. Dissecting the molecular details of this perturbation will greatly increase our understanding and may help to explain other aspects of E5 function such as the down-regulation of MHCI.

In summary, recent work has emphasized how little we understand about this difficult to study oncoprotein, and reinforced how important E5 is both for the virus life cycle and for transformation. Our goal over the coming years will be to resolve its function at the molecular level and to translate these findings into potential therapeutic benefits.

Acknowledgements

The authors would like to thank Dr Nicola Stonehouse for critical reading of the manuscript and useful comments. LFW is funded by an MRC studentship, AM is a recipient of an RCUK Academic Fellowship and research in his laboratory is funded by Yorkshire Cancer Research (PP015, L339 and LPP041) and the Royal Society.

References

Alonso, A., and Reed, J. (2002). Modelling of the human papillomavirus type 16 E5 protein. Biochim. Biophys. Acta *1601*, 9–18.

Andresson, T., Sparkowski, J., Goldstein, D.J., and Schlegel, R. (1995). Vacuolar H(+)-ATPase mutants transform cells and define a binding site for the papillomavirus E5 oncoprotein. J. Biol. Chem. *270*, 6830–6837.

Annaert, W.G., Becker, B., Kistner, U., Reth, M., and Jahn, R. (1997). Export of cellubrevin from the endoplasmic reticulum is controlled by BAP31. J. Cell. Biol. *139*, 1397–1410.

Ashby, A.D., Meagher, L., Campo, M.S., and Finbow, M.E. (2001). E5 transforming proteins of papillomaviruses do not disturb the activity of the vacuolar H(+)-ATPase. J. Gen. Virol. *82*, 2353–2362.

Ashrafi, G.H., Brown, D.R., Fife, K.H., and Campo, M.S. (2006a). Down-regulation of MHC class I is a property common to papillomavirus E5 proteins. Virus Res. *120*, 208–211.

Ashrafi, G.H., Haghshenas, M., Marchetti, B., and Campo, M.S. (2006b). E5 protein of human papillomavirus 16 down-regulates HLA class I and interacts with the heavy chain via its first hydrophobic domain. Int. J. Cancer *119*, 2105–2112.

Authier, F., and Chauvet, G. (1999). *In vitro* endosome-lysosome transfer of dephosphorylated EGF receptor and Shc in rat liver. FEBS Lett. *461*, 25–31.

Auvinen, E., Alonso, A., and Auvinen, P. (2004). Human papillomavirus type 16 E5 protein colocalizes with the antiapoptotic Bcl-2 protein. Arch. Virol. *149*, 1745–1759.

Boiko, I.V., Mitchell, M.F., Hu, W., Pandey, D.K., Mathevet, P., Malpica, A., and Hittelman, W.N. (1998). Epidermal growth factor receptor expression in cervical intraepithelial neoplasia and its modulation during an alpha-difluoromethylornithine chemoprevention trial. Clin. Cancer Res. *4*, 1383–1391.

Bontkes, H.J., van Duin, M., de Gruijl, T.D., Duggan-Keen, M.F., Walboomers, J.M., Stukart, M.J., Verheijen, R.H., Helmerhorst, T.J., Meijer, C.J., Scheper, R.J., *et al.* (1998). HPV16 infection and progression of cervical intra-epithelial neoplasia: analysis of HLA polymorphism and HPV16 E6 sequence variants. Int. J. Cancer *78*, 166–171.

Borzacchiello, G., Roperto, F., Campo, M.S., and Venuti, A. (2010). 1st International Workshop on Papillomavirus E5 Oncogene – a report. Virology *408*, 135–137.

Bouvard, V., Matlashewski, G., Gu, Z.M., Storey, A., and Banks, L. (1994). The human papillomavirus type 16 E5 gene cooperates with the E7 gene to stimulate proliferation of primary cells and increases viral gene expression. Virology *203*, 73–80.

Bravo, I.G., and Alonso, A. (2004). Mucosal human papillomaviruses encode four different E5 proteins whose chemistry and phylogeny correlate with malignant or benign growth. J. Virol. *78*, 13613–13626.

Briggs, M.W., Adam, J.L., and McCance, D.J. (2001). The human papillomavirus type 16 E5 protein alters vacuolar H(+)-ATPase function and stability in *Saccharomyces cerevisiae*. Virology *280*, 169–175.

Bubb, V., McCance, D.J., and Schlegel, R. (1988). DNA sequence of the HPV16 E5 ORF and the structural conservation of its encoded protein. Virology *163*, 243–246.

Campo, M.S., Graham, S.V., Cortese, M.S., Ashrafi, G.H., Araibi, E.H., Dornan, E.S., Miners, K., Nunes, C., and Man, S. (2010). HPV16 E5 down-regulates expression of surface HLA class I and reduces recognition by CD8 T cells. Virology *407*, 137–142.

Cartin, W., and Alonso, A. (2003). The human papillomavirus HPV2a E5 protein localizes to the Golgi apparatus and modulates signal transduction. Virology *314*, 572–579.

Chen, S.L., and Mounts, P. (1990). Transforming activity of E5a protein of human papillomavirus type 6 in NIH 3T3 and C127 cells. J. Virol. *64*, 3226–3233.

Chen, S.L., Lin, Y.K., Li, L.Y., Tsao, Y.P., Lo, H.Y., Wang, W.B., and Tsai, T.C. (1996). E5 proteins of human papillomavirus types 11 and 16 transactivate the c-fos promoter through the NF1 binding element. J. Virol. *70*, 8558–8563.

Chen, S.L., Lin, S.T., Tsai, T.C., Hsiao, W.C., and Tsao, Y.P. (2007). ErbB4 (JM-b/CYT-1)-induced expression and phosphorylation of c-Jun is abrogated by human papillomavirus type 16 E5 protein. Oncogene *26*, 42–53.

Chen, Y.F., Lin, C.W., Tsao, Y.P., and Chen, S.L. (2004). Cytotoxic-T-lymphocyte human papillomavirus type 16 E5 peptide with CpG-oligodeoxynucleotide can eliminate tumour growth in C57BL/6 mice. J. Virol. *78*, 1333–1343.

Conrad, M., Bubb, V.J., and Schlegel, R. (1993). The human papillomavirus type 6 and 16 E5 proteins are membrane-associated proteins which associate with the 16-kilodalton pore-forming protein. J. Virol. *67*, 6170–6178.

Conrad, M., Goldstein, D., Andresson, T., and Schlegel, R. (1994). The E5 protein of HPV-6, but not HPV16, associates efficiently with cellular growth factor receptors. Virology *200*, 796–800.

Crusius, K., Auvinen, E., and Alonso, A. (1997). Enhancement of EGF- and PMA-mediated MAP kinase activation in cells expressing the human papillomavirus type 16 E5 protein. Oncogene *15*, 1437–1444.

Crusius, K., Auvinen, E., Steuer, B., Gaissert, H., and Alonso, A. (1998). The human papillomavirus type 16 E5-protein modulates ligand-dependent activation of the EGF receptor family in the human epithelial cell line HaCaT. Exp. Cell Res. *241*, 76–83.

Crusius, K., Kaszkin, M., Kinzel, V., and Alonso, A. (1999). The human papillomavirus type 16 E5 protein modulates phospholipase C-gamma-1 activity and phosphatidyl inositol turnover in mouse fibroblasts. Oncogene *18*, 6714–6718.

Crusius, K., Rodriguez, I., and Alonso, A. (2000). The human papillomavirus type 16 E5 protein modulates ERK1/2 and p38 MAP kinase activation by an EGFR-independent process in stressed human keratinocytes. Virus Genes *20*, 65–69.

DiMaio, D., and Mattoon, D. (2001). Mechanisms of cell transformation by papillomavirus E5 proteins. Oncogene *20*, 7866–7873.

DiMaio, D., Guralski, D., and Schiller, J.T. (1986). Translation of open reading frame E5 of bovine papillomavirus is required for its transforming activity. Proc. Natl. Acad. Sci. U.S.A. *83*, 1797–1801.

Disbrow, G.L., Sunitha, I., Baker, C.C., Hanover, J., and Schlegel, R. (2003). Codon optimization of the HPV16 E5 gene enhances protein expression. Virology *311*, 105–114.

Disbrow, G.L., Hanover, J.A., and Schlegel, R. (2005). Endoplasmic reticulum-localized human papillomavirus type 16 E5 protein alters endosomal pH but not trans-Golgi pH. J. Virol. *79*, 5839–5846.

Doorbar, J., Parton, A., Hartley, K., Banks, L., Crook, T., Stanley, M., and Crawford, L. (1990). Detection of novel splicing patterns in a HPV16-containing keratinocyte cell line. Virology *178*, 254–262.

Durst, M., Glitz, D., Schneider, A., and zur Hausen, H. (1992). Human papillomavirus type 16 (HPV16) gene expression and DNA replication in cervical neoplasia: analysis by *in situ* hybridization. Virology *189*, 132–140.

Ettenberg, S.A., Magnifico, A., Cuello, M., Nau, M.M., Rubinstein, Y.R., Yarden, Y., Weissman, A.M., and Lipkowitz, S. (2001). Cbl-b-dependent coordinated degradation of the epidermal growth factor receptor signalling complex. J. Biol. Chem. *276*, 27677–27684.

Faccini, A.M., Cairney, M., Ashrafi, G.H., Finbow, M.E., Campo, M.S., and Pitts, J.D. (1996). The bovine papillomavirus type 4 E8 protein binds to ductin and causes loss of gap junctional intercellular communication in primary fibroblasts. J. Virol. *70*, 9041–9045.

Fehrmann, F., Klumpp, D.J., and Laimins, L.A. (2003). Human papillomavirus type 31 E5 protein supports cell cycle progression and activates late viral functions upon epithelial differentiation. J. Virol. *77*, 2819–2831.

Flores, E.R., and Lambert, P.F. (1997). Evidence for a switch in the mode of human papillomavirus type 16 DNA replication during the viral life cycle. J. Virol. *71*, 7167–7179.

Flores, E.R., Allen-Hoffmann, B.L., Lee, D., and Lambert, P.F. (2000). The human papillomavirus type 16 E7 oncogene is required for the productive stage of the viral life cycle. J. Virol. *74*, 6622–6631.

Futter, C.E., Pearse, A., Hewlett, L.J., and Hopkins, C.R. (1996). Multivesicular endosomes containing internalized EGF-EGF receptor complexes mature and then fuse directly with lysosomes. J. Cell. Biol. *132*, 1011–1023.

Genther, S.M., Sterling, S., Duensing, S., Munger, K., Sattler, C., and Lambert, P.F. (2003). Quantitative role of the human papillomavirus type 16 E5 gene during the productive stage of the viral life cycle. J. Virol. *77*, 2832–2842.

Genther Williams, S.M., Disbrow, G.L., Schlegel, R., Lee, D., Threadgill, D.W., and Lambert, P.F. (2005). Requirement of epidermal growth factor receptor for hyperplasia induced by E5, a high-risk human papillomavirus oncogene. Cancer Res. *65*, 6534–6542.

Gieswein, C.E., Sharom, F.J., and Wildeman, A.G. (2003). Oligomerization of the E5 protein of human papillomavirus type 16 occurs through multiple hydrophobic regions. Virology *313*, 415–426.

Gill, D.K., Bible, J.M., Biswas, C., Kell, B., Best, J.M., Punchard, N.A., and Cason, J. (1998). Proliferative T-cell responses to human papillomavirus type 16 E5 are decreased amongst women with high-grade neoplasia. J. Gen. Virol. *79 (Pt 8)*, 1971–1976.

Gruener, M., Bravo, I.G., Momburg, F., Alonso, A., and Tomakidi, P. (2007). The E5 protein of the human papillomavirus type 16 down-regulates HLA-I surface expression in calnexin-expressing but not in calnexin-deficient cells. Virol. J. *4*, 116.

Hu, L., and Ceresa, B.P. (2009). Characterization of the plasma membrane localization and orientation of HPV16 E5 for cell–cell fusion. Virology *393*, 135–143.

Hummel, M., Hudson, J.B., and Laimins, L.A. (1992). Differentiation-induced and constitutive transcription of human papillomavirus type 31b in cell lines containing viral episomes. J. Virol. *66*, 6070–6080.

Hwang, E.S., Nottoli, T., and Dimaio, D. (1995). The HPV16 E5 protein: expression, detection, and stable complex formation with transmembrane proteins in COS cells. Virology *211*, 227–233.

Kabsch, K., and Alonso, A. (2002a). The human papillomavirus type 16 (HPV16) E5 protein sensitizes human keratinocytes to apoptosis induced by osmotic stress. Oncogene *21*, 947–953.

Kabsch, K., and Alonso, A. (2002b). The human papillomavirus type 16 E5 protein impairs TRAIL- and

FasL-mediated apoptosis in HaCaT cells by different mechanisms. J. Virol. *76*, 12162–12172.

Kabsch, K., Mossadegh, N., Kohl, A., Komposch, G., Schenkel, J., Alonso, A., and Tomakidi, P. (2004). The HPV16 E5 protein inhibits TRAIL- and FasL-mediated apoptosis in human keratinocyte raft cultures. Intervirology *47*, 48–56.

Kell, B., Jewers, R.J., Cason, J., Pakarian, F., Kaye, J.N., and Best, J.M. (1994). Detection of E5 oncoprotein in human papillomavirus type 16-positive cervical scrapes using antibodies raised to synthetic peptides. J. Gen. Virol. *75 (Pt 9)*, 2451–2456.

Kersemaekers, A.M., Fleuren, G.J., Kenter, G.G., Van den Broek, L.J., Uljee, S.M., Hermans, J., and Van de Vijver, M.J. (1999). Oncogene alterations in carcinomas of the uterine cervix: overexpression of the epidermal growth factor receptor is associated with poor prognosis. Clin. Cancer Res. *5*, 577–586.

Kim, S.H., Oh, J.M., No, J.H., Bang, Y.J., Juhnn, Y.S., and Song, Y.S. (2009). Involvement of NF-kappaB and AP-1 in COX-2 upregulation by human papillomavirus 16 E5 oncoprotein. Carcinogenesis *30*, 753–757.

Krawczyk, E., Hanover, J.A., Schlegel, R., and Suprynowicz, F.A. (2008a). Karyopherin beta3: a new cellular target for the HPV16 E5 oncoprotein. Biochem. Biophys. Res. Commun. *371*, 684–688.

Krawczyk, E., Suprynowicz, F.A., Liu, X., Dai, Y., Hartmann, D.P., Hanover, J., and Schlegel, R. (2008b). Koilocytosis: a cooperative interaction between the human papillomavirus E5 and E6 oncoproteins. Am J. Pathol. *173*, 682–688.

Krawczyk, E., Suprynowicz, F.A., Sudarshan, S.R., and Schlegel, R. (2010). Membrane orientation of the human papillomavirus type 16 E5 oncoprotein. J. Virol. *84*, 1696–1703.

Kulkarni, S., Rader, J.S., Zhang, F., Liapis, H., Koki, A.T., Masferrer, J.L., Subbaramaiah, K., and Dannenberg, A.J. (2001). Cyclooxygenase-2 is overexpressed in human cervical cancer. Clin. Cancer Res. *7*, 429–434.

Lazarczyk, M., Pons, C., Mendoza, J.A., Cassonnet, P., Jacob, Y., and Favre, M. (2008). Regulation of cellular zinc balance as a potential mechanism of EVER-mediated protection against pathogenesis by cutaneous oncogenic human papillomaviruses. J. Exp. Med. *205*, 35–42.

Leechanachai, P., Banks, L., Moreau, F., and Matlashewski, G. (1992). The E5 gene from human papillomavirus type 16 is an oncogene which enhances growth factor-mediated signal transduction to the nucleus. Oncogene *7*, 19–25.

Lewis, C., Baro, M.F., Marques, M., Gruner, M., Alonso, A., and Bravo, I.G. (2008). The first hydrophobic region of the HPV16 E5 protein determines protein cellular location and facilitates anchorage-independent growth. Virol. J. *5*, 30.

McCance, D.J., Kopan, R., Fuchs, E., and Laimins, L.A. (1988). Human papillomavirus type 16 alters human epithelial cell differentiation *in vitro*. Proc. Natl. Acad. Sci. U.S.A. *85*, 7169–7173.

Mattoon, D., Gupta, K., Doyon, J., Loll, P.J., and DiMaio, D. (2001). Identification of the transmembrane dimer interface of the bovine papillomavirus E5 protein. Oncogene *20*, 3824–3834.

Maufort, J.P., Williams, S.M., Pitot, H.C., and Lambert, P.F. (2007). Human papillomavirus 16 E5 oncogene contributes to two stages of skin carcinogenesis. Cancer Res. *67*, 6106–6112.

Maufort, J.P., Shai, A., Pitot, H.C., and Lambert, P.F. (2010). A role for HPV16 E5 in cervical carcinogenesis. Cancer Res. *70*, 2924–2931.

Mayer, T.J., and Meyers, C. (1998). Temporal and spatial expression of the E5a protein during the differentiation-dependent life cycle of human papillomavirus type 31b. Virology *248*, 208–217.

Miura, S., Kawana, K., Schust, D.J., Fujii, T., Yokoyama, T., Iwasawa, Y., Nagamatsu, T., Adachi, K., Tomio, A., Tomio, K., *et al.* (2010). CD1d, a sentinel molecule bridging innate and adaptive immunity, is down-regulated by the human papillomavirus (HPV) E5 protein: a possible mechanism for immune evasion by HPV. J. Virol. *84*, 11614–11623.

Nicol, C., Bunka, D.H., Blair, G.E., and Stonehouse, N.J. (2011). Effects of single nucleotide changes on the binding and activity of RNA aptamers to human papillomavirus 16 E7 oncoprotein. Biochem. Biophys. Res. Commun. *405*, 417–421.

Oelze, I., Kartenbeck, J., Crusius, K., and Alonso, A. (1995). Human papillomavirus type 16 E5 protein affects cell-cell communication in an epithelial cell line. J. Virol. *69*, 4489–4494.

Oh, J.M., Kim, S.H., Cho, E.A., Song, Y.S., Kim, W.H., and Juhnn, Y.S. (2010). Human papillomavirus type 16 E5 protein inhibits hydrogen-peroxide-induced apoptosis by stimulating ubiquitin-proteasome-mediated degradation of Bax in human cervical cancer cells. Carcinogenesis *31*, 402–410.

Orth, G. (2008). Host defenses against human papillomaviruses: lessons from epidermodysplasia verruciformis. Curr. Top. Microbiol. Immunol. *321*, 59–83.

Ozsaran, A.A., Ates, T., Dikmen, Y., Zeytinoglu, A., Terek, C., Erhan, Y., Ozacar, T., and Bilgic, A. (1999). Evaluation of the risk of cervical intraepithelial neoplasia and human papilloma virus infection in renal transplant patients receiving immunosuppressive therapy. Eur. J. Gynaecol. Oncol. *20*, 127–130.

Pater, M.M., and Pater, A. (1985). Human papillomavirus types 16 and 18 sequences in carcinoma cell lines of the cervix. Virology *145*, 313–318.

Pedroza-Saavedra, A., Lam, E.W., Esquivel-Guadarrama, F., and Gutierrez-Xicotencatl, L. (2010). The human papillomavirus type 16 E5 oncoprotein synergizes with EGF-receptor signalling to enhance cell cycle progression and the down-regulation of p27(Kip1). Virology *400*, 44–52.

Pfeiffer, D., Stellwag, B., Pfeiffer, A., Borlinghaus, P., Meier, W., and Scheidel, P. (1989). Clinical implications of the epidermal growth factor receptor in the squamous cell carcinoma of the uterine cervix. Gynecol. Oncol. *33*, 146–150.

Pim, D., Collins, M., and Banks, L. (1992). Human papillomavirus type 16 E5 gene stimulates the transforming activity of the epidermal growth factor receptor. Oncogene *7*, 27–32.

Regan, J.A., and Laimins, L.A. (2008). Bap31 is a novel target of the human papillomavirus E5 protein. J. Virol. *82*, 10042–10051.

Rodriguez, M.I., Finbow, M.E., and Alonso, A. (2000). Binding of human papillomavirus 16 E5 to the 16 kDa subunit c (proteolipid) of the vacuolar H^+-ATPase can be dissociated from the E5-mediated epidermal growth factor receptor overactivation. Oncogene *19*, 3727–3732.

Ruesch, M.N., and Laimins, L.A. (1998). Human papillomavirus oncoproteins alter differentiation-dependent cell cycle exit on suspension in semisolid medium. Virology *250*, 19–29.

Schapiro, F., Sparkowski, J., Adduci, A., Suprynowicz, F., Schlegel, R., and Grinstein, S. (2000). Golgi alkalinization by the papillomavirus E5 oncoprotein. J. Cell. Biol. *148*, 305–315.

Schmitt, M., and Pawlita, M. (2011). The HPV transcriptome in HPV16 positive cell lines. Mol. Cell. Probes. *25*, 108–13.

Schneider, D.L. (1987). The proton pump ATPase of lysosomes and related organelles of the vacuolar apparatus. Biochim. Biophys. Acta *895*, 1–10.

Schnell, J.R., and Chou, J.J. (2008). Structure and mechanism of the M2 proton channel of influenza A virus. Nature *451*, 591–595.

Schrevel, M., Gorter, A., Kolkman-Uljee, S.M., Trimbos, J.B., Fleuren, G.J., and Jordanova, E.S. (2011). Molecular mechanisms of epidermal growth factor receptor overexpression in patients with cervical cancer. Mod. Pathol. *24*, 720–8.

Schwarz, E., Freese, U.K., Gissmann, L., Mayer, W., Roggenbuck, B., Stremlau, A., and zur Hausen, H. (1985). Structure and transcription of human papillomavirus sequences in cervical carcinoma cells. Nature *314*, 111–114.

Stoler, M.H., and Broker, T.R. (1986). *In situ* hybridization detection of human papillomavirus DNAs and messenger RNAs in genital condylomas and a cervical carcinoma. Hum. Pathol. *17*, 1250–1258.

Straight, S.W., Hinkle, P.M., Jewers, R.J., and McCance, D.J. (1993). The E5 oncoprotein of human papillomavirus type 16 transforms fibroblasts and effects the down-regulation of the epidermal growth factor receptor in keratinocytes. J. Virol. *67*, 4521–4532.

Straight, S.W., Herman, B., and McCance, D.J. (1995). The E5 oncoprotein of human papillomavirus type 16 inhibits the acidification of endosomes in human keratinocytes. J. Virol. *69*, 3185–3192.

Sudarshan, S.R., Schlegel, R., and Liu, X. (2010). The HPV16 E5 protein represses expression of stress pathway genes XBP-1 and COX-2 in genital keratinocytes. Biochem. Biophys. Res. Commun. *399*, 617–622.

Suprynowicz, F.A., Disbrow, G.L., Krawczyk, E., Simic, V., Lantzky, K., and Schlegel, R. (2008). HPV16 E5 oncoprotein upregulates lipid raft components caveolin-1 and ganglioside GM1 at the plasma membrane of cervical cells. Oncogene *27*, 1071–1078.

Suprynowicz, F.A., Krawczyk, E., Hebert, J.D., Sudarshan, S.R., Simic, V., Kamonjoh, C.M., and Schlegel, R. (2010). The human papillomavirus type 16 E5 oncoprotein inhibits epidermal growth factor trafficking independently of endosome acidification. J. Virol. *84*, 10619–10629.

Surti, T., Klein, O., Aschheim, K., DiMaio, D., and Smith, S.O. (1998). Structural models of the bovine papillomavirus E5 protein. Proteins *33*, 601–612.

Thomsen, P., van Deurs, B., Norrild, B., and Kayser, L. (2000). The HPV16 E5 oncogene inhibits endocytic trafficking. Oncogene *19*, 6023–6032.

Tsao, Y.P., Li, L.Y., Tsai, T.C., and Chen, S.L. (1996). Human papillomavirus type 11 and 16 E5 represses p21(WafI/SdiI/CipI) gene expression in fibroblasts and keratinocytes. J. Virol. *70*, 7535–7539.

Valle, G.F., and Banks, L. (1995). The human papillomavirus (HPV)-6 and HPV16 E5 proteins co-operate with HPV16 E7 in the transformation of primary rodent cells. J. Gen. Virol. *76 (Pt 5)*, 1239–1245.

Vinokurova, S., Wentzensen, N., Kraus, I., Klaes, R., Driesch, C., Melsheimer, P., Kisseljov, F., Durst, M., Schneider, A., and von Knebel Doeberitz, M. (2008). Type-dependent integration frequency of human papillomavirus genomes in cervical lesions. Cancer Res. *68*, 307–313.

Yang, D.H., Wildeman, A.G., and Sharom, F.J. (2003). Overexpression, purification, and structural analysis of the hydrophobic E5 protein from human papillomavirus type 16. Protein Expr. Purif. *30*, 1–10.

Yarden, Y. (2001). The EGFR family and its ligands in human cancer. signalling mechanisms and therapeutic opportunities. Eur. J. Cancer *37 (Suppl 4)*, S3–8.

Zhang, B., Spandau, D.F., and Roman, A. (2002). E5 protein of human papillomavirus type 16 protects human foreskin keratinocytes from UV B-irradiation-induced apoptosis. J. Virol. *76*, 220–231.

Zhang, B., Srirangam, A., Potter, D.A., and Roman, A. (2005). HPV16 E5 protein disrupts the c-Cbl–EGFR interaction and EGFR ubiquitination in human foreskin keratinocytes. Oncogene *24*, 2585–2588.

Zimmermann, M., Zouhair, A., Azria, D., and Ozsahin, M. (2006). The epidermal growth factor receptor (EGFR) in head and neck cancer: its role and treatment implications. Radiat. Oncol. *1*, 11.

E6 Oncoproteins: Structure and Associations

5

Scott B. Vande Pol

Abstract

Papillomavirus E6 oncoproteins are small zinc-binding proteins with a bewildering array of biological activities, including modulation of apoptosis, cellular transcription, host cell differentiation, growth factor dependence, DNA damage responses, and cell cycle progression. How can such a tiny protein do so much? This review examines insights from studies of oncogenic human papillomavirus E6 and bovine papillomavirus E6 to illuminate the mechanism by which E6 proteins interact with cellular binding partners. The origins of E6 and the history of its investigation are presented with the discovery of the major interaction partners that mediate E6 effects on DNA damage responses, cellular transcription, and modulation of keratinocyte differentiation.

Introduction to papillomavirus biology

What are papillomaviruses?

Papillomaviruses are small, encapsidated, DNA viruses with double-stranded circular genomes, that induce benign squamous epithelial neoplasms (papillomas) in vertebrates and replicate within the differentiating cell layers of the papilloma. Some papillomas are so small as to be invisible to the eye (such as human β-papillomavirus lesions in normal hosts) while others can make kilogram size lesions (such as bovine papillomavirus type 1 (BPV-1). Although all papillomas are initially benign, some may evolve over time to produce malignancies, most notably human anogenital and upper respiratory carcinomas. That subset of viruses associated with anogenital mucosal cancers are referred to as 'high-risk' HPV types (HR HPV), and the related mucosal viruses that do not cause malignancies are called 'low risk'.

The infectious cycle begins with an injury to the squamous epithelium, giving the virus access to the basal layer and basement membrane at the base of the epithelial sheet. The initial virus-infected cells have low copy numbers of episomal viral DNA in proliferative basal epithelial cells. After cell division in the basal cell layer, progeny cells are pushed up off the basement membrane into the spinous cell layer, and a subset of these spinous cells, under the influence of viral oncogenes, re-enter the cell cycle to amplify viral DNAs to high copy number (Fig. 5.1 and reviewed in Chow and Broker, 1994, and Stubenrauch and Laimins, 1999). As cells with amplified viral DNA copy number move even higher within the stratified epithelium, a subset of the cells that have amplified genomes express the two late gene capsid proteins within the granular cell layers, encapsidate viral DNA, and are finally shed from the surface of the papilloma as desquamated epithelial cells containing infectious virus. This is unlike normal uninfected squamous epithelium where cells divide in the basal layer, move into the spinous cell layer, but do not re-enter the cell cycle and commit to a terminal differentiation pathway.

Papillomaviruses express eight early gene products, with additional variation in these proteins produced by splicing, proteolysis, and post-translational modifications (reviewed in (Moody and Laimins, 2010). The early gene products are

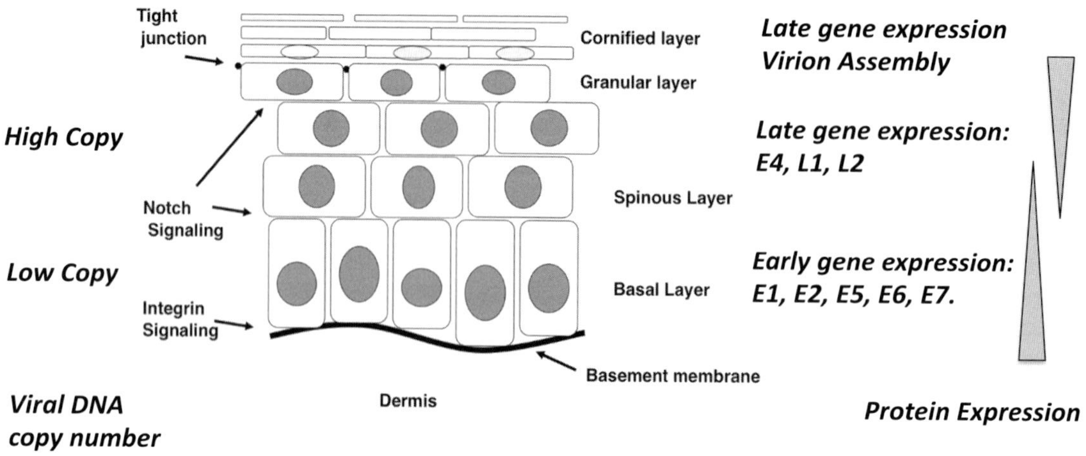

Figure 5.1 Squamous epithelial structure and differentiation. Papillomaviruses infect keratinocytes in the basal layer of the epithelium. Upon infection, the viral genomes exist as low-copy episomes in the basal layer; upon cell division, a daughter cell will move up from the basal layer and undergo differentiation in a Notch-signalling dependent fashion. Differentiation induces the productive phase of the viral life cycle under the expression of viral oncoproteins, inducing re-entry of cells into S phase. Only a subset of cells that amplified the viral genome express the late-phase L1 and L2 capsid proteins in the last layers of the epithelium, encapsidate viral genomes, and are shed as 'bags o' virus'.

named by the size of the open reading frames, with E1 and E2 being the two largest open reading frames encoding early gene products that regulate viral replication and transcription. Three early open reading frames (ORFs) encode proteins that stimulate cell proliferation, survival, and modulate keratinocyte differentiation called E5, E6, and E7. These are the papillomavirus oncoproteins, because in the prototype papillomaviruses (HR HPV16 and BPV-1), these ORFs have activity in classic cell transformation assays, although in other papillomaviruses, such transforming activity can be difficult to demonstrate. The E5 gene is located 3′ to the E1 and E2 replication genes. In high-risk HPV-induced invasive anogenital cancers, the virus is usually found integrated into the host chromosome such that only the E6 and E7 ORFs are still expressed. In most cases, random integration of the HPV genome into a host chromosome would be expected to be unproductive, as the loss of the early transcript polyA site would destabilize viral transcripts of E6 and E7, and further, loss of the E5 oncoprotein would be expected to put cells with integrated genomes at a disadvantage. However, in cervical cancers a cellular splice acceptor and polyaddenylaiton site serve to stabilize viral messages which only express the intact E6 and E7 ORFs (Fig. 5.2). It is possible that the location of E5 in the HPV genome selects against cells in which the virus has integrated into the host chromosome (thereby yielding no virus) because in such integrants, E5 expression is no longer produced from the viral early promoter. In HPV-associated cancers, E6 and E7 expression is necessary for the continued cancer phenotype. This has been demonstrated by repression of the HPV early promoter (by re-expression of the viral E2 protein or by RNAi-mediated repression of viral expression); in both cases, cancer cell lines withdraw from the cell cycle and initiate a differentiation process (Thierry and Yaniv, 1987; Francis *et al.*, 2000; Goodwin and DiMaio, 2000). The subject of this review, E6, has been the subject of other excellent reviews recently that focus upon physiological consequences of E6 expression (Fan and Chen, 2004; Li *et al.*, 2005; Narisawa-Saito and Kiyono, 2007; Liu and Baleja, 2008; Tungteakkhun and Duerksen-Hughes, 2008; Wise-Draper and Wells, 2008). This review is more focused upon the structure of E6, the known interaction

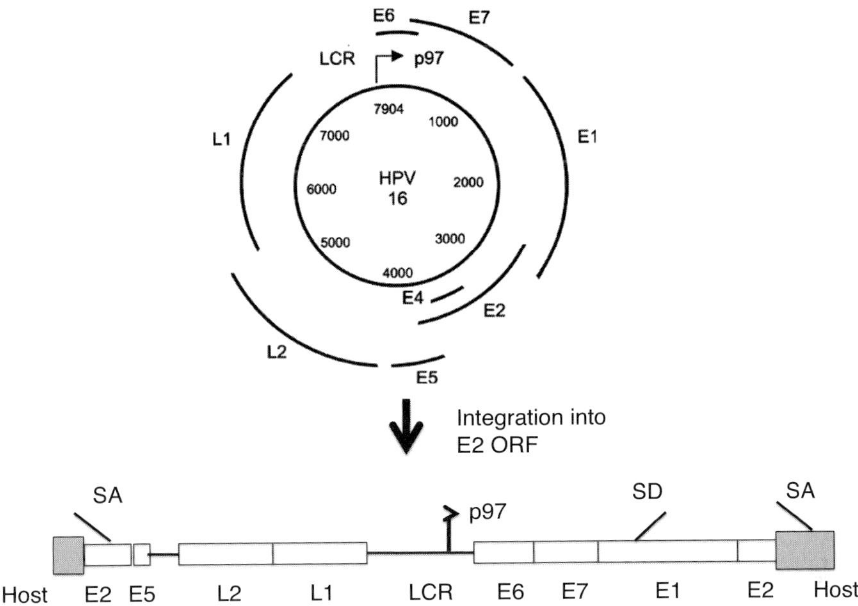

Figure 5.2 The HPV16 genome organization and integration into cellular chromosomes during cancer progression. The circular HPV16 genome is shown with illustrated open reading frames and location of the major early promoter p97. In the transition from *in situ* carcinoma to invasive carcinomas, integration of the viral genome within the E2 gene results in loss of the E2 and E5 proteins and loss of the viral early splice acceptor. In productive integration events, a cellular splice acceptor is utilized, giving rise to expression of the E6 and E7 oncoproteins.

partners of the model E6 proteins and the direct consequence of those interactions rather than the downstream physiologic functions of E6.

The evolutionary origins of E6

Infections by a particular papillomavirus type are typically species and site specific, with distinct groups of viruses specialized to a given species and further specialized to either mucosal or squamous epithelia in that species. This indicates a long association between the virus and the host and suggests that each virus coevolved and co-speciated with its host; this notion is supported by taxonomy of papillomavirus DNA and protein sequences (Chan *et al.*, 1992; Bernard, 1994; Tachezy *et al.*, 2002). Since birds and reptiles harbour papillomaviruses, this implies that papillomaviruses have been co-evolving with their hosts for at least 300 million years. However, not everything in the virus appears to have evolved together, in that the evolutionary history of the replication and encapsidation genes appears to be distinct from the evolutionary history of the viral oncoproteins, and taxonomy charts prepared from each of these groups of genes differ (Garcia-Vallve *et al.*, 2005). Papillomaviruses are classified on the basis of the most conserved L1 open reading frame, but there are two areas in the papillomavirus genome where divergence is enhanced: in the hinge region of the E2 open reading frame that overlaps with E4, and in the E6-E7 oncogene region, (Garcia-Vallve *et al.*, 2005). Recent protein sequence analysis has shown evidence for recombination between species (Shah *et al.*, 2010) which may have contributed to this finding of sequence divergence. The end result of this evolutionary process is an amazing diversity of papillomavirus species. For many years, workers in the field have been pleased to point out that there are 'over one hundred different human papillomaviruses', but recent molecular approaches to make an unbiased survey of papillomaviruses in the skin have discovered this estimate to be a gross underestimation (Christopher Buck, personal communication).

Although all papillomaviruses contain the replication and structural genes E1, E2, L1 and L2, there is more variation among the oncogenes encoded by papillomaviruses. Almost all current papillomaviruses contain an E7 gene (except genital porpoise papillomavirus; Van Bressem *et al.*, 2007), but several papillomaviruses have no E6 gene (BPV types 3, 4, 6, and HPV101 and 103 and two avian papillomaviruses (Tachezy *et al.*, 2002; Terai *et al.*, 2002; Villiers *et al.*, 2004; Chen *et al.*, 2007) contain an E6 gene with only one zinc finger instead of two (Van Doorslaer *et al.*, 2009). The E5 region is difficult to assess because the region can encode hydrophobic peptides that are not well conserved between different papillomavirus types. These considerations give rise to a model in which a progenitor papillomavirus genome with E1 and E2 early proteins and L1 and L2 capsid proteins somehow acquired an E7 oncoprotein with a single zinc-binding region. The zinc-binding region of E7 then may have duplicated and diverged, giving rise to a single zinc finger E6 protein similar to that found in avian species today, and then an additional early duplication of that E6 domain in reptiles (as seen within turtles) giving rise to the E6 protein most commonly observed today with two zinc fingers (recently reviewed in Garcia-Vallve *et al.*, 2005, and Shah *et al.*, 2010). The notion that E7 is the origin of E6 is further suggested by the strong conservation in the spacing between the two CXXC zinc-binding motifs common to both E6 and E7. Since the spacing between the two CXXC motifs is conserved, this suggests some common structural requirement for the conservation, thus suggesting that the protein folds for E6 and E7 would be related. However, this evolutionary speculation is not strongly supported by the comparison of the structure of E6 zinc binding fold with the solved structure of the E7 zinc-binding domain; the folds of the two proteins are distinct (Liu *et al.*, 2006; Nomine *et al.*, 2006).

It is unclear why papillomaviruses have separated their oncogenes into three separate open reading frames instead of combining the functions of E6 and E7 together into a single polypeptide. Presumably, this reflects some requirement for separate spatial or temporal expression that remains undefined. While the E7 oncoprotein from HR HPVs will immortalize keratinocytes at low frequency, the E6 oncoprotein alone does not, but the combination of hrE6 and hrE7 immortalizes keratinocytes at high frequency (Ma *et al.*, 1987; Munger *et al.*, 1989a).

What is the E6 protein for?

Since the E6 and E7 proteins are expressed from a common early transcript, their functions need to be considered together. E7 proteins (with the rare exception of BPV-1 E7) contain LXCXE peptide motifs that bind to and target members of the retinoblastoma family of proteins that regulate E2F family transcription factors (recently reviewed in McLaughlin-Drubin and Munger, 2009). While E7 oncoproteins that are associated with anogenital cancers ('high risk') are themselves oncogenic, E7 from non-cancer-associated viruses are weakly oncogenic and either show little oncogenic activity directly, or only co-operative activity when co-expressed with additional oncogenes. This has been recently explained by the observation that while all E7 proteins associate with and target the degradation of the p130 RB2 protein that regulates G0 to G1 transition in the cell cycle, only the oncogenic E7 proteins additionally target the degradation of p105 RB that controls G1 to S transition as well (Zhang *et al.*, 2006). The targeted degradation of p105 RB by oncogenic E7 proteins sets in play the stabilization of the p53 tumour suppressor and the sensitization of E7 expressing cells to apoptosis (Eichten *et al.*, 2002). The finding that the E6 oncoproteins of oncogenic HPVs targeted the degradation of p53 led to a satisfying paradigm, where the 'purpose' of E7 was to transform target cells (so that differentiated epithelial cells will re-enter the cell cycle to amplify viral DNAs in the spinous layer of squamous epithelium) and the 'purpose' of E6 was to deal with the consequences of E7 transformation, by blocking the function of p53 and inhibiting apoptosis. Indeed, all the examined high-risk human papillomaviruses where E7 targets the degradation of RB also contain E6 proteins that at some level target the degradation of p53. However, this is a myopic view of the world of papillomaviruses, where the targeted degradation of RB is not necessarily linked to the degradation of p53; while the E7 oncoprotein of cotton-tailed papillomavirus reduces RB expression levels in rabbit keratinocytes, its E6 proteins do not target the degradation of p53, and p53 is still inducible by mitomycin C in the presence of E6 (Ganzenmueller *et al.*, 2008), and the BPV-1 E7 oncoprotein does not contain a LXCXE RB family docking site The cellular proteins that associate with E6 have only been closely studied in the genital high-risk HPV types and BPV-1 E6, so it is premature to assert that the purpose of E6 proteins overall is to 'clean up' after the consequences of E7 expression. Outside of cancer-associated HPV E6 proteins, very little is understood about the mechanism by which E6 supports the viral life cycle of any other human or animal papillomavirus.

The principal E6 model systems: hrE6 and BPV-1 E6

Over the last two decades, two E6 proteins have been the subject of substantial study, the 'high-risk' E6 proteins (hrE6), principally from HPV16 and HPV18 (termed hereafter as 16E6 and 18E6), and BPV-1 E6 (termed hereafter as BE6). The reasons for this are both programmatic (with the intense interest and focus upon the carcinogenic role of HR HPVs) and practical, in that the non-cancer-associated HPV E6 proteins did not initially show many compelling phenotypes when expressed in cells in culture. However the exception to this rule is BE6. The first physiologic function for any E6 protein was the transformation of cells in tissue culture by BPV-1 E6 (Schiller *et al.*, 1984). This quantitative transformation assay was a tremendous help because it provided a physiologically relevant assay against which to gage the phenotypes of E6 mutants and thereby the significance of E6-associated proteins. However, despite this early advantage of BE6 studies, the first biochemical functions for an E6 protein was the association of hrE6 with p53 (Werness *et al.*, 1990) and shortly thereafter the targeted degradation of p53 by hrE6 in rabbit reticulocyte lysate (Scheffner *et al.*, 1990). These observations with hrE6 proteins provided a similar quantitative and relevant function that sustained early studies of hrE6 proteins. With time and the development of further insights into the papillomavirus life cycle, additional quantitative assays for cell survival, transcriptional modulation, and signal transduction have opened up new opportunities for the study of other HPV and animal papillomavirus E6 proteins.

E6 association with cellular proteins

The central role of E6 binding to LXXLL peptides

Biochemical fractionation of reticulocyte lysates that supported the degradation of p53 by 16E6 led to the identification of a cellular factor termed E6AP (for E6-associated protein, now also referred to as the product of the UBE3A gene) that associated with E6 and p53 and supported the ubiquitination and degradation of p53 (Huibregtse *et al.*, 1993a). Analysis of E6AP defined a 20 amino acid peptide that bound to E6 and was subsequently recognized to contain an LXXLL peptide motif (Huibregtse *et al.*, 1993b). Subsequent work on BPV-1 E6 identified the BE6-associated proteins paxillin and AP1 (Tong and Howley, 1997; Tong *et al.*, 1998; Vande Pol *et al.*, 1998). It was recognized that BE6-binding peptides from paxillin and E6AP shared similar LXXLL peptides, identifying LXXLL motifs as E6 docking sites (Vande Pol *et al.*, 1998). Yeast two-hybrid selection of 16E6-associated peptides identified a fragment of the LXXLL motif (termed an ELLG motif) as the E6 docking peptide (Elston *et al.*, 1998); mutagenesis of the 20 amino acid E6AP peptide that bound 16E6 and mutagenesis of the paxillin peptide that bound BE6 more clearly defined the binding sequence as an acidic LXXLL peptide (Chen *et al.*, 1998; Bohl *et al.*, 2000). More recently, selection of random peptides from a large phage display library has identified the full BE6 binding peptide as a 10 amino acid peptide with the consensus sequence $\Phi_1X_2D_3L_4D_5(D/E)\ {}_{6}L_7(F/L)_8X_9(D/E)_{10}$. An alignment of BE6-binding peptides from BE6-associated proteins is shown in Fig. 5.3, which agrees quite nicely with the phage display results. The hydrophobic first residue corresponds to the methionine residue in the LD1 motif of paxillin and a methionine in the BE6 binding protein MAML1. The strongest conservation is observed for the hydrophobic positions L_4, L_7 and $(F/L)_8$,

```
260-    T R E L D E L M A S    Paxillin LD4
1-      M D D L D A L L A D    Paxillin LD1
142-    L S E L D R L L L E    Paxillin LD2
1-      M E D L D E L L S E    HIC5 LD1
1005-   M S D L D D L L G S    MAML1
406     E L T L Q E L L G E    E6AP
```

Figure 5.3 Known LXXLL binding motifs in BE6-associated proteins. The number refers to the first amino acid in the motif. Not shown are potential LXXLL docking sites from the adapter protein complex 1, whose BE6 docking sites have not been mapped.

i.e. LXXLL. Four positions in the motif (3, 5, 6 and 10) show preferences for negative residues. Clearly, there is some variation in ligands that can be bound by E6 with some substitutions of the hydrophobic residues (especially position 8), but we will here generically refer to E6 binding peptides as acidic LXXLL motifs. It is probable that other E6 proteins also interact with an extended 10 amino acid acidic LXXLL peptide motif, but this has not yet been demonstrated.

There are four lines of evidence that E6 docking to LXXLL peptides on target proteins is essential. First, mutants of BE6 that fail to bind to LXXLL containing target proteins fail to transform cells, and 16E6 mutants that fail to bind to E6AP fail to target the degradation of p53 (Vande Pol *et al.*, 1998; Das *et al.*, 2000; Cooper *et al.*, 2003). Of course, some uncertainty in this result arises from the likelihood that such mutants may have global defects in proper folding or that *in vitro* and *in vivo* assays reveal different phenotypes for the same mutant. A second line of evidence is that deletion of the LXXLL motif in E6AP prevents E6 from directing E6AP-mediated p53 degradation (Huibregtse *et al.*, 1993b), and deletion of the BE6 binding motifs of paxillin prevent BE6 transformation (Wade *et al.*, 2008). And third, fusion of a LXXLL motif to the amino-terminus of BE6 blocked cellular transformation by BE6, and upon mutation of the LXXLL motif, transformation was restored; this indicates that the fused LXXLL motif was binding to BE6 in cis and preventing BE6 association with cellular proteins in trans, and avoided potential artefacts owing to mutation of BE6 itself (Bohl *et al.*, 2000). Finally, peptides or drugs that competitively block 16E6 interactions with LXXLL targets inhibit *in vitro* and *in vivo* p53 degradation by 16E6 (Liu *et al.*, 2004; Sterlinko Grm *et al.*, 2004; Baleja *et al.*, 2006).

Why the selection of acidic amphipathic binding peptides by divergent E6 proteins?

Both BE6 and 16E6 associate with similar acidic LXXLL motifs on paxillin and E6AP respectively, and HPV11 E6 (11E6) also binds the LXXLL motif of E6AP (Brimer *et al.*, 2007). Recent studies of BE6, HPV1 and HPV8 E6 show that all three of these E6 proteins bind to the same acidic LXXLL motif of the MAML1 co-activator (see below). Thus these divergent E6 proteins so far studied all associate with similar acidic LXXLL motifs. Is there some underlying biological reason for this? While as yet obscure, it is noteworthy that LXXLL peptides are used as docking sites for the interaction of nuclear hormone receptor transcription factors and their co-activators and co-repressors, and are referred to as signature motifs, charged leucine motifs or LXXLL motifs (Savkur and Burris, 2004). Upon binding the hormone ligand, nuclear hormone receptors open a cleft and interact with LXXLL peptides on co-activators and co-repressors. However, unlike E6-binding LXXLL motifs, the motifs that associate with nuclear hormone receptors are typically basic (Heery *et al.*, 1997), while the LXXLL motif of MAML1, E6AP, and paxillin are acidic. The association of E6 proteins from HPV types 1, 8, 11, 16 and BPV-1 with acidic LXXLL motifs implies a conserved biological importance that is as yet unappreciated.

The structure of E6 bound to LXXLL peptide

Studies of the structure of E6 have been hampered for two decades owing to the poor solubility of E6. When expressed in bacteria and concentrated, E6 proteins aggregate into filaments (Zanier *et al.*,

2007). However as noted above, when an LXXLL peptide from paxillin is fused to the amino-terminus of BPV-1 E6, the fused peptide binds to E6 and blocks transformation by BE6 (Bohl *et al.*, 2000). Fortunately, the LXXLL–BE6 fusion also results in a protein that can be overexpressed and concentrated. This approach has recently allowed for the crystallization of LXXLL-BE6 fused to a crystallization prone variant of maltose binding protein in the laboratories of Gilles Trave and Jean Cavarelli. Since this work is currently under review (Charbonnier *et al.*, 2010), it can only summarized below (see Fig. 5.4).

BPV-1 E6 contains two zinc domains with a conserved fold connected by a helical linker

Each of the two BE6 zinc-binding domains has a conserved and similar overall fold to what was previously solved for the 16E6 C-terminal motif (Nomine *et al.*, 2006). The two zinc domains, together with a helical inter-domain linker, form a deep pocket in which the LXXLL peptide makes close and specific contacts. The paxillin LXXLL peptide (MDDLDALLAD) adopts an alpha-helical conformation with an amphipathic character: the hydrophobic methionine and

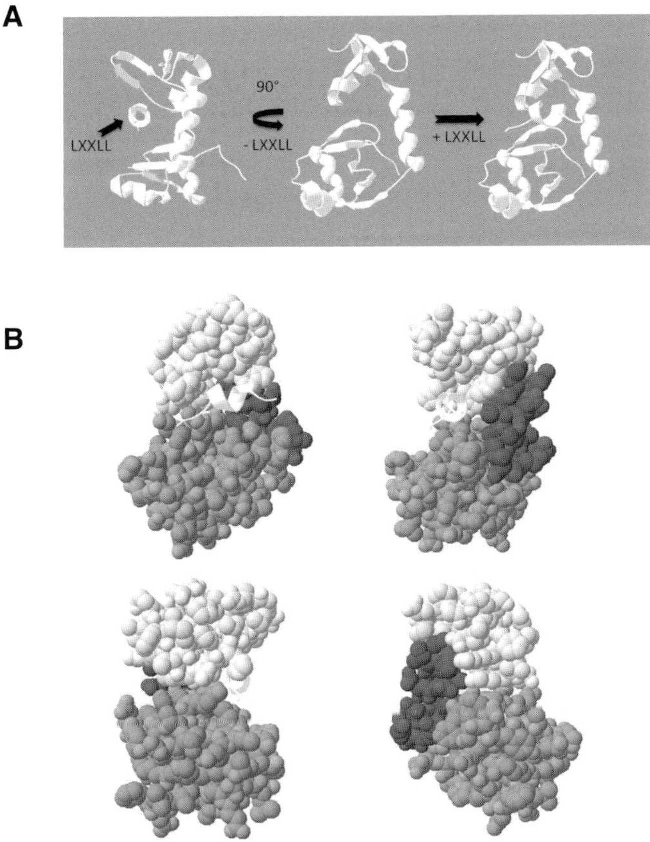

Figure 5.4 BE6 bound to an LXXLL motif. (A) Ribbon diagram of BE6 bound to an LXXLL motif. In the left panel, the amino-terminal zinc-binding domain is shown at the top and the helical LXXLL peptide is viewed on end through its axis. The LXXLL peptide resides a deep pocket. The interdomain connecting helix is not well seen in the left panel, but is clearly seen to the right side of the middle panel, where the LXXLL peptide has been removed. The right panel is the same view as the middle panel but with the LXXLL peptide inserted into the pocket. (B) Space filling views of the BE6 protein bound to the LXXLL peptide. In the upper left panel, the same view as the right panel of the preceding figure is shown with the amino-terminal zinc binding domain at the top in light grey, the C-terminal zinc-binding domain in medium grey at the bottom, the interdomain connecting helix in dark grey and the LXXLL peptide as a white ribbon. The additional views are all 90° sequential rotations.

leucine side chains are oriented to one side into the pocket, opposite from the negatively charged aspartic acids of the LXXLL peptide, which face outward and make charge interactions with E6. The E6 pocket that contains the LXXLL amphipathic alpha helix has a complementary surface. The interdomain linker is anchored at each end of the helix to each of the zinc domains. The main differences between the solved BE6 structure and the proposed model of 16E6 from the 2006 study (Nomine *et al.*, 2006) is the helical interdomain linker and the specific interdomain contacts.

In summary, an amphipathic negatively charged LXXLL alpha helical peptide found on E6-binding cellular proteins fits tightly into a deep complementary pocket in E6. One recent observation suggests that E6 protein stability in cells favours that E6 be bound to an appropriate LXXLL binding partner: for the high cancer risk HPV E6 proteins, E6 expression is reduced in the absence of its principal binding partner E6AP, such that when E6AP protein is knocked down by RNAi, E6 protein levels decline (Tomaic *et al.*, 2009b). This suggests that E6 proteins may exist in cells primarily bound to LXXLL peptides on target cellular proteins. Indeed, for 16E6, E6AP is required for the induction of cervical cancer (Shai *et al.*, 2010), but 16E6 retained oncogenic activity in the skin of E6AP null mice, indicating either LXXLL independent functions or as yet undefined additional important LXXLL binding partners for 16E6 (Shai *et al.*, 2007).

How does E6 interact with secondary associated proteins?

E6 proteins bind to more than just LXXLL motifs: amino-terminal interactions

Although the interaction of E6 with LXXLL motifs on cellular proteins seems to be the initial binding function of E6, it is not everything. E6 proteins have amino-terminal sequences in the E6N domain beyond what is required for interaction with the LXXLL motif (Cooper *et al.*, 2003). As mentioned above, in the case of BE6, there are ten amino acids that can be deleted before the ability of BE6 to bind to LXXLL motifs is abolished. For most papillomaviruses, the amount of 'extra' amino-terminal sequences is more substantial than BE6, ranging from 23 to 25 amino acids for the cancer-associated E6 proteins, to 34 amino acids for the cutaneous HPV5 E6, and an entire additional zinc domain for the long form of cotton-tailed rabbit papillomavirus (Meyers *et al.*, 1992). Before E6 interaction with LXXLL motifs was appreciated, the first described interaction between E6 and a cellular protein was the interaction of p53 with cancer-associated HPV E6 proteins (Werness *et al.*, 1990). Subsequent binding experiments *in vitro* and in yeast demonstrated that the first eight amino acids of 16E6 could be deleted, ablating p53 binding but without altering LXXLL interactions (Kao *et al.*, 2000; Cooper *et al.*, 2003). Thus, while a central core region of E6 (corresponding to BE6 amino acids 11–132) is required for LXXLL interactions, additional amino-terminal sequences seem to be required for additional interactions. Much work remains to understand these interactions, since in only a single instance (cancer-associated HPV E6 and its p53 interaction) have any such interactions been demonstrated and mapped to the amino-terminal surface of E6.

Interaction of hrE6 with LXXLL peptides reshapes E6 to interact with p53 via amino-terminal E6 sequences

As mentioned above, early studies that mapped the domains of 16E6 that interact with E6AP mapped the E6 binding domain to a central 20 amino acid peptide of E6AP containing a LXXLL motif. Additional sequences of E6AP were thought to be necessary for interaction with p53; the model being that p53 interactions were with a combination of E6AP and E6 (Huibregtse *et al.*, 1993b). In yeast expression systems however, LXXLL peptide interactions with 16E6 are sufficient for recruitment of p53 to 16E6 in the absence of additional E6AP sequences (Brimer and Vande Pol, 2011). This demonstrates that all of the binding specificity for p53 resides within E6 and not in E6AP, and gives rise to a more simple model for how E6 proteins recognize secondary substrates: upon binding to LXXLL peptides,

E6 undergoes a conformational change that then allows the interaction of a secondary substrate such as p53 requiring amino-terminal sequences of E6. It is possible that other E6 proteins that have amino-terminal sequences beyond what is required for interaction with LXXLL (such as β-papillomaviruses or anogenital low-risk human papillomaviruses) will act in a similar manner, although with secondary substrates other than p53 that remain to be discovered.

Carboxy-terminal PDZ ligands on hrE6 proteins interact with a subset of PDZ-containing proteins implicated in signal transduction and polarity

A long-standing observation in the biology of DNA tumour viruses is the association between the targeting of p105RB by viral oncoproteins such as Adenovirus E1A or high-risk HPV E7 and the additional association with cellular proteins containing PDZ domains or the targeting of cell polarity proteins by other viral oncoproteins of the same virus. For example, Adenovirus E1A interacts with p105RB and the E4ORF4 with PDZ proteins. High-risk papillomavirus E7 targets p105RB for degradation and hrE6 associates with a subset of PDZ proteins. SV40 TAg associates with RB and small T antigen disrupts the integrity of cellular tight junctions (reviewed in Thomas *et al.*, 2008, and Javier, 2008). A recent exception to the rules in papillomaviruses underscored the importance of this association: the rhesus monkey papillomavirus E7 protein both targets RB and has a PDZ ligand at the carboxy-terminus that can interact with DLG1, a PDZ binding protein in common with hrE6 (Tomaic *et al.*, 2009a). In contrast, low-risk mucosal HPVs whose oncoproteins target RB family member p130 but do not target the degradation of RB itself have as yet shown no direct E6 associations with PDZ proteins.

The PDZ domain (named for the proteins PSD95, DLG, and ZO1) is a small folded domain that binds to peptide ligands on target proteins. The ligands can be internal, but are most typically carboxy-terminal peptide ligands with a consensus sequence XX(S/T/Y) X(V/L/M). Adapter proteins can contain multiple PDZ domains, resulting in large complexes built through the association of multiple PDZ domain proteins and their binding partners together. Affinities of PDZ–ligand interaction are typically in the low micromolar range, and can be modulated by phosphorylation of the PDZ ligand or the PDZ domain (reviewed in Lee and Zheng, 2010). There are over 200 different PDZ domain containing human proteins, and they have diverse functions in signal transduction and the building of protein complexes.

There are several phenotypes for E6 interactions with PDZ proteins that are revealed by comparison of wild-type E6 to mutants deleted of the PDZ ligand at the carboxy-terminus of E6. In mice, 16E6 expression in the skin or eye cause aberrant differentiation; this phenotype requires the PDZ ligand of 16E6 (Nguyen *et al.*, 2003a,b). In the context of the HPV genome, deletion of the E6 PDZ ligand in the context of the entire HPV31 genome results in loss of the HPV episomal plasmid upon cell passage and enhanced epithelial differential of organotypic raft cultures compared with wild-type HPV31 (Lee and Laimins, 2004). More confusing and controversial phenotypes have been described in tissue culture systems where E6 function was assessed in cancer cell lines or under overexpression conditions. Under low expression conditions, E6 reduces growth factor dependence in human keratinocytes, and this requires the PDZ ligand of E6 (Jing *et al.*, 2007). In SV40 immortalized keratinocytes, the E6 PDZ ligand function promotes epithelial to mesenchymal transitions (Watson *et al.*, 2003), and in human keratinocytes the E6 PDZ ligand promotes co-operation with ras and anchorage independent colony formation (Spanos *et al.*, 2008a).

So which PDZ associations with E6 mediate these phenotypes? Only a handful of the total number of known cellular PDZ proteins associate with high-risk E6 proteins through PDZ interactions including multi-PDZ adapter proteins (DLG1 (Lee *et al.*, 1997), PSD95/DLG4 (Handa *et al.*, 2007), MUPP1 (Lee *et al.*, 2000), MAGI 1, 2, and 3 (Glaunsinger *et al.*, 2000; Thomas *et al.*, 2002), PADJ (Storrs and Silverstein, 2007), scribble (Nakagawa and Huibregtse, 2000)), and the PDZ-containing tyrosine phosphatases (PTPN3 and PTP13) (Jing *et al.*, 2007; Topffer *et al.*, 2007; Spanos *et al.*, 2008b). Of note, all of these associations were initially published as resulting in the targeted degradation of the PDZ protein, usually in an E6AP and proteasome-dependent manner, analogous to the previously described and widely replicated targeted degradation of p53 by hrE6 proteins. However, the targeted overall degradation of some of the PDZ proteins by hrE6 has more recently been challenged (Kranjec and Banks, 2010). DLG, which is degraded by hrE6 *in vitro* does not show reduced expression or relocalization in the context of E6 expressed from episomal genomes in primary keratinocytes (Lee and Laimins, 2004). Some studies have found that only certain subcellular fractions of hrE6-associated PDZ proteins are degraded (Massimi *et al.*, 2004, 2006), but again, these experiments involve expression levels higher than produced by episomal genomes. More recently, subcellular phosphorylated forms of DLG1 have been shown to be preferentially targeted for degradation (Narayan *et al.*, 2009). Further, there is disagreement as to the mechanism by which E6 reduces expression of PDZ proteins, with some studies showing E6AP and proteasome dependence (Handa *et al.*, 2007; Jing *et al.*, 2007; Kuballa *et al.*, 2007), and others seeing neither ubiquitin nor proteasome dependence (Grm and Banks, 2004; Ainsworth *et al.*, 2008). While more than one study has observed loss of PSD95 expression by HPV18 E6, no study has shown a common target of all the high-risk E6 types (although a recent study found a special significance for MAGI1 interactions (Kranjec and Banks, 2010). Further complication arises from the possible phosphorylation of the hrE6 PDZ ligand (Massimi *et al.*, 2001), which could require particular culture conditions for PDZ interactions to occur *in vivo*. In summary, although the importance of the interaction between hrE6 and PDZ proteins is clear, the critical cellular target(s), and the mechanisms or consequences of the interactions that are necessary for the HPV life cycle remain unresolved.

Other interactions of cellular proteins with E6 have been described but have not been mapped to domains of E6, in part because of the uncertainties of E6 mutants. Because many E6 mutants have not been shown to associate with LXXLL motifs, it is uncertain if they retain the overall fold of the central LXXLL binding region of E6, and may in fact thus be globally defective. Careful mapping of these mutants onto the structure of BE6 or onto models of hrE6 based either upon BE6 or possible future structures of hrE6 proteins will allow for a reinterpretation of prior studies that have utilized mutants of E6 to ascertain the significance of E6 associations. Some associations with E6 are probably LXXLL and PDZ independent because they do not contain these motifs, and include FADD (Filippova *et al.*, 2004), hADA3 (Kumar *et al.*, 2002), TIP60 (Jha *et al.*, 2010), p300/CBP (Patel *et al.*, 1999; Zimmermann *et al.*, 1999), NFX1 (Gewin *et al.*, 2004), procaspase 8 (Filippova *et al.*, 2007) and Gps2 (Degenhardt and Silverstein, 2001). These interactions could be direct or indirect with E6.

E6* proteins

E6* is a shorter portion of the full-length E6 protein that is produced by a splice internal to the E6 ORF. The splice donor site is highly conserved, expressing the first 42 or 44 amino acids of 16E6 or 18E6 respectively and then non-conserved sequence derived from the splice acceptor. The polypeptide contains the first CXXC zinc-binding motif of E6 and overall, the sequence conservation among the hrE6* proteins is 50–60%. Functions for this polypeptide in the context of the viral infection have long been obscure, because mutation of the splice donor site in the context of the viral genome could effect expression of E7 and E1. It is unclear if this short polypeptide has the potential to adopt a stable fold on its own or must form associations with other polypeptides. It seems implausible that it could either alone or as a multimer form an LXXLL-binding pocket. Yet early studies of E6* showed that when E6* was expressed as a glutathione *S*-transferase (GST) fusion protein, it bound to *in vitro* translated 16E6, 18E6, E6AP, and inhibited E6-mediated p53 degradation (Pim *et al.*, 1997). More recently, this same group has shown that overexpression of 18E6* alone promotes the proteasome dependent degradation of a variety of proteins that are the target of full-length E6 (such as DLG1) or proteins that are not the target of full-length E6 (such as AKT) (Pim *et al.*, 2009). These results might be dismissed as *in vitro* or overexpression artefacts, but recently 16E6* and 18E6* proteins have been shown in stable retroviral expression in keratinocytes to target the degradation of the TIP60 acetyltransferase (Jha *et al.*, 2010). How such a short polypeptide could affect the stability of so many targets remains an enigma, but it seems probable that E6* will prove to alter cellular global effectors of protein stability. As yet, direct cellular interaction targets of E6* have not been described.

E6 interaction with p53

P53 was discovered as a cellular protein that associates with the SV40 T antigen by the immune precipitation of radiolabelled cell lysates with serum from hamsters bearing SV40 induced tumours (Lane and Crawford, 1979). Later, similar techniques identified p53 as a Adenovirus E1B55-associated protein (Sarnow *et al.*, 1982). It was thought that because a 53 kDa protein could be often observed in Coomassie stained gels of tumour cells but not in the corresponding normal tissues, that p53 might be an induced cellular oncoprotein. Initial molecular clones of p53 derived from tumour cell lines supported this interpretation, because these versions of p53 could cooperate with oncogenic ras genes in the transformation of primary cells (Eliyahu *et al.*, 1984; Parada *et al.*, 1984). However, groups that had cloned p53 from normal tissues could not replicate these findings, and a comparison of the sequences from the differing groups led to the conclusion that p53 had acquired oncogenic properties through cancer-associated mutations (Finlay *et al.*, 1989; Hinds *et al.*, 1989). It was in the midst of this evolving story of p53 and its association with viral oncogenes that a parallel story about the association of papovavirus and adenovirus oncogenes with members of the retinoblastoma (RB) family developed: first adenovirus E1a (Whyte *et al.*, 1988) and then SV40 T antigen (DeCaprio *et al.*, 1988) were found to associate with RB. The hypothesis that high-risk papillomaviruses might transform cells through associations with the same set of cellular proteins occurred as if through the collective subconscious, and it was soon found that hrE7 proteins associate with RB as well (Dyson *et al.*, 1989; Munger *et al.*, 1989b). The correlations between RB and p53 associations in adenovirus and SV40 prompted a specific search for hrE6 association with p53, which was demonstrated by the *in vitro* association of radiolabelled E6 from reticulocyte lysates with p53 in cellular lysates (Werness *et al.*, 1990). Observation of the gels in that study showed some high molecular weight E6 in a smear at the top of the gel, which prompted the hypothesis that E6 and p53 might be involved in poly ubiquitin mediated conjugation, which was confirmed shortly thereafter (Scheffner *et al.*, 1990). That three different oncogenic viruses all targeted a common set of cellular proteins was a seminal observation in cancer biology, and preceded the subsequent clinical correlation that most spontaneous malignancies are mutated in some manner in both the p53 and RB signalling pathways. Although targeted degradation of p53 is an essential function for high-risk HPV genome replication (Park and Androphy, 2002), E6 mutants that fail to target p53 still immortalize cells demonstrating additional critical functions of E6 (Kiyono *et al.*, 1998; Liu *et al.*, 1999).

How does E6 associate with p53?

As noted above, hrE6 interaction with p53 is minimal in the absence of E6AP, but the LQELL peptide of E6AP alone can induce hrE6 to interact with p53 which is dependent upon amino-terminal sequences in E6 that are not required for interaction with the LXXLL peptide (Cooper *et al.*, 2003)). Although p53 has a LXXLL motif in the amino-terminus that is required for full p53 transactivation (LWKLL), E6 has not been shown to associate with this motif *in vitro*. *In vitro* binding experiments have shown that E6 associates with the core DNA-binding domain of p53 when p53 is in a native conformation, but does not associate with the DNA-binding domain of most (but not all) p53 cancer-associated mutants (Scheffner *et al.*, 1992). Thus, there is a good correlation between a wild-type p53 conformation that is competent for DNA binding with the ability to associate with hrE6.

A second modality of E6 association with p53 was defined using bacterially expressed E6 proteins, where GST-E6 protein from both high- and low-risk papillomavirus types associates with the p53 oligomerization domain at the carboxy-terminus of p53 (Li and Coffino, 1996). To this author it remains unclear if this is a biologically meaningful result or an artefact of aggregated and detergent treated E6. As an aside, E6 *in vitro* binding assays are notoriously difficult. Bacterially expressed E6 preparations invariably contain aggregated E6 proteins that have high non-specific binding. To decrease non-specific binding, non-ionic detergents are added, which (in our hands at least) strongly degrades specific interactions between hrE6 and LXXLL peptides. If a hrE6 preparation does not specifically interact with LXXLL motifs, then it is by definition not in a native conformation. LXXLL interactions with E6 are most easily performed by yeast 2-hybrid assays (Elston *et al.*, 1998; Vande Pol *et al.*, 1998; Cooper *et al.*, 2003). In contrast to hrE6, BPV-1 E6 is not inhibited by non-ionic detergents, which has allowed for robust *in vitro* binding assays (Das *et al.*, 2000). LXXLL binding assays with hrE6 proteins are possible, but require careful preparation and purification of hrE6 proteins (Nomine *et al.*, 2001).

How does the formation of the E6–E6AP–p53 complex result in the degradation of p53?

The interaction of 16E6 with E6AP induces the dimerization and ubiquitination of E6AP (Nuber *et al.*, 1998), although it is unclear in those studies the mechanisms involved. Interestingly, Lipari and co-workers found that the E6 amino-terminal zinc binding domain (which could be expressed alone in soluble form) could dimerize *in vitro* (Lipari *et al.*, 2001), suggesting the possibility that E6–E6 interactions could result in the trans-ubiquitination of E6AP or other components in the complex including p53. In this light, a fascinating 16E6 mutant in the amino-terminal domain, F47R, has been described that when expressed forms a stable complex with E6AP and p53 without degradation (Ristriani *et al.*, 2009); this mutant when expressed in HeLa cells (an HPV18 hrE6 expressing cell line) can induce senescence, presumably through a dominant negative interaction with E6AP and p53. A detailed analysis of this mutant could elucidate the mechanism by which the formation of the E6–E6AP–p53 complex initiates targeted degradation of p53.

Related to this question is how does E6 escape being the target of E6AP ubiquitination and degradation? A recent important study demonstrated that E6 purified from cells by tandem affinity chromatography is in a complex with the ubiquitin specific protease USP15 (Vos *et al.*, 2009). RNAi knockdown of USP15 resulted in the reduction of E6 expression, giving rise to the exciting possibility that these enzymes could be therapeutic targets for high-risk HPV caused malignancies. If USP15 was sufficiently inhibited, p53 could presumably accumulate and have a therapeutic role in the malignancy. However, while knockdown of USP15 reduced the expression of E6, p53 was not induced in cervical cancer cell lines, indicating that further development of this area is an important research goal (Vos *et al.*, 2009).

In addition to hrE6 plus E6AP, a variety of other E3-ligases have also been shown to promote p53 degradation including Mdm2 (Honda *et al.*, 1997) (murine double minute 2), COP1 (constitutively photomorphogenic 1) (Dornan *et al.*, 2004), and PIRH2 (p53-induced-RING-H2) (Leng *et al.*, 2003). p53 protein levels are induced by a variety of genotoxic and other stressful stimuli that result in the post-translational modification of the p53 amino-terminus, blocking the association of p53 with Mdm2, and resulting in the accumulation of p53 protein levels. Because the hrE6 + E6AP ubiquitin ligase complex does not interact with the amino terminus but rather with the core DNA binding domain, p53 does not accumulate in cells that express E6 upon genotoxic stimuli, but interestingly, it does accumulate in cells where p53 is induced by hypoxic stress (Alarcon *et al.*, 1999). Because E6 targets the degradation of p53, cervical cancer cell lines have loss of p53 function while often retaining wild-type p53 genes (Crook *et al.*, 1991; Scheffner *et al.*, 1991).

While hrE6 targets p53 degradation, p53 is often not fully degraded in hrE6 expressing cells although p53 dependent transcription, checkpoint control and p53-induced apoptosis is blocked. Further, in cells expressing hrE6, low-risk mucosal HPV E6 genes and β-papillomavirus E6 (both of which fail to target p53 degradation) all block some p53-induced transcription (Giampieri *et al.*, 2004); one important mechanism involves modulation of protein acetylation (discussed below). How the low-risk and β-papillomavirus E6 proteins gain access to target p53 has not been full resolved. hrE6 degradation of p53 is blocked by inhibitors of nuclear export (Freedman and Levine, 1998), indicating that p53 degradation occurs in cytoplasmic and not nuclear proteasomes.

The role of hrE6–E6AP associations

Discovery of the E6AP ubiquitin ligase

Prior observations that p53 associates with SV40 T antigen (Lane and Crawford, 1979; Linzer and Levine, 1979) and adenovirus E1B 55 kDa protein (Sarnow *et al.*, 1982) prompted an examination of the ability of E6 to associate with p53. *In vitro* translated E6 proteins from cancer-associated HPV16 and 18 were shown to associate with p53 in tumour cell lysates while the non-cancer-associated E6 proteins from HPV types 6 and 11 and BPV-1 E6 did not (Werness *et al.*, 1990). In that study, high molecular weight radiolabelled E6 was present at the top of the gel, indicating the possibility of polyubiquitination. This, and the prior observation that normal fibroblasts and HeLa cells contained translatable mRNA for p53 but no detectable protein (Matlashewski *et al.*, 1986), suggested the possibility that E6 might target the degradation of p53, which proved to be the case (Scheffner *et al.*, 1990, 1994). Biochemical fractionation of reticulocyte lysates that supported the degradation of p53 by 16E6 led to the identification of a cellular factor termed E6AP (for E6-associated protein, the product of the UBE3A gene) that associated with E6 and p53 and supported the ubiquitination and degradation of p53 (Huibregtse *et al.*, 1993a). Domain analysis of E6AP defined three E6AP regions required for p53 degradation together with E6: first, a 20 amino acid peptide that bound to E6 containing an LQELLGEE peptide motif, second, a larger region in the amino-terminus required to associate with p53, and third, a domain in the carboxy-terminus required for ubiquitination (Huibregtse *et al.*, 1993b) (Fig. 5.5). It was determined that E6 bound to E6AP, and that the resulting complex recruited p53, leading to the transfer of ubiquitin from a thio-ester cysteine bond in the ubiquitination domain to p53 (Scheffner *et al.*, 1993). The carboxy-terminal ubiquitination domain was found present in a family of similar ubiquitin ligases now termed HECT domain ubiquitin ligases (for homologous to E6AP carboxy-terminus) of which E6AP is the prototype (Huibregtse *et al.*, 1995). E6AP expression is imprinted, and loss of E6AP or mutation with loss of ubiquitin ligase activity is the cause of Angelman syndrome, a complex neurodevelopmental disorder (Kishino *et al.*, 1997; Matsuura *et al.*, 1997). Mutation of the UBE3A gene that encodes E6AP in mice creates a phenotype that bears close resemblance to Angelman syndrome in humans and served as a source of E6AP null cells for *in vitro* studies (Jiang *et al.*, 1998; Cheron *et al.*, 2005). How loss of E6AP ubiquitin ligase activity results in Angelman syndrome remains poorly understood.

Protein ubiquitination by E6AP is an enzymatic cascade where ubiquitin is transferred from an initial E1 charging enzyme to an E6AP-associated E2 enzyme, and finally to a cysteine residue in the HECT domain of E6AP and has been recently reviewed (Beaudenon and Huibregtse, 2008). Multiple different E2 enzymes can associate with and function with E6AP (Nuber *et al.*, 1996), but the affinity between the E2 enzymes and E6AP is low, with detection most easily observed by yeast two-hybrid assays (Kumar *et al.*, 1997). Protein ubiquitination is complex and can have diverse activities in the cell. A single ubiquitin may be linked at a specific site (mono-ubiquitination), or multiple ubiquitins can be linked. Each of the seven lysine residues of ubiquitin can be conjugated to form oligo and polyubiquitin chains, with linear, branched, homogeneous or heterogeneous linkages resulting in various biological effects. How polyubiquitin chains are formed and the linkages within ubiquitin that form the chains is important because different linkages between ubiquitin molecules in the chain signal different biochemical outcomes (Ikeda and Dikic, 2008). Polyubiquitination at ubiquitin K29 and K48 result in proteasomal degradation. Ubiquitination of p53 by E6 + E6AP results in a K48-linked polyubiquitination (Wang and Pickart, 2005),

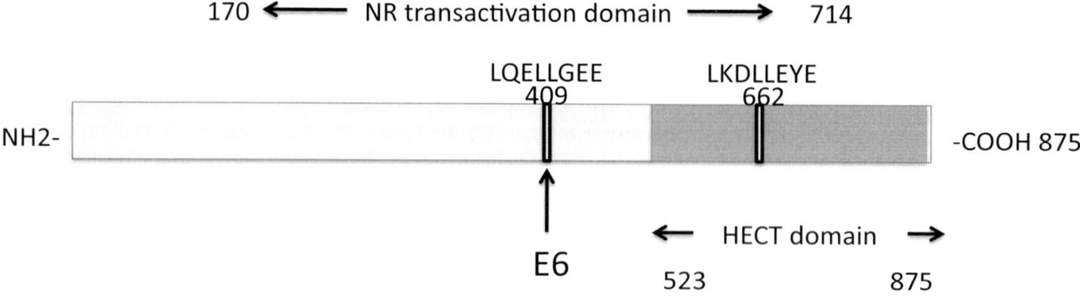

Figure 5.5 Diagram of the functional domains of E6AP. The locations of the nuclear receptor co-activation region, E6 LXXLL-binding domain and HECT domain are illustrated. See text for details and references.

resulting in proteasomal degradation. However, K63 polyubiquitin regulates membrane trafficking pathways (Acconcia *et al.*, 2009; Nathan and Lehner, 2009; Rotin and Kumar, 2009), and more recent studies implicate protein ubiquitination in the regulation of many diverse cellular functions. The ubiquitin chain linkage specificity is a function of the c-terminal lobe of the E3 HECT domain, and not of the E2 component (Kim and Huibregtse, 2009). Finally, β-papillomaviruses also associate with E6AP, presumably through the same LXXLL mechanism as cancer-associated genital HPV types (Bedard *et al.*, 2008).

Why E6AP?

The targeted degradation of p53 by E6 is exquisitely analogous to the association of p53 with the oncoproteins of other small DNA tumour viruses, yet the association of mucosal HPV E6 proteins, both high risk and low risk, with E6AP remains enigmatic. If one assumes that targeted degradation of p53 has particular advantages to the viral life cycle compared with the formation of inactivating complexes, why was the association of E6 with E6AP selected rather than any of the other 28 HECT domain ubiquitin ligases (recently reviewed in Rotin and Kumar, 2009), or the approximately 95% of ubiquitin ligases that belong to the RING family of ligases? The neurodevelopmental phenotype of Angelman syndrome offers few clues, but presumably, the role of E6AP within genital mucosal epithelium plays some important role, and the identification of E6-independent substrates of E6AP could give insight into why E6 'chose' E6AP. The interaction of 16E6 with E6AP induces the dimerization and ubiquitination of E6AP (Nuber *et al.*, 1998), and these authors thus proposed that self-ubiquitinization by E6AP could regulate the abundance of E6AP. Along this line of argument, in cervical cancers it appears that E6 causes a reduction of E6AP half-life and expression level (Kao *et al.*, 2000). In that paper, E6AP had a shorter half-life in HPV positive cervical cancer cell lines compared with non-HPV transformed C33A cells, but all the mechanistic experiments were either *in vitro* or in transient transfections so the effect of E6 upon the abundance and half life of E6AP is still not fully defined. In contrast, in a recent study of transgenic mice that express E6 from the keratin 14 promoter, no clear reduction of E6AP in tissues expressing E6 compared with non-E6-expressing cells was observed by immunohistochemistry (Shai *et al.*, 2010). If it is possible that one consequence of E6 interaction with E6AP is to modulate the abundance and half-life of E6AP, which could thereby modulate the normal function of E6AP to the advantage of the viral life cycle, some consideration of the normal cellular substrates of E6AP is indicated.

E6-independent substrates of E6AP

Nuclear hormone receptors

Because of the essential role of oestrogen in the development of cervical cancers in mice that express hrE6 and hrE7, a consideration of the role of E6AP in hormone responses is warranted. In

1999, Nawaz *et al.* found that E6AP associated with the progesterone receptor in a yeast two-hybrid hunt (Nawaz *et al.*, 1999). They further demonstrated that E6AP could also serve as a co-activator for oestrogen, gluccocorticoid, androgen, thyroid hormone and retinoic acid receptors in transient transfections assays. Nuclear hormone receptors change conformation upon the binding of the hormone, resulting in a binding site for LXXLL motifs found on co-activators and co-repressors. It would be attractive to hypothesize that the interaction of E6 with the LXXLL motif of E6AP evolved from nuclear hormone receptor interactions with LXXLL motifs of co-activators, but this is not the case. The ligand binding domains of the nuclear hormone receptors are helical bundles and interact with the LXXLL motif by superficial contacts (Savkur and Burris, 2004), while BE6 and presumably hrE6 binds the LXXLL motif of paxillin in a deep pocket formed by the amino-terminal and carboxy-terminal zinc-binding motifs and the connecting helix (discussed above). However, the nuclear hormone coactivator function of E6AP resembles traditional coactivators such as NcoA1 or NcoA3 in that the transactivation function resides within the amino terminus of E6AP containing the LQELL E6-binding motif, but the ubiquitin ligase function of E6-AP was not required in these initial studies (Nawaz *et al.*, 1999). Subsequent studies demonstrated that E6AP had a physiological role as a steroid receptor coactivator. E6AP is recruited to the promoter of the androgen-responsive PSA gene (Khan *et al.*, 2006), and to the oestrogen-responsive pS2 promoter in a hormone responsive manner (Reid *et al.*, 2003).

But what about the role of E6AP as a ubiquitin ligase in hormone responses? Proteasome inhibitors MG132 and lactacystin diminish hormone induced transcriptional activation by numerous hormone receptors (Lonard *et al.*, 2000), demonstrating that proteasomal activity plays a key role in modulating steroid hormone mediated transcription, and thus suggesting E6AP ubiquitin ligase activity might play a role as well. E6AP was shown to be able to target the degradation of the oestrogen receptor and progesterone receptor (Li *et al.*, 2006). Interestingly, E6AP null mice and mice expressing an E6AP mutant defective for ubiquitin ligase activity have mammary glands with enhanced lateral branching and alveologenesis; these phenotypes correlate with the altered progesterone receptor protein levels (Ramamoorthy *et al.*, 2010). These authors also showed that E6AP directly regulated PR-B protein levels (Ramamoorthy *et al.*, 2010). Finally, E6AP has been shown to play a role in mediating the degradation of the NcoA3 nuclear receptor co-activator (Mani *et al.*, 2006) under conditions of serum starvation or cell culture confluence. NcoA3 is particularly intriguing because it is overexpressed in numerous cancers, promotes neoplastic changes in transgenic mice, (Torres-Arzayus *et al.*, 2004), and loss of NcoA3 reduces Ha-*ras*-induced tumorigenesis in mice (Kuang *et al.*, 2004). Although E6AP associated with NcoA3, the ubiquitination and degradation of NcoA3 by E6AP could be indirect (Mani *et al.*, 2006).

Thus, altered expression of E6AP via E6 interaction could cause altered expression dynamics of nuclear hormone receptors and NcoA3 which could alter cell proliferation and survival in the hormone responsive tissues in which the mucosal HPV's replicate in the female reproductive tract. This of course leaves unanswered what role this might play in the male reproductive tract where mucosal HPV's also replicate, or the role in anal and upper respiratory tissues. The role of E6AP as a nuclear receptor co-activator has been recently reviewed (Ramamoorthy and Nawaz, 2008).

Other E6-independent E6AP substrates

HHR23, the human homologue of the yeast DNA repair protein rad23, was identified in a yeast two hybrid hunt as an E6AP interacting protein (Kumar *et al.*, 1999). At present, the precise roles of HHR23 proteins in repair of UV-damaged DNA is not clear, but the carboxy-terminal ubiquitin-like domain is thought to play a role in the transport of ubiquitinated proteins to the proteasome. HHR23 binds to polyubiquitylated p53, thereby inhibiting p53 deubiquitylation, and RNAi against HHR23 results in the accumulation of p53 (Glockzin *et al.*, 2003).

The Rho-GEF protein ECT2 was identified in a proteomics screen in Drosophila as a potential E6AP substrate (Reiter *et al.*, 2006). Ect2 is the antagonist of p190Rho-Gap and activates Rho and thus the ROCK kinase. Therefore, E6AP could alter the balance between Rho and Rac signalling in the cell, promoting Rac1 signalling, and thus inhibiting keratinocyte differentiation and promoting keratinocyte proliferation (McMullan *et al.*, 2003). E6 might activate Rac1 through reducing the expression or half-life of E6AP, which thereby could, through ECT2, alter the balance between Rho and Rac1 signalling pathways.

Trihydrophobin 1 (TH1) is a subunit of the human negative transcription elongation factor (NELF) complex that associates with RNA polymerase II complexes and represses *in vitro* RNA polymerase II transcriptional elongation. Using a TH1 bait, E6AP was identified in a yeast two hybrid screen (Yang *et al.*, 2007). Overexpression of E6AP enhanced TH1 degradation while whereas RNAi against endogenous E6AP or expression of E6APc-833a (defective in ubiquitin ligase activity) enhanced TH1 expression. An *in vitro* ubiquitination assay also demonstrated that TH1 can be ubiquitinated by E6AP. Thus, interaction of E6 with E6AP could alter transcriptional elongation through altered degradation of TH1 with potentially widespread effects upon transcription.

A GST-E6AP binding experiment using cellular lysate identified annexin A1 as a potential E6AP substrate by proteomic analysis (Shimoji *et al.*, 2009). Ectopic expression of E6AP enhanced the degradation of annexin A1 *in vivo* and RNAi-mediated repression of endogenous E6AP increased the levels of endogenous annexin A1 protein, and annexin A1 could be ubiquitinated by E6AP *in vitro* (Shimoji *et al.*, 2009). Annexin A1 is a calcium dependent membrane-binding protein that is a phospholipase A2 inhibitor. Because PLA2 generates precursors for pro-inflammatory lipid molecules and secreted Annexin A1 has potent anti-inflammatory properties, modulation of annexin A1 levels by E6AP or E6 + E6AP could modulate inflammatory responses to HPV infected cells (reviewed in Perretti and D'Acquisto, 2009). Finally, for consideration of the HPV replication cycle, how might E6 expression modulate the activity of E6AP towards its non-E6-dependent substrates? This has not been reported!

What is the role of E6AP in E6-induced cancers?

Two recent studies have examined the role of E6AP in the production of skin hyperplasias and cervical cancer in mice that express 16E6 in squamous epithelia from the keratin 14 promoter (K14E6 mice). K14E6 mice produce skin hyperplasias, which these authors previously showed depended upon the PDZ ligand of E6 (Nguyen *et al.*, 2003a). K14E6 mice also produce cervical cancers with prolonged latency when also treated with oestrogen; in this system, K14E6 enhances the tumorigenicity of oestrogen treatment upon cervical and vaginal neoplasms, and loss of E6AP ablated this enhancement (Shai *et al.*, 2010). Interestingly, loss of E6AP resulted in an enhanced incidence of cancer in oestrogen treated animals without E6 (Shai *et al.*, 2010). As expected, cell cycle arrest in irradiated cells was ablated by E6, but this did not require E6AP, indicating that an E6AP-independent mechanism for E6 ablation of p53 functions exists in the mouse, despite the increase of p53 expression in K14E6-E6AP$^{-/-}$ mice compared with K14E6 mice (Shai *et al.*, 2010). E6AP null mouse cells have been reported to target the degradation of p53, but the mechanisms by which this occurs have yet to be elucidated (Massimi *et al.*, 2008).

E6 and the global manipulation of cellular transcription

Microarray analysis of E6 expressing keratinocytes has demonstrated both p53 dependent and p53 independent alterations of global cellular transcription by hrE6 proteins (Duffy *et al.*, 2003; Garner-Hamrick *et al.*, 2004; Kuner *et al.*, 2007; Mendoza-Villanueva *et al.*, 2008). Although the manipulation of cellular signal transduction by hrE6 (such as through the effects of E6 upon cellular PDZ proteins) could in part explain these effects, specific interactions of E6 with cellular transcriptional regulators that have global effects are the probable cause. Histone acetyltransferases (HATs) are components of eukaryotic transcription complexes. Apart from acetylating histones to enable chromosomal remodelling, several

HATs (p300, CBP, PCAF, TIP60, and hMOF) acetylate p53 and function as p53 co-activators. hrE6 proteins target the degradation of both the TIP60 acetyltransferase, and the ADA3 adapter protein that is part of the PCAF acetyltransferase complex.

ADA3 (for the yeast alteration/deficiency in activation protein) is a component of yeast HAT complexes (Balasubramanian *et al.*, 2002). Mammalian Ada3 is a transcription co-activator for p53 and other cellular transcription facts such as RXR-alpha that is targeted for degradation by 16E6 and E6AP (Kumar *et al.*, 2002; Shamanin *et al.*, 2008; Hu *et al.*, 2009). RNAi knockdown of hAda3 blocks the acetylation of lysine 382 in p53, inhibits p53 stabilization, and attenuates p14ARF-induced senescence (Nag *et al.*, 2007; Sekaric *et al.*, 2007). Ada3 is also a co-activator for the oestrogen receptor with RNAi knockdown of Ada3 attenuating oestrogen induction of target genes and E6 having the same effect (Meng *et al.*, 2004). Thus the E6-mediated degradation of ada3 blocks p53 transcription and could serve as a feedback repression on oestrogen effects in HPV infected cervical cells.

The TIP60 tumour suppressor is a histone acetyltransferase involved in checkpoint activation and p53-directed pro-apoptotic pathways. Both high- and low-risk mucosal HPV E6 proteins destabilize TIP60 *in vitro* and *in vivo* in both transient transfections, stably transduced keratinocytes, and cervical cancer cell lines (Jha *et al.*, 2010). TIP60 is present at the HPV major early promoter and through acetylation of histone H4 recruits Brd4, a repressor of the HPV early promoter. Degradation of TIP60, therefore, allows low- and high-risk HPV to promote cell proliferation and cell survival. Correlating with the above study, EP400, a component of the NuA4/TIP60 histone acetyltransferase complex, was identified in a genome wide RNAi scan for factors that are necessary for E2-mediated repression of the high-risk HPV16 early promoter. Thus, E6 is implicated in a feed-forward regulation of both basal transcription from the early promoter and possibly E2 mediated repression (Smith *et al.*, 2010). Destabilization of TIP60 by HPV E6 also relieves cellular promoters from TIP60-initiated repression and blocks p53-mediated activation of transcription.

Like Ada3 and TIP60, p300 acetylates p53, thus augmenting p53 DNA binding and activation of p53-responsive genes (Gu and Roeder, 1997 and reviewed in Grossman, 2001). Given the central role of p300 associations with Adenovirus E1a, a search for similar interactions showed *in vitro* translated E6 proteins to associate *in vitro* to GST fusions of p300 fragments (Patel *et al.*, 1999) or GST-E6 proteins to associate *in vitro* with partially purified p300 preparations from HeLa cells (Zimmermann *et al.*, 1999). A similar *in vitro* binding experiment showed association with BE6 (Zimmermann *et al.*, 2000). In all of these studies, the association was related to the inhibition of p53 transcriptional activation independent of p53 degradation, and loss of p53 acetylation. More recently, CRPV E6 and the β-papillomavirus HPV38 E6 association with p300 was shown to correlate with the blockage of p53 acetylation by p300; mutants that failed to associate with p300 were defective for tumorigenesis (Muench *et al.*, 2010). It is unclear if the E6 association with p300 is indirect or indirect. *In vitro* reconstituted chromatin templates allow for the sequential formation of transcription complexes using fractionated nuclear extracts and purified E6 proteins; this system demonstrated that both high- and low-risk E6 proteins could repress p53 transcription through inhibition of p300 dependent histone acetylation, thus converting p53 transcription complexes from transcriptional activation to repressor complexes (Thomas and Chiang, 2005). They also observed that the low-risk HPV11 E6 protein could also ablate *in vitro* and *in vivo* p53-responsive transcription in a similar manner. In contrast, keratinocytes transduced with low-risk E6 and E7 genes are unaltered in their checkpoint response to radiation (Demers *et al.*, 1994).

An important transcription target of E6, the hTERT promoter, has been reviewed recently; E6 associations that mediate the induction of telomerase activity, in particular the targeted degradation of NFX-91 by E6 and E6AP (Gewin *et al.*, 2004), are nicely reviewed there (Howie *et al.*, 2009).

E6 interactions that modulate apoptosis: Bak

The E6 oncoproteins from both high- and low-risk mucosal HPVs and β-papillomavirus cutaneous HPVs all target the degradation of BAK. Initial studies in immortalized cells showed constitutive hrE6 + E6AP association with BAK, and constitutive proteasome dependent degradation of BAK (Thomas and Banks, 1998). Subsequent studies in primary keratinocytes showed little effect of E6 upon BAK expression in the absence of UV treatment, but upon such treatment, while BAK expression increases in vector transduced keratinocytes, it declines in a proteasome dependent manner in cells expressing hrE6 or β-papillomavirus E6 (Jackson and Storey, 2000; Jackson *et al.*, 2000; Underbrink *et al.*, 2008). It is probable that the difference between these two sets of results reflects altered signalling in transformed cells versus primary keratinocytes. While some studies have shown E6AP dependence by shRNA knockdown (Underbrink *et al.*, 2008), others, also using β-papillomaviruses, do not (Simmonds and Storey, 2008). Mutational mapping of BAK degradation in HPV5 E6 has been somewhat controversial, because mutants that lost BAK degradation had an unexpected gain of function for the degradation of p53, which is not a property of β-papillomavirus wild-type E6 proteins (Simmonds and Storey, 2008). How E6 proteins interact with BAK and how signalling to BAK modifies the association of BAK with E6 remain to be elucidated. The broad and conserved targeting of BAK by alpha and beta papillomaviruses implies a role for a conserved structural feature of these E6 proteins that has yet to be unambiguously demonstrated.

hrE6 interaction with components of the TNF receptor signalling is another mechanism by which E6 regulates apoptotic responses and has been the subject of recent excellent reviews (Garnett and Duerksen-Hughes, 2006; Tungteakkhun and Duerksen-Hughes, 2008).

E6 and the manipulation of keratinocyte differentiation

Although the preoccupation with HR HPV's interactions with the pRB and p53 tumour suppressors is understandable, it overlooks the common biology of papillomaviruses. Most papillomaviruses do not cause cancer. BPV-1 shares the same essential life cycle as HR HPVs in stratified squamous epithelium, yet its E7 oncoproteins does not have an RB-family LXCXE docking site, and its E6 protein does not target the degradation of p53. And yet, it lives! How does it do it, and what is the role of BE6?

The BPV-1 E6 (BE6) protein can interact with the cellular proteins E6AP, paxillin, the AP-1 adaptin complex (Tong and Howley, 1997; Tong *et al.*, 1998; Vande Pol *et al.*, 1998) and the transcriptional coactivator MAML1 (Brimer *et al.*, 2011). While transformation by BE6 requires paxillin expression and LXXLL docking sites on paxillin (Wade *et al.*, 2008), the mechanism by which BE6 transforms at least in part through interaction with paxillin remains unresolved. Although the comparative binding of BE6 mutants to E6AP and paxillin more strongly supported a role of paxillin in transformation by BE6, more rigorous studies (such as the use of E6AP knockdown cells) have not yet been reported, so a potential role for BE6 association with E6AP in transformation has not yet been eliminated. Pull-down experiments with GST-BE6 identified the clathrin adaptor complex AP1 (Tong *et al.*, 1998). BE6 preferentially associated with membrane-bound AP-1 and was further shown to be recruited to isolated Golgi membranes and to co purify with clathrin-coated vesicles; the E6 interaction with AP-1 was required for its association with clathrin-coated vesicles. Although this association implicates BE6 in the possible modulation of membrane trafficking, the physiologic consequences of this interaction for transformation or the virus life cycle remain unknown.

BE6 manipulates the Notch tumour suppressor pathway in keratinocyte differentiation

The MAML1 coactivator is most well known for its function in Notch signalling. Notch signalling between adjacent cells affects developmental choices of neighbouring cells. Thus, Notch functions in development to link the differentiation fate of a given cell to that of its adjacent neighbour. Notch1 and 2 are expressed in the first spinous cell layer and the Notch ligand Jagged2 is expressed in the basal layer; Notch1 signalling in the spinous cell layer then drives early squamous epithelial differentiation as well as terminal epithelial differentiation (Rangarajan *et al.*, 2001b; Blanpain *et al.*, 2006; and reviewed in Watt *et al.*, 2008). Upon canonical Notch signalling, the Notch receptor is cleaved by the intramembranous gamma-secretase protease, liberating the Notch intracellular domain that forms a complex with the RBP-J DNA-binding protein. This displaces a repressor–histone-deacetylase complex and recruits the MAML1 coactivator, thus converting the RBP-J complex from a transcriptional repressor to an activator (Fig. 5.6 and reviewed in Tanigaki and Honjo, 2010). BE6 binds to an acidic LXXLL transactivation motif at the carboxy-terminus of MAML1 (Fig. 5.3) and thereby represses Notch-induced transcription. BE6, and the HPV types 1E6 and 8E6 can all associate with the c-terminal Notch LXXLL ligand and repress the transcriptional activity of MAML1, although the role of this interaction for 1E6 and 8E6 in Notch signalling remains to be determined (Brimer *et al.*, 2011).

It is fair to say that Notch is to squamous cell differentiation what p53 is to DNA damage responses, so it is a given that all papillomaviruses have developed strategies that in some way manipulate Notch signalling. One characteristic feature is that modest differences in Notch signalling have developmental consequences, classically revealed in genetic haploinsufficiency syndromes, where upon haploid mutation of Notch, Notch ligands, and components of the Notch signalling complexes developmental abnormalities ensue (reviewed in Artavanis-Tsakonas and Muskavitch, 2010). Complete disruption of Notch signalling in the squamous epithelium of transgenic mice by tissue specific Notch deletion (Nicolas *et al.*, 2003; Dotto, 2008), epithelial deletion of the RBP-J DNA-binding subunit of the Notch transcription complex (Blanpain *et al.*, 2006), or expression of a dominant negative MAML1 (Proweller *et al.*,

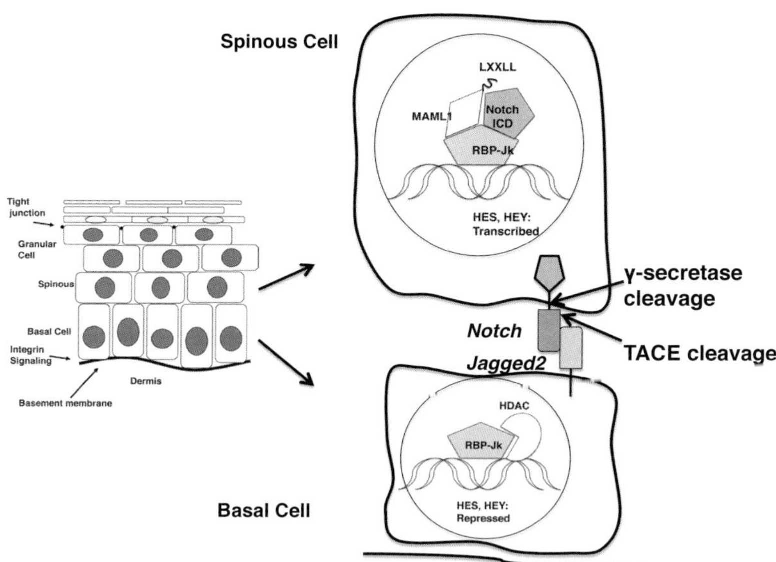

Figure 5.6 Notch signalling in squamous epithelium. Notch signalling is initiated between basal cells expressing Notch ligands and Notch expressed in suprabasal cells, resulting in two successive proteolytic cleavages of Notch. The first cleavage within the extracellular domain is mediated by TACE, while the second cleavage is mediated by the γ-secretase activity of presenilin protein complexes. The cleaved Notch ICD translocates to the nucleus where it associates with MAML1 and RBP-Jk, displacing HDAC. This converts the RBP-J complex from a transcriptional repressor to a transcriptional activator, activating transcription of basis helix–loop–helix transcription factors that are the Notch effectors.

2006) results in loss of differentiation and squamous cell carcinomas, demonstrating that Notch signalling is a tumour suppressor in squamous epithelium. However, Notch signalling in different tissues can drive either terminal differentiation or cellular proliferation, making this subject wickedly complex. For example, in numerous cancers Notch receptors are constitutively activated and drive cancer cell proliferation (Maliekal *et al.*, 2008). In T-cell leukaemia, Notch1 is itself mutated, resulting in constitutive cleavage of Notch and constitutive Notch signalling (Ellisen *et al.*, 1991), and activated Notch can cooperate with high-risk E6 + E7 oncoproteins to transform HaCat cells (Rangarajan *et al.*, 2001a). In contrast, expression of activated Notch in HPV expressing cervical cancers results in growth arrest (Talora *et al.*, 2005). The role of Notch expression in combination with other signalling pathways in the papillomavirus life cycle and oncogenesis remains to be elucidated.

A longstanding observation with HR HPV is the ability of hrE6 to promote the growth of colonies of keratinocytes that fail to stratify when cell cultures are switched from low to high calcium media (Sherman and Schlegel, 1996). It is unclear what interactions with E6 are responsible for this phenomenon as it is not observed in pooled colonies that are transduced by hrE6.

Conclusions

The explosion of knowledge about E6 proteins and the effect of E6 upon cell physiology leaves one shocked: how can such a small protein do so much? How can such a small protein modulate almost every aspect of cell biology from global transcription, protein synthesis, signal transduction, cell polarity, cell survival, cell cycle progression, DNA damage responses, host cell differentiation, immune system responses, and cell adhesion and migration? At the same time, a blizzard of interactions have been reported, many of which were not explored here; certainly such a small protein cannot have even a fraction of all the reported interaction partners! The answer is that E6 targets proteins that are themselves at nodal control points of cell physiology. Sorting out which of the effects of E6 are direct and which are indirect is still an ongoing challenge. Compounding this, the analysis of the phenotypes of E6 mutants has until now been quite difficult, and many E6 mutants used in the past will turn out to have been globally defective for core functions of E6 such as LXXLL interactions, making conclusions drawn from the use of these mutants now subject to new interpretation. The solved structure of E6 proteins, BE6 and structures still to come, will now allow a precise mapping of the functions of E6 to its structure and thereby to its direct interaction partners, opening a new era of understanding.

Acknowledgements

This review is indebted to my longstanding collaboration with the laboratories of Gilles Trave and Jean Cavarelli at the CNRS in Strasbourg France on the structure of E6, and in whose labs the structure of BE6 was solved. Embedded in that subject are numerous conversations with Gilles Trave and the members of the Trave Lab, especially Sebastian Charbonnier and Katia Zanier. I am also grateful for the communication of unpublished observations from Christopher Buck (NIH) regarding the diversity of HPV types in skin, and Karl Munger (Harvard) regarding the association of cutaneous E6 proteins with MAML1. This work was supported by NIH grants (CA120352 and CA08093) to S. V. P, and institutional support from the University of Virginia.

References

Acconcia, F., Sigismund, S., and Polo, S. (2009). Ubiquitin in trafficking: the network at work. Exp. Cell. Res. *315*, 1610–1618.

Ainsworth, J., Thomas, M., Banks, L., Coutlee, F., and Matlashewski, G. (2008). Comparison of p53 and the PDZ domain containing protein MAGI-3 regulation by the E6 protein from high-risk human papillomaviruses. Virol. J. *5*, 67.

Alarcon, R., Koumenis, C., Geyer, R.K., Maki, C.G., and Giaccia, A.J. (1999). Hypoxia induces p53 accumulation through MDM2 down-regulation and inhibition of E6-mediated degradation. Cancer Res. *59*, 6046–6051.

Artavanis-Tsakonas, S., and Muskavitch, M.A. (2010). Notch: the past, the present, and the future. Curr. Top. Dev. Biol. *92*, 1–29.

Balasubramanian, R., Pray-Grant, M.G., Selleck, W., Grant, P.A., and Tan, S. (2002). Role of the Ada2 and Ada3 transcriptional coactivators in histone acetylation. J. Biol. Chem. *277*, 7989–7995.

Baleja, J.D., Cherry, J.J., Liu, Z., Gao, H., Nicklaus, M.C., Voigt, J.H., Chen, J.J., and Androphy, E.J. (2006). Identification of inhibitors to papillomavirus type 16 E6 protein based on three-dimensional structures of interacting proteins. Antiviral Res. *72*, 49–59.

Beaudenon, S., and Huibregtse, J.M. (2008). HPV E6, E6AP and cervical cancer. BMC Biochem. *9 (Suppl 1)*, S4.

Bedard, K.M., Underbrink, M.P., Howie, H.L., and Galloway, D.A. (2008). The E6 oncoproteins from human betapapillomaviruses differentially activate telomerase through an E6AP-dependent mechanism and prolong the lifespan of primary keratinocytes. J. Virol. *82*, 3894–3902.

Bernard, H.U. (1994). Coevolution of papillomaviruses with human populations. Trends Microbiol. *2*, 140–143.

Blanpain, C., Lowry, W.E., Pasolli, H.A., and Fuchs, E. (2006). Canonical notch signalling functions as a commitment switch in the epidermal lineage. Genes Dev. *20*, 3022–3035.

Bohl, J., Das, K., Dasgupta, B., and Vande Pol, S.B. (2000). Competitive binding to a charged leucine motif represses transformation by a papillomavirus E6 oncoprotein. Virology *271*, 163–170.

Brimer, N., Lyons, C., and Vande Pol, S.B. (2007). Association of E6AP (UBE3A) with human papillomavirus type 11 E6 protein. Virology *358*, 303–310.

Brimer, N., Lyons, C., Wallberg, A.E., and Vande Pol, S.B. (2011). Papillomavirus E6 oncoproteins associate with MAML1 to repress transactivation and Notch signalling. Submitted for publication.

Brimer, N., and Vande Pol, S. (2011). Manuscript in preparation.

Chan, S.Y., Bernard, H.U., Ong, C.K., Chan, S.P., Hofmann, B., and Delius, H. (1992). Phylogenetic analysis of 48 papillomavirus types and 28 subtypes and variants: a showcase for the molecular evolution of DNA viruses. J. Virol. *66*, 5714–5725.

Charbonnier, S., Brimer, N., Ould Sidi, A., Luck, K., Zanier, K., Babah, O., Ansari, T., Muller, I., Poussin, P., Cura, V., *et al.* (2010). Structural basis for hijacking of cellular LxxLL motifs by a papillomavirus E6 oncoprotein. Submitted for publication.

Chen, J.J., Hong, Y., Rustamzadeh, E., Baleja, J.D., and Androphy, E.J. (1998). Identification of an alpha helical motif sufficient for association with papillomavirus E6. J. Biol. Chem. *273*, 13537–13544.

Chen, Z., Schiffman, M., Herrero, R., Desalle, R., and Burk, R.D. (2007). Human papillomavirus (HPV) types 101 and 103 isolated from cervicovaginal cells lack an E6 open reading frame (ORF) and are related to gamma-papillomaviruses. Virology *360*, 447–453.

Cheron, G., Servais, L., Wagstaff, J., and Dan, B. (2005). Fast cerebellar oscillation associated with ataxia in a mouse model of Angelman syndrome. Neuroscience *130*, 631–637.

Chow, L.T., and Broker, T.R. (1994). Papillomavirus DNA replication. Intervirology *37*, 150–158.

Cooper, B., Schneider, S., Bohl, J., Jiang, Y., Beaudet, A., and Vande Pol, S. (2003). Requirement of E6AP and the features of human papillomavirus E6 necessary to support degradation of p53. Virology *306*, 87–99.

Crook, T., Wrede, D., and Vousden, K.H. (1991). p53 point mutation in HPV negative human cervical carcinoma cell lines. Oncogene *6*, 873–875.

Das, K., Bohl, J., and Vande Pol, S.B. (2000). Identification of a second transforming region in bovine papillomavirus type 1 E6 and the role of E6 interaction with paxilin, E6BP, and E6AP. J. Virol. *74*, 812–816.

DeCaprio, J.A., Ludlow, J.W., Figge, J., Shew, J.Y., Huang, C.M., Lee, W.H., Marsilio, E., Paucha, E., and Livingston, D.M. (1988). SV40 large tumor antigen forms a specific complex with the product of the retinoblastoma susceptibility gene. Cell *54*, 275–283.

Degenhardt, Y.Y., and Silverstein, S.J. (2001). Gps2, a protein partner for human papillomavirus E6 proteins. J. Virol. *75*, 151–160.

Demers, G.W., Foster, S.A., Halbert, C.L., and Galloway, D.A. (1994). Growth arrest by induction of p53 in DNA damaged keratinocytes is bypassed by human papillomavirus 16 E7. Proc. Natl. Acad. Sci. U.S.A. *91*, 4382–4386.

Dornan, D., Wertz, I., Shimizu, H., Arnott, D., Frantz, G.D., Dowd, P., O'Rourke, K., Koeppen, H., and Dixit, V.M. (2004). The ubiquitin ligase COP1 is a critical negative regulator of p53. Nature *429*, 86–92.

Dotto, G.P. (2008). Notch tumour suppressor function. Oncogene *27*, 5115–5123.

Duffy, C.L., Phillips, S.L., and Klingelhutz, A.J. (2003). Microarray analysis identifies differentiation-associated genes regulated by human papillomavirus type 16 E6. Virology *314*, 196–205.

Dyson, N., Howley, P.M., Munger, K., and Harlow, E. (1989). The human papilloma virus-16 E7 oncoprotein is able to bind to the retinoblastoma gene product. Science *243*, 934–937.

Eichten, A., Westfall, M., Pietenpol, J.A., and Munger, K. (2002). Stabilization and functional impairment of the tumour suppressor p53 by the human papillomavirus type 16 E7 oncoprotein. Virology *295*, 74–85.

Eliyahu, D., Raz, A., Gruss, P., Givol, D., and Oren, M. (1984). Participation of p53 cellular tumour antigen in transformation of normal embryonic cells. Nature *312*, 646–649.

Ellisen, L.W., Bird, J., West, D.C., Soreng, A.L., Reynolds, T.C., Smith, S.D., and Sklar, J. (1991). TAN-1, the human homolog of the Drosophila notch gene, is broken by chromosomal translocations in T lymphoblastic neoplasms. Cell *66*, 649–661.

Elston, R.C., Napthine, S., and Doorbar, J. (1998). The identification of a conserved binding motif within human papillomavirus type 16 E6 binding peptides, E6AP and E6BP. J. Gen. Virol. *79*, 371–374.

Fan, X., and Chen, J.J. (2004). Regulation of cell cycle progression and apoptosis by the papillomavirus E6 oncogene. Crit. Rev. Eukaryot. Gene Expr. *14*, 183–202.

Filippova, M., Johnson, M.M., Bautista, M., Filippov, V., Fodor, N., Tungteakkhun, S.S., Williams, K., and

Duerksen-Hughes, P.J. (2007). The large and small isoforms of human papillomavirus type 16 E6 bind to and differentially affect procaspase 8 stability and activity. J. Virol. *81*, 4116–4129.

Filippova, M., Parkhurst, L., and Duerksen-Hughes, P.J. (2004). The human papillomavirus 16 E6 protein binds to Fas-associated death domain and protects cells from Fas-triggered apoptosis. J. Biol. Chem. *279*, 25729–25744.

Finlay, C.A., Hinds, P.W., and Levine, A.J. (1989). The p53 proto-oncogene can act as a suppressor of transformation. Cell *57*, 1083–1093.

Francis, D.A., Schmid, S.I., and Howley, P.M. (2000). Repression of the integrated papillomavirus E6/E7 promoter is required for growth suppression of cervical cancer cells. J. Virol. *74*, 2679–2686.

Freedman, D.A., and Levine, A.J. (1998). Nuclear export is required for degradation of endogenous p53 by MDM2 and human papillomavirus E6. Mol. Cell Biol. *18*, 7288–7293.

Ganzenmueller, T., Matthaei, M., Muench, P., Scheible, M., Iftner, A., Hiller, T., Leiprecht, N., Probst, S., Stubenrauch, F., and Iftner, T. (2008). The E7 protein of the cottontail rabbit papillomavirus immortalizes normal rabbit keratinocytes and reduces pRb levels, while E6 cooperates in immortalization but neither degrades p53 nor binds E6AP. Virology *372*, 313–324.

Garcia-Vallve, S., Alonso, A., and Bravo, I.G. (2005). Papillomaviruses: different genes have different histories. Trends Microbiol. *13*, 514–521.

Garner-Hamrick, P.A., Fostel, J.M., Chien, W.M., Banerjee, N.S., Chow, L.T., Broker, T.R., and Fisher, C. (2004). Global effects of human papillomavirus type 18 E6/E7 in an organotypic keratinocyte culture system. J. Virol. *78*, 9041–9050.

Garnett, T.O., and Duerksen-Hughes, P.J. (2006). Modulation of apoptosis by human papillomavirus (HPV) oncoproteins. Arch. Virol. *151*, 2321–2335.

Gewin, L., Myers, H., Kiyono, T., and Galloway, D.A. (2004). Identification of a novel telomerase repressor that interacts with the human papillomavirus type-16 E6/E6–AP complex. Genes Dev. *18*, 2269–2282.

Giampieri, S., Garcia-Escudero, R., Green, J., and Storey, A. (2004). Human papillomavirus type 77 E6 protein selectively inhibits p53-dependent transcription of proapoptotic genes following UV-B irradiation. Oncogene *23*, 5864–5870.

Glaunsinger, B.A., Lee, S.S., Thomas, M., Banks, L., and Javier, R. (2000). Interactions of the PDZ-protein MAGI-1 with adenovirus E4-ORF1 and high-risk papillomavirus E6 oncoproteins. Oncogene *19*, 5270–5280.

Glockzin, S., Ogi, F.X., Hengstermann, A., Scheffner, M., and Blattner, C. (2003). Involvement of the DNA repair protein hHR23 in p53 degradation. Mol. Cell Biol. *23*, 8960–8969.

Goodwin, E.C., and DiMaio, D. (2000). Repression of human papillomavirus oncogenes in HeLa cervical carcinoma cells causes the orderly reactivation of dormant tumour suppressor pathways. Proc. Natl. Acad. Sci. U.S.A. *97*, 12513–12518.

Grm, H.S., and Banks, L. (2004). Degradation of hDlg and MAGIs by human papillomavirus E6 is E6-AP-independent. J. Gen. Virol. *85*, 2815–2819.

Grossman, S.R. (2001). p300/CBP/p53 interaction and regulation of the p53 response. Eur. J. Biochem. *268*, 2773–2778.

Handa, K., Yugawa, T., Narisawa-Saito, M., Ohno, S., Fujita, M., and Kiyono, T. (2007). E6AP-dependent degradation of DLG4/PSD95 by high-risk human papillomavirus type 18 E6 protein. J. Virol. *81*, 1379–1389.

Heery, D.M., Kalkhoven, E., Hoare, S., and Parker, M.G. (1997). A signature motif in transcriptional co-activators mediates binding to nuclear receptors [see comments]. Nature *387*, 733–736.

Hinds, P., Finlay, C., and Levine, A.J. (1989). Mutation is required to activate the p53 gene for cooperation with the ras oncogene and transformation. J. Virol. *63*, 739–746.

Honda, R., Tanaka, H., and Yasuda, H. (1997). Oncoprotein MDM2 is a ubiquitin ligase E3 for tumour suppressor p53. FEBS Lett. *420*, 25–27.

Howie, H.L., Katzenellenbogen, R.A., and Galloway, D.A. (2009). Papillomavirus E6 proteins. Virology *384*, 324–334.

Hu, Y., Ye, F., Lu, W., Hong, D., Wan, X., and Xie, X. (2009). HPV16 E6-induced and E6AP-dependent inhibition of the transcriptional coactivator hADA3 in human cervical carcinoma cells. Cancer Invest. *27*, 298–306.

Huibregtse, J.M., Scheffner, M., and Howley, P.M. (1993a). Cloning and expression of the cDNA for E6-AP, a protein that mediates the interaction of the human papillomavirus E6 oncoprotein with p53. Mol. Cell Biol. *13*, 775–784.

Huibregtse, J.M., Scheffner, M., and Howley, P.M. (1993b). Localization of the E6-AP regions that direct human papillomavirus E6 binding, association with p53, and ubiquitination of associated proteins. Mol. Cell Biol. *13*, 4918–4927.

Huibregtse, J.M., Scheffner, M., Beaudenon, S., and Howley, P.M. (1995). A family of proteins structurally and functionally related to the E6-AP ubiquitin-protein ligase. Proc. Natl. Acad. Sci. U.S.A. *92*, 2563–2567.

Ikeda, F., and Dikic, I. (2008). Atypical ubiquitin chains: new molecular signals. 'Protein Modifications: Beyond the Usual Suspects' review series. EMBO Rep. *9*, 536–542.

Jackson, S., and Storey, A. (2000). E6 proteins from diverse cutaneous HPV types inhibit apoptosis in response to UV damage. Oncogene *19*, 592–598.

Jackson, S., Harwood, C., Thomas, M., Banks, L., and Storey, A. (2000). Role of Bak in UV-induced apoptosis in skin cancer and abrogation by HPV E6 proteins. Genes Dev. *14*, 3065–3073.

Javier, R.T. (2008). Cell polarity proteins: common targets for tumorigenic human viruses. Oncogene *27*, 7031–7046.

Jha, S., Vande Pol, S., Banerjee, N.S., Dutta, A.B., Chow, L.T., and Dutta, A. (2010). Destabilization of TIP60 by human papillomavirus E6 results in attenuation of TIP60-dependent transcriptional regulation and apoptotic pathway. Mol. Cell *38*, 700–711.

Jiang, Y.H., Armstrong, D., Albrecht, U., Atkins, C.M., Noebels, J.L., Eichele, G., Sweatt, J.D., and Beaudet, A.L. (1998). Mutation of the Angelman ubiquitin ligase in mice causes increased cytoplasmic p53 and deficits of contextual learning and long-term potentiation [see comments]. Neuron *21*, 799–811.

Jing, M., Bohl, J., Brimer, N., Kinter, M., and Vande Pol, S.B. (2007). Degradation of tyrosine phosphatase PTPN3 (PTPH1) by association with oncogenic human papillomavirus E6 proteins. J. Virol. *81*, 2231–2239.

Kao, W.H., Beaudenon, S.L., Talis, A.L., Huibregtse, J.M., and Howley, P.M. (2000). Human papillomavirus type 16 E6 induces self-ubiquitination of the E6AP ubiquitin-protein ligase. J. Virol. *74*, 6408–6417.

Khan, O.Y., Fu, G., Ismail, A., Srinivasan, S., Cao, X., Tu, Y., Lu, S., and Nawaz, Z. (2006). Multifunction steroid receptor coactivator, E6-associated protein, is involved in development of the prostate gland. Mol. Endocrinol. *20*, 544–559.

Kim, H.C., and Huibregtse, J.M. (2009). Polyubiquitination by HECT E3s and the determinants of chain type specificity. Mol. Cell Biol. *29*, 3307–3318.

Kishino, T., Lalande, M., and Wagstaff, J. (1997). UBE3A/E6-AP mutations cause Angelman syndrome. Nat. Genet. *15*, 70–73.

Kiyono, T., Foster, S.A., Koop, J.I., McDougall, J.K., Galloway, D.A., and Klingelhutz, A.J. (1998). Both Rb/p16INK4a inactivation and telomerase activity are required to immortalize human epithelial cells. Nature *396*, 84–88.

Kranjec, C., and Banks, L. (2010). A systematic analysis of human papillomavirus (HPV) E6 PDZ substrates identifies MAGI-1 as a major target of HPV type 16 (HPV16) and HPV-18 whose loss accompanies disruption of tight junctions. J. Virol. *85*, 1757–1764.

Kuang, S.Q., Liao, L., Zhang, H., Lee, A.V., O'Malley, B.W., and Xu, J. (2004). AIB1/SRC-3 deficiency affects insulin-like growth factor I signalling pathway and suppresses v-Ha-ras-induced breast cancer initiation and progression in mice. Cancer Res. *64*, 1875–1885.

Kuballa, P., Matentzoglu, K., and Scheffner, M. (2007). The role of the ubiquitin ligase E6-AP in human papillomavirus E6-mediated degradation of PDZ domain-containing proteins. J. Biol. Chem. *282*, 65–71.

Kumar, A., Zhao, Y., Meng, G., Zeng, M., Srinivasan, S., Delmolino, L.M., Gao, Q., Dimri, G., Weber, G.F., Wazer, D.E., *et al.* (2002). Human papillomavirus oncoprotein E6 inactivates the transcriptional coactivator human ADA3. Mol. Cell Biol. *22*, 5801–5812.

Kumar, S., Kao, W.H., and Howley, P.M. (1997). Physical interaction between specific E2 and Hect E3 enzymes determines functional cooperativity. J. Biol. Chem. *272*, 13548–13554.

Kumar, S., Talis, A.L., and Howley, P.M. (1999). Identification of HHR23A as a substrate for E6-associated protein-mediated ubiquitination. J. Biol. Chem. *274*, 18785–18792.

Kuner, R., Vogt, M., Sultmann, H., Buness, A., Dymalla, S., Bulkescher, J., Fellmann, M., Butz, K., Poustka, A., and Hoppe-Seyler, F. (2007). Identification of cellular targets for the human papillomavirus E6 and E7 oncogenes by RNA interference and transcriptome analyses. J. Mol. Med. *85*, 1253–1262.

Lane, D.P., and Crawford, L.V. (1979). T antigen is bound to a host protein in SV40-transformed cells. Nature *278*, 261–263.

Lee, C., and Laimins, L.A. (2004). Role of the PDZ domain-binding motif of the oncoprotein E6 in the pathogenesis of human papillomavirus type 31. J. Virol. *78*, 12366–12377.

Lee, H.J., and Zheng, J.J. (2010). PDZ domains and their binding partners: structure, specificity, and modification. Cell. Commun. Signal. *8*, 8.

Lee, S.S., Weiss, R.S., and Javier, R.T. (1997). Binding of human virus oncoproteins to hDlg/SAP97, a mammalian homolog of the Drosophila discs large tumour suppressor protein. Proc. Natl. Acad. Sci. U.S.A. *94*, 6670–6675.

Lee, S.S., Glaunsinger, B., Mantovani, F., Banks, L., and Javier, R.T. (2000). MultiPDZ domain protein MUPP1 is a cellular target for both adenovirus E4-ORF1 and high-risk papillomavirus type 18 E6 oncoproteins. J. Virol. *74*, 9680–9693.

Leng, R.P., Lin, Y., Ma, W., Wu, H., Lemmers, B., Chung, S., Parant, J.M., Lozano, G., Hakem, R., and Benchimol, S. (2003). Pirh2, a p53-induced ubiquitin-protein ligase, promotes p53 degradation. Cell *112*, 779–791.

Li, L., Li, Z., Howley, P.M., and Sacks, D.B. (2006). E6AP and calmodulin reciprocally regulate estrogen receptor stability. J. Biol. Chem. *281*, 1978–1985.

Li, T.T., Zhao, L.N., Liu, Z.G., Han, Y., and Fan, D.M. (2005). Regulation of apoptosis by the papillomavirus E6 oncogene. World J. Gastroenterol. *11*, 931–937.

Li, X., and Coffino, P. (1996). High-risk human papillomavirus E6 protein has two distinct binding sites within p53, of which only one determines degradation. J. Virol. *70*, 4509–4516.

Linzer, D.I., and Levine, A.J. (1979). Characterization of a 54K dalton cellular SV40 tumour antigen present in SV40-transformed cells and uninfected embryonal carcinoma cells. Cell *17*, 43–52.

Lipari, F., McGibbon, G.A., Wardrop, E., and Cordingley, M.G. (2001). Purification and biophysical characterization of a minimal functional domain and of an N-terminal $Zn2^+$-binding fragment from the human papillomavirus type 16 E6 protein. Biochemistry *40*, 1196–1204.

Liu, X., Clements, A., Zhao, K., and Marmorstein, R. (2006). Structure of the human papillomavirus E7 oncoprotein and its mechanism for inactivation of the retinoblastoma tumour suppressor. J. Biol. Chem. *281*, 578–586.

Liu, Y., and Baleja, J.D. (2008). Structure and function of the papillomavirus E6 protein and its interacting proteins. Front. Biosci. *13*, 121–134.

Liu, Y., Chen, J.J., Gao, Q., Dalal, S., Hong, Y., Mansur, C.P., Band, V., and Androphy, E.J. (1999). Multiple functions of human papillomavirus type 16 E6 contribute to the immortalization of mammary epithelial cells. J. Virol. *73*, 7297–7307.

Liu, Y., Liu, Z., Androphy, E., Chen, J., and Baleja, J.D. (2004). Design and characterization of helical peptides that inhibit the E6 protein of papillomavirus. Biochemistry *43*, 7421–7431.

Lonard, D.M., Nawaz, Z., Smith, C.L., and O'Malley, B.W. (2000). The 26S proteasome is required for estrogen receptor-alpha and coactivator turnover and for efficient estrogen receptor-alpha transactivation. Mol. Cell *5*, 939–948.

Ma, H., Kunes, S., Schatz, P.J., and Botstein, D. (1987). Plasmid construction by homologous recombination in yeast. Gene *58*, 201–216.

Maliekal, T.T., Bajaj, J., Giri, V., Subramanyam, D., and Krishna, S. (2008). The role of Notch signalling in human cervical cancer: implications for solid tumors. Oncogene *27*, 5110–5114.

Mani, A., Oh, A.S., Bowden, E.T., Lahusen, T., Lorick, K.L., Weissman, A.M., Schlegel, R., Wellstein, A., and Riegel, A.T. (2006). E6AP mediates regulated proteasomal degradation of the nuclear receptor coactivator amplified in breast cancer 1 in immortalized cells. Cancer Res. *66*, 8680–8686.

Massimi, P., Pim, D., Kuhne, C., and Banks, L. (2001). Regulation of the human papillomavirus oncoproteins by differential phosphorylation. Mol. Cell. Biochem. *227*, 137–144.

Massimi, P., Gammoh, N., Thomas, M., and Banks, L. (2004). HPV E6 specifically targets different cellular pools of its PDZ domain-containing tumour suppressor substrates for proteasome-mediated degradation. Oncogene *23*, 8033–8039.

Massimi, P., Narayan, N., Cuenda, A., and Banks, L. (2006). Phosphorylation of the discs large tumour suppressor protein controls its membrane localisation and enhances its susceptibility to HPV E6-induced degradation. Oncogene *25*, 4276–4285.

Massimi, P., Shai, A., Lambert, P., and Banks, L. (2008). HPV E6 degradation of p53 and PDZ containing substrates in an E6AP null background. Oncogene *27*, 1800–1804.

Matlashewski, G., Banks, L., Pim, D., and Crawford, L. (1986). Analysis of human p53 proteins and mRNA levels in normal and transformed cells. Eur. J. Biochem. *154*, 665–672.

Matsuura, T., Sutcliffe, J.S., Fang, P., Galjaard, R.J., Jiang, Y.H., Benton, C.S., Rommens, J.M., and Beaudet, A.L. (1997). *De novo* truncating mutations in E6-AP ubiquitin-protein ligase gene (UBE3A) in Angelman syndrome. Nat. Genet. *15*, 74–77.

McLaughlin-Drubin, M.E., and Munger, K. (2009). The human papillomavirus E7 oncoprotein. Virology *384*, 335–344.

McMullan, R., Lax, S., Robertson, V.H., Radford, D.J., Broad, S., Watt, F.M., Rowles, A., Croft, D.R., Olson, M.F., and Hotchin, N.A. (2003). Keratinocyte differentiation is regulated by the Rho and ROCK signalling pathway. Curr. Biol. *13*, 2185–2189.

Mendoza-Villanueva, D., Diaz-Chavez, J., Uribe-Figueroa, L., Rangel-Escareao, C., Hidalgo-Miranda, A., March-Mifsut, S., Jimenez-Sanchez, G., Lambert, P., and Gariglio, P. (2008). Gene expression profile of cervical and skin tissues from human papillomavirus type 16 E6 transgenic mice. BMC Cancer *8*, 347.

Meng, G., Zhao, Y., Nag, A., Zeng, M., Dimri, G., Gao, Q., Wazer, D.E., Kumar, R., Band, H., and Band, V. (2004). Human ADA3 binds to estrogen receptor (ER) and functions as a coactivator for ER-mediated transactivation. J. Biol. Chem. *279*, 54230–54240.

Meyers, C., Harry, J., Lin, Y., and Wettstein, F. (1992). Identification of three transforming proteins encoded by cottontail rabbit papillomavirus. J. Virol. *66*, 165–164.

Moody, C.A., and Laimins, L.A. (2010). Human papillomavirus oncoproteins: pathways to transformation. Nat. Rev. Cancer *10*, 550–560.

Muench, P., Probst, S., Schuetz, J., Leiprecht, N., Busch, M., Wesselborg, S., Stubenrauch, F., and Iftner, T. (2010). Cutaneous papillomavirus E6 proteins must interact with p300 and block p53-mediated apoptosis for cellular immortalization and tumorigenesis. Cancer Res. *70*, 6913–6924.

Munger, K., Phelps, W.C., Bubb, V., Howley, P.M., and Schlegel, R. (1989a). The E6 and E7 genes of the human papillomavirus type 16 together are necessary and sufficient for transformation of primary human keratinocytes. J. Virol. *63*, 4417–4421.

Munger, K., Werness, B.A., Dyson, N., Phelps, W.C., Harlow, E., and Howley, P.M. (1989b). Complex formation of human papillomavirus E7 proteins with the retinoblastoma tumor suppressor gene product. Embo J. *8*, 4099–4105.

Nag, A., Germaniuk-Kurowska, A., Dimri, M., Sassack, M.A., Gurumurthy, C.B., Gao, Q., Dimri, G., Band, H., and Band, V. (2007). An essential role of human Ada3 in p53 acetylation. J. Biol. Chem. *282*, 8812–8820.

Nakagawa, S., and Huibregtse, J.M. (2000). Human scribble (Vartul) is targeted for ubiquitin-mediated degradation by the high-risk papillomavirus E6 proteins and the E6AP ubiquitin-protein ligase. Mol. Cell Biol. *20*, 8244–8253.

Narayan, N., Subbaiah, V.K., and Banks, L. (2009). The high-risk HPV E6 oncoprotein preferentially targets phosphorylated nuclear forms of hDlg. Virology *387*, 1–4.

Narisawa-Saito, M., and Kiyono, T. (2007). Basic mechanisms of high-risk human papillomavirus-induced carcinogenesis: roles of E6 and E7 proteins. Cancer Sci. *98*, 1505–1511.

Nathan, J.A., and Lehner, P.J. (2009). The trafficking and regulation of membrane receptors by the RING-CH ubiquitin E3 ligases. Exp. Cell Res. *315*, 1593–1600.

Nawaz, Z., Lonard, D.M., Smith, C.L., Lev-Lehman, E., Tsai, S.Y., Tsai, M.J., and O'Malley, B.W. (1999). The Angelman syndrome-associated protein, E6-AP, is a coactivator for the nuclear hormone receptor superfamily. Mol. Cell Biol. *19*, 1182–1189.

Nguyen, M.L., Nguyen, M.M., Lee, D., Griep, A.E., and Lambert, P.F. (2003a). The PDZ ligand domain of the human papillomavirus type 16 E6 protein is required for E6's induction of epithelial hyperplasia *in vivo*. J. Virol. *77*, 6957–6964.

Nguyen, M.M., Nguyen, M.L., Caruana, G., Bernstein, A., Lambert, P.F., and Griep, A.E. (2003b). Requirement of PDZ-containing proteins for cell cycle regulation and differentiation in the mouse lens epithelium. Mol. Cell Biol. *23*, 8970–8981.

Nicolas, M., Wolfer, A., Raj, K., Kummer, J.A., Mill, P., van Noort, M., Hui, C.C., Clevers, H., Dotto, G.P., and Radtke, F. (2003). Notch1 functions as a tumour suppressor in mouse skin. Nat. Genet. *33*, 416–421.

Nomine, Y., Masson, M., Charbonnier, S., Zanier, K., Ristriani, T., Deryckere, F., Sibler, A.P., Desplancq, D., Atkinson, R.A., Weiss, E., *et al.* (2006). Structural and functional analysis of E6 oncoprotein: insights in the molecular pathways of human papillomavirus-mediated pathogenesis. Mol. Cell *21*, 665–678.

Nomine, Y., Ristriani, T., Laurent, C., Lefevre, J.F., Weiss, E., and Trave, G. (2001). A strategy for optimizing the monodispersity of fusion proteins: application to purification of recombinant HPV E6 oncoprotein. Protein Eng. *14*, 297–305.

Nuber, U., Schwarz, S., Kaiser, P., Schneider, R., and Scheffner, M. (1996). Cloning of human ubiquitin-conjugating enzymes UbcH6 and UbcH7 (E2-F1) and characterization of their interaction with E6-AP and RSP5. J. Biol. Chem. *271*, 2795–2800.

Nuber, U., Schwarz, S.E., and Scheffner, M. (1998). The ubiquitin-protein ligase E6-associated protein (E6-AP) serves as its own substrate. Eur. J. Biochem. *254*, 643–649.

Parada, L.F., Land, H., Weinberg, R.A., Wolf, D., and Rotter, V. (1984). Cooperation between gene encoding p53 tumour antigen and ras in cellular transformation. Nature *312*, 649–651.

Park, R.B., and Androphy, E.J. (2002). Genetic analysis of high-risk e6 in episomal maintenance of human papillomavirus genomes in primary human keratinocytes. J. Virol. *76*, 11359–11364.

Patel, D., Huang, S.M., Baglia, L.A., and McCance, D.J. (1999). The E6 protein of human papillomavirus type 16 binds to and inhibits co-activation by CBP and p300. Embo J. *18*, 5061–5072.

Perretti, M., and D'Acquisto, F. (2009). Annexin A1 and glucocorticoids as effectors of the resolution of inflammation. Nat. Rev. Immunol. *9*, 62–70.

Pim, D., Massimi, P., and Banks, L. (1997). Alternatively spliced HPV-18 E6* protein inhibits E6 mediated degradation of p53 and suppresses transformed cell growth. Oncogene *15*, 257–264.

Pim, D., Tomaic, V., and Banks, L. (2009). The human papillomavirus (HPV) E6* proteins from high-risk, mucosal HPVs can direct degradation of cellular proteins in the absence of full-length E6 protein. J. Virol. *83*, 9863–9874.

Proweller, A., Tu, L., Lepore, J.J., Cheng, L., Lu, M.M., Seykora, J., Millar, S.E., Pear, W.S., and Parmacek, M.S. (2006). Impaired notch signalling promotes *de novo* squamous cell carcinoma formation. Cancer Res. *66*, 7438–7444.

Ramamoorthy, S., and Nawaz, Z. (2008). E6-associated protein (E6-AP) is a dual function coactivator of steroid hormone receptors. Nucl. Recept. Sig. *6*, e006.

Ramamoorthy, S., Dhananjayan, S.C., Demayo, F.J., and Nawaz, Z. (2010). Isoform-specific degradation of PR-B by E6-AP is critical for normal mammary gland development. Mol. Endocrinol. *24*, 2099–2113.

Rangarajan, A., Syal, R., Selvarajah, S., Chakrabarti, O., Sarin, A., and Krishna, S. (2001a). Activated Notch1 signalling cooperates with papillomavirus oncogenes in transformation and generates resistance to apoptosis on matrix withdrawal through PKB/Akt. Virology *286*, 23–30.

Rangarajan, A., Talora, C., Okuyama, R., Nicolas, M., Mammucari, C., Oh, H., Aster, J.C., Krishna, S., Metzger, D., Chambon, P., *et al.* (2001b). Notch signalling is a direct determinant of keratinocyte growth arrest and entry into differentiation. Embo J. *20*, 3427–3436.

Reid, G., Hubner, M.R., Metivier, R., Brand, H., Denger, S., Manu, D., Beaudouin, J., Ellenberg, J., and Gannon, F. (2003). Cyclic, proteasome-mediated turnover of unliganded and liganded ERalpha on responsive promoters is an integral feature of estrogen signalling. Mol. Cell *11*, 695–707.

Reiter, L.T., Seagroves, T.N., Bowers, M., and Bier, E. (2006). Expression of the Rho-GEF Pbl/ECT2 is regulated by the UBE3A E3 ubiquitin ligase. Hum. Mol. Genet. *15*, 2825–2835.

Ristriani, T., Fournane, S., Orfanoudakis, G., Trave, G., and Masson, M. (2009). A single-codon mutation converts HPV16 E6 oncoprotein into a potential tumour suppressor, which induces p53-dependent senescence of HPV-positive HeLa cervical cancer cells. Oncogene *28*, 762–772.

Rotin, D., and Kumar, S. (2009). Physiological functions of the HECT family of ubiquitin ligases. Nat. Rev. Mol. Cell. Biol. *10*, 398–409.

Sarnow, P., Ho, Y.S., Williams, J., and Levine, A.J. (1982). Adenovirus E1b-58kd tumour antigen and SV40 large tumour antigen are physically associated with the same 54 kd cellular protein in transformed cells. Cell *28*, 387–394.

Savkur, R.S., and Burris, T.P. (2004). The coactivator LXXLL nuclear receptor recognition motif. J. Pept. Res. *63*, 207–212.

Scheffner, M., Werness, B.A., Huibregtse, J.M., Levine, A.J., and Howley, P.M. (1990). The E6 oncoprotein encoded by human papillomavirus types 16 and 18 promotes the degradation of p53. Cell *63*, 1129–1136.

Scheffner, M., Munger, K., Byrne, J.C., and Howley, P.M. (1991). The state of the p53 and retinoblastoma genes in human cervical carcinoma cell lines. Proc. Natl. Acad. Sci. U.S.A. *88*, 5523–5527.

Scheffner, M., Takahashi, T., Huibregtse, J.M., Minna, J.D., and Howley, P.M. (1992). Interaction of the human papillomavirus type 16 E6 oncoprotein with wild-type and mutant human p53 proteins. J. Virol. *66*, 5100–5105.

Scheffner, M., Huibregtse, J.M., Vierstra, R.D., and Howley, P.M. (1993). The HPV16 E6 and E6–AP complex functions as a ubiquitin-protein ligase in the ubiquitination of p53. Cell *75*, 495–505.

Scheffner, M., Huibregtse, J.M., and Howley, P.M. (1994). Identification of a human ubiquitin-conjugating enzyme that mediates the E6-AP-dependent ubiquitination of p53. Proc. Natl. Acad. Sci. U.S.A. *91*, 8797–8801.

Schiller, J.T., Vass, W.C., and Lowy, D.R. (1984). Identification of a second transforming region in bovine papillomavirus. Proc. Natl. Acad. Sci. U.S.A. *81*, 7880–7884.

Sekaric, P., Shamanin, V.A., Luo, J., and Androphy, E.J. (2007). hAda3 regulates p14ARF-induced p53 acetylation and senescence. Oncogene *26*, 6261–6268.

Shah, S.D., Doorbar, J., and Goldstein, R.A. (2010). Analysis of host-parasite incongruence in papillomavirus evolution using importance sampling. Mol. Biol. Evol. *27*, 1301–1314.

Shai, A., Nguyen, M.L., Wagstaff, J., Jiang, Y.H., and Lambert, P.F. (2007). HPV16 E6 confers p53-dependent and p53-independent phenotypes in the epidermis of mice deficient for E6AP. Oncogene *26*, 3321–3328.

Shai, A., Pitot, H.C., and Lambert, P.F. (2010). E6-associated protein is required for human papillomavirus type 16 E6 to cause cervical cancer in mice. Cancer Res. *70*, 5064–5073.

Shamanin, V.A., Sekaric, P., and Androphy, E.J. (2008). hAda3 degradation by papillomavirus type 16 E6 correlates with abrogation of the p14ARF-p53 pathway and efficient immortalization of human mammary epithelial cells. J. Virol. *82*, 3912–3920.

Sherman, L., and Schlegel, R. (1996). Serum- and calcium-induced differentiation of human keratinocytes is inhibited by the E6 oncoprotein of human papillomavirus type 16. J. Virol. *70*, 3269–3279.

Shimoji, T., Murakami, K., Sugiyama, Y., Matsuda, M., Inubushi, S., Nasu, J., Shirakura, M., Suzuki, T., Wakita, T., Kishino, T., *et al.* (2009). Identification of annexin A1 as a novel substrate for E6AP-mediated ubiquitylation. J. Cell. Biochem. *106*, 1123–1135.

Simmonds, M., and Storey, A. (2008). Identification of the regions of the HPV-5 E6 protein involved in Bak degradation and inhibition of apoptosis. Int. J. Cancer *123*, 2260–2266.

Smith, J.A., White, E.A., Sowa, M.E., Powell, M.L., Ottinger, M., Harper, J.W., and Howley, P.M. (2010). Genome-wide siRNA screen identifies SMCX, EP400, and Brd4 as E2-dependent regulators of human papillomavirus oncogene expression. Proc. Natl. Acad. Sci. U.S.A. *107*, 3752–3757.

Spanos, W.C., Geiger, J., Anderson, M.E., Harris, G.F., Bossler, A.D., Smith, R.B., Klingelhutz, A.J., and Lee, J.H. (2008a). Deletion of the PDZ motif of HPV16 E6 preventing immortalization and anchorage-independent growth in human tonsil epithelial cells. Head Neck *30*, 139–147.

Spanos, W.C., Hoover, A., Harris, G.F., Wu, S., Strand, G.L., Anderson, M.E., Klingelhutz, A.J., Hendriks, W., Bossler, A.D., and Lee, J.H. (2008b). The PDZ binding motif of human papillomavirus type 16 E6 induces PTPN13 loss, which allows anchorage-independent growth and synergizes with ras for invasive growth. J. Virol. *82*, 2493–2500.

Sterlinko Grm, H., Weber, M., Elston, R., McIntosh, P., Griffin, H., Banks, L., and Doorbar, J. (2004). Inhibition of E6-induced degradation of its cellular substrates by novel blocking peptides. J. Mol. Biol. *335*, 971–985.

Storrs, C.H., and Silverstein, S.J. (2007). PATJ, a tight junction-associated PDZ protein, is a novel degradation target of high-risk human papillomavirus E6 and the alternatively spliced isoform 18 E6. J. Virol. *81*, 4080–4090.

Stubenrauch, F., and Laimins, L.A. (1999). Human papillomavirus life cycle: active and latent phases. Semin. Cancer Biol. *9*, 379–386.

Tachezy, R., Rector, A., Havelkova, M., Wollants, E., Fiten, P., Opdenakker, G., Jenson, B., Sundberg, J., and Van Ranst, M. (2002). Avian papillomaviruses: the parrot *Psittacus erithacus* papillomavirus (PePV) genome has a unique organization of the early protein region and is phylogenetically related to the chaffinch papillomavirus. BMC Microbiol. *2*, 19.

Talora, C., Cialfi, S., Segatto, O., Morrone, S., Kim Choi, J., Frati, L., Paolo Dotto, G., Gulino, A., and Screpanti, I. (2005). Constitutively active Notch1 induces growth arrest of HPV-positive cervical cancer cells via separate signalling pathways. Exp. Cell Res. *305*, 343–354.

Tanigaki, K., and Honjo, T. (2010). Two opposing roles of RBP-J in Notch signalling. Curr. Top. Dev. Biol. *92*, 231–252.

Terai, M., DeSalle, R., and Burk, R.D. (2002). Lack of canonical E6 and E7 open reading frames in bird papillomaviruses: *Fringilla coelebs* papillomavirus and *Psittacus erithacus* timneh papillomavirus. J. Virol. *76*, 10020–10023.

Thierry, F., and Yaniv, M. (1987). The BPV1-E2 trans-acting protein can be either an activator or a repressor of the HPV18 regulatory region. EMBO J. *6*, 3391–3397.

Thomas, M., and Banks, L. (1998). Inhibition of Bak-induced apoptosis by HPV-18 E6. Oncogene *17*, 2943–2954.

Thomas, M., Laura, R., Hepner, K., Guccione, E., Sawyers, C., Lasky, L., and Banks, L. (2002). Oncogenic human papillomavirus E6 proteins target the MAGI-2 and MAGI-3 proteins for degradation. Oncogene *21*, 5088–5096.

Thomas, M., Narayan, N., Pim, D., Tomaic, V., Massimi, P., Nagasaka, K., Kranjec, C., Gammoh, N., and Banks, L. (2008). Human papillomaviruses, cervical cancer and cell polarity. Oncogene *27*, 7018–7030.

Thomas, M.C., and Chiang, C.M. (2005). E6 oncoprotein represses p53-dependent gene activation via inhibition of protein acetylation independently of inducing p53 degradation. Mol. Cell *17*, 251–264.

Tomaic, V., Gardiol, D., Massimi, P., Ozbun, M., Myers, M., and Banks, L. (2009a). Human and primate tumour viruses use PDZ binding as an evolutionarily conserved mechanism of targeting cell polarity regulators. Oncogene *28*, 1–8.

Tomaic, V., Pim, D., and Banks, L. (2009b). The stability of the human papillomavirus E6 oncoprotein is E6AP dependent. Virology 393, 7–10.

Tong, X., and Howley, P.M. (1997). The bovine papillomavirus E6 oncoprotein interacts with paxillin and disrupts the actin cytoskeleton. Proc. Natl. Acad. Sci. U.S.A. *94*, 4412–4417.

Tong, X., Boll, W., Kirchhausen, T., and Howley, P.M. (1998). Interaction of the bovine papillomavirus E6 protein with the clathrin adaptor complex AP-1. J. Virol. 72, 476–482.

Topffer, S., Muller-Schiffmann, A., Matentzoglu, K., Scheffner, M., and Steger, G. (2007). Protein tyrosine phosphatase H1 is a target of the E6 oncoprotein of high-risk genital human papillomaviruses. J. Gen. Virol. *88*, 2956–2965.

Torres-Arzayus, M.I., Font de Mora, J., Yuan, J., Vazquez, F., Bronson, R., Rue, M., Sellers, W.R., and Brown, M. (2004). High tumour incidence and activation of the PI3K/AKT pathway in transgenic mice define AIB1 as an oncogene. Cancer Cell. *6*, 263–274.

Tungteakkhun, S.S., and Duerksen-Hughes, P.J. (2008). Cellular binding partners of the human papillomavirus E6 protein. Arch. Virol. *153*, 397–408.

Underbrink, M.P., Howie, H.L., Bedard, K.M., Koop, J.I., and Galloway, D.A. (2008). E6 proteins from multiple human betapapillomavirus types degrade Bak and protect keratinocytes from apoptosis after UVB irradiation. J. Virol. *82*, 10408–10417.

Van Bressem, M.F., Cassonnet, P., Rector, A., Desaintes, C., Van Waerebeek, K., Alfaro-Shigueto, J., Van Ranst, M., and Orth, G. (2007). Genital warts in Burmeister's porpoises: characterization of *Phocoena spinipinnis* papillomavirus type 1 (PsPV-1) and evidence for a second, distantly related PsPV. J. Gen. Virol. *88*, 1928–1933.

Van Doorslaer, K., Sidi, A.O., Zanier, K., Rybin, V., Deryckere, F., Rector, A., Burk, R.D., Lienau, E.K., van Ranst, M., and Trave, G. (2009). Identification of unusual E6 and E7 proteins within avian papillomaviruses: cellular localization, biophysical characterization, and phylogenetic analysis. J. Virol. *83*, 8759–8770.

Vande Pol, S.B., Brown, M.C., and Turner, C.E. (1998). Association of Bovine Papillomavirus Type 1 E6 oncoprotein with the focal adhesion protein paxillin through a conserved protein interaction motif. Oncogene *16*, 43–52.

Vos, R.M., Altreuter, J., White, E.A., and Howley, P.M. (2009). The ubiquitin-specific peptidase USP15 regulates human papillomavirus type 16 E6 protein stability. J. Virol. *83*, 8885–8892.

Wade, R., Brimer, N., and Vande Pol, S. (2008). Transformation by bovine papillomavirus type 1 E6 requires paxillin. J. Virol. *82*, 5962–5966.

Wang, M., and Pickart, C.M. (2005). Different HECT domain ubiquitin ligases employ distinct mechanisms of polyubiquitin chain synthesis. EMBO J. *24*, 4324–4333.

Watson, R.A., Thomas, M., Banks, L., and Roberts, S. (2003). Activity of the human papillomavirus E6 PDZ-binding motif correlates with an enhanced morphological transformation of immortalized human keratinocytes. J. Cell Sci. *116*, 4925–4934.

Watt, F.M., Estrach, S., and Ambler, C.A. (2008). Epidermal Notch signalling: differentiation, cancer and adhesion. Curr. Opin. Cell. Biol. *20*, 171–179.

Werness, B.A., Levine, A.J., and Howley, P.M. (1990). Association of human papillomavirus types 16 and 18 E6 proteins with p53. Science *248*, 76–79.

Whyte, P., Buchkovich, K.J., Horowitz, J.M., Friend, S.H., Raybuck, M., Weinberg, R.A., and Harlow, E. (1988). Association between an oncogene and an anti-oncogene: the adenovirus E1A proteins bind to the retinoblastoma gene product. Nature *334*, 124–129.

Wise-Draper, T.M., and Wells, S.I. (2008). Papillomavirus E6 and E7 proteins and their cellular targets. Front. Biosci. *13*, 1003–1017.

de Villiers, E.M., Fauquet, C., Broker, T.R., Bernard, H.U., and zur Hausen, H. (2004). Classification of papillomaviruses. Virology *324*, 17–27.

Yang, Y., Liu, W., Zou, W., Wang, H., Zong, H., Jiang, J., Wang, Y., and Gu, J. (2007). Ubiquitin-dependent proteolysis of trihydrophobin 1 (TH1) by the human papilloma virus E6-associated protein (E6-AP). J. Cell. Biochem. *101*, 167–180.

Zanier, K., Nomine, Y., Charbonnier, S., Ruhlmann, C., Schultz, P., Schweizer, J., and Trave, G. (2007). Formation of well-defined soluble aggregates upon fusion to MBP is a generic property of E6 proteins from various human papillomavirus species. Protein Expr. Purif. *51*, 59–70.

Zhang, B., Chen, W., and Roman, A. (2006). The E7 proteins of low- and high-risk human papillomaviruses share the ability to target the pRB family member p130 for degradation. Proc. Natl. Acad. Sci. U.S.A. *103*, 437–442.

Zimmermann, H., Degenkolbe, R., Bernard, H.U., and O'Connor, M.J. (1999). The human papillomavirus type 16 E6 oncoprotein can down-regulate p53 activity by targeting the transcriptional coactivator CBP/p300. J. Virol. *73*, 6209–6219.

Zimmermann, H., Koh, C.H., Degenkolbe, R., O'Connor, M.J., Muller, A., Steger, G., Chen, J.J., Lui, Y., Androphy, E., and Bernard, H.U. (2000). Interaction with CBP/p300 enables the bovine papillomavirus type 1 E6 oncoprotein to down-regulate CBP/p300-mediated transactivation by p53. J. Gen. Virol. *81*, 2617–2623.

Biochemical and Structure–Function Analyses of the HPV E7 Oncoprotein

Leonardo G. Alonso, Lucía B. Chemes, María L. Cerutti, Karina I. Dantur and Gonzalo de Prat-Gay

Abstract

The human papillomavirus E7 oncoprotein is the main transforming agent of this important pathogen. Although its primary action is binding to and inactivating the retinoblastoma tumour suppressor protein, over two decades of research has shown a much more complex mode of action where multiple cellular partners and cellular events take place before ultimate progression to carcinogenesis. In this chapter we describe the HPV16 E7 protein in biochemical terms in an attempt to understand some of the various interactions in which this protein participates. We describe its multiple equilibria and conformational species in solution, show that these can explain some of its puzzling promiscuous binding activities and we review the few interactions that have been addressed by biochemical and mechanistic approaches to date. We finally discuss the cellular localization of E7 conformers, how they influence its antigenic capacity, and how they can be exploited in therapeutic applications.

Introduction

Small DNA tumour viruses, as stated within their definition, have small genomes coding for a handful of protein products. Thus, they rely on few proteins in order to replicate and complete their life cycles. Since these viruses carry information for very few enzymatic activities, the main one being DNA replication, viral amplification requires that they make use of the host cell machinery, for which they must force the cells into S phase. These viruses use, in order to do so, a combination of biochemical activities, mainly protein-protein and protein–DNA interactions, with the exception of the DNA helicase activity. These activities are distributed in different protein products depending on the virus and, for some of the detrimental effects inflicted on the host cell proteins, these viruses also rely on alternatively spliced products.

Most of the aspects related to these important viruses' life cycle are reviewed by experts throughout the chapters of this book and, not surprisingly, the larger body of information, although not the first chronologically, comes from HPV since the discovery of its implication in cervical and related cancers of high impact on human health (Schwarz *et al.*, 1985; zur Hausen, 1994, 2009). The oncogenic activity of HPV is concentrated on the E6 and E7 oncoproteins and, to a minor extent, on the enigmatic E5 protein. E6 and E7 have been studied not only since the HPV-cancer link era, but also as model instruments for understanding the function of two key cell cycle regulators, p53 and the retinoblastoma (Rb) proteins, together with other relevant mechanisms (Moody and Laimins, 2010; Pim and Banks, 2010).

The present chapter focuses on the biochemistry and structure–function analysis of the HPV E7 oncoprotein, the protein that constitutes the major transforming activity of the virus, and information is based largely on studies on the high-risk HPV16 E7 protein. Some key features for understanding the biochemical basis of the biological activity of this protein have been its initial description as an ORF (Chen *et al.*, 1982; Giri *et al.*, 1985; Seedorf *et al.*, 1985; Cole and Danos, 1987), its ability to interact and promote degradation of Rb as the primary event of cell transformation (Dyson *et al.*, 1989; Heck *et al.*, 1992; Phelps *et al.*, 1992; Boyer *et al.*, 1996), its recombinant expression and characterization (Pahel *et al.*, 1993; Alonso *et al.*, 2002), recent structural characterization of its C- and N-terminal domains (Liu *et al.*, 2006; Ohlenschlager *et al.*, 2006; Garcia-Alai *et al.*, 2007a), and interestingly, characterization of its non-globular intrinsically disordered nature (Alonso *et al.*, 2002; Daughdrill, 2005; Garcia-Alai *et al.*, 2007a), a probable explanation of its well-known target promiscuity. The existence of multiple equilibria and conformers and their possible cell functions and localization are also discussed.

E7 functions

The E7 protein is a small polypeptide (~100 aa) that is able to bind to multiple cellular and viral protein targets (McCance, 2006; Wise-Draper and Wells, 2008; McLaughlin-Drubin and Munger, 2009; Moody and Laimins, 2010; Pim and Banks, 2010) (Table 6.1). E7 lacks any known enzymatic activity and, in view of the multiple effects associated to E7 expression, it is generally accepted that the E7 protein mediates its functions through the establishment of a complex network of protein–protein interactions. However, as it will be discussed throughout this chapter, the biochemical properties of E7 make it prone to interact non-specifically with many proteins, making it difficult to define its bona fide targets from the large list of proteins to which E7 has been reported to associate.

As opposed to other proteins whose function is defined, for example by an enzymatic activity, the biological function of E7 is given by its interaction with host proteins and, therefore, it is usually described indirectly through the function of these protein targets. However, phenotypes observed upon E7 expression are complex and not so straightforward to infer from the individual interactions. Next, we discuss those activities that are currently accepted as E7-mediated.

HPV replicates within a differentiating epithelium, and the virus lacks the enzymatic machinery necessary for autonomous DNA synthesis. Thus, induction of S-phase re-entry and uncoupling of keratinocyte replication and differentiation are prerequisites for completion of the virus life cycle, an activity dependent on E7 expression. This activity is mediated mainly through the interaction of E7 with the retinoblastoma tumour suppressor protein (Rb) and with the pocket protein family members p107 and p130 (Genovese *et al.*, 2008; Wise-Draper and Wells, 2008) (Table 6.1). Complex formation between E7 and Rb leads to abrogation of the inhibitory effects of Rb on cell cycle progression and thus, induces S-phase progression through two mechanisms (Helt and Galloway, 2003). Hypo-phosphorylated Rb binds to and inhibits the transcription factor E2F, keeping cells arrested in the G1 phase. The interaction between E7 and Rb leads to dissociation of the Rb–E2F complex, releasing the transcriptionally active E2F and inducing the expression of genes necessary for S-phase progression. On the other hand, E7 induces proteasome-mediated Rb degradation, further contributing to abrogating Rb function (Helt and Galloway, 2003).

In addition to the Rb-mediated effects of E7, other effects of E7 on the cell cycle are Rb independent. E7 targets many others proteins involved in cell cycle regulation (Table 6.1), probably allowing for fine-tuning of the cellular cycle. Downstream of the G1/S checkpoint, E7 can interfere directly with the cyclin-kinase inhibitors p21 and p27, activate cdk2-Cyclin A and E kinase complexes and modulate HDAC activity (see below) in a coordinate manner.

E7 has also been found to associate to several proteins involved in transcriptional regulation, such as the TATA binding protein (TBP) and the TBP-associated factor 110 (TAF-110), the forkhead transcription factor family MPP2 and the basal transcription factor AP-1 (Table 6.1), and to proteins involved in histone modification such as histone acetytransferase p300 and the histone deacetylase (HDAC) recruiting factor Mi-2 (Table 6.1). Finally, it is known that E7 is able to interfere with the IFN-mediated host antiviral response through its interaction with the IRF-1 and p48 proteins (Barnard and McMillan, 1999).

Taking into consideration the complexity of the E7 interaction network and the large number of targets described, biochemical and quantitative

Table 6.1 Reported E7 interaction partners

Protein	Reported effect	References
Cell cycle control		
pRb	Proteasome-induced degradation of Rb Rb–E2F complex dissociation	Dyson *et al.* (1989), Chellappan *et al.* (1992), Boyer *et al.* (1996)
p107, p130	Cell cycle deregulation during epithelial differentiation	Dyson *et al.* (1992), Davies *et al.* (1993), Genovese *et al.* (2008)
p21, p27	Abrogation of p21- and p27-mediated inhibition of cyclin-dependent kinases Cell cycle deregulation during epithelial differentiation	Zerfass-Thome *et al.* (1996), Jones *et al.* (1997)
E2F1	Rb-independent increase in E2F1 transcriptional activity	Hwang *et al.* (2002)
E2F6	Prevent repression by E2F6. S-phase induction	McLaughlin-Drubin *et al.* (2008)
Cyclin A/CDK2 Cyclin E/CDK2	Activation of cyclin/cdk complexes	Tommasino *et al.* (1993), McIntyre *et al.* (1996)
Cullin 1 and 2	Regulation of E7 and Rb turnover	Oh *et al.* (2004), Huh *et al.* (2007)
Chromatin modification		
Mi2β-HDAC	Modulation of histone deacetylase activity Episomal viral genome maintenance	Brehm *et al.* (1998, 1999), Longworth *et al.* (2005)
NuMA-1	Delocalization of dynein from mitotic spindles Disruption of mitotic events	Nguyen and Munger, (2009)
Transcriptional control regulation		
TBP	Modulation of basal transcription machinery	Phillips and Vousden (1997)
TAF-110	Modulation of basal transcription machinery	Mazzarelli *et al.* (1995)
AP-1	Modulation of transcription factor activity	Antinore *et al.* (1996)
p300/CBP	Modulation of transcriptional coactivator and acetyltransferase activity	Bernat *et al.* (2003)
MPP2	Modulation of transcription factor activity	Luscher-Firzlaff *et al.* (1999)
Other functions		
HPV E2	Regulation of HPV E7 protein stability	Gammoh *et al.* (2006, 2009), Smal *et al.* (2009)
p600	Regulation of Anoikis. Contribution to anchorage-independent growth and cellular transformation	Huh *et al.* (2005)
IRF-1	Repression of interferon beta promoter through recruitment of HDAC. Immune evasion	Park *et al.* (2000)
S4 proteasome subunit	Rb-degradation?	Berezutskaya and Bagchi (1997)
M2-PK (pyruvate kinase)	Modulation of enzymatic activity	Zwerschke *et al.* (1999)
Alpha-glucosidase	Modulation of enzymatic activity	Zwerschke *et al.* (2000)

studies of the different functional interactions, *in vitro* and *in vivo* become essential tools towards understanding the hierarchy of protein complexes established within cells and their contribution to different scenarios such as natural infections or pathogenic overexpression.

Modular architecture of HPV E7

Fig. 6.1 shows the E7 protein domain organization. E7 proteins from most papillomavirus types sequenced to date show similar length, ranging from 95 to 105 residues (Bernard *et al.*, 2010) (Fig. 6.1A). As it will be discussed throughout the following sections, both the sequence conservation pattern as well as structural and biochemical evidence indicate that E7 is organized into two modular domains, an architecture that is conserved across papillomaviruses and which includes human papillomavirus E7 proteins from 'high'- and 'low'-risk viral types (Bernard *et al.*, 2010). The N-terminal domain of HPV16 E7 (E7N) spans residues 1–40, while the C-terminal domain (E7C) spans residues 41–98 (Fig. 6.1A). It is worth mentioning that three conserved motifs (CR1, CR2 and CR3) were initially identified on the basis of the sequence homology found between HPV E7 and the DNA tumour virus proteins adenovirus E1A (AdE1A) and simian virus 40 large T antigen (SV40-LT) (Phelps *et al.*, 1988; Barbosa *et al.*, 1990; Ikeda and Nevins, 1993) (Fig. 6.1B). Both CR1 and CR2 are contained within the E7N domain: CR1 spans residues 2 to 15, while CR2 spans residues 16 to 40. Conserved region 3, or CR3, corresponds to the globular E7C domain spanning residues 40 to 98.

The E7 N-terminal domain

E7 is an acidic protein (theoretical PI of 4.6 for HPV16 E7) mainly owing to the high proportion of negatively charged residues present in its E7N domain. As mentioned, this domain contains the conserved regions CR1 and CR2. A closer look at the sequence alignment of the E7N domain reveals that sequence conservation is high across the entire domain, with short stretches of poorly conserved residues separating several highly conserved linear motifs (Fig. 6.2A). These include a helix-forming motif located in the CR1 region that shows high sequence identity with residues of the AdE1A CR1 region (Fig. 6.1B). This region in AdE1A is known to interact with the RbAB domain forming an alpha helix (Liu and Marmorstein, 2007), an interaction required for AdE1A-mediated E2F displacement (Ikeda and Nevins, 1993). Intriguingly, as will be discussed below, the helix-forming motif of E7 also shows a low affinity interaction with RbAB (Chemes

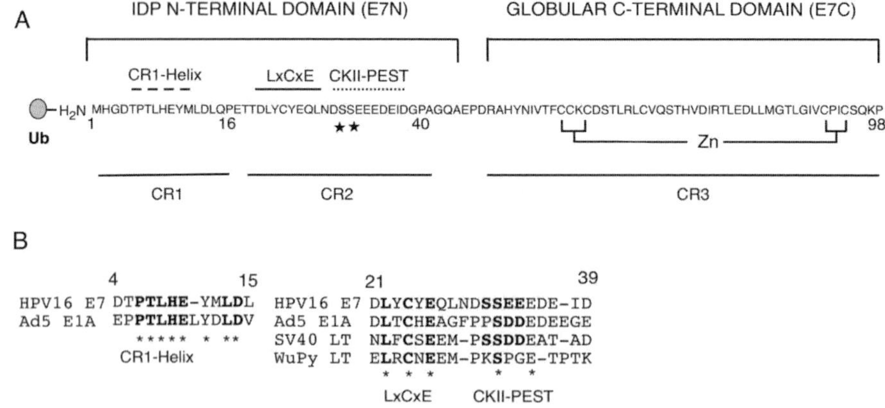

Figure 6.1 Domain organization of the papillomavirus E7 oncoprotein. (A) Schematic representation of the E7 protein and its N- and C-terminal domains. Shown is the sequence of the HPV16 E7 protein, marking the positions of conserved regions 1, 2 and 3 (CR1, CR2, and CR3), the linear motifs within the E7N domain (CR1-Helix, LxCxE and CKII-PEST) and the cysteine residues responsible for Zn^{2+} coordination. Ub: ubiquitin conjugation site. *, phosphorylation site. (B) Sequence conservation among DNA tumour virus proteins in the CR1 (left panel) and CR2 (right panel) regions. Numbers above the sequence correspond to the amino acid position of the HPV16 E7 protein. Ad5 E1A, Adenovirus type 5 E1A protein; SV40-LT, Simian virus 40 large T antigen; WuPy-LT, Wu polyomavirus large T antigen. Strictly conserved positions are marked by an asterisk.

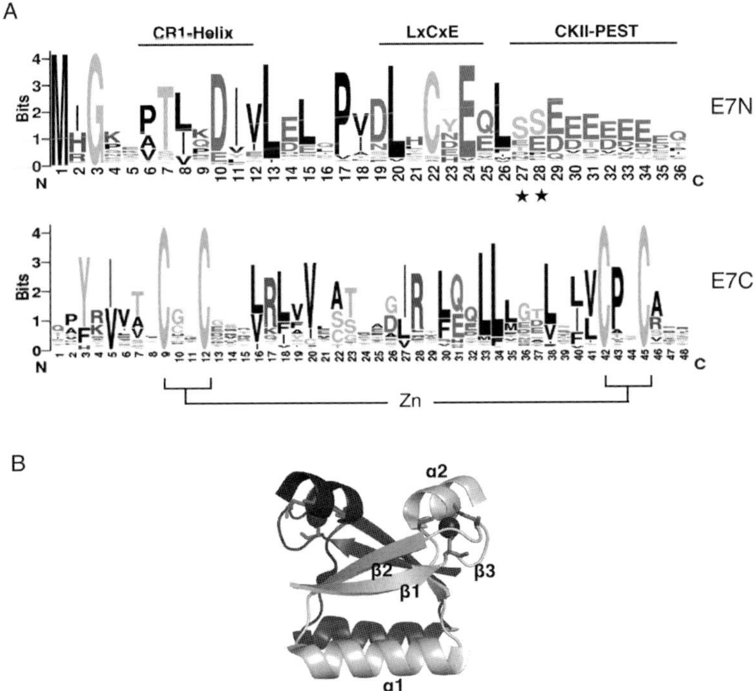

Figure 6.2 Sequence conservation and structure of the E7N and E7C domains. (A) Sequence logos of the E7N (upper panel) and E7C (lower panel) domains obtained from the alignment of 184 (E7N) or 196 (E7C) papillomavirus E7 sequences. The height of the letters in each position is correlated to their conservation. *, phosphoylation site. CR1-Helix:CR1 helix-forming region, LxCxE: LxCxE motif, CKII-PEST: CKII-PEST region. The cysteine residues involved in Zn coordination are marked. (B) Structure of the HPV45 E7C domain (PDB ID: 2F8B). Each monomer forming the dimer is shown in dark and light grey respectively. Grey ball: Zn atom. The four cysteine residues of each monomer involved in Zn coordination are shown in stick representation.

et al., 2010). This motif is also required for the interaction between E7 and several target proteins including p600, p300, IRF-1 and Cullin-2 (Table 6.1). Deletion of this region has been shown to abrogate the transforming activity of E7 from the high-risk HPV16 type(Phelps *et al.*, 1992) and to be required for E7-mediated Rb degradation, highlighting its functional importance.

The CR2 region of E7N contains the high-affinity LxCxE motif, required for the interaction between E7 and Rb, a conserved stretch of acidic residues scoring as a PEST degradation signal (Rechsteiner and Rogers, 1996) and one or two casein kinase II (CKII) phosphorylation sites (S/TxxD/E) (Fig. 6.2A). PEST motifs have been shown to regulate turnover of short-lived cellular proteins (Lin *et al.*, 1996b). We will refer to the conserved acidic region containing the phosphorylation site as the CKII-PEST region. These motifs, located in the CR2 region, are also found in the related DNA tumour virus proteins AdE1A and SV40-LTAg (Fig. 6.1B). Also, an acidic region is found following the LxCxE motif in cellular Rb targets such as HDAC, RBP1 and RIZ, and a CKII phosphorylation site is present in one third of the cellular Rb targets harbouring the LxCxE motif, including HDAC and CtIP. Additionally to the interaction with Rb, several E7–target interactions have been mapped to the CR2 region, including other members of the pocket protein family (p107 and p130), the kinase inhibitor p21^{CIP1}, and TBP among others (Table 6.1). Therefore, the E7N domain contains several linear motifs required for specific protein targeting activities of E7, which mediate different functional effects. As will be discussed in detail below, E7N has been found to be an intrinsically disordered domain of the E7 protein.

The E7 C-terminal domain

Early biochemical studies showed that the E7 protein could exist as a dimer in solution (Clements *et al.*, 2000) and was able to bind one zinc (Zn) atom per protein monomer. The Zn atom is coordinated by four conserved cysteine residues that form a CxxC-29-CxxC Zn binding motif (Barbosa *et al.*, 1989) (Fig. 6.2A). To date, two high-resolution structures of the E7C protein have been solved. One corresponds to a crystal structure from the low-risk HPV1A E7 protein (Liu *et al.*, 2006) and the other corresponds to an NMR structure of the high-risk HPV45 E7C domain (Ohlenschlager *et al.*, 2006). The E7C domain presents a $\beta1\beta2\alpha1\beta3\alpha2$ topology and forms a dimer both under crystallization conditions and in solution (Fig. 6.2B). In E7, Zn coordination is required for protein folding and for dimerization (McIntyre *et al.*, 1993; Alonso *et al.*, 2002). However, the Zn coordination site is proximal but separate from the dimerization interface. It should be noted that both the Zn coordination site and the dimer interface residues are highly conserved throughout E7 evolution, pointing to their functional relevance. Recently, it has been proposed that the fold found in E7 and E6 could be derived from an ancestral treble-cleft fold present in cellular PHD and RING finger domains, and also in LIM domains (de Souza *et al.*, 2009). Interestingly, the proteins containing these domains are involved in interactions with proteins from the chromatin remodelling and ubiquitination machinery, and are able to oligomerize, properties shared by the E7 protein.

Early studies established the functional importance of the E7C domain, showing that its mutation yielded E7 proteins defective in transformation activity (Phelps *et al.*, 1992). Many protein–protein interactions have been mapped to the E7C domain, including those of the cellular kinase inhibitors $p21^{KIP1}$ and $p27^{CIP1}$, the TATA box binding protein (TBP), the S4 subunit of the proteasome and E2F family proteins, among others (Table 6.1). Nevertheless, the structure of the E7C domain reveals that most probably many of the mutations or deletions tested in early studies disrupt the folded structure of this domain, making it difficult to interpret the results in terms of structure–function relationships. Future rational mutagenesis studies targeting residues that do not affect the fold of the E7C domain will be required in order to understand their role in the multiple known E7 interactions. To date, the binding sites for $p21^{CIP1}$ and for the RbC domain have been mapped to an overlapping conserved surface patch of E7C (Liu *et al.*, 2006; Ohlenschlager *et al.*, 2006).

Biochemical properties of the E7 protein in solution

E7 is an extended dimer

Well before the first two structures of the E7C domain were solved, several experimental studies provided evidence that E7 presented an anomalous electrophoretic mobility in denaturing reducing polyacrylamide gel electrophoresis (SDS-PAGE), migrating as a ca. 20 kDa polypeptide despite the fact that its actual molecular weight is 11 kDa (Smotkin and Wettstein, 1987). Accordingly, E7 has an anomalous hydrodynamic behaviour, presenting an elution profile corresponding to a 44 kDa globular protein in size exclusion chromatography (Alonso *et al.*, 2002). Yet, techniques such as multiangle light scattering reveal that this species of E7 has a molecular weight of 22 kDa, indicating that E7 exists in solution as an extended dimer (Alonso *et al.*, 2002) (Fig. 6.3A). Dynamic light scattering measurements reveal a hydrodynamic diameter of 6.8 nm, twice the one expected for a 22 kDa globular dimer (Smal, 2010). The dimer formed by E7 is rather weak, as inferred by sedimentation equilibrium experiments that yield a dissociation constant of about 1 μM (Clements *et al.*, 2000) (Fig. 6.4). The extended HPV16 E7 protein can undergo compaction in mild denaturing conditions, a conformational change that could involve interactions between the E7N and E7C domains (Alonso *et al.*, 2002). Therefore, E7 has a non-globular and extended conformation in solution, and can associate to form a weak dimer at concentrations at or above 1 μM.

The HPV16 E7 dimer is highly thermostable and it was shown to undergo a conformational

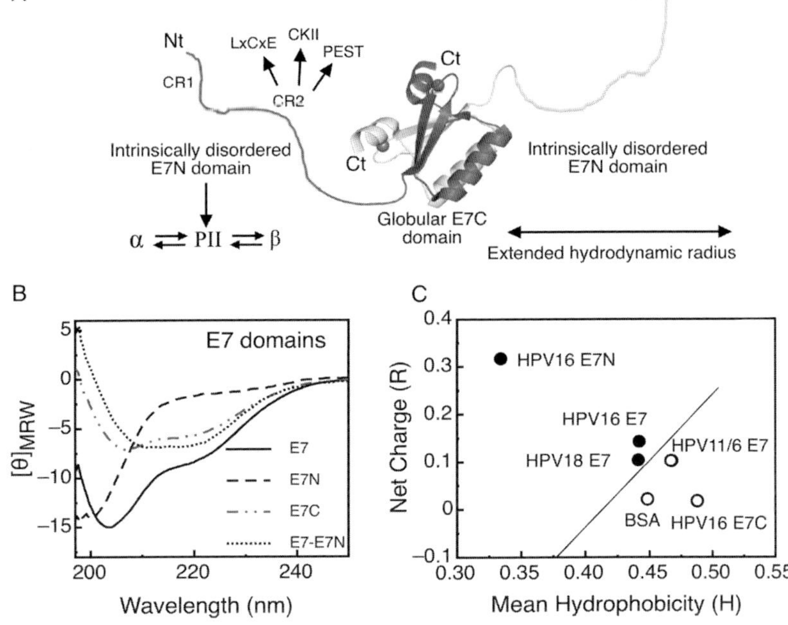

Figure 6.3 Intrinsically disordered nature of HPV16 E7. (A) Schematic representation of the HPV E7 protein. E7 is represented as an extended dimer. Each monomer consists of a globular E7C domain and an intrinsically disordered E7N domain. Conformational equilibria demonstrated for E7N are indicated. CR1, 2: conserved regions 1 and 2. LxCxE, Rb binding motif; CKII, casein kinase II site; PEST, PEST degradation region; Nt, N-terminus, Ct: C-terminus. (B) CD spectrum of the full-length E7 protein and the E7N and E7C domains. The similarity between the E7-E7N difference spectrum and the E7C domain spectrum indicates the modular character of both domains of the E7 protein. (C) Charge-hydrophobicity plot for E7 proteins. The line represents a theoretical prediction obtained from the analysis of a large set of proteins, which separates intrinsically disordered proteins (to the left of the line) from globular proteins (to the right of the line) (Uversky *et al.*, 2000). The graph shows the charge–hydrophobicity ratio of different E7 proteins and domains.

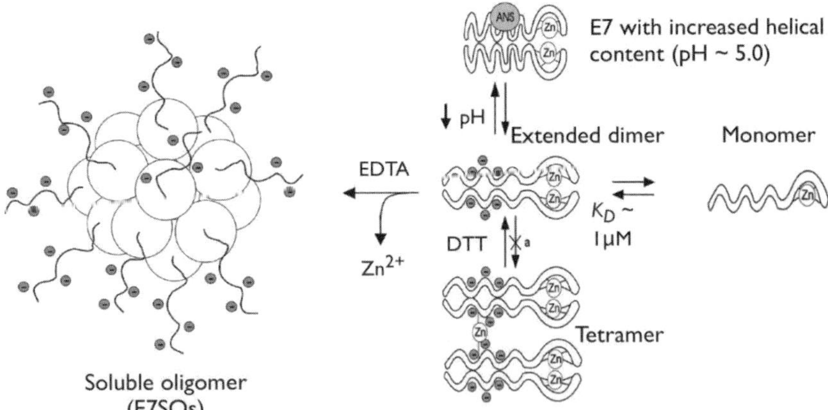

Figure 6.4 E7 conformational equilibria in solution. Schematic representation of conformational transitions of the HPV16 E7 protein. E7 exists as an extended dimer in solution, which can undergo compaction and is in equilibrium with a monomeric species. Excess Zn leads to the formation of tetrameric species, and Zn removal with EDTA leads to the formation of E7 soluble spheric oligomeric species (E7SOs). Adapted from Alonso *et al.*, (2002, 2004).

transition between pH 5.0 and 7.0, with an increase in alpha-helix content and solvent accessibility of hydrophobic regions based on 1-anilino-8-naphtalene-sulphonate (ANS) binding (Fig. 6.4). It was then proposed that these plastic conformational properties could be related to its ability to interact with different targets (Alonso *et al.*, 2002).

Evidence of intrinsic disorder in E7

Recently, it has been recognized that many functional proteins or long protein segments are devoid of stable secondary and/or tertiary structure and exist as a very dynamic ensemble of conformations, as opposed to globular folded proteins (Uversky, 2002). These so-called intrinsically disordered proteins or IDPs are recognized as important players in key regulatory pathways such as cell cycle control (Fink, 2005). In 2006, it was proposed that high-risk E7 and E6 transforming proteins present increased amounts of intrinsic disorder, as judged by disorder-prediction algorithms (Uversky *et al.*, 2006) (Fig. 6.3C). In addition, experimental evidence suggested that E7 has properties resembling those of both IDPs and globular folded proteins. E7 has an extended hydrodynamic radius and a low overall secondary structure content similar to IDPs but, besides, shows tertiary structure and cooperative folding, similar to globular proteins (Alonso *et al.*, 2002; Garcia-Alai *et al.*, 2007a).

This could be explained on the basis of the modular nature of the E7 protein: while the isolated E7C domain is compact (Ohlenschlager *et al.*, 2006) and does not have anomalous hydrodynamic properties (Smal, 2010), the E7N domain is responsible for the extended and non-globular behaviour of the full-length protein (Garcia-Alai *et al.*, 2007a). Spectroscopic studies using circular dichroism (CD) showed that E7N presents a CD spectrum characteristic of a disordered polypeptide, being devoid of stable secondary and tertiary structure (Garcia-Alai *et al.*, 2007a) (Fig. 6.3B). Additionally, it was shown that IDPs have a high mean charge/hydrophobicity ratio (Uversky *et al.*, 2000). Fig. 6.3C shows that both the full-length HPV16 E7 protein and particularly the E7N domain fall within the intrinsically disordered region of a charge-hydrophobicity plot.

Several regions within the IDP E7N domain show dynamic secondary structure, and are able to undergo coil-helix, coil-beta sheet and coil-polyproline type II (PII) conformational transitions upon changes in the chemical environment (Garcia-Alai *et al.*, 2007a) (Fig. 6.3A). Moreover, acidification leads to charge neutralization within E7N and induces compaction of the E7 protein (Alonso *et al.*, 2002) (Fig. 6.4). Within E7N, charge neutralization favours alpha helical structures, whereas at neutral pH where E7N exists as a negatively charged polypeptide, PII conformations are favoured (Garcia-Alai *et al.*, 2007a). Thermodynamic analyses suggest that separate residues are involved in PII structure and in nucleation of the alpha helix. In addition, CD studies of the isolated CR1 and CR2 regions show that while CR1 is able to acquire alpha helix structure upon addition of Trifluoroethanol (TFE), CR2 has an extended conformation and presents PII structure (Chemes *et al.*, 2010). Moreover, detergents such as SDS can induce alpha helix above the critical micellar concentration (CMC) or beta sheet conformations (below the CMC) (Garcia-Alai *et al.*, 2007a), similarly to what is observed in other IDP proteins such as alpha synuclein (Woods *et al.*, 2007).

The E7N domain can finely tune these different conformations depending on the chemical environments *in vitro*, which can mirror the environment encountered by this domain upon interaction with other proteins or with cellular membranes *in vivo*. This high structural plasticity of E7N is similar to the one seen in the protein α-synuclein, where α-helix and β-sheet conformational transitions correlate with specific target protein binding activities acting as a molecular switch (Ferreon *et al.*, 2009a). These properties provide a plausible explanation for the broad protein binding activity of E7 (Chemes, 2011a). The extended nature of E7N, the lack of stable secondary and tertiary structure and its non-random conformational properties classify this domain as an intrinsically disordered domain. Interestingly the AdE1A protein, which shares sequence homology with E7 in both the CR1 and CR2 regions, is also considered as being intrinsically disordered (Pelka *et al.*, 2008). The presence of disorder in viral proteins is thought to endow them with a high functional versatility necessary for interfering with cellular regulatory networks (Tokuriki *et al.*, 2009).

The classical protein domain definition refers to an independent and cooperatively folded unit, which adopts a compact structure, and is not based solely on sequence alignment. The evidence presented shows that HPV16 E7N is a bona fide domain, albeit intrinsically disordered, providing conformational plasticity to the oncoprotein (Garcia-Alai *et al.*, 2007a), and that this domain is responsible for the intrinsically disordered nature of the E7 protein (Uversky *et al.*, 2006).

E7 equilibria in solution

A salient biochemical feature of the E7 protein is it ability to exist in several conformational and oligomerization states (Fig. 6.4). The E7 monomer (11 kDa) is able to self-associate into different oligomeric species including dimers (22 kDa), tetramers (44 kDa) and high-molecular-weight soluble oligomers (E7SOs, >790 kDa) (McIntyre *et al.*, 1993; Chinami *et al.*, 1994; Clemens *et al.*, 1995; Zwerschke *et al.*, 1996; Clements *et al.*, 2000; Alonso *et al.*, 2002, 2004).

The measured monomer–dimer equilibrium dissociation constant of 1 μM obtained by analytical centrifugation (Clements *et al.*, 2000) and chemical unfolding experiments (Alonso *et al.*, 2002) indicate that E7 forms a weak dimer in solution. Although solution experiments provide evidence that a monomer-dimer equilibrium is present, the conformation of the free E7 monomer is currently unknown, and the dissociation reaction could be coupled to partial unfolding of the E7 monomer. A tetrameric form of E7 was observed when refolding of recombinant HPV16 E7 was performed in the presence of excess of Zn and in low reductant conditions (Alonso *et al.*, 2002).

On the other hand, Zn chelation by EDTA of a dimeric E7 sample *in vitro* leads to the irreversible formation of high molecular weight (ca. 790 kDa) spherical oligomers showing a diameter of 50 nm (Chinami *et al.*, 1994; Alonso *et al.*, 2004). The E7C Zn binding domain is responsible for the oligomerization properties of E7 (McIntyre *et al.*, 1993; Clemens *et al.*, 1995; Zwerschke *et al.*, 1996; Alonso *et al.*, 2004). The oligomerization process involves a substantial secondary structure reorganization that yields a highly stable macromolecular assembly. Many dyes that are able to bind to repetitive beta-sheet structures have been shown to bind to the E7SOs species, and this evidence, in addition to its CD spectrum, indicates that the E7SOs are rich in beta-sheet structure.

In the light of the recent structural evidence (Liu *et al.*, 2006; Ohlenschlager *et al.*, 2006), it becomes clear that both the cysteine residues involved in metal coordination and the hydrophobic residues involved in monomer–monomer interactions are fundamental in determining the folding state of the E7 protein and its oligomerization state. This information will hopefully guide future studies aimed at elucidating the key determinants of E7 conformational stability and conformational equilibria *in vitro* and *in vivo*, along with its functional consequences. This recent evidence also puts a note of caution on early functional data obtained from sequence deletion or non-conservative residue replacement within E7C, since many of these deletions/substitutions are likely to disrupt E7 structure altogether, as explained above. Using the two-hybrid system (Clemens *et al.*, 1995), and more recently through the use of conformation-specific antibodies that are able to recognize oligomeric E7 species in cervical cancer cell lines and transfected cells (Dantur *et al.*, 2009), it was shown that HPV16 E7 oligomers exist inside the cell, suggesting that the oligomerized state is functionally relevant.

Structural and functional correlates of E7 phosphorylation

E7 proteins from all human and most other animal papillomavirus types present one to two serine or threonine phosphorylation sites in the CR2 region following the LxCxE motif (Fig. 6.2A). Studies performed on the E7 protein from the high-risk HPV16 and HPV18 types showed that CKII is able to recognize the S/TxxD/E consensus sites and phosphorylate these residues both *in vitro* and *in vivo* (Barbosa *et al.*, 1990). Phosphorylation is required for E7 transforming activity in high-risk HPV types (Firzlaff *et al.*, 1991) and for induction of S-phase re-entry in infected keratinocytes in high- and low-risk HPV types (Banerjee *et al.*, 2006; Genovese *et al.*, 2008). Recently, it was described that E7 phosphorylation has a moderate effect on pocket protein binding affinity (Genovese *et al.*, 2008; Chemes *et al.*, 2010). Additionally, E7 phosphorylation was shown to vary throughout the cell cycle and to increase affinity for the TATA binding protein (Massimi and Banks, 2000). Phosphorylation of the conserved CKII site regulates the transforming potential and Rb-binding affinity of AdE1A (Whalen *et al.*, 1996), and modulates nuclear-cytoplasmic shuttling of SV40-LTAg (Rihs *et al.*, 1991), highlighting the functional importance of this post-translational modification in related DNA tumour virus proteins.

An alternative means of regulating E7 function is through modulation of E7 protein stability. In this regard, it was found that the CKII-PEST region of E7 presents dynamic PII structure, and that phosphorylation increases the PII content of this region (Garcia-Alai *et al.*, 2007a; Chemes *et al.*, 2010). It was previously found that, in BPV E2, the phosphorylation of the 'hinge' interdomain region decreases the PII content of a similar CKII-PEST site, leading to unstable polypeptides that are susceptible to intracellular degradation (Garcia-Alai *et al.*, 2006). Therefore, E7 phosphorylation may act as a conformational switch, regulating E7 protein stability and constituting an additional mechanism for the regulation of E7 function.

Different oligomeric species of E7 show differential susceptibility to CKII phosphorylation. E7SOs cannot be phosphorylated *in vitro*; however, the phosphorylated dimeric E7 can form E7SOs, indicating that the CKII phosphorylation site is inaccessible in the oligomeric structure (Alonso *et al.*, 2004).

Regulation of protein turnover: ubiquitination and degradation of E7 and Rb

The transforming and immortalization potential of the E7 protein strongly depends on the ubiquitin-proteasome pathway (Boyer *et al.*, 1996; Berezutskaya and Bagchi, 1997). At present, a large body of experimental evidence supports the view that the ubiquitin-proteasome pathway regulates two major aspects of E7 biology: First, E7-mediated degradation of the retinoblastoma tumour suppressor protein (Boyer *et al.*, 1996; Berezutskaya and Bagchi, 1997; Gonzalez *et al.*, 2001; Wang *et al.*, 2001) and second the regulation of E7 steady-state levels and protein half-life (Reinstein *et al.*, 2000).

E7-mediated proteasome-dependent degradation of Rb

The first strong evidence demonstrating that the proteasome pathway was involved in E7-mediated Rb degradation came from the observation that HPV16 E7-immortalized mammary epithelial cells (MECs) showed drastically decreased levels of the Rb protein (Boyer *et al.*, 1996) while maintaining normal Rb mRNA levels, indicating that the decrease was due to an enhanced proteolysis. Moreover, the treatment of these HPV E7-immortalized cells with proteasome inhibitors led to the restoration of normal Rb levels (Boyer *et al.*, 1996). It was further demonstrated that HPV16 positive CaSki cervical cancer cells, which naturally express the HPV16 E7 protein, show low or non-detectable Rb protein levels that are restored by treatment with proteasome inhibitors (Aviel *et al.*, 2000; Wang *et al.*, 2001).

Ubiquitin conjugation is necessary, except for some exceptions (Breitschopf *et al.*, 1998; Aviel *et al.*, 2000), for recognition of protein substrates by the 26S proteasome subunit of the proteolytic machinery (Hershko and Ciechanover, 1998). Ubiquitination involves at least three enzymes: E1 ubiquitin-activating enzyme that activates and transfers ubiquitin to E2 ubiquitin-conjugating enzyme. E2 that, in turn, cooperates with E3 ubiquitin ligase to transfer ubiquitin to the substrate. E3 ubiquitin ligases are a group of proteins that provide specificity to the degradation process by recognizing a restricted subset of substrates (Kornitzer and Ciechanover, 2000). Different experimental evidences support the involvement of the ubiquitin conjugation machinery in E7-mediated Rb degradation. First, experiments performed on an HPV16 E7 transfected cell line expressing a thermo-sensitive E1 ubiquitin-activating enzyme showed that E7-mediated degradation of Rb requires E1 and is, therefore, ubiquitin dependent (Boyer *et al.*, 1996). Second, Munger and co-workers (Munger *et al.*, 1991) reported that HPV16 E7 associates with the Cullin-2 ubiquitin ligase (E3) complex and that this interaction contributes to Rb degradation (Huh *et al.*, 2007).

The Rb protein is actively degraded through a process that requires a direct protein–protein interaction between E7 and Rb. Abrogation of the E7–Rb interaction through disruption of the high affinity LxCxE Rb-binding motif using point mutations such as C24S and E26Q (Wang *et al.*, 2001) or partial deletions such as Δ21–24 (Gonzalez *et al.*, 2001) leads to inhibition of E7-mediated Rb degradation (Fig. 6.5A). Additional information obtained with E7 Δ6–10 indicates that the CR1 region, which is necessary for E7-mediated transformation but not for Rb binding, is strictly necessary for proteasomal degradation of Rb (Gonzalez *et al.*, 2001) (Fig. 6.5A). The CR1 region of HPV16 E7 was shown to contribute to the interaction with the Cullin-2 ubiquitin ligase complex, as judged by experiments using the Δ6–10 and H2P mutants (Huh *et al.*, 2007). It should be mentioned that binding of the related DNA tumour virus proteins E1A and SV40-LTAg to Rb does not lead to Rb degradation (Helt and Galloway, 2003), indicating that Rb binding is necessary but not sufficient for inducing Rb degradation.

It has been proposed that E7-induced Rb proteasomal degradation involves a two step process in which the E7 protein first binds to full-length Rb (residues 1 to 928) and activates the calcium-dependent cysteine protease calpain, which in turn cleaves Rb at position 810 yielding the

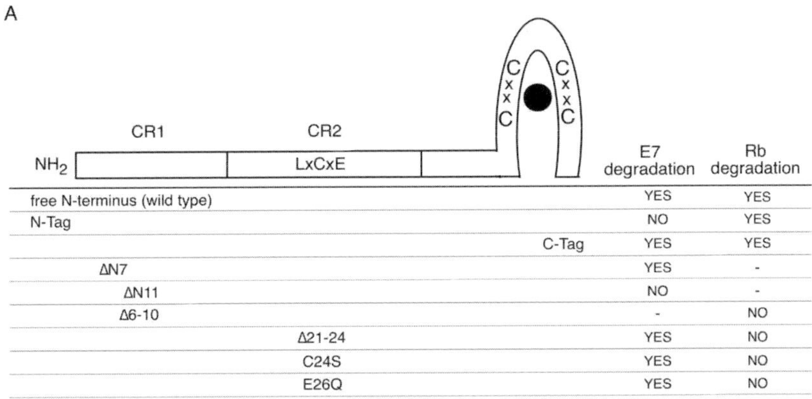

	E7 degradation	Rb degradation
free N-terminus (wild type)	YES	YES
N-Tag	NO	YES
C-Tag	YES	YES
ΔN7	YES	-
ΔN11	NO	-
Δ6-10	-	NO
Δ21-24	YES	NO
C24S	YES	NO
E26Q	YES	NO

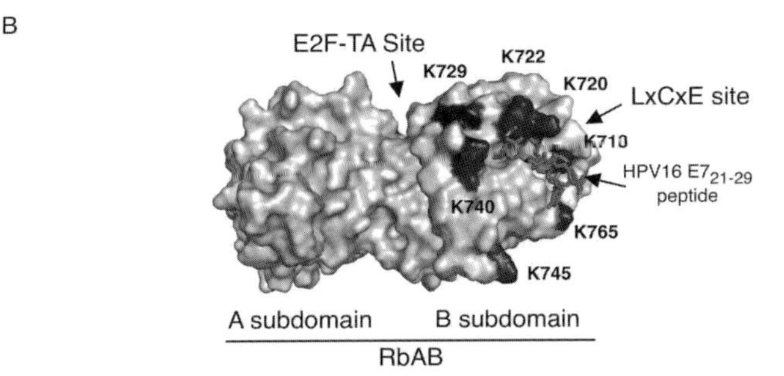

Figure 6.5 Main sequence determinants for HPV E7 degradation and E7-induced Rb degradation and binding. (A) Schematic representation of the HPV16 E7 monomer and E7 mutants used to assess the region involved in E7 degradation and in E7-induced Rb degradation. See text for details. The table on the right indicates whether the mutants were able or impaired in E7 or Rb degradation. YES: efficient degradation; NO: impaired for degradation. (B) Structure of the HPV16 $E7_{21-29}$ peptide (containing the LxCxE motif) in complex with the RbAB domain (PDB ID: 1GUX). The conserved RbAB lysine residues involved in binding to the E7 LxCxE motif are marked in grey. The E7 peptide is depicted in stick representation.

shorter form Rb (1–810). The cleaved Rb form is unable to promote cell cycle arrest, suggesting that the E2F–Rb complex has been disrupted. Furthermore, Rb (1–810) but not full-length Rb is promptly degraded through the proteasome pathway (Darnell *et al.*, 2007).

Proteasome mediated degradation of E7

E7 proteins from high-risk HPV types 16 and 18 are short lived, with a half-life that has been reported to be 30–40 min (Smotkin and Wettstein, 1987; Selvey *et al.*, 1994; Reinstein *et al.*, 2000). The degradation pathway has been studied in detail for the HPV16 E7 protein, and was shown to involve the ubiquitin-proteasome pathway (Reinstein *et al.*, 2000). It was shown that the addition of proteasome inhibitors to CaSki cells leads to the accumulation of highly insoluble poly ubiquitinated E7 species, which were also observed in cell free assays (Reinstein *et al.*, 2000; Oh *et al.*, 2004).

Most proteins are ubiquitinated at lysine epsilon-amino groups. An outstanding feature of E7 is that poly ubiquitination occurs at its N-terminal free amino group (Fig. 6.1A), a feature shared with a small subset of proteins including MyoD and the EBV transforming Latent Membrane Protein 1 (LMP1) (Breitschopf *et al.*, 1998; Aviel *et al.*, 2000; Reinstein *et al.*, 2000). This mechanism was elegantly proven by showing that both a naturally occurring lysine-less E7 protein from the HPV58 strain (Ben-Saadon *et al.*, 2004) and an HPV16 E7 protein in which internal lysines had been mutated (Reinstein *et al.*, 2000; Gonzalez *et al.*, 2001) could be ubiquitinated and degraded. Furthermore, the introduction of an N-terminal but not a C-terminal tag prevented E7 ubiquitination and degradation, increasing E7 half-life and steady state levels (Reinstein *et al.*, 2000; Gonzalez *et al.*, 2001; Huh *et al.*, 2007) (Fig. 6.5A).

The deletion of the first 11 residues in HPV16 E7 leads to a substantial stabilization of the protein while the deletion of only 7 residues does not affect the degradation rate (Fig. 6.5A). This indicates that E7 ubiquitination requires a free amino-terminus and also sequence determinants located in the CR1 region, probably necessary for interaction with a specific E3 ligase complex (Reinstein *et al.*, 2000). Interestingly, it has been shown that the CR1 region of E7 is required for the interaction with the Cullin-2 ubiquitin ligase complex, which mediates Rb degradation (Huh *et al.*, 2007). E7 ubiquitin-proteasome degradation requires the E2 ubiquitin conjugating enzyme UbcH7 and a Cullin1-containing E3 ligase complex (Oh *et al.*, 2004), but the specific binding site for this complex has not yet been identified.

Despite the fact that both E7 degradation and E7-induced Rb degradation are proteasome-dependent, they are two independent processes. Several lines of evidence support this hypothesis: first, N-terminally tagged E7 (TAG-E7) protein that is neither ubiquitinated nor efficiently degraded by the proteasome is as efficient in prompting Rb for proteasomal degradation as untagged E7 (Gonzalez *et al.*, 2001; Huh *et al.*, 2007) (Fig. 6.5). On the other hand, E7 Δ(21–24), which is impaired in its ability to bind and degrade Rb, is efficiently degraded by the proteasome (Gonzalez *et al.*, 2001) (Fig. 6.5A). Finally, it has been reported that the E3 ubiquitin–ligase complex involved in E7 degradation and in E7-induced Rb degradation is not the same: Bagchi and co-workers showed that HPV16 E7 is ubiquitinated by UbcH7 and Cullin-1/skp2-containing E3 ligase (Oh *et al.*, 2004) while Munger and co-workers showed that E7-induced Rb degradation requires the Cullin-2 ubiquitin ligase complex (Huh *et al.*, 2007). In addition, the E7 protein itself is able to establish a direct interaction with the S4 subunit of the proteasome (Berezutskaya and Bagchi, 1997). Therefore, experimental evidence strongly suggests that E7 uses multiple interactions with different components of the degradation machinery in order to regulate its own degradation or degradation of cell cycle regulators, such as Rb.

Protein–protein interaction mechanisms of E7

Cell cycle progression is a tightly regulated process that involves many different checkpoints and a large network of interacting proteins (DeGregori, 2004). However, papillomaviruses are able to override this robust control mechanism by using just a few proteins coded in the HPV genome

(Moody and Laimins, 2010). Papillomaviruses lack their own replication, transcription and translation machinery and, hence, use cell enzymes to replicate their genome, transcribe their genes, and translate their proteins. In this scenario E7 plays a key role in the normal virus life cycle, reprogramming the host cell cycle through the interaction with a broad repertoire of cellular proteins involved in cell cycle control such as Rb, HDAC, and p21, among others (Table 6.1). From the viewpoint of the virus life cycle, cancer development is a rare 'unproductive' event with a dead end since it precludes the formation of infective particles. Most infective events do not lead to cancer, and, instead, the virus replicates in the epithelium in a process that requires uncoupling of cell differentiation and cell cycle arrest, allowing the production of infective virions (Howley, 2007). This process requires fine-tuning of the numerous pathways involved in cell cycle control. E7 performs its cell cycle reprogramming activity through a complex network of protein–protein interactions (Table 6.1), leading to diverse effects on its targets such as the dissociation of protein complexes, the activation of enzymes, or the induction of protein degradation. Despite the apparent complexity of E7-mediated protein–protein interactions, the multiple effects of E7 on different target proteins appear to act in a coordinate manner ultimately leading to cell cycle reprogramming.

Many of these protein–protein interactions are shared by both high- and low-risk HPV strains, indicating their central importance in papillomavirus biology. Interactions are complex and in most cases involve several contact sites in the interacting proteins. We will review here selected examples of interactions that have been thoroughly characterized by biochemical or biophysical methods, for example through *in vitro* studies with pure components in solution, or interactions for which high resolution structures are available for the protein complexes.

The E7–Rb interaction

The retinoblastoma protein (Rb) was the first cellular target identified for the E7 protein (Dyson *et al.*, 1989). Binding of E7 to Rb is required for the transforming properties of the high-risk HPV16 E7 protein in transfected cells (Munger and Phelps, 1993), and it is the best-studied interaction related to E7-mediated cellular transformation. The interaction between E7 and Rb leads to dissociation of the Rb–E2F complex (Chellappan *et al.*, 1992) and induces progression of G1-arrested cells into the S-phase of the cell cycle. E7 from prototypical high-risk HPV types (16 and 18) has also been found to induce Rb degradation, contributing to functional inactivation of the retinoblastoma control pathway. Although E7 has several Rb-independent functions (McLaughlin-Drubin and Munger, 2009), the E7–Rb interaction is central to E7 transforming properties. Studies of natural HPV infection in keratinocyte raft models showed that induction of S-phase re-entry by E7 also required the interaction between E7 and other 'pocket protein' family proteins, such as p107 and p130 (Banerjee *et al.*, 2006). Therefore, the E7–Rb interaction plays a key role in viral replication during natural infection and also for HPV-mediated transformation. The E7 protein from prototypical high-risk HPV types (HPV16 and HPV18) binds to Rb with higher affinity than the E7 protein from prototypical low-risk types (HPV6 and HPV11) (Munger *et al.*, 1991). These differences in affinity are thought to contribute to the stronger transforming properties of prototypical high-risk E7 proteins, even though Rb affinity has been measured for only a handful of the known HPV types (Jones *et al.*, 1990; Munger *et al.*, 1991; Dong *et al.*, 2001; Chemes *et al.*, 2010). Moreover, it is not known whether Rb binding affinity correlates strongly with E7 transforming properties beyond the prototypical high- and low-risk HPV types 16 and 18, and 11 and 6 respectively.

The conserved LxCxE motif of E7 is required for the interaction with Rb (Edmonds and Vousden, 1989; Phelps *et al.*, 1992), and it is the main determinant of binding for E7 proteins from both high- and low-risk HPV types, contributing up to 90% of the interaction free energy (Jones *et al.*, 1990; Munger *et al.*, 1991; Chemes *et al.*, 2010). The dissociation constant in solution was recently measured accurately to be 5 nM (Chemes *et al.*, 2010). The crystal structure of a nine-residue peptide containing the LxCxE motif from HPV16 E7 complexed with the RbAB domain (residues 372–787) shows that this linear motif binds in an extended conformation to a conserved and shallow surface groove located in the B subdomain of RbAB (Lee *et al.*, 1998) (Fig. 6.5B). Even when the LxCxE motif in E7 constitutes the main determinant for Rb binding, several lower affinity interaction sites were shown to contribute to binding and functional inactivation of Rb by E7. Dissociation of the Rb–E2F complex by E7 is a two-step process requiring a high-affinity interaction mediated by the LxCxE motif that does not compete for E2F binding to Rb, and a second lower affinity interaction between the E7C domain and the RbC domain responsible for displacement of E2F (Patrick *et al.*, 1994). It has been shown that the RbC domain binds to a conserved surface groove formed by the β1 strand and the α1 helix in E7C (Liu *et al.*, 2006) (Fig. 6.2B). On the other hand, binding of full-length E7 to the RbAB domain increases the interaction affinity by 10-fold with respect to peptide fragments containing the LxCxE motif, indicating that additional surfaces of the RbAB domain are being targeted by the E7C domain, potentially interfering with additional protein–protein interactions (Chemes *et al.*, 2010). Interestingly, in canine and gamma papillomaviruses the main determinant for Rb binding is not the LxSxE motif present in these proteins. Instead, both Rb binding and Rb degradation are mediated by the E7C domain and occur independently of the LxSxE motif (Wang *et al.*, 2010). In the AdE1A protein, the CR1 region is strictly required for E2F displacement, since it binds to the so-called E2F-TA site in RbAB competing for E2F binding (Liu and Marmorstein, 2007) (Fig. 6.5B). In HPV E7, the CR1 region has been shown to contribute to E2F displacement in cooperation with the CR2 region (Helt and Galloway, 2001) and to bind to RbAB with micromolar affinity (Chemes *et al.*, 2010), which suggests that the CR1 region may also play a role in Rb binding and in E2F displacement by the E7 dimer. This likeliness highlights the structural and functional similarity between the E7 and AdE1A proteins (Ferreon *et al.*, 2009b).

Kinetic and thermodynamic studies show that binding of E7 to Rb is guided by the LxCxE motif and by electrostatic interactions, which favour association. This electrostatic component arises from the complementarity between the negatively charged residues from the LxCxE motif and adjacent CKII-PEST region in E7 (Chemes, 2011b) (Fig. 6.1) and the positively charged lysine patch surrounding the RbAB cleft (Dick and Dyson, 2002) (Fig. 6.5B). Early studies showed that the differences in Rb affinity between prototypical high- and low-risk E7 proteins were explained mainly by substitutions at a position adjacent to the LxCxE motif, where high-risk HPV16 and 18 E7 proteins have an aspartic acid, and low-risk HPV11 and 6 E7 proteins have a glycine (position D21 in HPV16 E7; Fig. 6.1B) (Munger *et al.*, 1991). Recently, it was shown that the lower affinity of the low-risk HPV11 E7 protein is due to a lower half-life of the E7–Rb complex (Chemes, 2011b), which would allow for faster ligand exchange between E7–Rb and cellular–Rb complexes.

Phosphorylation also stabilizes the E7–Rb complex, probably owing to additional direct interactions established between the phosphate groups in E7 and positively charged residues in RbAB (Chemes, 2011b). Since not all cellular LxCxE-containing proteins that bind to Rb contain a phosphorylation site like that found in E7, this post-translational modification can contribute to fine-tuning of the interaction affinity and kinetics in favour of the viral protein(Chemes, 2011b).

E7 binding to histone deacetylase

The cell cycle-inhibitory activity of the Rb protein requires the direct binding and inhibition of the E2F protein and also the recruitment of histone deacetylase protein (HDAC), which, in turn, deacetylates histones and cooperates to locally repress gene transcription (Brehm *et al.*, 1998). This deacetylation uncovers a positively charged lysine side chain, allowing for optimal interaction of histones with DNA. Subsequently, this tight binding leads to heterochromatin formation and repression of transcription. HDAC binds the RbAB pocket domain and HPV16 E7 is able to dissociate the Rb–HDAC complex in a process that requires the E7 LxCxE motif (Brehm *et al.*, 1998, 1999).

In addition to dissociating the Rb–HDAC complex through an LxCxE-mediated interaction, E7 proteins from HPV16 and HPV31 were shown to bind class I HDAC through their C-terminal domain (Brehm *et al.*, 1999; Longworth and Laimins, 2004) acting in a coordinate manner with the dissociation of the Rb–HDAC complex to further promote cell cycle deregulation (Brehm *et al.*, 1999). These effects of E7 depend on two separate and independent binding sites for Rb and HDAC. In differentiating keratinocytes, the HPV31 E7 protein specifically up-regulates the transcription of E2F2 by precluding the interaction of HDAC with the E2F2 promoter (Longworth *et al.*, 2005). This effect depends on an interaction between E7 and HDAC, since it was shown that the HPV31 E7 L67R mutant, which does not form a complex with HDAC, is unable to inhibit binding of HDAC to the E2F2 promoter. Moreover, this effect is specific for a subset of HDAC-regulated promoters, since HDAC-regulated transcriptional activity of the cdc6 promoter is unaffected by E7 expression (Longworth *et al.*, 2005). In the case of HPV16 E7, the interaction with HDAC was shown to occur indirectly through binding of E7 to the Mi2 protein, a component of the NuRD chromatin remodelling complex (Brehm *et al.*, 1999).

The effect of E7 on HDAC-mediated epigenetic regulation includes a complex and coordinated strategy involving gene activation, as in the case of E2F2, as well as gene repression. As an example, it has been shown that HPV16 E7 directly binds to the interferon regulatory factor-1 (IRF-1) through its N-terminal domain (Park *et al.*, 2000). This interaction abrogates the transactivation function of IRF-1 by recruiting HDAC to the IFN promoter, which leads to the subversion of the interferon antiviral response (Moody and Laimins, 2010).

It was reported that mutation of the C-terminal Zn-binding motif of the HPV31 E7 protein that impairs HDAC binding affects viral episome maintenance, suggesting that the E7–HDAC interaction is important for the viral life cycle (Longworth and Laimins, 2004).

E7 binding to cyclin-dependent kinase (CDK) inhibitors p21 and p27

In addition to the previously described interactions, the E7 protein targets the cdk-2 inhibitors p21 and p27, which contributes to overriding the growth arrest induced during epithelial differentiation (Zerfass-Thome *et al.*, 1996; Funk *et al.*, 1997; Jones *et al.*, 1997). The interaction between E7 and p21 blocks p21-mediated inhibition of cyclin A and E-associated kinases, increasing cdk-2 activity in differentiated keratinocytes and, thus, uncoupling differentiation and proliferation (Jones *et al.*, 1997). It has been shown that the C-terminal region of p21 (residues 139–164) interacts with a conserved surface groove formed by the β1 strand and the α1 helix in E7 (Ohlenschlager *et al.*, 2006) (Fig. 6.2B), which overlaps with the RbC binding site in E7C (Liu *et al.*, 2006). E7 from HPV16 can also associate to the p27 cdk inhibitor abrogating the G0/G1 arrest induced in response to both serum withdrawal and loss of cell adhesion (Zerfass-Thome *et al.*, 1996). As in the case of p21, E7 binds p27 through its C-terminal domain.

The E7 protein from high-risk types 16, 18 and 31 and low-risk HPV6b directly interacts with the cdk2-cyclin A and E complexes, increasing their kinase activity and supporting an additional mechanism to further control the cellular replication machinery (Tommasino *et al.*, 1993; McIntyre *et al.*, 1996; He *et al.*, 2003). Residues 9 to 38 in the HPV16 E7N domain, which span both the CR1 and CR2 regions, are necessary and sufficient for the association with the cdk2–cyclin complexes and for increasing kinase activity. These cyclin–kinase complexes mediate phosphorylation of both histone H1 and Rb, but E7 produces a specific upregulation of the histone H1 kinase activity of these cyclin–kinase complexes, without affecting the Rb phosphorylation rate (He *et al.*, 2003). It is worth mentioning that the LxCxE Rb-binding motif and the CKII-PEST sites are not required for this functional effect on cdk2–cyclin complexes, since the E7Δ22–26 and E7Δ31–32 mutants that individually eliminate these sites, are able to increase kinase activity of the cdk2–cyclin complex (He *et al.*, 2003).

Interaction of E7 with the HPV E2 transcriptional regulator

The HPV E2 protein regulates gene transcription and viral genome replication (Hamid *et al.*, 2009; Thierry, 2009) and can strongly repress transcription of the E6/E7 ORFs. Integration of the viral DNA genome into the host genome is part of the carcinogenic process. In fact, Kalantari *et al.* showed that there is a correlation between the progress of the cancerous lesion and DNA integration (Kalantari *et al.*, 1998). The E2 open reading frame is often disrupted upon viral DNA integration, and the E2-mediated repression of E6 and E7 transcription is released, increasing E6 and E7 expression levels (Stoler *et al.*, 1992; Thierry, 2009). An interaction between E2 and E7 has been reported *in vivo*, where E2 was capable of inhibiting E7-ras cooperation in cell transformation and affected E7 localization and stability (Gammoh *et al.*, 2006, 2009).

HPV E2 proteins are *ca.* 400 amino acid polypeptides consisting of an N-terminal transactivation domain and a C-terminal DNA binding and dimerization domain (E2C), separated by a non-conserved 'hinge' region. E2C DNA binding domains from all strains analysed to date share a unique folding topology, namely, the dimeric β-barrel (Hegde, 2002). The N-terminal domain of E7 interacts with E2C with a K_D of 0.1 μM. The stretch of residues 25–40 of E7, encompassing both the PEST motif and the two phosphorylation sites (S31 and S32) (Fig. 6.1A), was mapped as the primary binding site (Smal *et al.*, 2009). The DNA-binding helix of E2C is involved in the protein–protein interface, as shown by site-directed mutagenesis of the helix and by the fact that the E7–E2C complex can be displaced by an E2 site DNA duplex (Smal *et al.*, 2009).

Depending on the relative concentration of each protein in solution, the interaction of the HPV16 E7 protein with the HPV16 E2 DNA-binding domain leads to the formation of soluble (12, 40 and 115 nm diameter) and insoluble complexes (Smal *et al.*, 2009). The possibility of a finely tuned mechanism for regulating the relative availability of the E2 and E7 protein levels emerges from these results. Through this mechanism, E7 could block its own repression through interaction with the E2 repressor. In this way, the relative change in the protein levels, resulting from the loss of E2 upon integration of the viral genome, could generate a drastic increase in both the E7 and E6 oncoprotein levels. In addition, E7 could sequester E2 by forming oligomeric species in the cytosol prior to integration, thus, increasing E7 expression levels in cells bearing episomal viral DNA.

HPV16 E7 chaperone holdase activity

E7SOs were shown to prevent incorrect folding and aggregation of two model chaperone substrates, citrate synthase and luciferase, at substoichiometric concentrations *in vitro* (Alonso *et al.*, 2006). E7SOs bind to these model substrates while being in a partly unfolded conformation, holding them in a folding-competent conformational state with significant overall secondary and tertiary structure. Fully folded proteins are not bound by E7SOs, and ATP is not required for this activity which defines E7SOs as a *chaperone holdase* (Alonso *et al.*, 2006).

This chaperone activity could play an important role in Rb degradation since E7SOs bind strongly to full-length native RbAB. This is not unexpected, given the fact that the high affinity LxCxE Rb binding motif is located in the E7N intrinsically disordered domain and that this domain is accessible for protein–protein interactions in the E7SOs (Alonso *et al.*, 2006). The ~70 molecules of E7 that form E7SOs (Alonso *et al.*, 2004) provide a high avidity molecular assembly that could sequester Rb from the Rb–E2F complex or target it for proteasomal degradation. Although the E7N intrinsically disordered domain does not interfere with E7SOs chaperone holdase activity, it is able to outcompete E7SOs for Rb binding (Alonso *et al.*, 2006).

An interaction between the SV40-LT J domain and the hsc70 cellular chaperone is required for displacement of the Rb–E2F complex (Sullivan *et al.*, 2000). Hence, it seems that for both the E7 and the SV40-LT viral proteins, a chaperone activity (coded directly by E7 and coded in a separate protein in the case of SV40-LT) is required for Rb–E2F complex displacement. On the other hand, in the case of the AdE1A protein, Rb-E2F displacement does not require a chaperone activity (Helt and Galloway, 2003).

Cellular localization of HPV E7

The E7 oncoprotein has been identified in carcinoma biopsies, in cervical cancer cell lines CaSki, SiHa (HPV16), HeLa (HPV18) and in E7-transfected cells, by immunochemical and biochemical assays (Greenfield *et al.*, 1991; Fiedler *et al.*, 2004). Different authors have reported that HPV16 E7 localizes to the nucleus in the cervical cancer cell line CaSki and in other human cell lines transiently expressing E7 (Greenfield *et al.*, 1991; Fiedler *et al.*, 2004). Moreover, in cervical cancer biopsies, E7 was predominantly detected in the nucleus of the tumour cells and also, to a minor extent, in the cytoplasm (Greenfield *et al.*, 1991; Fiedler *et al.*, 2004). Angeline *et al.* reported that E7 can enter the nucleus via a Ran-dependent pathway (Angeline *et al.*, 2003), while other authors have shown that at least a part of the E7 protein is localized and might function within the cytoplasmic compartment (Oltersdorf *et al.*, 1987; Kanda *et al.*, 1991; Fujikawa *et al.*, 1994; Rey *et al.*, 2000). It is possible that at least under some conditions, a certain fraction of E7 may reside in the cytoplasm, as initially found by Smotkin and Wettstein (1987).

To date, it has been difficult to determine which of the conformational species of E7 (monomers, dimers or oligomers) exist in different subcellular compartments, mainly because of the difficulty in obtaining antibodies able to differentially identify E7 conformers. The preferential identification of different conformers by the used antibodies may account for the discrepancies found in the literature regarding the localization of E7 in cells.

The generation of conformation-specific antibodies for E7 using purified conformers revealed that monomeric forms of these proteins are localized preferentially to the nucleus of model HPV transformed cell lines and HPV E7 transfected cells (Dantur *et al.*, 2009). On the other hand, oligomeric forms that share conformational epitopes with the *in vitro* characterized oligomers are localized to the cytoplasm (Dantur *et al.*, 2009) (Fig. 6.6A). The oligomeric E7 species found in the cytosol colocalized with an amyloid stain (Dantur *et al.*, 2009), a property described for the soluble purified E7SOs (Alonso *et al.*, 2004).

Cytosolic oligomeric E7 appeared as the most abundant species in all cell systems tested. Nuclear E7 levels were shown to be replenished dynamically from the cytosolic pools and not to arise from *de novo* protein synthesis (Dantur *et al.*, 2009). Then, long-term events related to de-repression of E7 may cause accumulation of excess

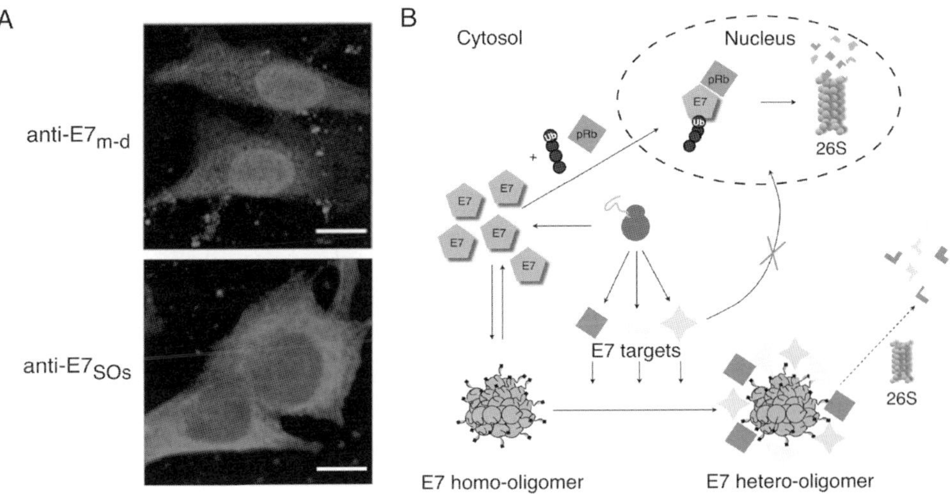

Figure 6.6 Cellular localization of E7 conformers. (A) Subcellular localization of E7 species tested by using conformation-specific antibodies. Upper panel: CaSKi cells stained with an antibody raised against monomeric and dimeric E7 species (anti-$E7_{m-d}$). Lower panel: CaSKi cells stained with an antibody raised against oligomeric E7 species (anti-$E7_{SOs}$). White bar: 10 µm. (B) Model for the localization and translocation of monomeric-dimeric and oligomeric E7 species within cells. Cellular E7 targets can be bound by E7 oligomers and degraded by the 26S proteasome. The oligomeric E7 species is in equilibrium with monomeric-dimeric species (pentagon), which can translocate to the nucleus where Rb binding and degradation take place. Adapted from Dantur *et al.* (2009).

E7 as oligomeric species in the cytosol (Fig. 6.6B). These facts, together with the known target promiscuity of E7, strongly suggest that interactions with many of the non-pRb dependent targets described, even if the specificity or affinity is low, may take place. Moreover, the natural occurrence of the oligomeric species is further supported by the detection of E7 oligomers in the cytosol of cancerous cells from tissue biopsies (Dantur *et al.*, 2009).

Among potential E7 interacting proteins are pocket proteins, transcription factors, chromatin remodelling proteins, negative regulators of the cell cycle, components of the innate immune response and cytoskeletal proteins (Howley *et al.*, 1991; Rey *et al.*, 2000; Munger and Howley, 2002; Campo, 2005) (Table 6.1). Many of these proteins are located in the cytosol, such as those involved in destabilization of centrosomes and genome instability (Duensing and Munger, 2002). Transcription factors or proteins that act inside the nucleus need to be imported to this compartment after synthesis. Accumulated high levels of cytosolic E7 are likely to interfere with their transport to the nucleus, thus, interfering with their key gene regulatory functions (Fig. 6.6B). The fact that different conformers of E7 exist in cells raises new questions related to the underlying mechanisms operating in tissues that would go beyond the promotion of pRb and pocket protein degradation. Interestingly, a similar situation of monomeric-oligomeric conformer distribution is observed for endogenous HPV16 and -18 E6 proteins in model CaSki and HeLa cells (Garcia-Alai *et al.*, 2007b). It should be important to establish which E7 (or E6) species are antigenic and which are 'invisible' to immune surveillance, given the need for therapeutic vaccines (Hung *et al.*, 2008).

E7 protein as an immunogen

Papillomaviruses have co-evolved with their vertebrate hosts over millions of years, elaborating a non-lytic replicative cycle that causes asymptomatic infections ensuring long-term spreading. During the natural course of HPV infection, no cytolysis or concomitant inflammation is produced, avoiding activation of professional antigen presenting cells and subsequent initiation of the immune response, and keeping HPV antigens away from immunologic surveillance.

Initial HPV infection is largely neutralized by humoral immunity, which prevents viral spreading by the production of neutralizing antibodies against the capsid proteins L1 and L2. On the other hand, oncogenic viruses such as papillomaviruses, Epstein–Barr or hepatitis B virus have built up complex mechanisms to evade host immune surveillance during the initiation of cell transformation and tumour progression. E6 and E7 oncoproteins, mostly from high-risk HPV types, have been shown to down-regulate the expression of several IFN-inducible genes in differentiating cervical keratinocytes: interferon regulatory factor-3 (IRF-3), IFNα, IFNβ and IFN-transcriptional activator STAT-1, among others (Ronco *et al.*, 1998; Nees *et al.*, 2001). E7 from HPV11, -16 and -18 was shown to down-regulate the expression of components of the antigen processing and presentation pathway, like the major histocompatibility complex (MHC) class I heavy chain, TAP1 and LMP2 genes (Georgopoulos *et al.*, 2000; Vambutas *et al.*, 2001; Bottley *et al.*, 2008). HPV16 E7 impairs IFN-γ mediated antigen processing and T-cell lysis (Zhou *et al.*, 2011) and overcomes tumour necrosis factor (TNF)-alpha cytostatic effect on cervical keratinocytes (Boccardo *et al.*, 2010).

The constitutive expression of HPV16 E6 and E7 proteins in HPV-cervical neoplasias has been extensively reported. Several serological assays consistently demonstrated that antibodies against HPV16 E7 can be found at a higher frequency in cervical cancer patients than in age-matched non-cancer controls (Jochmus-Kudielka *et al.*, 1989; Galloway, 1992; Jha *et al.*, 1993; Nindl *et al.*, 1996). Unexpectedly, no significant differences in antibodies against the cell cycle regulator p16, the current biomarker for detection of high-risk human papillomavirus in Papanicolaou (Pap) smears, have been found (Reuschenbach *et al.*, 2008). On the other hand, HPV typing and E7 oncoprotein immunohistochemistry showed the regular expression of the high-risk HPV E7 proteins in biopsies of cervical cancers, although no correlation between expression levels and tumour stages was found (Ressler *et al.*, 2007). Recently, the presence of E7 in HPV16 cervical adenocarcinomas *in situ* and adenocarcinoma samples was also reported, denoting the potential application of this protein as biomarker in these malignancies (Dreier *et al.*, 2011).

Prophylactic vaccines against HPV are based on the virus-like particle assembly of the L1 major capsid protein and, although with a demonstrated high efficacy in generating a protective (humoral) immunity against viral infection, these vaccines are unlikely to have a therapeutic effect on pre-existing neoplastic processes; thus, they have no immediate impact on the incidence of cervical cancer (Kols *et al.*, 2006; Leggatt and Frazer, 2007). Clearly, therapeutic vaccines should ideally be developed to treat established persistent HPV infections, having an immediate effect on the prevalence of HPV-associated malignancies. The HPV E6 and E7 oncoproteins represent ideal targets for immunotherapy, since their constitutive expression in transformed cells reduces the chance of immune evasion through antigen loss. Besides, as both oncoproteins constitute foreign antigens, immune tolerance issues commonly associated to other cancer vaccines might be circumvented (Hung *et al.*, 2008).

In recent years, several HPV16 E7-based vaccine candidates have been tested at preclinical and clinical stages. These vaccines cover a wide spectrum of technical approaches: DNA plasmids or recombinant live-vectors expressing the E7 gene, purified E7 protein, peptides comprising HLA-E7 epitopes, or E7-loaded dendritic cells (DCs). Many of these platforms also include fusing the E7 protein, peptide or gene to another molecule known to enhance inflammation or mediate antigen processing and presentation. A current compendium of therapeutic vaccines can be found in Cid-Arregui *et al.* (2009) and Hung *et al.* (2008). The efficacy of therapeutic HPV vaccine candidates is generally tested in the cervical cancer murine TC-1 model. The TC-1 cell line derives from primary lung epithelial cells of C57BL/6 mice, which were immortalized with HPV16 E6 and E7 genes and transformed with the cellular c-Ha-*ras* oncogene (Lin *et al.*, 1996a). This co-transformation produced a tumorigenic cell line expressing the E6 and E7 oncoproteins, mimicking the natural sequence of cervical cancer progression in which these oncoproteins immortalize cells. Further mutations in cellular proto-oncogenes transform them in tumour cells with metastatic potential. Protective immunity against tumour challenge or therapeutic eradication of the E7-expressing tumour cells provides the validity of this cell line as a murine model for immunotherapeutic intervention (Lin *et al.*, 1996a; Chu *et al.*, 2000; van der Burg *et al.*, 2001; Kim *et al.*, 2002; Michel *et al.*, 2002).

It has been previously shown that removal of the zinc atom from HPV16 E7 induces its oligomerization into highly stable, soluble and homogeneous spherical particles, named E7SOs (Alonso *et al.*, 2004). Recently, the potential of a further chemically stabilized version of the E7SOs (E7SOx) as an immunotherapeutic agent for HPV16-associated lesions was evaluated at a pre-clinical stage with promising results (Cerutti *et al.*, 2011). Notwithstanding the extraordinary improvements that took place in the last decade on DNA-based vaccine technology and the promising results achieved with experimental models, both the scarce immunogenicity and ethical issues are still their main drawbacks.

Compilations of preclinical and clinical studies of HPV therapeutic vaccine candidates highlight the limited efficacy these agents have in high-grade CINs or cervical cancer patients. This indicates that future clinical protocols might involve combined schedules of immunotherapy with chemotherapy or radiotherapy to achieve better results. Clearly, different formulations, in particular those involving the E7 protein as immunogen, are promising candidates for developing effective and safe therapeutic vaccines.

Concluding remarks

Since the first descriptions of E7 as a transforming agent (Vousden *et al.*, 1988; Yutsudo *et al.*, 1988), E7 has been a paradigmatic example of a viral oncogene, in particular, within the small DNA tumour virus group. Following the first molecular explanation for its transforming action, i.e. targeting the Rb tumour suppressor for degradation, a vast number of cellular targets have been described, together with many different targeted pathways. It is hard to envisage how a 98 amino acid protein can cope with all these interactions simultaneously, which indicates that there are many factors such as protein expression levels, post-translational modifications, cellular compartmentalization, timing of expression and so on, that dictate the biological outcome. Given the complexity of all of the competing interactions, the biochemical and quantitative study of the relative affinities and kinetics of different E7 targets becomes essential for understanding the hierarchy of interactions established in a given cell type and at a given point in the cell differentiation or transformation process. Thus, in order to reach a complete understanding and discrimination among all possible biological targets and functions or 'effects' exerted on cell transformation, structure–function studies and elucidation of biochemical mechanisms using recombinant species and biophysical approaches should be established.

Besides its relevance to HPV-linked cancers, as a member of a group of transforming proteins from other tumour viruses, and as a model protein targeting Rb and instrumental for understanding its many roles, E7 emerges as an example of an intrinsically disordered protein (Alonso *et al.*, 2002; Daughdrill, 2005; Uversky *et al.*, 2006). These types of proteins display extended regions that can accommodate different linear motifs or modules, and are becoming more of a common feature than a rarity among proteomes, in particular among viral proteins that must be multifunctional, owing to their minimal number of genes (Tokuriki *et al.*, 2009; Xue *et al.*, 2010).

References

Alonso, L.G., Garcia-Alai, M.M., Nadra, A.D., Lapena, A.N., Almeida, F.L., Gualfetti, P., and Prat-Gay, G.D. (2002). High-risk (HPV16) human papillomavirus E7 oncoprotein is highly stable and extended, with conformational transitions that could explain its multiple cellular binding partners. Biochemistry *41*, 10510–10518.

Alonso, L.G., Garcia-Alai, M.M., Smal, C., Centeno, J.M., Iacono, R., Castano, E., Gualfetti, P., and de Prat-Gay, G. (2004). The HPV16 E7 viral oncoprotein self-assembles into defined spherical oligomers. Biochemistry *43*, 3310–3317.

Alonso, L.G., Smal, C., Garcia-Alai, M.M., Chemes, L., Salame, M., and de Prat-Gay, G. (2006). Chaperone holdase activity of human papillomavirus E7 oncoprotein. Biochemistry *45*, 657–667.

Angeline, M., Merle, E., and Moroianu, J. (2003). The E7 oncoprotein of high-risk human papillomavirus type 16 enters the nucleus via a nonclassical Ran-dependent pathway. Virology *317*, 13–23.

Antinore, M.J., Birrer, M.J., Patel, D., Nader, L., and McCance, D.J. (1996). The human papillomavirus type 16 E7 gene product interacts with and trans-activates the AP1 family of transcription factors. EMBO J. *15*, 1950–1960.

Aviel, S., Winberg, G., Massucci, M., and Ciechanover, A. (2000). Degradation of the Epstein–Barr virus latent membrane protein 1 (LMP1) by the ubiquitin-proteasome pathway. Targeting via ubiquitination of the N-terminal residue. J. Biol. Chem. *275*, 23491–23499.

Banerjee, N.S., Genovese, N.J., Noya, F., Chien, W.M., Broker, T.R., and Chow, L.T. (2006). Conditionally activated E7 proteins of high-risk and low-risk human papillomaviruses induce S phase in postmitotic, differentiated human keratinocytes. J. Virol. *80*, 6517–6524.

Barbosa, M.S., Edmonds, C., Fisher, C., Schiller, J.T., Lowy, D.R., and Vousden, K.H. (1990). The region of the HPV E7 oncoprotein homologous to adenovirus E1a and Sv40 large T antigen contains separate domains for Rb binding and casein kinase II phosphorylation. EMBO J. *9*, 153–160.

Barbosa, M.S., Lowy, D.R., and Schiller, J.T. (1989). Papillomavirus polypeptides E6 and E7 are zinc-binding proteins. J. Virol. *63*, 1404–1407.

Barnard, P., and McMillan, N.A. (1999). The human papillomavirus E7 oncoprotein abrogates signalling mediated by interferon-alpha. Virology *259*, 305–313.

Ben-Saadon, R., Fajerman, I., Ziv, T., Hellman, U., Schwartz, A.L., and Ciechanover, A. (2004). The tumour suppressor protein p16(INK4a) and the human papillomavirus oncoprotein-58 E7 are naturally occurring lysine-less proteins that are degraded by the ubiquitin system. Direct evidence for ubiquitination at the N-terminal residue. J. Biol. Chem. *279*, 41414–41421.

Berezutskaya, E., and Bagchi, S. (1997). The human papillomavirus E7 oncoprotein functionally interacts with the S4 subunit of the 26 S proteasome. J. Biol. Chem. *272*, 30135–30140.

van der Burg, S.H., Kwappenberg, K.M., O'Neill, T., Brandt, R.M., Melief, C.J., Hickling, J.K., and Offringa, R. (2001). Pre-clinical safety and efficacy of TA-CIN, a recombinant HPV16 L2E6E7 fusion protein vaccine, in homologous and heterologous prime-boost regimens. Vaccine *19*, 3652–3660.

Bernard, H.U., Burk, R.D., Chen, Z., van Doorslaer, K., Hausen, H., and de Villiers, E.M. (2010). Classification of papillomaviruses (PVs) based on 189 PV types and proposal of taxonomic amendments. Virology *401*, 70–79.

Bernat, A., Avvakumov, N., Mymryk, J.S., and Banks, L. (2003). Interaction between the HPV E7 oncoprotein and the transcriptional coactivator p300. Oncogene *22*, 7871–7881.

Boccardo, E., Manzini Baldi, C.V., Carvalho, A.F., Rabachini, T., Torres, C., Barreta, L.A., Brentani, H., and Villa, L.L. (2010). Expression of human papillomavirus type 16 E7 oncoprotein alters keratinocytes expression profile in response to tumour necrosis factor-alpha. Carcinogenesis *31*, 521–531.

Bottley, G., Watherston, O.G., Hiew, Y.L., Norrild, B., Cook, G.P., and Blair, G.E. (2008). High-risk human papillomavirus E7 expression reduces cell-surface MHC class I molecules and increases susceptibility to natural killer cells. Oncogene *27*, 1794–1799.

Boyer, S.N., Wazer, D.E., and Band, V. (1996). E7 protein of human papilloma virus-16 induces degradation of retinoblastoma protein through the ubiquitin-proteasome pathway. Cancer Res. *56*, 4620–4624.

Brehm, A., Miska, E.A., McCance, D.J., Reid, J.L., Bannister, A.J., and Kouzarides, T. (1998). Retinoblastoma protein recruits histone deacetylase to repress transcription. Nature *391*, 597–601.

Brehm, A., Nielsen, S.J., Miska, E.A., McCance, D.J., Reid, J.L., Bannister, A.J., and Kouzarides, T. (1999). The E7 oncoprotein associates with Mi2 and histone deacetylase activity to promote cell growth. EMBO J. *18*, 2449–2458.

Breitschopf, K., Bengal, E., Ziv, T., Admon, A., and Ciechanover, A. (1998). A novel site for ubiquitination: the N-terminal residue, and not internal lysines of MyoD, is essential for conjugation and degradation of the protein. EMBO J. *17*, 5964–5973.

Campo, S. (2005). The essential transforming proteins of HPV: E5, E6 and E7. HPV Today *7*, 8–10.

Cerutti, M.L., Alonso, L.G., Tatti, S., and Prat-Gay, G.D. (2011). Long-lasting immunoprotective and therapeutic effects of a hyperstable E7 oligomer based vaccine in a murine HPV tumor model. Int. J. Cancer. DOI: 10.1002/ijc.26294.

Chellappan, S., Kraus, V.B., Kroger, B., Munger, K., Howley, P.M., Phelps, W.C., and Nevins, J.R. (1992). Adenovirus E1A, simian virus 40 tumour antigen, and human papillomavirus E7 protein share the capacity to disrupt the interaction between transcription factor E2F and the retinoblastoma gene product. Proc. Natl. Acad. Sci. U.S.A. *89*, 4549–4553.

Chemes, L.B., Sanchez, I.E., Smal, C., and de Prat-Gay, G. (2010). Targeting mechanism of the retinoblastoma tumour suppressor by a prototypical viral oncoprotein. Structural modularity, intrinsic disorder and phosphorylation of human papillomavirus E7. FEBS J. 277, 973–988.

Chemes, L.B., Sánchez, I.E., Alonso, L.G., and de Prat-Gay, G. (2011a). Intrinsic disorder in the human papillomavirus E7 protein. In Flexible Viruses: Structural Disorder within Viral Proteins, Longhi, S., Uversky, V.N., eds. (John Wiley and Sons). In press, pp. 313–347.

Chemes, L.B., Sanchez, I.E., and de Prat-Gay, G. (2011b). Kinetic recognition of the retinoblastoma tumour suppressor by a specific protein target. J. Mol. Biol. *412*, 267–284.

Chen, E.Y., Howley, P.M., Levinson, A.D., and Seeburg, P.H. (1982). The primary structure and genetic organization of the bovine papillomavirus type 1 genome. Nature *299*, 529–534.

Chinami, M., Sasaki, S., Hachiya, N., Yuge, K., Ohsugi, T., Maeda, H., and Shingu, M. (1994). Functional oligomerization of purified human papillomavirus types 16 and 6b E7 proteins expressed in *Escherichia coli*. J. Gen. Virol. *75 (Pt 2)*, 277–281.

Chu, N.R., Wu, H.B., Wu, T., Boux, L.J., Siegel, M.I., and Mizzen, L.A. (2000). Immunotherapy of a human papillomavirus (HPV) type 16 E7-expressing tumour by administration of fusion protein comprising *Mycobacterium bovis bacille* Calmette-Guerin (BCG) hsp65 and HPV16 E7. Clin. Exp. Immunol. *121*, 216–225.

Cid-Arregui, A. (2009). Therapeutic vaccines against human papillomavirus and cervical cancer. Open Virol. J. *3*, 67–83.

Clemens, K.E., Brent, R., Gyuris, J., and Munger, K. (1995). Dimerization of the human papillomavirus E7 oncoprotein *in vivo*. Virology *214*, 289–293.

Clements, A., Johnston, K., Mazzarelli, J.M., Ricciardi, R.P., and Marmorstein, R. (2000). Oligomerization properties of the viral oncoproteins adenovirus E1A and human papillomavirus E7 and their complexes with the retinoblastoma protein. Biochemistry *39*, 16033–16045.

Cole, S.T., and Danos, O. (1987). Nucleotide sequence and comparative analysis of the human papillomavirus type 18 genome. Phylogeny of papillomaviruses and repeated structure of the E6 and E7 gene products. J. Mol. Biol. *193*, 599–608.

Dantur, K., Alonso, L., Castano, E., Morelli, L., Centeno-Crowley, J.M., Vighi, S., and de Prat-Gay, G. (2009). Cytosolic accumulation of HPV16 E7 oligomers supports different transformation routes for the prototypic viral oncoprotein: the amyloid-cancer connection. Int. J. Cancer *125*, 1902–1911.

Darnell, G.A., Schroder, W.A., Antalis, T.M., Lambley, E., Major, L., Gardner, J., Birrell, G., Cid-Arregui, A., and Suhrbier, A. (2007). Human papillomavirus E7 requires the protease calpain to degrade the retinoblastoma protein. J. Biol. Chem. *282*, 37492–37500.

Daughdrill, G.W., Pielak, G.J., Uversky, V.N., Cortese, M.S., and Dunker A. K. (2005). Natively disordered proteins. In Protein Folding Handbook, Buchner, J., and Kiefhaber, T., ed. (Wiley-VCH, Weinheim, Germany), pp. 275–357.

Davies, R., Hicks, R., Crook, T., Morris, J., and Vousden, K. (1993). Human papillomavirus type 16 E7 associates with a histone H1 kinase and with p107 through sequences necessary for transformation. J. Virol. *67*, 2521–2528.

DeGregori, J. (2004). The Rb network. J. Cell Sci. *117*, 3411–3413.

Dick, F.A., and Dyson, N.J. (2002). Three regions of the pRB pocket domain affect its inactivation by human papillomavirus E7 proteins. J. Virol. *76*, 6224–6234.

Dong, W.L., Caldeira, S., Sehr, P., Pawlita, M., and Tommasino, M. (2001). Determination of the binding affinity of different human papillomavirus E7 proteins for the tumour suppressor pRb by a plate-binding assay. J. Virol. Methods *98*, 91–98.

Dreier, K., Scheiden, R., Lener, B., Ehehalt, D., Pircher, H., Muller-Holzner, E., Rostek, U., Kaiser, A., Fiedler, M., Ressler, S., *et al.* (2011). Subcellular localization of the human papillomavirus 16 E7 oncoprotein in CaSki cells and its detection in cervical adenocarcinoma and adenocarcinoma *in situ*. Virology *409*, 54–68.

Duensing, S., and Munger, K. (2002). Human papillomaviruses and centrosome duplication errors: modeling the origins of genomic instability. Oncogene *21*, 6241–6248.

Dyson, N., Howley, P.M., Munger, K., and Harlow, E. (1989). The human papilloma virus-16 E7 oncoprotein is able to bind to the retinoblastoma gene product. Science *243*, 934–937.

Dyson, N., Guida, P., Munger, K., and Harlow, E. (1992). Homologous sequences in adenovirus E1A and human papillomavirus E7 proteins mediate interaction with the same set of cellular proteins. J. Virol. *66*, 6893–6902.

Edmonds, C., and Vousden, K.H. (1989). A point mutational analysis of human papillomavirus type 16 E7 protein. J. Virol. *63*, 2650–2656.

Ferreon, A.C., Gambin, Y., Lemke, E.A., and Deniz, A.A. (2009a). Interplay of alpha-synuclein binding and conformational switching probed by single-molecule fluorescence. Proc. Natl. Acad. Sci. U.S.A. *106*, 5645–5650.

Ferreon, J.C., Martinez-Yamout, M.A., Dyson, H.J., and Wright, P.E. (2009b). Structural basis for subversion of cellular control mechanisms by the adenoviral E1A oncoprotein. Proc. Natl. Acad. Sci. U.S.A. *106*, 13260–13265.

Fiedler, M., Muller-Holzner, E., Viertler, H.P., Widschwendter, A., Laich, A., Pfister, G., Spoden, G.A., Jansen-Durr, P., and Zwerschke, W. (2004). High level HPV16 E7 oncoprotein expression correlates with reduced pRb-levels in cervical biopsies. FASEB J. *18*, 1120–1122.

Fink, A.L. (2005). Natively unfolded proteins. Curr. Opin. Struct. Biol. *15*, 35–41.

Firzlaff, J.M., Luscher, B., and Eisenman, R.N. (1991). Negative charge at the casein kinase II phosphorylation site is important for transformation but not for Rb protein binding by the E7 protein of human papillomavirus type 16. Proc. Natl. Acad. Sci. U.S.A. *88*, 5187–5191.

Fujikawa, K., Furuse, M., Uwabe, K., Maki, H., and Yoshie, O. (1994). Nuclear localization and transforming activity of human papillomavirus type 16 E7–beta-galactosidase fusion protein: characterization of the nuclear localization sequence. Virology *204*, 789–793.

Funk, J.O., Waga, S., Harry, J.B., Espling, E., Stillman, B., and Galloway, D.A. (1997). Inhibition of CDK activity and PCNA-dependent DNA replication by p21 is blocked by interaction with the HPV16 E7 oncoprotein. Genes Dev. *11*, 2090–2100.

Galloway, D.A. (1992). Serological assays for the detection of HPV antibodies. IARC Sci. Publ. 147–161.

Gammoh, N., Grm, H.S., Massimi, P., and Banks, L. (2006). Regulation of human papillomavirus type 16 E7 activity through direct protein interaction with the E2 transcriptional activator. J. Virol. *80*, 1787–1797.

Gammoh, N., Isaacson, E., Tomaic, V., Jackson, D.J., Doorbar, J., and Banks, L. (2009). Inhibition of HPV16 E7 oncogenic activity by HPV16 E2. Oncogene *28*, 2299–2304.

Garcia-Alai, M.M., Gallo, M., Salame, M., Wetzler, D.E., McBride, A.A., Paci, M., Cicero, D.O., and de Prat-Gay, G. (2006). Molecular basis for phosphorylation-dependent, PEST-mediated protein turnover. Structure *14*, 309–319.

Garcia-Alai, M.M., Alonso, L.G., and de Prat-Gay, G. (2007a). The N-terminal module of HPV16 E7 is an intrinsically disordered domain that confers conformational and recognition plasticity to the oncoprotein. Biochemistry *46*, 10405–10412.

Garcia-Alai, M.M., Dantur, K.I., Smal, C., Pietrasanta, L., and de Prat-Gay, G. (2007b). High-risk HPV E6 oncoproteins assemble into large oligomers that allow localization of endogenous species in prototypic HPV-transformed cell lines. Biochemistry *46*, 341–349.

Genovese, N.J., Banerjee, N.S., Broker, T.R., and Chow, L.T. (2008). Casein kinase II motif-dependent phosphorylation of human papillomavirus E7 protein promotes p130 degradation and S-phase induction in differentiated human keratinocytes. J. Virol. *82*, 4862–4873.

Georgopoulos, N.T., Proffitt, J.L., and Blair, G.E. (2000). Transcriptional regulation of the major histocompatibility complex (MHC) class I heavy chain, TAP1 and LMP2 genes by the human papillomavirus (HPV) type 6b, 16 and 18 E7 oncoproteins. Oncogene *19*, 4930–4935.

Giri, I., Danos, O., and Yaniv, M. (1985). Genomic structure of the cottontail rabbit (Shope) papillomavirus. Proc. Natl. Acad. Sci. U.S.A. *82*, 1580–1584.

Gonzalez, S.L., Stremlau, M., He, X., Basile, J.R., and Munger, K. (2001). Degradation of the retinoblastoma tumour suppressor by the human papillomavirus type 16 E7 oncoprotein is important for functional inactivation and is separable from proteasomal degradation of E7. J. Virol. *75*, 7583–7591.

Greenfield, I., Nickerson, J., Penman, S., and Stanley, M. (1991). Human papillomavirus 16 E7 protein is associated with the nuclear matrix. Proc. Natl. Acad. Sci. U.S.A. *88*, 11217–11221.

Hamid, N.A., Brown, C., and Gaston, K. (2009). The regulation of cell proliferation by the papillomavirus early proteins. Cell. Mol. Life Sci. *66*, 1700–1717.

zur Hausen, H. (1994). Molecular pathogenesis of cancer of the cervix and its causation by specific human papillomavirus types. Curr. Top. Microbiol. Immunol. *186*, 131–156.

zur Hausen, H. (2009). Papillomaviruses in the causation of human cancers – a brief historical account. Virology *384*, 260–265.

He, W., Staples, D., Smith, C., and Fisher, C. (2003). Direct activation of cyclin-dependent kinase 2 by human papillomavirus E7. J. Virol. *77*, 10566–10574.

Heck, D.V., Yee, C.L., Howley, P.M., and Munger, K. (1992). Efficiency of binding the retinoblastoma protein correlates with the transforming capacity of the E7 oncoproteins of the human papillomaviruses. Proc. Natl. Acad. Sci. U.S.A. *89*, 1442–1446.

Hegde, R.S. (2002). The papillomavirus E2 proteins: structure, function, and biology. Annu. Rev. Biophys. Biomol. Struct. *31*, 343–360.

Helt, A.M., and Galloway, D.A. (2001). Destabilization of the retinoblastoma tumour suppressor by human papillomavirus type 16 E7 is not sufficient to overcome cell cycle arrest in human keratinocytes. J. Virol. *75*, 6737–6747.

Helt, A.M., and Galloway, D.A. (2003). Mechanisms by which DNA tumour virus oncoproteins target the Rb family of pocket proteins. Carcinogenesis *24*, 159–169.

Hershko, A., and Ciechanover, A. (1998). The ubiquitin system. Annu. Rev. Biochem. *67*, 425–479.

Howley, P.M., Munger, K., Romanczuk, H., Scheffner, M., and Huibregtse, J.M. (1991). Cellular targets of the oncoproteins encoded by the cancer associated human papillomaviruses. Princess Takamatsu Symp. *22*, 239–248.

Howley, P.M., and Lowy, D.R. (2007). Papillomaviruses. In Fields Virology, Knipe, D.M., and Howley, P.M., eds (Lippincott, Williams and Wilkins, Philadelphia), pp. 2299–2354.

Huh, K., Zhou, X., Hayakawa, H., Cho, J.Y., Libermann, T.A., Jin, J., Harper, J.W., and Munger, K. (2007). Human papillomavirus type 16 E7 oncoprotein associates with the cullin 2 ubiquitin ligase complex, which contributes to degradation of the retinoblastoma tumor suppressor. J. Virol. *81*, 9737–9747.

Huh, K.W., DeMasi, J., Ogawa, H., Nakatani, Y., Howley, P.M., and Munger, K. (2005). Association of the human papillomavirus type 16 E7 oncoprotein with the 600-kDa retinoblastoma protein-associated factor, p600. Proc. Natl. Acad. Sci. U.S.A. *102*, 11492–11497.

Hung, C.F., Ma, B., Monie, A., Tsen, S.W., and Wu, T.C. (2008). Therapeutic human papillomavirus vaccines: current clinical trials and future directions. Expert Opin. Biol. Ther. *8*, 421–439.

Hwang, S.G., Lee, D., Kim, J., Seo, T., and Choe, J. (2002). Human papillomavirus type 16 E7 binds to E2F1 and activates E2F1-driven transcription in a retinoblastoma protein-independent manner. J. Biol. Chem. *277*, 2923–2930.

Ikeda, M.A., and Nevins, J.R. (1993). Identification of distinct roles for separate E1A domains in disruption of E2F complexes. Mol. Cell Biol. *13*, 7029–7035.

Jha, P.K., Beral, V., Peto, J., Hack, S., Hermon, C., Deacon, J., Mant, D., Chilvers, C., Vessey, M.P., Pike, M.C., *et al.* (1993). Antibodies to human papillomavirus and to other genital infectious agents and invasive cervical cancer risk. Lancet *341*, 1116–1118.

Jochmus-Kudielka, I., Schneider, A., Braun, R., Kimmig, R., Koldovsky, U., Schneweis, K.E., Seedorf, K., and Gissmann, L. (1989). Antibodies against the human papillomavirus type 16 early proteins in human sera: correlation of anti-E7 reactivity with cervical cancer. J. Natl. Cancer Inst. *81*, 1698–1704.

Jones, D.L., Alani, R.M., and Munger, K. (1997). The human papillomavirus E7 oncoprotein can uncouple cellular differentiation and proliferation in human keratinocytes by abrogating p21Cip1-mediated inhibition of cdk2. Genes Dev. *11*, 2101–2111.

Jones, R.E., Wegrzyn, R.J., Patrick, D.R., Balishin, N.L., Vuocolo, G.A., Riemen, M.W., Defeo-Jones, D., Garsky, V.M., Heimbrook, D.C., and Oliff, A. (1990). Identification of HPV16 E7 peptides that are potent antagonists of E7 binding to the retinoblastoma suppressor protein. J. Biol. Chem. *265*, 12782–12785.

Kalantari, M., Karlsen, F., Kristensen, G., Holm, R., Hagmar, B., and Johansson, B. (1998). Disruption of the E1 and E2 reading frames of HPV16 in cervical carcinoma is associated with poor prognosis. Int. J. Gynecol. Pathol. *17*, 146–153.

Kanda, T., Zanma, S., Watanabe, S., Furuno, A., and Yoshiike, K. (1991). Two immunodominant regions of the human papillomavirus type 16 E7 protein are masked in the nuclei of monkey COS-1 cells. Virology *182*, 723–731.

Kim, T.Y., Myoung, H.J., Kim, J.H., Moon, I.S., Kim, T.G., Ahn, W.S., and Sin, J.I. (2002). Both E7 and CpG-oligodeoxynucleotide are required for protective immunity against challenge with human papillomavirus 16 (E6/E7) immortalized tumor cells: involvement of CD4+ and CD8+ T cells in protection. Cancer Res. *62*, 7234–7240.

Kols, A., Sherris, J., Herdman, C., and Wittet, S. (2006). Current and Future HPV Vaccines: Promise and Challenges (PATH, Seattle, USA). pp 1–72.

Kornitzer, D., and Ciechanover, A. (2000). Modes of regulation of ubiquitin-mediated protein degradation. J. Cell Physiol. *182*, 1–11.

Lee, J.O., Russo, A.A., and Pavletich, N.P. (1998). Structure of the retinoblastoma tumour-suppressor pocket domain bound to a peptide from HPV E7. Nature *391*, 859–865.

Leggatt, G.R., and Frazer, I.H. (2007). HPV vaccines: the beginning of the end for cervical cancer. Curr. Opin. Immunol. *19*, 232–238.

Lin, K.Y., Guarnieri, F.G., Staveley-O'Carroll, K.F., Levitsky, H.I., August, J.T., Pardoll, D.M., and Wu, T.C. (1996a). Treatment of established tumors with a novel vaccine that enhances major histocompatibility class II presentation of tumor antigen. Cancer Res. *56*, 21–26.

Lin, R., Beauparlant, P., Makris, C., Meloche, S., and Hiscott, J. (1996b). Phosphorylation of IkappaBalpha in the C-terminal PEST domain by casein kinase II affects intrinsic protein stability. Mol. Cell Biol. *16*, 1401–1409.

Liu, X., Clements, A., Zhao, K., and Marmorstein, R. (2006). Structure of the human Papillomavirus E7 oncoprotein and its mechanism for inactivation of the retinoblastoma tumor suppressor. J. Biol. Chem. *281*, 578–586.

Liu, X., and Marmorstein, R. (2007). Structure of the retinoblastoma protein bound to adenovirus E1A reveals the molecular basis for viral oncoprotein inactivation of a tumor suppressor. Genes Dev. *21*, 2711–2716.

Longworth, M.S., and Laimins, L.A. (2004). The binding of histone deacetylases and the integrity of zinc finger-like motifs of the E7 protein are essential for the life cycle of human papillomavirus type 31. J. Virol. *78*, 3533–3541.

Longworth, M.S., Wilson, R., and Laimins, L.A. (2005). HPV-31 E7 facilitates replication by activating E2F2 transcription through its interaction with HDACs. EMBO J. *24*, 1821–1830.

Luscher-Firzlaff, J.M., Westendorf, J.M., Zwicker, J., Burkhardt, H., Henriksson, M., Muller, R., Pirollet, F., and Luscher, B. (1999). Interaction of the fork head domain transcription factor MPP2 with the human papilloma virus 16 E7 protein: enhancement of transformation and transactivation. Oncogene *18*, 5620–5630.

McCance, D.J. (2006). The biology of the E7 protein of HPV16. In Papillomavirus Research: From Natural History to Vaccines and Beyond, Saveria Campo, M., ed. (Caister Academic Press, Norfolk, England), pp. 133–144.

McIntyre, M.C., Frattini, M.G., Grossman, S.R., and Laimins, L.A. (1993). Human papillomavirus type 18 E7 protein requires intact Cys-X-X-Cys motifs for zinc binding, dimerization, and transformation but not for Rb binding. J. Virol. *67*, 3142–3150.

McIntyre, M.C., Ruesch, M.N., and Laimins, L.A. (1996). Human papillomavirus E7 oncoproteins bind a single form of cyclin E in a complex with cdk2 and p107. Virology *215*, 73–82.

McLaughlin-Drubin, M.E., and Munger, K. (2009). Oncogenic activities of human papillomaviruses. Virus Res. *143*, 195–208.

McLaughlin-Drubin, M.E., Huh, K.W., and Munger, K. (2008). Human papillomavirus type 16 E7 oncoprotein associates with E2F6. J. Virol. *82*, 8695–8705.

Massimi, P., and Banks, L. (2000). Differential phosphorylation of the HPV16 E7 oncoprotein during the cell cycle. Virology *276*, 388–394.

Mazzarelli, J.M., Atkins, G.B., Geisberg, J.V., and Ricciardi, R.P. (1995). The viral oncoproteins Ad5 E1A, HPV16 E7 and SV40 TAg bind a common region of the TBP-associated factor-110. Oncogene *11*, 1859–1864.

Michel, N., Ohlschlager, P., Osen, W., Freyschmidt, E.J., Guthohrlein, H., Kaufmann, A.M., Muller, M., and Gissmann, L. (2002). T cell response to human papillomavirus 16 E7 in mice: comparison of Cr release assay, intracellular IFN-gamma production, ELISPOT and tetramer staining. Intervirology *45*, 290–299.

Moody, C.A., and Laimins, L.A. (2010). Human papillomavirus oncoproteins: pathways to transformation. Nat. Rev. Cancer *10*, 550–560.

Munger, K., and Howley, P.M. (2002). Human papillomavirus immortalization and transformation functions. Virus Res. *89*, 213–228.

Munger, K., and Phelps, W.C. (1993). The human papillomavirus E7 protein as a transforming and transactivating factor. Biochim. Biophys. Acta *1155*, 111–123.

Munger, K., Yee, C.L., Phelps, W.C., Pietenpol, J.A., Moses, H.L., and Howley, P.M. (1991). Biochemical and biological differences between E7 oncoproteins of the high- and low-risk human papillomavirus types are determined by amino-terminal sequences. J. Virol. *65*, 3943–3948.

Nees, M., Geoghegan, J.M., Hyman, T., Frank, S., Miller, L., and Woodworth, C.D. (2001). Papillomavirus type 16 oncogenes down-regulate expression of interferon-responsive genes and upregulate proliferation-associated and NF-kappaB-responsive genes in cervical keratinocytes. J. Virol. *75*, 4283–4296.

Nguyen, C.L., and Munger, K. (2009). Human papillomavirus E7 protein deregulates mitosis via an association with nuclear mitotic apparatus protein 1. J. Virol. *83*, 1700–1707.

Nindl, I., Gissmann, L., Fisher, S.G., Bribiesca, L.B., Berumen, J., and Muller, M. (1996). The E7 protein of human papillomavirus (HPV) type 16 expressed by recombinant vaccinia virus can be used for detection of antibodies in sera from cervical cancer patients. J. Virol. Methods *62*, 81–85.

Oh, K.J., Kalinina, A., Wang, J., Nakayama, K., Nakayama, K.I., and Bagchi, S. (2004). The papillomavirus E7 oncoprotein is ubiquitinated by UbcH7 and Cullin 1- and Skp2-containing E3 ligase. J. Virol. *78*, 5338–5346.

Ohlenschlager, O., Seiboth, T., Zengerling, H., Briese, L., Marchanka, A., Ramachandran, R., Baum, M., Korbas, M., Meyer-Klaucke, W., Durst, M., *et al.* (2006). Solution structure of the partially folded high-risk human papilloma virus 45 oncoprotein E7. Oncogene *25*, 5953–5959.

Oltersdorf, T., Seedorf, K., Rowekamp, W., and Gissmann, L. (1987). Identification of human papillomavirus type 16 E7 protein by monoclonal antibodies. J. Gen. Virol. *68 (Pt 11)*, 2933–2938.

Pahel, G., Aulabaugh, A., Short, S.A., Barnes, J.A., Painter, G.R., Ray, P., and Phelps, W.C. (1993). Structural and functional characterization of the HPV16 E7 protein expressed in bacteria. J. Biol. Chem. *268*, 26018–26025.

Park, J.S., Kim, E.J., Kwon, H.J., Hwang, E.S., Namkoong, S.E., and Um, S.J. (2000). Inactivation of interferon regulatory factor-1 tumor suppressor protein by HPV E7 oncoprotein. Implication for the E7-mediated immune evasion mechanism in cervical carcinogenesis. J. Biol. Chem. *275*, 6764–6769.

Patrick, D.R., Oliff, A., and Heimbrook, D.C. (1994). Identification of a novel retinoblastoma gene product binding site on human papillomavirus type 16 E7 protein. J. Biol. Chem. *269*, 6842–6850.

Pelka, P., Ablack, J.N., Fonseca, G.J., Yousef, A.F., and Mymryk, J.S. (2008). Intrinsic structural disorder in adenovirus E1A: a viral molecular hub linking multiple diverse processes. J. Virol. *82*, 7252–7263.

Phelps, W.C., Yee, C.L., Munger, K., and Howley, P.M. (1988). The human papillomavirus type 16 E7 gene encodes transactivation and transformation functions similar to those of adenovirus E1A. Cell *53*, 539–547.

Phelps, W.C., Munger, K., Yee, C.L., Barnes, J.A., and Howley, P.M. (1992). Structure–function analysis of the human papillomavirus type 16 E7 oncoprotein. J. Virol. *66*, 2418–2427.

Phillips, A.C., and Vousden, K.H. (1997). Analysis of the interaction between human papillomavirus type 16 E7 and the TATA-binding protein, TBP. J. Gen. Virol. *78 (Pt 4)*, 905–909.

Pim, D., and Banks, L. (2010). Interaction of viral oncoproteins with cellular target molecules: infection with high-risk vs low-risk human papillomaviruses. APMIS *118*, 471–493.

Rechsteiner, M., and Rogers, S.W. (1996). PEST sequences and regulation by proteolysis. Trends Biochem Sci *21*, 267–271.

Reinstein, E., Scheffner, M., Oren, M., Ciechanover, A., and Schwartz, A. (2000). Degradation of the E7 human papillomavirus oncoprotein by the ubiquitin-proteasome system: targeting via ubiquitination of the N-terminal residue. Oncogene *19*, 5944–5950.

Ressler, S., Scheiden, R., Dreier, K., Laich, A., Muller-Holzner, E., Pircher, H., Morandell, D., Stein, I., Viertler, H.P., Santer, F.R., *et al.* (2007). High-risk human papillomavirus E7 oncoprotein detection in cervical squamous cell carcinoma. Clin. Cancer Res. *13*, 7067–7072.

Reuschenbach, M., Waterboer, T., Wallin, K.L., Einenkel, J., Dillner, J., Hamsikova, E., Eschenbach, D., Zimmer, H., Heilig, B., Kopitz, J., *et al.* (2008). Characterization of humoral immune responses against p16, p53, HPV16 E6 and HPV16 E7 in patients with HPV-associated cancers. Int. J. Cancer *123*, 2626–2631.

Rey, O., Lee, S., Baluda, M.A., Swee, J., Ackerson, B., Chiu, R., and Park, N.H. (2000). The E7 oncoprotein of human papillomavirus type 16 interacts with F-actin *in vitro* and *in vivo*. Virology *268*, 372–381.

Rihs, H.P., Jans, D.A., Fan, H., and Peters, R. (1991). The rate of nuclear cytoplasmic protein transport is determined by the casein kinase II site flanking the nuclear localization sequence of the SV40 T-antigen. EMBO J. *10*, 633–639.

Ronco, L.V., Karpova, A.Y., Vidal, M., and Howley, P.M. (1998). Human papillomavirus 16 E6 oncoprotein binds to interferon regulatory factor-3 and inhibits its transcriptional activity. Genes Dev. *12*, 2061–2072.

Schwarz, E., Freese, U.K., Gissmann, L., Mayer, W., Roggenbuck, B., Stremlau, A., and zur Hausen, H. (1985). Structure and transcription of human papillomavirus sequences in cervical carcinoma cells. Nature *314*, 111–114.

Seedorf, K., Krammer, G., Durst, M., Suhai, S., and Rowekamp, W.G. (1985). Human papillomavirus type 16 DNA sequence. Virology *145*, 181–185.

Selvey, L.A., Dunn, L.A., Tindle, R.W., Park, D.S., and Frazer, I.H. (1994). Human papillomavirus (HPV) type 18 E7 protein is a short-lived steroid-inducible phosphoprotein in HPV-transformed cell lines. J. Gen. Virol. *75 (Pt 7)*, 1647–1653.

Smal, C. (2010). Oligomerización de la oncoproteína E7 del papilomavirus humano y su interacción con el regulador de la transcripción y replicación viral E2. In Physics Department, PhD Thesis (University of Buenos Aires, Argentina).

Smal, C., Wetzler, D.E., Dantur, K.I., Chemes, L.B., Garcia-Alai, M.M., Dellarole, M., Alonso, L.G., Gaston, K., and de Prat-Gay, G. (2009). The human papillomavirus E7–E2 interaction mechanism *in vitro* reveals a finely tuned system for modulating available E7 and E2 proteins. Biochemistry *48*, 11939–11949.

Smotkin, D., and Wettstein, F.O. (1987). The major human papillomavirus protein in cervical cancers is a cytoplasmic phosphoprotein. J. Virol. *61*, 1686–1689.

de Souza, R.F., Iyer, L.M., and Aravind, L. (2009). Diversity and evolution of chromatin proteins encoded by DNA viruses. Biochim. Biophys. Acta *1799*, 302–318.

Stoler, M.H., Rhodes, C.R., Whitbeck, A., Wolinsky, S.M., Chow, L.T., and Broker, T.R. (1992). Human papillomavirus type 16 and 18 gene expression in cervical neoplasias. Hum. Pathol. *23*, 117–128.

Sullivan, C.S., Cantalupo, P., and Pipas, J.M. (2000). The molecular chaperone activity of simian virus 40 large T antigen is required to disrupt Rb-E2F family complexes by an ATP-dependent mechanism. Mol. Cell Biol. *20*, 6233–6243.

Thierry, F. (2009). Transcriptional regulation of the papillomavirus oncogenes by cellular and viral transcription factors in cervical carcinoma. Virology *384*, 375–379.

Tokuriki, N., Oldfield, C.J., Uversky, V.N., Berezovsky, I.N., and Tawfik, D.S. (2009). Do viral proteins possess unique biophysical features? Trends Biochem. Sci. *34*, 53–59.

Tommasino, M., Adamczewski, P.J., Carlotti, F., Barth, C.F., Manetti, R., Contorni, M., Cavalieri, F., Hunt, T., and Crawdford, L. (1993). HPV16 E7 protein associates with the protein kinase p33CDK2 and cyclin A. Oncogene *8*, 195–202.

Uversky, V.N. (2002). What does it mean to be natively unfolded? Eur. J. Biochem. *269*, 2–12.

Uversky, V.N., Gillespie, J.R., and Fink, A.L. (2000). Why are 'natively unfolded' proteins unstructured under physiologic conditions? Proteins *41*, 415–427.

Uversky, V.N., Roman, A., Oldfield, C.J., and Dunker, A.K. (2006). Protein intrinsic disorder and human papillomaviruses: increased amount of disorder in E6 and E7 oncoproteins from high risk HPVs. J. Proteome Res. *5*, 1829–1842.

Vambutas, A., DeVoti, J., Pinn, W., Steinberg, B.M., and Bonagura, V.R. (2001). Interaction of human papillomavirus type 11 E7 protein with TAP-1 results in the reduction of ATP-dependent peptide transport. Clin. Immunol. *101*, 94–99.

Vousden, K.H., Doniger, J., DiPaolo, J.A., and Lowy, D.R. (1988). The E7 open reading frame of human papillomavirus type 16 encodes a transforming gene. Oncogene Res. *3*, 167–175.

Wang, J., Sampath, A., Raychaudhuri, P., and Bagchi, S. (2001). Both Rb and E7 are regulated by the ubiquitin proteasome pathway in HPV-containing cervical tumor cells. Oncogene *20*, 4740–4749.

Wang, J., Zhou, D., Prabhu, A., Schlegel, R., and Yuan, H. (2010). The canine papillomavirus and gamma HPV E7 proteins use an alternative domain to bind and destabilize the retinoblastoma protein. PLoS Pathog. *6*, e1001089.

Whalen, S.G., Marcellus, R.C., Barbeau, D., and Branton, P.E. (1996). Importance of the Ser-132 phosphorylation site in cell transformation and apoptosis induced by the adenovirus type 5 E1A protein. J. Virol. *70*, 5373–5383.

Wise-Draper, T.M., and Wells, S.I. (2008). Papillomavirus E6 and E7 proteins and their cellular targets. Front. Biosci. *13*, 1003–1017.

Woods, W.S., Boettcher, J.M., Zhou, D.H., Kloepper, K.D., Hartman, K.L., Ladror, D.T., Qi, Z., Rienstra, C.M., and George, J.M. (2007). Conformation-specific binding of alpha-synuclein to novel protein partners detected by phage display and NMR spectroscopy. J. Biol. Chem. *282*, 34555–34567.

Xue, B., Williams, R.W., Oldfield, C.J., Goh, G.K., Dunker, A.K., and Uversky, V.N. (2010). Viral disorder or disordered viruses: do viral proteins possess unique features? Protein Pept. Lett. *17*, 932–951.

Yutsudo, M., Okamoto, Y., and Hakura, A. (1988). Functional dissociation of transforming genes of human papillomavirus type 16. Virology *166*, 594–597.

Zerfass-Thome, K., Zwerschke, W., Mannhardt, B., Tindle, R., Botz, J.W., and Jansen-Durr, P. (1996). Inactivation of the cdk inhibitor p27KIP1 by the human papillomavirus type 16 E7 oncoprotein. Oncogene *13*, 2323–2330.

Zhou, F., Leggatt, G.R., and Frazer, I.H. (2011). Human papillomavirus 16 E7 protein inhibits interferon-gamma-mediated enhancement of keratinocyte antigen processing and T-cell lysis. FEBS J. *278*, 955–63.

Zwerschke, W., Joswig, S., and Jansen-Durr, P. (1996). Identification of domains required for transcriptional activation and protein dimerization in the human papillomavirus type-16 E7 protein. Oncogene *12*, 213–220.

Zwerschke, W., Mazurek, S., Massimi, P., Banks, L., Eigenbrodt, E., and Jansen-Durr, P. (1999). Modulation of type M2 pyruvate kinase activity by the human papillomavirus type 16 E7 oncoprotein. Proc. Natl. Acad. Sci. U.S.A. *96*, 1291–1296.

Zwerschke, W., Mannhardt, B., Massimi, P., Nauenburg, S., Pim, D., Nickel, W., Banks, L., Reuser, A.J., and Jansen-Durr, P. (2000). Allosteric activation of acid alpha-glucosidase by the human papillomavirus E7 protein. J. Biol. Chem. *275*, 9534–9541.

7 Replication and Maintenance of Viral Genomes by Association with Host Chromatin

Koenraad Van Doorslaer, Vandana Sekhar, Jameela Khan and Alison A. McBride

Abstract

Papillomaviruses persistently infect dividing epithelial cells. This cellular environment presents papillomaviruses with the challenge of having to replicate and retain their genome in proliferating cells. Papillomaviruses have evolved to maintain their genome as an extra-chromosomal element. The viral E2 protein binds to the viral genome and tethers it to the host chromosomes. This chapter will review some of the recent work showing how different papillomaviruses have evolved variations on this theme. Where appropriate, parallels will be drawn with other persistent, double-stranded DNA viruses.

Introduction

Unlike many acute viruses, persistent viruses have evolved alongside their host, resulting in a symbiotic relationship. From the viewpoint of an evolutionary biologist, the *Papillomaviridae* family (PVs for the remainder of this chapter) is one of the more successful groups of DNA viruses. This ubiquitous group of viruses are the causative agents of cutaneous and mucosotropic infections in almost all studied amniotes (Bernard *et al.*, 2010). PVs are classified into different genera based on nucleotide similarity across the highly conserved L1 structural gene. These genera are named according to the Greek alphabet (Bernard *et al.*, 2010). In this review we will mainly focus on the alpha, beta and delta genera. The *Alphapapillomavirus* (alpha-PV) genus contains mainly mucosotropic viruses (e.g. HPV16), while the viruses in the beta-PV genus infect mainly the skin. The delta-PV genus contains PVs that infect ruminants. The delta-PV, bovine PV type 1 (BPV1) has been historically studied because of the relative ease in obtaining virus and because of its ability to readily transform and replicate in murine fibroblasts. Hence, BPV1 became an early molecular prototype in the study of viral replication, transcription and transformation.

Infections with PVs span the range from asymptomatic to invasive carcinomas. Persistent infection with a specific subset of HPVs is responsible for the aetiology of cervical cancer (Walboomers *et al.*, 1999; Bouvard *et al.*, 2010). One of the most challenging problems for a persistent virus is the need for a robust mechanism for viral genome maintenance. This poses a set of problems for a virus that maintains its genome extrachromosomally throughout infection. This chapter will focus on the methods used by PVs (and other double-stranded DNA viruses) to physically link their genomes to the host chromosome allowing for viral genome partitioning and nuclear retention following cellular division.

Papillomavirus life cycle

Despite their ubiquitous presence in nature (Antonsson *et al.*, 2000; Antonsson and Hansson, 2002), PVs are highly species specific. In addition, PVs have specialized to infect particular anatomical niches in the cutaneous and mucosal epithelia of their host, thus strictly limiting the cellular targets available for infection. Despite this species and niche specificity, the life cycle of all PVs progresses in a specific order of events. These steps are tightly associated with the differentiation state of the infected host keratinocyte (reviewed in Doorbar, 2006). The current model of PV infection proposes that the virus reaches the basal cell layer following epithelial wounding. Binding of the virus to heparin sulfate proteoglycans (HSPG) on the exposed basement membrane triggers a conformational change, which in turn exposes the viral L2 protein for furin cleavage (Schiller *et al.*, 2010). This allows unmasking of an L1 epitope and binding to an, as yet unknown, secondary receptor on the migrating keratinocytes at the leading edge of the wound. Two to four hours following cell surface binding, the virus is internalized. Following entry, uncoating of the viral capsid is initiated in the late endosomes (Kamper *et al.*, 2006). Following uncoating, the minor capsid protein, L2 and the viral genome are trafficked towards the nucleus along the microtubule network (Florin *et al.*, 2006). The viral genome may be actively imported into the nucleus in complex with the L2 protein (Bordeaux *et al.*, 2006) or may enter during nuclear membrane breakdown in mitosis (Pyeon *et al.*, 2009). In the nucleus the viral genome accumulates, with the L2 protein, in the vicinity of promyelocytic leukaemia oncogenic (PML) domains and a role for this subnuclear structure for successful viral establishment has been suggested (Day *et al.*, 2004).

Following nuclear entry, replication of the viral genome can be divided into three main stages. An initial amplification phase, where the incoming genome is amplified to a low copy number, is followed by maintenance replication. In the maintenance stage the viral genome replicates in synchrony with host DNA synthesis and is maintained at a constant copy number. Finally, when the infected cells migrate upwards and begin to differentiate, a switch is made towards vegetative replication, and the viral genomes are amplified to a high copy number. Expression of the late major and minor capsid antigens (L1 and L2) results in packaging of the progeny genomes. These newly packaged genomes are released into the environment as viral-laden squames, when the superficial cells slough-off the surface of the epithelium. The cues responsible for switching replication modes are not well understood, but a link to the differentiation state of the host epithelium is clear. It has been proposed that this strategy allows the virus to avoid detection, as the host cells that support vegetative replication are terminally differentiating and therefore not under stringent immune surveillance.

Viral persistence

Some viruses are able to establish persistent infections. In order to persist, the virus needs to subvert a wide array of antiviral tactics utilized by the host, including cellular apoptotic pathways and the host's immunological surveillance. Notably, many persistent viruses have evolved similar mechanisms to overcome these hurdles. A key feature adapted by most persistent viruses is the ability to establish a latent or low level, chronic infection. In this state no, or very limited amounts of, viral proteins are expressed in cells harbouring the viral genome. The selection of target cells is crucial for this strategy. For example, following the initial infection of keratinocytes, the *Alphaherpesvirinae*, such as herpes simplex (HSV-1) and varicella zoster virus (VZV), enter a latent phase of their lifecycle in long-lived, non-dividing neuronal cells. Neurons that are latently infected with HSV or VZV express only a latency-associated transcript, whose function remains elusive, until reactivation of the lytic lifecycle (Bloom *et al.*, 2010).

However, other viruses enter the latent phase of their lifecycle in actively dividing cells, which adds an additional level of complexity in maintaining the viral genome. Members of the *Gammaherpesvirinae*, such as Epstein–Barr virus (EBV) and Kaposi's sarcoma-associated herpesvirus (KSHV) infect both epithelial and lymphoid cells. Following initial infection and viral genome amplification these viruses establish a latent lifecycle in a subset of CD19 positive B cells. During the latent phase of infection, which lasts for the lifetime of the host, the number of infected cells and viral load is typically very low.

A common feature of the gammaherpesviruses and the papillomaviruses is the ability to maintain a long-term latent or low-level persistent infection and maintain their genomes as extrachromosomal circular genomes in dividing cells.

Viral genome replication and partitioning

Modes of papillomavirus replication

As mentioned above, three modes of replication exist during the PV lifecycle, and it is assumed that for all three modes of replication, the origin of replication (minimally containing binding sites for E1 and E2) and the viral E1 and E2 proteins (Mohr *et al.*, 1990; Ustav *et al.*, 1991; Ustav and Stenlund, 1991) are required for initiation. While the E1 helicase is the primary replication protein, the E2 protein is required to increase the specificity and affinity of the E1 protein for the viral origin.

The PV E1 protein is an ATP-dependent helicase that initiates replication by binding specifically to the origin (Yang *et al.*, 1993). E1 consists of four domains; an N-terminal domain, a sequence specific DNA binding domain, an oligomerization domain and a helicase domain. The N-terminal domain contains both nuclear import and export signals that are regulated by cellular kinases. E1 binds to DNA both specifically and non-specifically using two different DNA binding regions. While DNA binding to the replication origin is mediated by the specific DNA-binding domain (DBD) (Sarafi and McBride, 1995; Chen and Stenlund, 1998) the helicase domain is also able to bind DNA with low sequence specificity in the region flanking the E1BS (Schuck and Stenlund, 2006). The helicase domain of E1 interacts with the transactivation domain (TAD) of the E2 protein (Sarafi and McBride, 1995; Sedman *et al.*, 1997) and binding of the E2 dimer to its origin binding site results in cooperative recruitment of E1. This is followed by dissociation of E2 and the formation of a double hexameric ring of E1 proteins (Sanders and Stenlund, 1998; Sedman and Stenlund, 1998). The solved X-ray crystallographic structure of the E1 helicase associated with its DNA target shows that single-stranded DNA is threaded through the hexamer channel, implying melting of the origin either before or during hexamer ring formation (Enemark and Joshua-Tor, 2006). The E1 and E2 proteins also recruit many cellular proteins that synthesize viral DNA in a bidirectional manner (reviewed in McBride, 2008).

In addition to their role in replication, the E2 proteins are the primary regulators of viral transcription. The E2 protein consists of an N-terminal domain important for transcriptional regulation and interaction with the E1 protein and a C-terminal DNA binding and dimerization domain. A non-conserved linker region connects the N- and the C-terminal domains (Fig. 7.1). The full-length E2 protein is often named E2-TA

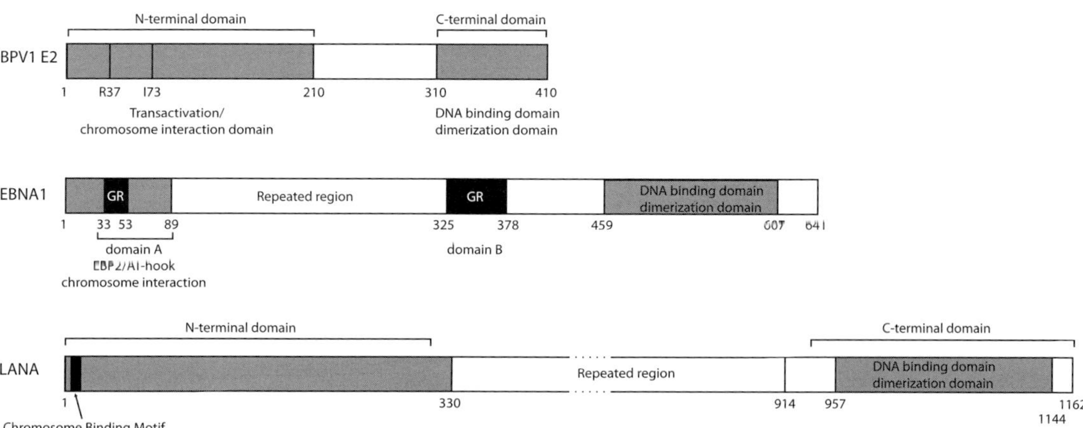

Figure 7.1 Functional domains of the viral tethering proteins. Despite the lack of primary sequence similarity, the viral tethering proteins have a similar modular build. The C-terminal domain is important for binding to canonical binding sites within the viral genome, as well as dimerization of the proteins. The N-terminal domains in turn are implicated in interaction with the host chromosomes. These domains are connected with an unfolded linker. The folded domains are indicated by grey boxes, while the white boxes represent unfolded regions in the proteins. Black boxes indicate the locations of mapped regions involved in chromosomal interaction.

(E2 transactivator). Shorter repressor proteins, the results of alternative promoter usage and/or alternative splicing (E2-TR or E8^E2) contain the C-terminal DNA-binding domain fused to sequences from the hinge or upstream open reading frames. The different E2 proteins regulate the activity of viral promoters by binding to specific binding motifs in the viral enhancer regions. Both the E2-TA and E2 repressor proteins regulate transcription by recruiting cellular transcription factors, coactivators and repressors (Wu and Chiang, 2007; Powell *et al.*, 2010; Smith *et al.*, 2010).

General strategy of genome maintenance and partitioning

As described above, the E2 protein is essential during viral replication where it acts as a loading factor for the E1 helicase. However, in BPV1 it became clear that E2 also played a key role during the maintenance phase of viral replication. It was noted that extra E2 binding sites (in addition to the E2BS in the viral origin) were essential for long-term maintenance of the viral episome (Piirsoo *et al.*, 1996). An explanation for this observation was provided when it was noted that the E2 protein and the viral episomes localized to discrete spots on the cellular mitotic chromosomes (Skiadopoulos and McBride, 1998), raising the possibility of a physical interaction between these molecules. Further experiments confirmed that the TAD of the BPV1 E2 protein physically interacted with the host chromosomes and the E2 DBD tethered the viral genome through the E2BSs (Lehman and Botchan, 1998; Ilves *et al.*, 1999; Bastien and McBride, 2000) (Fig. 7.2).

The interaction of BPV1 E2 with a cellular protein, Brd4, and their colocalization on human chromosomes suggested that this complex might tether the viral genome to the chromosomes (You *et al.*, 2004; Baxter *et al.*, 2005; McPhillips *et al.*, 2005). Initially, it was assumed that most, if not all, PVs used the interaction with Brd4 to maintain their viral genome. However, it was subsequently shown that, although all PVs interact with Brd4, this interaction is required for E2's role in transcription and not tethering (Stubenrauch *et al.*, 1998a; McPhillips *et al.*, 2006; Senechal *et al.*, 2007). It appears that different PVs have evolved

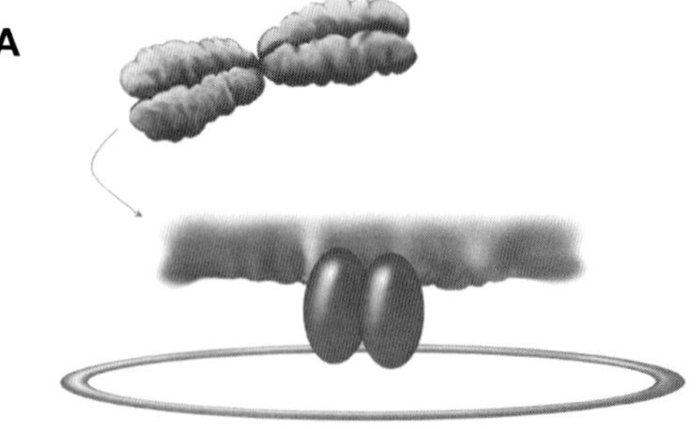

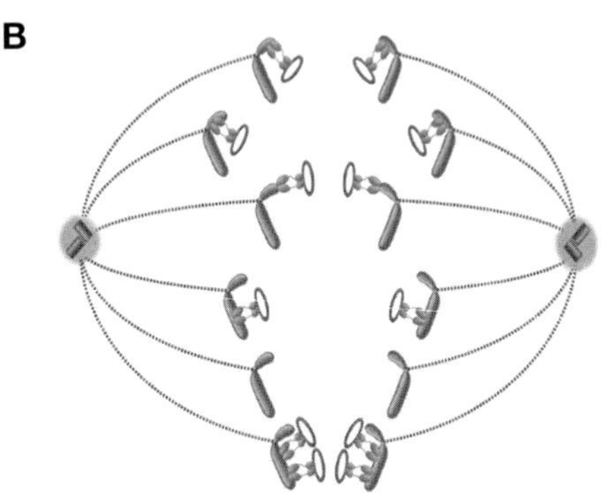

Figure 7.2 Plasmid maintenance model. (A) A virally encoded protein recognizes canonical binding sites in the viral genome. The viral protein completes the link with the host chromosome. This association can be established through protein–protein interactions or direct interaction with the hosts' DNA. (B) During mitosis, tethering associates the viral genome with the host chromosome. This allows for efficient partitioning of the viral genome to the daughter cells.

to use alternative targets to tether their genomes to the host chromosomes (see below).

A potential explanation for this observation could be that viruses infecting the same tissues attempt to avoid competition for the same target. However, double infection (i.e. infection of a single cell with two different viral types) appears to be rare. In addition, one would then expect that viruses infecting similar tissues (e.g. betaPVs infecting the skin) would use divergent tethering strategies, however this does not appear to be the case (Sekhar *et al.*, 2010). Since viral genomes are, like the host chromosomes, packaged into chromatin (Favre *et al.*, 1977) subnuclear localization of the viral genome might also be critical for viral gene expression. It is conceivable that different groups of viruses target specific, functional regions of the chromosomes, such as euchromatic or heterochromatic regions, to fine-tune viral gene expression and thus the viral lifecycle.

Similar to papillomaviruses, gammaherpesviruses infect dividing cells (i.e. a subset of B cells), and likewise attach their genomes to the chromosomes of the host. While latent, these viruses express only a limited number of proteins but this always includes a tethering protein analogous to the papillomavirus E2 protein. The best-studied gammaherpesvirus tethering proteins are EBNA1 and LANA (encoded by EBV and KSHV, respectively). Similar to the situation observed in E2, the viral DNA is bound through the DNA-binding domain of EBNA1/LANA. The viral protein then completes the link with the host chromosome (Fig. 7.2). An interesting difference between the gammaherpesviruses and PVs lies in the differences between numbers of canonical binding sites on the viral genome. The EBV origin (oriP) contains two functional elements separated by a long (~1 kb) spacer region. The 'family of repeats' (FR) consists of 20 repeats of 30 bp in length and each of these repeats contains an EBNA1 binding site. Similarly, the KSHV genome has ~35 binding sites for LANA (Fig. 7.3A). By comparison, BPV1 contains 'only' 17 E2 binding sites, eight of which are essential for plasmid maintenance (Piirsoo *et al.*, 1996). Many PVs have only four E2BS in their genome (Fig. 7.3B) and it is possible that for many PVs this number may not be sufficient for viral genome maintenance. Fewer sites (when compared with BPV1) may suffice for some PVs, however additional tethering mechanisms may be employed to enhance or substitute for E2-mediated tethering. It is conceivable that PV genomes contain additional cis-elements that bind other viral or cellular proteins that can help to retain the genomes in dividing cells.

The complexity of genome tethering is further illustrated by a recent study of the BPV1 E2 partitioning process. Silla *et al.* demonstrated that

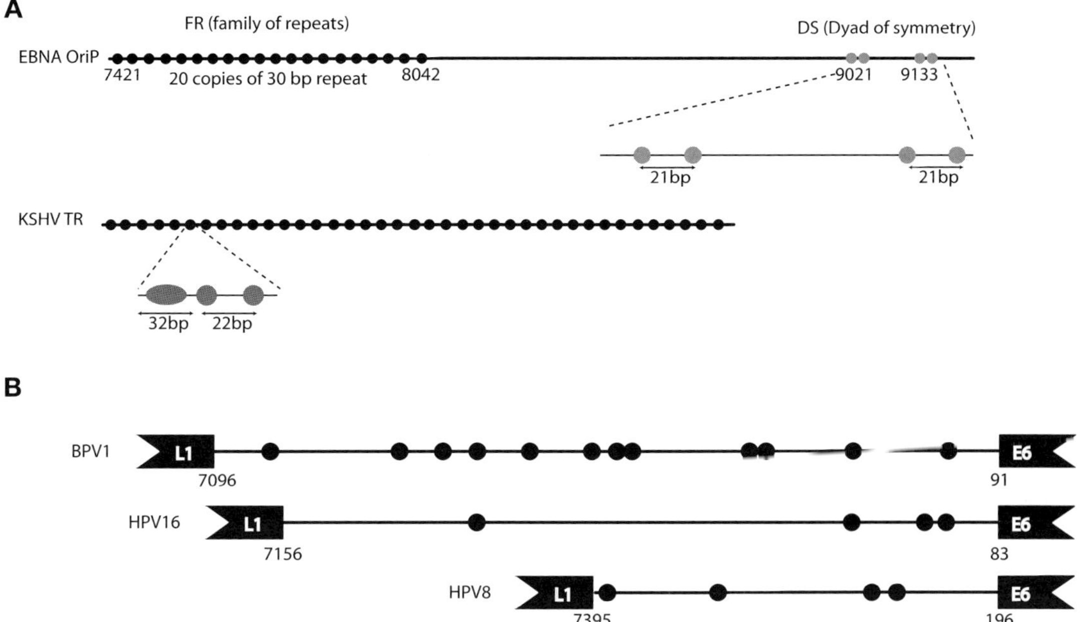

Figure 7.3 Comparison of location and number of consensus binding sites surrounding the viral origin of replication. (A) EBV: A functional OriP contains the DS and FR elements. Both elements are bound by EBNA1. While binding to the four EBNA1 binding sites in the DS is implicated in viral replication, tethering is dependent on the FR repeats. KSHV: Every individual TR contains a putative origin of replication. Two LANA binding sites (grey circles) in combination with the replication element (grey oval) form the minimal replicator. The array of TRs is likely involved in tethering. (B) Three different PV types are shown. Each circle represents a consensus E2BS located in the upstream regulatory region of the viral genome. While BPV1 has 12 sites, the HPVs have only four such sites. The implications of the reduced number of E2BS are not completely understood.

simple linking of plasmids to mitotic chromosomes is not sufficient for efficient partitioning. The tethering and transactivation functions of E2 are required to function in concert to form a segregation competent complex (Silla *et al.*, 2010).

Genome tethering proteins: structure and function

The tethering protein of each virus plays a key role in the viral life cycle. In addition to participating in genome maintenance and tethering, these proteins are involved in initiation of DNA replication and regulation of viral transcription. For papillomaviruses, the E2 protein functions as the tethering protein. The EBNA1 protein of EBV and the LANA protein of KSHV are the best characterized gammaherpesvirus genome maintenance proteins. Other members of the Lymphocryptoviruses (Callitrichine herpes virus 3, Macacine herpesvirus 4 and Papiine herpes virus 1) encode EBNA1 homologues. Rhadinoviruses and Macaviruses encode LANA homologues known as ORF73 (Blake, 2010). The domain structures of the PV E2, EBV EBNA1 and KSHV LANA proteins are shown in Fig. 7.1.

Depending on the PV type, the E2 gene is translated into a protein of 350–500 amino acids. It can be divided into three domains, with the N-terminal trans-activation domain (TAD) playing key roles in replication and regulation of transcription. The TAD is linked to the C-terminal domain through a flexible linker. The C-terminal domain is about 85 to 100 aa in length and is responsible for binding to the viral DNA (Fig. 7.3). While the N- and C-terminal domains are highly conserved across the different genera, the hinge or linker region is highly variable in sequence and length. Variations in length of the flexible linker are responsible for the differences in length of the full-length protein. It was initially thought that the hinge region functioned only as a flexible linker, but auxiliary elements, conserved only within each papillomavirus genus, have been mapped to the hinge. For example, the hinge contains regions that regulate nuclear localization in HPV11 E2 (Lai *et al.*, 1999; Zou *et al.*, 2000), proteasomal degradation in BPV1 E2 (Penrose and McBride, 2000) and transcriptional regulation and chromosome binding in HPV8 E2 (Steger *et al.*, 2002; Poddar *et al.*, 2009a; Sekhar *et al.*, 2010). The DNA-binding domain of E2 forms a core dimeric beta barrel structure with two surface alpha helices (Hegde *et al.*, 1992) (Fig. 7.4). One of these helices makes direct contact with the 12 bp E2 recognition sequence. The structure of the transactivation domain has also been solved; it forms a cashew shaped structure consisting of a bundle of alpha helices at the N-terminus linked by a fulcrum region to a beta sheet region (Antson *et al.*, 2000).

The EBNA1 protein of EBV also consists of N-terminal and C-terminal domains linked by a repeat domain composed of glycine and alanine residues (Fig. 7.1). The C-terminal domain encodes a DNA binding and dimerization domain that, remarkably, has a very similar structure to that of the PV E2 proteins (Bochkarev *et al.*, 1995) (Fig. 7.4). Two short regions of EBNA1 that are rich in arginine and glycine residues have been shown to associate with mitotic chromosomes. In addition to chromosomal association, EBNA1 is involved in transcriptional regulation and transactivation of the viral promoter.

The LANA protein is encoded by ORF73 in KSHV. Its C-terminal DNA binding and dimerization domain is thought to fold in a very similar manner to those of the E2 and EBNA1 proteins (Grundhoff and Ganem, 2003; Han *et al.*, 2010). Fig. 7.4 shows the structure of the E2 and EBNA1 DNA binding domains and a homology model of the LANA protein based on the solved structure of EBNA1. The central portion of LANA also consists of a long repetitive region that is not well conserved among herpesvirus ORF73 proteins. The N-terminal domain of LANA contains the chromosomal binding motif.

Viral genomic elements required for chromosome association

Binding sites for viral tethering proteins

The E2 proteins from all PV types specifically recognize a palindromic consensus sequence (ACCgNNNNcGGT) known as the E2 binding

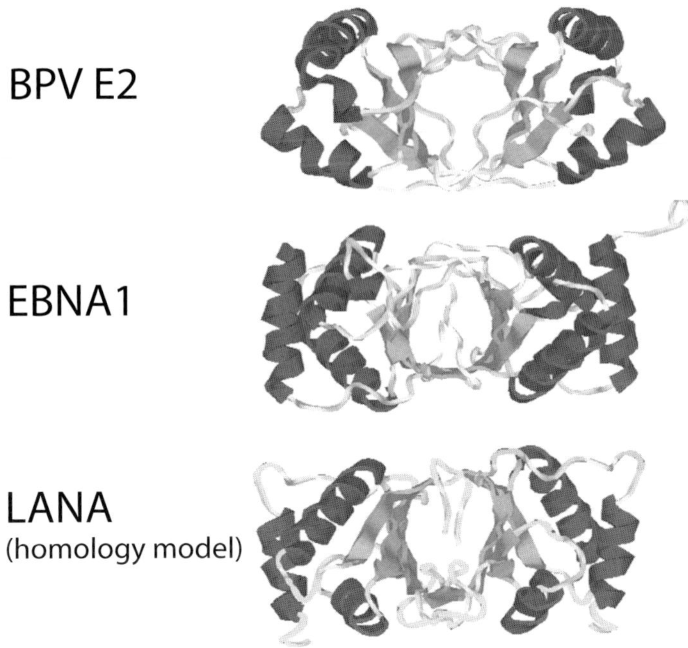

Figure 7.4 Structural similarity between the DNA binding domains of the viral tethering proteins Despite the lack of primary sequence similarity, the tertiary structure of the viral DNA binding domains is highly similar. The beta-barrel fold is a unique feature to these proteins, and has, to date, not been seen in other proteins. The LANA structure shown is a homology model based on the solved EBNA1 structure.

site (E2BS). Small case letters are, while preferred, not essential for recognition by the E2 protein. Although the length of the spacer region ('NNNN') is absolutely conserved across PVs, the composition varies with viral type. BPV1 has 11 consensus recognition sites and an additional six mapped by DNA footprinting (Li *et al.*, 1989). In HPV16, four E2BS are located just upstream from the viral early promoter, flanking the binding site for the E1 helicase (Fig. 7.3B). It appears that PV types have potentially important variability in the nature of the electrostatic potential of their E2 DNA binding surfaces. For example, HPV16 and HPV18 show differences in net charge as well as in the distribution of these charges on their DNA binding domains (Hegde, 2002).

During binding to the E2BS, the E2 protein inserts its recognition helices into the major grooves of the DNA, making direct contact with the consensus residues of the E2BS (Hegde *et al.*, 1992). Binding to the DNA does not significantly alter the structure of the E2 protein; however, depending on the viral type, significant bending of the DNA is observed. Most of the deformation appears to be absorbed by the linker region of the E2BS and the spacer plays an important role for the affinity of E2 for a particular E2BS. In the case of HPV16, there is a specific preference for AA[a/t]N. This preference is not dependent on AT richness, as a 'TTAA' spacer has lower affinity than an 'AATT' spacer (Hines *et al.*, 1998). Similar preferences are observed for HPV18 (Kim *et al.*, 2000). The different spacer sequences give rise to different conformation flexibility with AT rich spacers having an intrinsic propensity to bend towards the major groove (Brown *et al.*, 2011).

It is difficult to determine the role of individual E2 binding sites in maintenance DNA replication, because these sites are also required for transcriptional regulation and initiation of DNA replication of the viral genome. For BPV1, a minimum of eight E2 binding sites are required for episomal maintenance (Piirsoo *et al.*, 1996). In HPV31, three of the four E2 binding sites are required for maintenance replication. Interestingly, the specific arrangement of E2 binding sites appears to be more important than the exact number of sites (Stubenrauch *et al.*, 1998b).

Episomal maintenance of the EBV genome depends on the presence of EBNA1 and the viral cis-element *oriP. OriP* contains the dyad symmetry (DS) element, which contains four 30 bp repeats containing low affinity EBNA1 binding sites and the family of repeats (FR) that consists of 20 repeats of high affinity EBNA1 binding sites (Fig. 7.3A). While replication is initiated in the DS region, long-term maintenance of EBV genomes is dependent on the FR domains (Lupton and Levine, 1985; Reisman *et al.*, 1985; Harrison *et al.*, 1994; Aiyar *et al.*, 1998; Chaudhuri *et al.*, 2001; Schepers *et al.*, 2001). Seven high affinity EBNA1 binding sites are minimally required for stable plasmid maintenance (Wysokenski and Yates, 1989). The FR region from the closely related virus, Herpesvirus papio (HVP), contains just eight EBNA1 binding sites. Moreover, the binding sites in EBV are separated by 14 bp while those in HVP are separated by just 10 bp. The EBV binding sites function optimally in plasmid maintenance when they are separated by their natural spacing, which would place them on the same face of the DNA helix (Hebner *et al.*, 2003).

In KSHV, the LANA protein binds to an 18 bp palindromic DNA sequence within the terminal repeats (TR) of the viral genome (Fig. 7.3A). The KSHV genome contains ~35 copies of these repeats but LANA may also bind to additional sites outside of the TR repeats (Komatsu *et al.*, 2001; Cotter *et al.*, 2001). The number of binding sites varies depending on the KSHV isolate. Only a few TR repeats are required for stable LANA-dependent plasmid replication and it is not clear why there are such a large number of repeats in the genome (Grundhoff and Ganem, 2003).

Additional *cis*-elements important for genome maintenance

Although E2 has the potential to specifically localize viral DNA to mitotic chromosomes, it may not be the only player involved in tethering. Additional cis-elements, bound by other viral or cellular proteins, could augment and stabilize the plasmid maintenance function of E2. In addition to the E2BS, a plethora of additional factor binding sites have been identified in the viral genome. This includes matrix attachment regions (MARs), binding sites for HMG proteins, topoisomerase II, the telomere repeat binding protein and CENP-B (Pittayakhajonwut and Angeletti, 2008; Thain *et al.*, 1996). In addition, many transcription factors specifically interact with the viral regulatory region and all of these proteins might be able to direct the viral genome to specific areas of the nucleus. Additional evidence for an E2 independent form of viral maintenance lies in the observation that certain E2BS are methylated in proliferating cells, thereby interfering with the binding of E2 to these sites (Kim *et al.*, 2003).

Matrix attachment regions (MARs) are attractive candidates for auxiliary viral genome maintenance elements in PV genomes. Episomal vectors that contain MAR elements are able to replicate and persist long term in dividing cells through association with the nuclear matrix and interaction with mitotic chromosomes (Baiker *et al.*, 2000; Jenke *et al.*, 2004). The nuclear matrix is thought to be a highly dynamic structure that is essential for nuclear architecture. The discovery that DNA can specifically interact with the nuclear matrix was a key step in the understanding of the functional and structural organization of the eukaryotic nucleus (Mirkovitch *et al.*, 1984). It was shown that specific stretches of DNA (designated MARs) mediated this interaction. Tethering of viral genomes to the nuclear matrix could occur through proteins involved in transcription and replication, or through specific anchoring proteins. For example, topoisomerase II is tightly bound to the nuclear matrix (Razin *et al.*, 1991) and has consensus binding-sites within the PV genome. Biochemical analyses have shown that the genomes of several PV types can bind the nuclear matrix *in vitro* (Tan *et al.*, 1998; Stunkel *et al.*, 2000). The HPV16 genome appears to contain several potential MARs located within the epithelial cell-specific enhancer, the E6 promoter, the replication origin, and the early late intergenic region (Tan *et al.*, 1998). BPV1 also has been shown to have a potential MAR next to the ori (Adom *et al.*, 1992).

In addition to the origin of replication and enhancer regions EBV utilizes a MAR to associate with the nuclear matrix and this interaction is required for efficient replication of the viral genome (White *et al.*, 2001). Association of LANA with the nuclear matrix fraction is also important for latent replication (Ohsaki *et al.*, 2009). Although LANA is primarily responsible for tethering of the KSHV genome, LANA-depleted cells continue to maintain viral genomes. Furthermore, recombinant plasmids containing cis-elements isolated from the KSHV genome are maintained efficiently in the absence of LANA. Thus, viral elements independent from the LANA BS are involved in partitioning and tethering (Verma *et al.*, 2007b).

Cellular targets important for viral tethering

Various persistent DNA tumour viruses have adopted a common mechanism of using virally encoded proteins to tether the viral genome to host chromatin during mitosis, for the purpose of viral genome partitioning and nuclear retention. The interaction of different viral tethering proteins with host chromatin could be either direct, through binding to specific sequences on the host DNA, or could be mediated by association with cellular proteins. Identifying different cellular protein partners that mediate the chromosomal binding functions of the various tethering proteins has been an area of intense research interest, especially because of their potential to be used as targets for novel antiviral therapeutics.

Targets of PV E2 proteins

Several studies have shown that, although the E2 proteins from different PVs associate with host mitotic chromosomes to efficiently partition the viral genome to daughter cells, they interact with distinct chromosomal targets. This is mainly reflected by the fact that the E2 proteins from different PVs show different patterns of binding on the mitotic chromosomes (Van Tine *et al.*, 2004; McPhillips *et al.*, 2006; Oliveira *et al.*, 2006; Poddar *et al.*, 2009b).

Brd4

The first clue that BPV1 E2 proteins might interact with chromosomes came from the observation that the BPV1 E2 protein dotted the arms of the host mitotic chromosomes (Skiadopoulos and McBride, 1998). It was further shown that the BPV1 E2 mediated mitotic tethering was carried out in association with a cellular partner identified as the Brd4 protein (You *et al.*, 2004). Brd4 is a double bromodomain protein that binds to acetylated lysine residues on histones H3 and H4 (Dey *et al.*, 2003), Brd4 normally forms a diffuse coat around the mitotic chromosomes, but in the presence of E2 both proteins colocalize on mitotic chromosomes in punctuate dots all over the mitotic chromosomes. Moreover, the E2 protein stabilizes the interaction of Brd4 with host chromosomes in interphase and mitosis (McPhillips *et al.*, 2005). Two conserved residues, R37 and I73, in the N-terminal domain of the E2 protein interact with the C-terminal domain of Brd4 (You *et al.*, 2004). Mutations in the N-terminal domain of E2 protein that abrogate Brd4 interaction also compromise E2's mitotic binding ability (Baxter *et al.*, 2005). The C-terminal DNA-binding domain of E2 is not absolutely required for the interaction with Brd4, but the dimerization function of this domain greatly increases the affinity of Brd4 for chromatin, most probably by enabling the assembly of higher order E2–Brd4 complexes (Cardenas-Mora *et al.*, 2008).

Furthermore, in *Saccharomyces cerevisiae*, BPV1 E2 was observed to maintain plasmids containing E2 binding sites only in the presence of exogenously expressed Brd4 (Brannon *et al.*, 2005). Overexpression of a dominant negative C-terminal domain of Brd4 inhibits the E2–Brd4 interaction and results in inhibition of E2 mediated chromosome binding, viral genome maintenance, as well as BPV1's ability to induce cellular transformation (You *et al.*, 2005; Ilves *et al.*, 2006; McPhillips *et al.*, 2006). These studies clearly highlight the importance of cellular Brd4 protein in the BPV1 lifecycle as well as in the BPV1 E2 mediated genome maintenance and partitioning function.

Although Brd4 is critical for E2 mediated transcription of all PVs (McPhillips *et al.*, 2006; Wu *et al.*, 2006; Smith *et al.*, 2010), it is clearly not a cellular partner for genome partitioning in all the different viruses (Ilves *et al.*, 2006; McPhillips *et al.*, 2006). This is further supported by observations that R37A and I73A mutations that abrogate E2–Brd4 interactions, do not affect the mitotic localization of E2 proteins belonging to the alpha or the beta genus (McPhillips *et al.*, 2006; Poddar *et al.*, 2009b). Another line of evidence is the observation that the HPV31 genome carrying I73L mutation in the E2 gene is still maintained stably as an extrachromosomal genome and undergoes amplification in differentiated keratinocytes (Stubenrauch *et al.*, 1998a). Notably, the EBNA1 and LANA proteins also interact with Brd4 (You *et al.*, 2006; Lin *et al.*, 2008). While this interaction may play some role in tethering, it is probably because Brd4 is essential for many steps in viral transcription (Wu and Chiang, 2007).

rDNA

Unlike BPV1 E2, where the binding was detected as small speckles distributed all over the arms of all mitotic chromosomes, the E2 proteins from the beta-PVs show a distinct pattern of mitotic binding. The HPV8 and HPV5 E2 proteins bind to the pericentromeric region of chromosomes as large distinct foci (Oliveira *et al.*, 2006). Furthermore, HPV8 E2 protein has been shown to associate with the rDNA loci present on the short arms of the acrocentric chromosomes where it colocalizes with the upstream binding factor (UBF) in mitosis (Poddar *et al.*, 2009b). UBF is a transcription factor required for rDNA transcription by RNA polymerase I. Surprisingly, UBF remains bound to chromosomes during mitosis although transcription is silenced (Jantzen *et al.*, 1990; Roussel *et al.*, 1993). During interphase, the HPV8 and HPV5 E2 proteins are seen localized both in a granular nuclear pattern and in SC35 speckles, which exclude the nucleolus (Lai *et al.*, 1999; Sekhar *et al.*, 2010). However, during the cell cycle, as cells transition from interphase to mitosis, there is disassembly of the nucleolus and the E2 protein is free to localize to rDNA in mitosis. This spatial segregation during the different stages of the cell cycle would allow the E2 protein to actively participate in viral DNA replication and transcription during interphase and tether to host chromosomes for genome partitioning during mitosis. This separation might be essential considering the multitude of functions the E2 protein has to perform during these distinct phases of the cell cycle. It is tempting to speculate why rDNA is an attractive chromosomal tethering target for PVs. The rDNA loci contain about 400 tandem repeats of rDNA units. The presence of such large numbers of repetitive units greatly increases the local concentration of binding targets for the E2 protein. In addition, the rDNA locus has a specialized mechanism for replication, condensation and segregation.

In the case of the alpha-PVs, the mechanism of E2 mediated viral genome tethering has not been clearly elucidated. Contrary to the E2 proteins from BPV1 and HPV8, where under normal fixation conditions E2 is observed on chromosomes throughout mitosis, under similar fixation conditions the alpha E2s are only observed on mitotic chromosomes during prophase and telophase. However, with a brief pre-extraction step prior to fixation, the E2 proteins from alpha-PVs such as HPV11, HPV16 and HPV31 show a mitotic localization pattern similar to HPV8 E2, suggesting that this interaction could be unstable and dynamic (Oliveira *et al.*, 2006). These E2 proteins might require a specialized cellular environment, or additional cellular or viral factors to stabilize the binding of the E2 proteins to mitotic chromosomes.

Mitotic spindle

In addition to association with rDNA, HPV11, HPV16 and HPV18 E2 proteins have been reported to associate with the mitotic spindle during mitosis to enable partitioning of the viral plasmid (Van Tine *et al.*, 2004; Dao *et al.*, 2006). Although both the N-terminal and the C-terminal domains of HPV11 E2 can interact independently with microtubules, a short stretch of sequences in the C-terminal domain was found to be essential for this interaction (Dao *et al.*, 2006). This is highlighted by the fact that the E2 protein has also been observed localized to the mid-body during cytokinesis (Yu *et al.*, 2007). Finally, E2 proteins have also been observed localized to the centrosomes (Van Tine *et al.*, 2004; Donaldson *et al.*, 2007). It remains unclear how these associations would result in the maintenance of the viral genomes within the host nucleus.

TopBP1, ChlR1 and MKlp2

HPV16 E2 also interacts and colocalizes with the topoisomerase II-binding protein1 (TopBP1) on mitotic chromosomes during late telophase (Donaldson *et al.*, 2007). TopBP1 has been reported to be a transcriptional coactivator of HPV16 E2 and a potential cellular candidate for HPV16 E2 chromosomal association (Boner *et al.*, 2002).

The E2 proteins of BPV1, HPV11 and HPV16 interact with two proteins required for cellular mitotic chromosome segregation: ChlR1 (an ATP-dependent DNA helicase important for sister chromatid cohesion) (Parish *et al.*, 2006) and mitotic kinesin like protein, MKlp2 (a motor protein required for cytokinesis) (Yu *et al.*, 2007). BPV1 viral genomes that contain a mutation in E2 resulting in loss of interaction with ChlR1 are not maintained long term. Furthermore, depletion of ChlR1 abrogates E2 association with mitotic chromosomes. Although E2 and ChlR1 do not colocalize on mitotic chromosomes, it is believed that ChlR1 could load the E2 protein onto chromosomes (Parish *et al.*, 2006). On the other hand, the colocalization of MKlp2 and E2 occurs in the mid-body during late mitosis, suggesting that this association might be important for viral genome partitioning (Yu *et al.*, 2007).

It is clear that the interaction of E2 proteins with mitotic chromosomes is complex and may involve multiple different interacting partners to aid in viral genome maintenance (Table 7.1).

Targets of Herpesvirus tethering proteins

Similar to the situation in PVs, the *gammaherpesvirinae* have evolved to 'piggy-back' on cellular proteins for their interaction with the host genomes. Notably, the chromosomal binding pattern of LANA is different depending on whether the KSHV viral genome is present. In the absence of the viral genome, LANA exhibits a diffuse pattern in the interphase nuclei and all over the mitotic chromosomes. However, in virally infected cells, LANA can be observed together with the viral genome in punctuate dots during interphase and mitosis (Ballestas *et al.*, 1999; Cotter and Robertson, 1999). This suggests that LANA could be targeting different protein partners in the presence and absence of viral genome. However, it is also thought that the tethering proteins become concentrated by binding to the repetitive binding sites on the viral genome and this may also enable the formation of higher order segregation complexes.

LANA associates with several cellular proteins to bind to host chromatin (Table 7.1). LANA has been reported to associate with methyl CpG-binding protein 2 (MeCP2) and DEK to bind to mouse chromosomes (Krithivas *et al.*, 2002). MeCP2 binds to methylated CpG dinucleotides in internucleosomal linker DNA (Chandler *et al.*, 1999). DEK associates with H2A and H2B (Alexiadis *et al.*, 2000) and LANA has also been shown to bind directly to these histone proteins (Barbera *et al.*, 2006). In addition, LANA interacts with Brd4 and Brd2/Ring3 (Platt *et al.*, 1999; Mattsson *et al.*, 2002; Viejo-Borbolla *et al.*, 2005; Ottinger *et al.*, 2006; You *et al.*, 2006). While all of these cellular proteins are known to specifically localize to chromatin, their role in LANA-mediated chromosome tethering is not entirely clear. Notably, the Brd2/Ring3 complex is excluded from the mitotic chromosomes in KSHV uninfected cells and becomes weakly localized to chromosomes upon infection. In contrast, Brd4 binds to LANA on chromosomes in interphase and throughout mitosis (You *et al.*, 2006). Binding to histones and MeCP2 has been mapped to a short stretch in the N-terminal domain of LANA (Fig. 7.1); however, recent work has also illustrated the importance of the C-terminal DBD in interaction with cellular proteins, including MeCP2, DEK, Brd4, and Brd2/Ring3.

Interaction of the C-terminus of LANA with the nuclear mitotic apparatus protein (NuMA) contributes to KSHV genome maintenance and partitioning (Si and Robertson, 2006). Although NuMA was observed to associate with LANA during interphase, this interaction is lost in mitosis. Blocking the association of NuMA with dynein/dynactin and microtubules resulted in loss of viral episomes in the progeny cells. Thus, NuMA is speculated to play a role in efficient partitioning of KSHV genome to daughter cells.

Work from the same group has also recently revealed that LANA targets centromeric protein CENP-F on host chromosomes of KSHV-infected cells. However, as cells enter anaphase, CENP-F is known to dissociate from the kinetochore. Notably, another kinetochore-associated protein called Bub1 colocalizes with both LANA and the viral genome throughout the cell cycle from prophase to cytokinesis (Xiao *et al.*, 2010). This group also

Table 7.1 Targets of viral tethering proteins with a potential role in genome partitioning

Virus	Tethering protein	Putative target or accessory factor	References
Papillomaviruses			
BPV1	E2	Brd4	Baxter *et al.* (2005), You *et al.* (2004)
HPV8	E2	rDNA	Poddar *et al.* (2009a)
HPV11	E2	Mitotic spindle	Van Tine *et al.* (2004)
HPV16	E2	TopBP1	Donaldson *et al.* (2007)
BPV1, HPV11, HPV16	E2	ChLR1 MKlp2	Parish *et al.* (2006) Yu *et al.* (2007)
Gamma-herpesviruses			
Epstein–Barr virus (EBV)	EBNA1	hEBP2 (p40)	Kapoor and Frappier (2003)
		A-T hook	Sears *et al.* (2004)
		Brd4	Lin *et al.* (2008)
Kaposi's sarcoma-associated herpesvirus (HHV-8)	LANA	Brd2/Ring3	Platt *et al.* (1999), Viejo-Borbolla *et al.* (2005)
		Brd4	You *et al.* (2006)
		Histone H1	Cotter and Robertson (1999)
		Histones H2A/B	Barbera *et al.* (2006)
		DEK, MeCP2	Krithivas *et al.* (2002)
		NuMA	Si *et al.* (2008)
		Bub1/CENP-F	Xiao *et al.* (2010)
Herpesvirus saimiri	ORF 73	MeCP2	Calderwood *et al.* (2004), Griffiths and Whitehouse (2007)

showed that the knockdown of Bub1 in KSHV positive cell lines resulted in drastic loss of episomal KSHV genomes. Thus, the tethering model proposed by this group suggests that, although LANA binds Bub1 in complex with CENP-F and other cellular proteins (NuMA, DEK, Brd4, Brd2/Ring3 and histone proteins), all of the interactions with the exception of Bub1 are lost as cells enter anaphase. Thus, the Bub1–LANA interaction could be crucial for the partitioning and long-term maintenance of the KSHV genomes (Xiao *et al.*, 2010). It was recently suggested that the LANA N-and C-terminal domains cooperatively associate with the hosts chromosomes (Matsumura *et al.*, 2010). However, all of the studies so far seem to indicate that the mechanism of LANA mediated chromosome tethering function is complex and involves interactions between multiple protein partners that are required for efficient viral genome tethering and genome partitioning to the progeny cells at the end of cell division.

Similarly, EBNA1 interacts with a number of cellular proteins (Table 7.1). Using a yeast two-hybrid screening, EBNA1-binding protein 2 (EBP2) was initially identified as an interacting partner for EBNA1 (Shire *et al.*, 1999). The EBNA1 protein interacts with the EBP2 protein, which binds metaphase chromosomes through its coiled-coil domain (Wu *et al.*, 2000; Kapoor *et al.*, 2001). This interaction is essential for EBNA1 mediated partitioning of plasmids containing EBNA1 binding motifs, using a reconstituted yeast segregation assay (Kapoor *et al.*, 2001). Additionally, knockdown of EBP2 resulted in reduction of metaphase bound EBNA1 protein, which in turn, resulted in loss of *oriP* plasmids from the cells (Kapoor *et al.*, 2005). Nevertheless, many contradictory observations raise questions regarding the role of EBP2 in EBNA1-mediated chromosome binding. EBP2 is primarily localized in the nucleolus and during interphase there is very little colocalization between EBP2 and EBNA1 (Sears *et al.*, 2004). Moreover, although EBNA1 and EBP2 partially colocalize on metaphase chromosomes, much of the cellular EBP2 is primarily localized to the periphery of the chromosomes. It appears that EBNA1 binds to chromosomes in an EBP2 independent manner during interphase and as the cells transition to mitosis, EBP2 plays an auxiliary role of stabilizing the EBNA1 association with chromosomes (Nayyar *et al.*, 2009).

Others have proposed an alternative model to explain the EBNA1 tethering of EBV genomes to chromosomes (Sears *et al.*, 2004). According to this group, EBNA1 targets chromosomes through direct DNA binding. It is suggested that EBNA1 has regions similar to AT-hook motifs that can bind DNA directly by targeting AT-rich DNA sequences (Sears *et al.*, 2004).

Recent work using EBNA1 ChIP-seq experiments has identified high affinity EBNA1 binding sites on human chromosome 11. These regions are clusters of ~10 kb repetitive sequences that are situated between the promoters of the FAM55B and FAM55D genes. They have increased levels of histone H3 K9me3 modification suggesting these regions are mainly heterochromatic (Lu *et al.*, 2010). Since many of the consensus binding sites in this region were different from the known viral EBNA1 binding site consensus, it is possible that these regions are targeted by EBNA1 indirectly through interactions with various cellular proteins (Lu *et al.*, 2010). However, another study showed that EBNA1 could interact with cellular sequences through sites that differed from its consensus recognition motif (Dresang *et al.*, 2009).

Structural similarity between tethering proteins

The papillomavirus and gammaherpesvirus tethering proteins, E2, EBNA1 and LANA, play similar roles in the life cycles of their viruses. Each protein supports replication initiation and is required for transcriptional regulation and genome maintenance. However, despite no evolutionary relationship between these viruses and no strong sequence similarity between the proteins, the tethering proteins do share structural similarity in their DNA binding domains and some similarities in the regions shown to be important for association with host chromatin. A diagram of these regions is shown in Fig. 7.5.

DNA binding domains

The papillomavirus E2 protein, and isolated DNA binding domain, binds with high affinity to the E2BS in the viral genome with a K_D of around 5 nM (Alexander and Phelps, 1996; Dell *et al.*, 2003). The first solved structure of a PV E2 protein bound to its DNA target represented the prototype of a novel dimeric beta-barrel fold (Hegde *et al.*, 1992). Each half of the barrel contributes four anti-parallel beta-sheets. The topology of the beta-barrel is shown in Fig. 7.4. The dimer interface is formed by extensive hydrogen bonding between the B2 and B4 strands. The result of dimerization is that about 2000 Å^2 (depending on which structure is being analysed) of the dimerization domain is buried, contributing significantly to the stability of the dimeric protein (Mok *et al.*, 1996). Two alpha-helices are presented on the surface of the barrel, one of which contains most of the residues that contact viral DNA. Since the description of the original BPV1 structure, several additional E2 DBD structures have added to the understanding of the E2 fold, and how it affects DNA target recognition. Notably, although all of the solved structures have a highly similar tertiary structure, the orientation of the recognition helices is variable. This highlights the existence of two structural classes: HPV16 and HPV31 belong to one class and BPV1 and HPV18 belong to a second. It appears that the proteins orient their recognition helices to match the specific sequence that they recognize, which might be slightly different for each viral strain. This conservation of tertiary and quaternary structure is not obviously reflected in the primary sequence (Hegde, 2002). Phylogenetic analysis clusters HPV16, HPV31 and HPV18 together in the *Alphapapillomavirus* genus, while BPV1 belongs to the distantly related Delta-PV genus (Bernard *et al.*, 2010). However, it is of note that although HPV16 and HPV18 are closely related viruses, they cause different diseases. HPV16 is strongly associated with squamous cell carcinomas, while HPV18 causes mainly adenocarcinomas of the cervix (Burk *et al.*, 2003).

To date, the only non-E2 protein shown experimentally to fold as a dimeric beta-barrel is the EBV EBNA1 protein (Bochkarev *et al.*, 1995). Structurally, the EBNA1 protein consists of a dimeric beta-barrel core domain with two alpha helices positioned similar to those of the E2 protein (Fig. 7.4). However, the EBNA1 structure contains an additional flanking domain, which consists of an alpha-helix and an extended chain. Surprisingly, however, when the EBNA1 domain was crystallized with its DNA binding target, the flanking domain formed the bulk of the interactions with the viral DNA (Bochkarev *et al.*, 1998). It was not clear whether EBNA1 had a different interaction with its DNA binding site or whether the interaction was due to an artefact of crystallization. More recently, mutational analysis has implicated the recognition helices (analogous to those of the E2 protein) in DNA recognition (Cruickshank *et al.*, 2000; Fujita *et al.*, 2001). The authors suggested a possible role for two-step mechanism of target recognition. In this model the core-domain is responsible for the initial interaction with the target DNA after which the flanking domain is loaded into the minor groove. This mechanism would be complementary with the mechanism suggested for E2 protein target recognition. In a sequence independent manner the E2 protein and the viral DNA form an 'encounter complex' which is stabilized by electrostatic interactions. It is probable that E2 slides across the genome until the specific site is encountered. This results in solvent exclusion, allowing for the indirect readout of the nucleotides (Ferreiro and Prat-Gay, 2003).

Despite the absence of amino acid similarity, it was suggested that the LANA DBD might be homologous with the equivalent domain of EBNA1 (Grundhoff and Ganem, 2003) (Fig. 7.4). In support of this theory, LANA residues important for DNA binding correspond to those with equivalent roles in EBNA1 (Komatsu *et al.*, 2004; Kelley-Clarke *et al.*, 2007b). Although this observation provides experimental evidence for (partial) functional homology between the LANA and EBNA1 DBDs, an experimentally solved structure of the DBD is needed to conclusively answer these questions.

Chromosome interaction motifs

All PVs E2 proteins interact with the cellular Brd4 protein and this interaction is important for the transcriptional activation and repression functions of E2 (McPhillips *et al.*, 2006; Senechal *et al.*, 2007; Smith *et al.*, 2010). However, as described above, only some PVs use Brd4 for chromosomal tethering. Mapping studies illustrated the importance of the conserved R37 and I73 residues for this interaction (Baxter *et al.*, 2005; McPhillips *et al.*, 2006; Senechal *et al.*, 2007). Based on their dependence of Brd4 for viral genome tethering, PVs can be classified into at least two groups. Chromosomal tethering of 'BPV1-like viruses' such as AaPV1, HPV1a, OcPV1 and SfPV1 (the last two viruses were previously known as ROPV and CRPV, respectively; Bernard *et al.*, 2010) is sensitive to mutation of R37 and I73, indicating a requirement for interaction with Brd4. Chromosomal binding of the E2 proteins of viruses belonging to the second class (mainly alpha- and beta-PVs) is not affected by these mutations.

Further analysis of the beta-PV E2 proteins showed that the TAD is dispensable for chromosomal association. Mapping studies showed that, in addition to the DBD, a 16 aa region in the hinge of HPV8 was essential for association with mitotic chromosomes (Sekhar *et al.*, 2010). The authors went on to show that a highly conserved RXXS motif is required for this association (Sekhar *et al.*, 2010). Furthermore, the authors suggested a role for this motif in regulation of viral tethering.

In the case of EBV EBNA1, both the DBD and N-terminal domain are essential for successful partitioning of the viral genome. The importance of chromosomal attachment tethering was illustrated experimentally. It was shown that a fusion protein of the EBNA1 DBD with HMGA1 (a cellular protein known to specifically associate with mitotic chromosomes) will support stable partitioning of the viral genome (Hung *et al.*, 2001), illustrating a role independent of other functions performed by the N-terminal domain. EBNA1 contains two regions in the N-terminal domain that are involved in chromosome attachment (Fig. 7.1). These domains have been termed domain A (aa 33–89) and domain B (aa 328–378). Fusion proteins of domain A or B attached to GFP were observed as punctae on sister chromatids (Marechal *et al.*, 1999; Sears *et al.*, 2003). Based on the observation that both domains are rich in glycine and arginine (GR) repeats, their role as AT-hooks for association with DNA was suggested. The presence of functional AT-hooks has since been confirmed (Sears *et al.*, 2004). Despite similarities between both domains, EBNA1 proteins lacking domain B were unable to support long term partitioning of plasmids and long term partitioning was not affected in EBNA1 proteins lacking domain A (Shire *et al.*, 1999). The authors suggested that this difference was due to the interaction of domain B with the cellular protein, EBP2.

Similar to EBNA1, KSHV LANA uses a stretch of residues close to its N-terminus (aa 5 to 22) to bind to host chromosomes (Fig. 7.1). Full-length LANA proteins with specific mutations in the N-terminal domain cannot associate with mitotic chromosomes (Barbera *et al.*, 2004). An alanine scanning approach was used to illustrate the key importance of residue 5 to 13 (the chromatin-binding motif, CBM). GFP-fusion proteins of CBM coat the host chromosomes throughout the cell cycle, implying the use of abundant targets.

In the absence of the N-terminal halve, the C-terminal domain of LANA interacts with the pericentromeric and telomeric regions of mitotic chromosomes. It was shown that the DBD interacts with MeCP2, independent of the N-terminal domain. This interacting region overlaps with the region important for viral DNA binding (Kelley-Clarke *et al.*, 2007a; Matsumura *et al.*, 2010). In support of this observation, it was shown that point mutations allow for the separation of regions in the C-terminal domain required for chromosome binding and specific DNA binding (Kelley-Clarke *et al.*, 2007b, 2009).

Another member of the Rhadinoviruses (Herpesvirus saimiri, HVS) causes asymptomatic but persistent infections in squirrel monkeys. Despite its close phylogenetic relationship to KSHV, the N-terminal domain of HSV LANA was shown to be dispensable for chromosome interaction (Calderwood *et al.*, 2004; Griffiths *et al.*, 2008). Instead, the authors mapped two chromosome association sites (CAS1 and -2) within the C-terminal domain. The authors identified a PKK (Pro-Lys-Lys) motif in CAS1 that is conserved in several gammaherpesvirus ORF73s (Griffiths *et al.*, 2008). These PKK motifs have been shown to bind to the minor groove of DNA (Churchill and Suzuki, 1989). It is interesting that, like PVs, closely related herpesviridae have evolved to use different parts of their respective tethering proteins to interact with the host chromosome.

Finally, a motif similar to the RXXS kinase motif identified in HPV8 E2 is conserved in the chromosome-binding region of the LANA and EBNA1 tethering proteins (Fig. 7.5) (Sekhar *et al.*, 2010). In EBNA1, several of the RXXS motifs overlap with AT-hook regions implied in chromosome binding (Sears *et al.*, 2004) and in LANA, residues overlapping the conserved RXXS motif disrupt episomal persistence (Barbera *et al.*, 2006). As shown in Fig. 7.5, all identified chromosome-binding regions are rich in GR motifs. These conserved motifs (RXXS and GR) are post-translationally modified as a way to regulate chromosomal interaction.

Regulation of chromosome binding functions of the different tethering proteins

While the details have not been elucidated, it is clear that post-translational modifications of the viral tethering proteins are important in the regulation of viral transcription, replication and genome maintenance.

For the BPV1 E2 protein, a CK2 phosphorylation site has been mapped to serine residue 301 within the hinge region, which triggers a conformational switch targeting the E2 protein for protein degradation through the ubiquitin-proteosome pathway (Skiadopoulos and McBride, 1998; Penrose and McBride, 2000; Penrose *et al.*, 2004). Alanine substitution of serine 301 interferes with this control step and increases the half-life of the E2 protein resulting in a very high viral DNA copy number (McBride and Howley, 1991). This is supported by the observation that when the E2 protein half-life increased, significantly more viral genomes are found to be tethered to the mitotic chromosomes (Skiadopoulos and McBride, 1998).

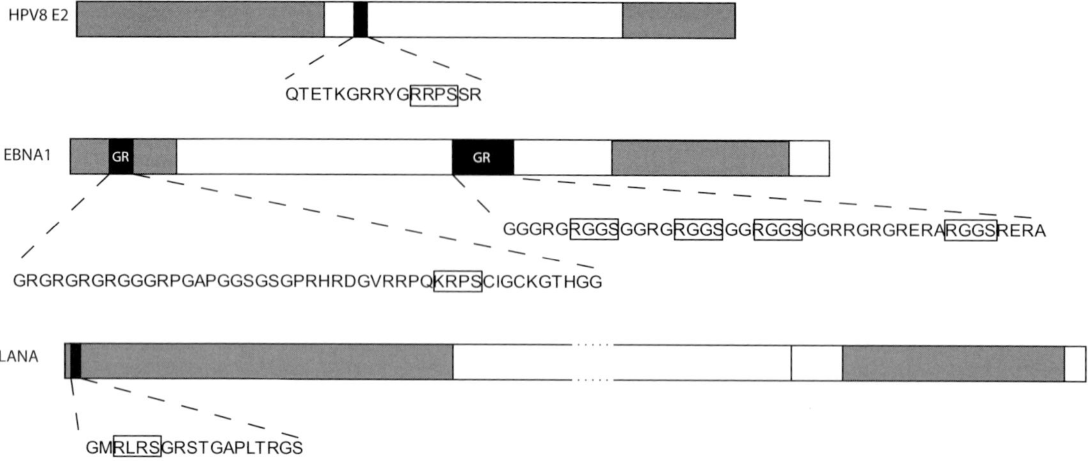

Figure 7.5 Location of conserved RXXS motifs in the viral tethering proteins. Sequence analysis revealed the presence of RXXS motifs in the three viral tethering proteins suggesting a common mechanism of regulation of chromosome tethering.

The E2 proteins from the distantly related beta-PVs are also phosphorylated (Sekhar *et al.*, 2010). Two residues crucial for chromosomal association, arginine 250 and serine 253, are contained in the chromosome binding region of the HPV8 E2 protein and are important for phosphorylation of the protein (Sekhar *et al.*, 2010). These residues lie within a RXXS kinase motif that is conserved in all the beta-PVs. These findings implicate phosphorylation in regulating the chromosome association function of beta-PV E2 proteins.

In the case of EBNA1, the 325–376 Gly Arg-rich region is modified by both serine phosphorylation and arginine methylation (Shire *et al.*, 2006). Although phosphorylation of multiple serine residues within the Gly Arg-rich region is important for the partitioning function of EBNA1, each serine appears to play a redundant role. The serine residues in this region are predicted to be target sites of calmodulin-dependent kinase II, and it has been suggested that phosphorylation of these residues regulates EBNA1 binding to the EBP2 protein, thereby controlling genome partitioning (Shire *et al.*, 2006). Arginine methylation of EBNA1 by PRMT1 and PRMT5 was suggested to be important for its nucleolar localization. Inhibition of methylation altered the localization of EBNA1 resulting in the formation of a ring of EBNA1 protein restricted to around the outer portion of the nucleoli (Shire *et al.*, 2006). While the functional significance of this observation is unknown, methylation may be important for EBNA1 interaction with nucleolar proteins or RNA. Furthermore, the association between EBNA1, EBP2 and the hosts' chromosomes is probably regulated by the Aurora kinase B. Although no Aurora kinase consensus motifs were identified on EBP2, down-regulation of the Aurora kinase B resulted in decreased EBP2 binding to mitotic chromosomes. This suggests that though EBP2 is not a direct target of the kinase, another chromosomal protein might be phosphorylated by the kinase, which could then be bound by EBP2.

There have been reports of phosphorylation of LANA by cellular kinases such as GSK-3β, Pim-1 and a kinase that is recruited by Ring3/Brd2. However, as yet, no studies have illustrated the importance of phosphorylation in the direct regulation of chromosome binding or genome partitioning functions.

The presence of RXXS kinase motifs in the chromosome binding regions of all three tethering proteins (Fig. 7.5) suggests that a common mechanism, such as phosphorylation, might regulate the chromosome binding function of these proteins (Sekhar *et al.*, 2010).

The presence of truncated repressor forms

of the E2 proteins introduces another layer of regulation of genome copy number and partitioning. The E2 repressor proteins do not contain the TAD but consist of the C-terminal DNA binding and dimerization domain fused to upstream sequences. Many viruses encode a protein, E8^E2, which consists of a short stretch of amino acids from E8 ORF fused to the DNA binding domain. In addition, BPV-1 encodes the E2-TR protein that is initiated from an internal promoter within the E2 ORF. In HPV31, the E8^E2 protein regulates genome copy number and is important for extrachromosomal maintenance of HPV31 in keratinocytes (Zobel *et al.*, 2003). In the case of BPV1, mutations that eliminate expression of the E2-TR protein replicate at greatly increased copy number (Lambert *et al.*, 1990). This regulation of copy number by repressor proteins could be either by forming inactive heterodimers with the full-length E2 protein or by directly competing for E2 binding sites within the viral genome (Lim *et al.*, 1998).

Varying expression levels of other viral proteins during the different stages of the viral life cycle could contribute to regulation of genome partitioning. In BPV1, expression of the viral replication protein E1 interferes with the E2 dependent tethering of the viral genome to the mitotic chromosomes (Voitenleitner and Botchan, 2002). When the levels of the E1 protein greatly exceeded that of the E2 protein, the E2 protein was inhibited from binding to mitotic chromosomes (Voitenleitner and Botchan, 2002). This interplay between the different viral proteins could, in turn be further modulated by different post-translational modifications affecting each of the proteins involved.

Replication licensing and regulation of maintenance replication

Cellular DNA is strictly licensed so that each DNA molecule is only replicated once per cell cycle. Viruses that replicate and maintain their genomes in dividing cells can also be licensed by similar mechanisms or can replicate randomly so that some genomes replicate several times and others not at all. In combination with a viral genome tethering mechanism, either of these approaches could conceivably allow for stable maintenance of the genome.

When extrachromosomal genomes are maintained at a high copy number, there is no need for a specific partitioning mechanism since there is a high probability that daughter nuclei will contain viral genomes after cell division. Furthermore, without replication licensing viral genomes could reamplify to reach an optimal copy number. However, with low copy number genomes, the once-per-cycle approach would not suffice for genome maintenance in the absence of tethering, as only a few viral genomes might be located in the nucleus following cellular division. Random replication would allow this limited number of viral genomes to be amplified but strict replication licensing would result in eventual loss of the genomes. Both papillomavirus and herpesvirus genomes are maintained at low numbers in the host. Using *in situ* techniques, it is very difficult to detect papillomavirus genomes undergoing maintenance replication in the basal cells of an infected lesion. Based on this observation it is estimated to be less than twenty copies per cell (Evans *et al.*, 2003).

Initial experiments that addressed the mode of papillomavirus replication were carried out in BPV1 infected mouse fibroblasts that maintained extrachromosomal genomes at a constant copy number. These studies concluded that BPV1 replicated by a 'random-choice' mechanism (Gilbert and Cohen, 1987; Roberts and Weintraub, 1988; Ravnan *et al.*, 1992). However, when HPV genomes were studied in human cell lines, the results were mixed and appeared to depend on the viral type and cell line used (Hoffmann *et al.*, 2006). The authors hypothesized that the replication mode might depend on whether the initial infected cell was a transit amplifying cells or a keratinocyte stem cell (Hoffmann *et al.*, 2006).

Gammaherpesvirus genomes use a strict once per cycle replication to maintain their genomes (Yates and Guan, 1991). This is thought to be because cellular, and not viral, replication machinery initiates replication from the latent origin of replication (OriP) (Aiyar *et al.*, 1998). Binding of EBNA1 recruits cellular ORC and MCM proteins, which are normally licensed, in a cell-cycle dependent manner (Chaudhuri *et al.*, 2001). Likewise, latent KSHV DNA replication is dependent on the host replication machinery and the ORC complex binds to TR in a LANA-dependent, once per cell cycle manner (Stedman *et al.*, 2004; Verma *et al.*, 2007a).

Since PVs encode their own initiator protein, E1, it is less probable that viral replication is licensed. However, there is some evidence to suggest that the E1 protein may have a different role during maintenance replication when compared with the initial amplification step. A temperature sensitive BPV1 E1 protein demonstrated that while E1 was required for establishment, it was not required for maintenance replication (Kim and Lambert, 2002). Notably, in experiments described by Hoffmann and colleagues, overexpression of the viral E1 protein converted a once-per-cell-cycle mechanism to the random approach (Hoffmann *et al.*, 2006). It is possible that when either no, or low, levels of E1 are available in the nucleus, cellular MCM proteins initiate replication of the viral genome resulting in licensing of replication. In the presence of higher E1 levels, such as in vegetative replication, origin firing might become random.

Nuclear targeting of the viral genomes

Despite the absence of separating membranes, the eukaryotic nucleus is divided into many separate domains where different factors are concentrated. It appears that specific protein domains target proteins into 'protein factories' where their respective functions are required (reviewed in Cardoso and Leonhardt, 1998)). The host chromosomes are located in individual chromosomal territories and active and silenced regions of chromatin are arranged in euchromatic and heterochromatic regions of the nucleus.

Viral tethering to mitotic chromosomes is an important mechanism employed to faithfully segregate the viral genomes to the daughter cells. However, tethering of viral genomes to specific regions of host chromatin, or nuclear compartments might not only be beneficial for genome partitioning. It is probable that the specific positioning of the viral genomes to a particular nuclear region is beneficial to the viral lifecycle and viruses might target their genomes to specific regions of the nucleus in an attempt to control viral gene expression. For example, using a genome-wide ChIP-on-chip analysis BPV1 E2 and Brd4 were found bound to most transcriptionally active promoters without affecting their activity (Jang *et al.*, 2009). The authors propose that this may be a way for the viral genome to remain in transcriptionally active regions of the nucleus to escape silencing.

The KSHV LANA protein functionally interacts with the heterochromatin protein 1 (HP1), which could explain the observed association of LANA with heterochromatin in interphase nuclei (Szekely *et al.*, 1999). It was hypothesized that targeting of the viral genome through interaction with LANA and HP1 might aid in maintaining a latent gene expression profile. Furthermore, the association of LANA with the nuclear matrix has been implicated to be important for replication during latency (Ohsaki *et al.*, 2009). Similarly, PVs have been suggested to target different subnuclear domains depending on the requirements imposed by their lifecycle. Upon viral infection, the L2 protein appears to target the viral genome towards PML bodies (Day *et al.*, 2004). Zou and colleagues suggested that residues in the hinge region of HPV11 E2 were required for nuclear matrix association of the viral genome (Zou *et al.*, 2000). Based on this interaction, a scenario was proposed where the E2 protein, through anchoring the viral genome to the nuclear matrix, increased the local concentration of host transcription and replication factors. This scenario would allow the virus optimal access to the hosts' resources. Sakakibara *et al* have recently shown that coexpression of PV E1 and E2 proteins results in nuclear replication foci that recruit DNA damage response proteins to facilitate viral DNA replication (Sakakibara *et al.*, 2011).

It is interesting to speculate that nuclear tethering might be dynamic and vary according to the requirements of the viral genome. Since components of transcription, replication and DNA repair are located to distinct areas of the nucleus, it is reasonable to hypothesize that the viral genome has evolved mechanisms to shuttle its genome towards these regions.

Conclusions and future perspectives

Partitioning as a target for therapeutic intervention

The introduction of the bivalent and quadrivalent prophylactic HPV vaccines is expected to significantly reduce the prevalence of infections and associated disease. However, because of the lack of immunological cross-reactivity these vaccines will not protect against infections with viral types other than the ones included in the vaccines (HPV6, HPV11 and/or HPV16 and HPV18). In addition, vaccine penetration in developing populations is likely to be limited. Finally, women currently infected with HPVs will gain little benefit from a prophylactic vaccine. Considering the prevalence of HPV infection and its proven association with malignancies, it is surprising that specific antiviral drugs are not available. The importance of the E2 protein in the PV life cycle makes it an attractive target for antiviral therapy.

Specific inhibition of E2 binding to its binding sites could interfere with the viral life cycle at different levels. The interaction of E2 with E2BS is essential for viral transcription, replication and genome partitioning. Therefore, compounds inhibiting E2BS binding could potentially interfere with oncogene production and result in curing of the viral episome. A class of polyamides specifically inhibits the interaction of HPV18 E2 with its E2BS *in vitro* ($K_i = 2\,\text{nM}$) (Schaal *et al.*, 2003). It appears that the minor groove binding polyamides prevented the bending of the viral DNA required for E2 binding (Bedrosian and Bastia, 1990; Schaal *et al.*, 2003; Zimmerman and Maher, III, 2003). However, bioavailability of these compounds is very limited, and the compounds were unable to reach the nucleus of cells in cell culture. Compounds that interfere with the interaction of E2 with mitotic chromosomes would be expected to result in loss of genomes thus curing of viral infection.

In addition, compounds have been developed that interfere with the interaction of the E1 and E2 proteins (Fradet-Turcotte and Archambault, 2007). However, it has to be noted that upon integration of the viral genome during tumorigenesis, the E2 ORF is usually lost. Therefore, the above-presented strategies are only beneficial prior to integration.

Use of papillomaviral-based vectors for gene therapy

The dangers of (random) integration of gene-therapy vectors have been described, and have severely limited advancement of the field. Therefore, vectors that are maintained extrachromosomally should be advantageous over integrating vectors. The use of PV and gammaherpesvirus based vectors have been described by several authors. The use of PV based vectors has mainly been limited to 'proof of principle' studies (Gadi *et al.*, 1999; Sverdrup *et al.*, 1999; Mannik *et al.*, 2003) although a recent HIV DNA vaccine capitalizes on the nuclear retention function of the E2 protein (Martinon *et al.*, 2009; Molder *et al.*, 2009). The use of EBNA1/OriP containing vectors is more widespread. Most studies have used EBV based vectors in actively proliferating cells derived from tumours and diseases of haematopoietic cells (Mazda and Kishida, 2009); however, EBNA1 based vectors appear to offer additional advantages over mere tethering of the therapeutic vector. Using an EBV based vector, Sclimenti and colleagues were able to show a 10- to 100-fold increased expression of a transgene in mice during the 8-month duration of the study (Sclimenti *et al.*, 2003). This study clearly shows the potential of these vectors for therapeutic applications.

Genome tethering, a conserved viral function

Certain viruses have evolved to persist inside actively dividing cells. In addition to avoiding the hosts' immune system this requires an active mechanism to maintain viral genomes inside these dividing cells. In this chapter we have illustrated how two families of unrelated viruses have converged to use a very similar strategy. Notably, this convergence has occurred at the single protein level, as EBNA1, LANA and the E2 proteins are the only proteins known, to date, to fold as a dimeric beta-barrel. Whether these proteins are true homologues (i.e. they share a common ancestor) or the result of convergent evolution remains an interesting question for evolutionary biologists. It is, however, clear that these proteins employ highly similar methods to tether their viral genomes to the host chromosome.

In addition, the interaction with the host genome is absolutely required for a successful infection. Increasing evidence suggests important roles for subnuclear targeting of viral DNA during infection. This raises a 'chicken-egg' situation. Did viruses initially tether their genomes to random locations within the nucleus, or was viral partitioning a 'lucky by-product' of subnuclear targeting? The role of subnuclear targeting promises to be a field of active research for some time to come.

Additional studies will improve our understanding of the importance of viral tethering. A better understanding of these processes will not only enable the development of targeted therapeutics, it might also allow for the development of improved vectors for gene therapy and vaccine applications.

Acknowledgements

The authors would like to acknowledge members of the McBride lab, past and present. The authors are funded by the Intramural Research Program of the NIAID, NIH.

References

Adom, J.N., Gouilleux, F., and Richard-Foy, H. (1992). Interaction with the nuclear matrix of a chimeric construct containing a replication origin and a transcription unit. Biochim. Biophys. Acta *1171*, 187–197.

Aiyar, A., Tyree, C., and Sugden, B. (1998). The plasmid replicon of EBV consists of multiple cis-acting elements that facilitate DNA synthesis by the cell and a viral maintenance element. EMBO J. *17*, 6394–6403.

Alexander, K.A., and Phelps, W.C. (1996). A fluorescence anisotropy study of DNA binding by HPV-11 E2C protein: a hierarchy of E2-binding sites. Biochemistry *35*, 9864–9872.

Alexiadis, V., Waldmann, T., Andersen, J., Mann, M., Knippers, R., and Gruss, C. (2000). The protein encoded by the proto-oncogene DEK changes the topology of chromatin and reduces the efficiency of DNA replication in a chromatin-specific manner. Genes Dev. *14*, 1308–1312.

Antonsson, A., and Hansson, B.G. (2002).Healthy skin of many animal species harbors papillomaviruses which are closely related to their human counterparts. J. Virol. *76*, 12537–12542.

Antonsson, A., Forslund, O., Ekberg, H., Sterner, G., and Hansson, B.G. (2000). The ubiquity and impressive genomic diversity of human skin papillomaviruses suggest a commensalic nature of these viruses. J. Virol. *74*, 11636–11641.

Antson, A.A., Burns, J.E., Moroz, O.V., Scott, D.J., Sanders, C.M., Bronstein, I.B., Dodson, G.G., Wilson, K.S., and Maitland, N.J. (2000). Structure of the intact transactivation domain of the human papillomavirus E2 protein. Nature *403*, 805–809.

Baiker, A., Maercker, C., Piechaczek, C., Schmidt, S.B., Bode, J., Benham, C., and Lipps, H.J. (2000). Mitotic stability of an episomal vector containing a human scaffold/matrix-attached region is provided by association with nuclear matrix. Nat. Cell Biol. *2*, 182–184.

Ballestas, M.E., Chatis, P.A., and Kaye, K.M. (1999). Efficient persistence of extrachromosomal KSHV DNA mediated by latency- associated nuclear antigen. Science *284*, 641–644.

Barbera, A.J., Ballestas, M.E., and Kaye, K.M. (2004). The Kaposi's sarcoma-associated herpesvirus latency-associated nuclear antigen 1 N terminus is essential for chromosome association, DNA replication, and episome persistence. J. Virol. *78*, 294–301.

Barbera, A.J., Chodaparambil, J.V., Kelley-Clarke, B., Joukov, V., Walter, J.C., Luger, K., and Kaye, K.M. (2006). The nucleosomal surface as a docking station for Kaposi's sarcoma herpesvirus LANA. Science *311*, 856–861.

Bastien, N., and McBride, A.A. (2000). Interaction of the papillomavirus E2 with mitotic chromosomes. Virology *270*, 124–134.

Baxter, M.K., McPhillips, M.G., Ozato, K., and McBride, A.A. (2005). The mitotic chromosome binding activity of the papillomavirus E2 protein correlates with interaction with the cellular chromosomal protein, Brd4. J. Virol. *79*, 4806–4818.

Bedrosian, C.L., and Bastia, D. (1990). The DNA-binding domain of HPV16 E2 protein interaction with the viral enhancer: Protein-induced DNA bending and role of the nonconserved core sequence in binding site affinity. Virology *174*, 557–575.

Bernard, H.U., Burk, R.D., Chen, Z., Van, D.K., Hausen, H., and de Villiers, E.M. (2010). Classification of papillomaviruses (PVs) based on 189 PV types and proposal of taxonomic amendments. Virology *401*, 70–79.

Blake, N. (2010). Immune evasion by gammaherpesvirus genome maintenance proteins. J. Gen. Virol. *91*, 829–846.

Bloom, D.C., Giordani, N.V., and Kwiatkowski, D.L. (2010). Epigenetic regulation of latent HSV-1 gene expression. Biochim. Biophys. Acta *1799*, 246–256.

Bochkarev, A., Barwell, J.A., Pfuetzner, R.A., Furey, W.J., Edwards, A.M., and Frappier, L. (1995). Crystal structure of the DNA-binding domain of the Epstein–Barr virus origin-binding protein EBNA 1. Cell *83*, 39–46.

Bochkarev, A., Bochkareva, E., Frappier, L., and Edwards, A.M. (1998). The 2.2 A structure of a permanganate-sensitive DNA site bound by the Epstein–Barr virus origin binding protein, EBNA1. J. Mol. Biol. *284*, 1273–1278.

Boner, W., Taylor, E.R., Tsirimonaki, E., Yamane, K., Campo, M.S., and Morgan, I.M. (2002). A Functional Interaction between the human papillomavirus 16 transcription/replication factor E2 and the DNA damage response protein TopBP1. J. Biol. Chem. *277*, 22297–22303.

Bordeaux, J., Forte, S., Harding, E., Darshan, M.S., Klucevsek, K., and Moroianu, J. (2006). The l2 minor capsid protein of low-risk human papillomavirus type 11 interacts with host nuclear import receptors and viral DNA. J. Virol. *80*, 8259–8262.

Bouvard, V., Baan, R., Straif, K., Grosse, Y., Secretan, B., El Ghissassi, F., Benbrahim-Tallaa, L., Guha, N., Freeman, C., Galichet, L., *et al.* (2010). A review of human carcinogens–Part B: biological agents. Lancet Oncol. *10*, 321–322.

Brannon, A.R., Maresca, J.A., Boeke, J.D., Basrai, M.A., and McBride, A.A. (2005). Reconstitution of papillomavirus E2-mediated plasmid maintenance in *Saccharomyces cerevisiae* by the Brd4 bromodomain protein. Proc. Natl. Acad. Sci. U.S.A. *102*, 2998–3003.

Brown, C., Campos-Leon, K., Strickland, M., Williams, C., Fairweather, V., Brady, R.L., Crump, M.P., and Gaston, K. (2011). Protein flexibility directs DNA recognition by the papillomavirus E2 proteins. Nucleic Acids Res. *39*, 2969–2980.

Burk, R.D., Terai, M., Gravitt, P.E., Brinton, L.A., Kurman, R.J., Barnes, W.A., Greenberg, M.D., Hadjimichael, O.C., Fu, L., McGowan, L., *et al.* (2003). Distribution of human papillomavirus types 16 and 18 variants in squamous cell carcinomas and adenocarcinomas of the cervix. Cancer Res. *63*, 7215–7220.

Calderwood, M.A., Hall, K.T., Matthews, D.A., and Whitehouse, A. (2004). The herpesvirus saimiri ORF73 gene product interacts with host-cell mitotic chromosomes and self-associates via its C terminus. J. Gen. Virol. *85*, 147–153.

Cardenas-Mora, J., Spindler, J.E., Jang, M.K., and McBride, A.A. (2008). Dimerization of the papillomavirus E2 protein is required for efficient mitotic chromosome association and Brd4 binding. J. Virol. *82*, 7298–7305.

Cardoso, M.C., and Leonhardt, H. (1998). Protein targeting to subnuclear higher order structures: a new level of regulation and coordination of nuclear processes. J. Cell Biochem. *70*, 222–230.

Chandler, S.P., Guschin, D., Landsberger, N., and Wolffe, A.P. (1999). The methyl-CpG binding transcriptional repressor MeCP2 stably associates with nucleosomal DNA. Biochemistry *38*, 7008–7018.

Chaudhuri, B., Xu, H., Todorov, I., Dutta, A., and Yates, J.L. (2001). Human DNA replication initiation factors, ORC and MCM, associate with oriP of Epstein–Barr virus. Proc. Natl. Acad. Sci. U.S.A. *98*, 10085–10089.

Chen, G., and Stenlund, A. (1998). Characterization of the DNA-binding domain of the bovine papillomavirus replication initiator E1. J. Virol. *72*, 2567–2576.

Churchill, M.E., and Suzuki, M. (1989). 'SPKK' motifs prefer to bind to DNA at A/T-rich sites. EMBO J. *8*, 4189–4195.

Cotter, M.A., and Robertson, E.S. (1999). The latency-associated nuclear antigen tethers the Kaposi's sarcoma-associated herpesvirus genome to host chromosomes in body cavity-based lymphoma cells. Virology *264*, 254–264.

Cotter, M.A., Subramanian, C., and Robertson, E.S. (2001). The Kaposi's sarcoma-associated herpesvirus latency-associated nuclear antigen binds to specific sequences at the left end of the viral genome through its carboxy-terminus. Virology *291*, 241–259.

Cruickshank, J., Shire, K., Davidson, A.R., Edwards, A.M., and Frappier, L. (2000). Two domains of the Epstein–Barr virus origin DNA-binding protein, EBNA1, orchestrate sequence-specific DNA binding. J. Biol. Chem. *275*, 22273–22277.

Dao, L.D., Duffy, A., Van Tine, B.A., Wu, S.Y., Chiang, C.M., Broker, T.R., and Chow, L.T. (2006). Dynamic localization of the human papillomavirus type 11 origin binding protein E2 through mitosis while in association with the spindle apparatus. J. Virol. *80*, 4792–4800.

Day, P.M., Baker, C.C., Lowy, D.R., and Schiller, J.T. (2004). Establishment of papillomavirus infection is enhanced by promyelocytic leukemia protein (PML) expression. Proc. Natl. Acad. Sci. U.S.A. *101*, 14252–14257.

Dell, G., Wilkinson, K.W., Tranter, R., Parish, J., Leo, B.R., and Gaston, K. (2003). Comparison of the structure and DNA-binding properties of the E2 proteins from an oncogenic and a non-oncogenic human papillomavirus. J. Mol. Biol. *334*, 979–991.

Dey, A., Chitsaz, F., Abbasi, A., Misteli, T., and Ozato, K. (2003). The double bromodomain protein Brd4 binds to acetylated chromatin during interphase and mitosis. Proc. Natl. Acad. Sci. U.S.A. *100*, 8758–8763.

Donaldson, M.M., Boner, W., and Morgan, I.M. (2007). TopBP1 regulates human papillomavirus type 16 E2 interaction with chromatin. J. Virol. *81*, 4338–4342.

Doorbar, J. (2006). Molecular biology of human papillomavirus infection and cervical cancer. Clin. Sci. (Lond) *110*, 525–541.

Dresang, L.R., Vereide, D.T., and Sugden, B. (2009). Identifying sites bound by Epstein–Barr virus nuclear antigen 1 (EBNA1) in the human genome: defining a position-weighted matrix to predict sites bound by EBNA1 in viral genomes. J. Virol. *83*, 2930–2940.

Enemark, E.J., and Joshua-Tor, L. (2006). Mechanism of DNA translocation in a replicative hexameric helicase. Nature *442*, 270–275.

Evans, M.F., Aliesky, H.A., and Cooper, K. (2003). Optimization of biotinyl-tyramide-based *in situ* hybridization for sensitive background-free applications on formalin-fixed, paraffin-embedded tissue specimens. BMC. Clin. Pathol. *3*, 2.

Favre, M., Breitburd, F., Croissant, O., and Orth, G. (1977). Chromatin-like structures obtained after alkaline disruption of bovine and human papillomaviruses. J. Virol. *21*, 1205–1209.

Ferreiro, D.U., and Prat-Gay, G. (2003). A protein-DNA binding mechanism proceeds through multistate or two-state parallel pathways. J. Mol. Biol. *331*, 89–99.

Florin, L., Becker, K.A., Lambert, C., Nowak, T., Sapp, C., Strand, D., Streeck, R.E., and Sapp, M. (2006). Identification of a dynein interacting domain in the papillomavirus minor capsid protein l2. J. Virol. *80*, 6691–6696.

Fradet-Turcotte, A., and Archambault, J. (2007). Recent advances in the search for antiviral agents against human papillomaviruses. Antiviral Ther. *12*, 431–451.

Fujita, T., Ikeda, M., Kusano, S., Yamazaki, M., Ito, S., Obayashi, M., and Yanagi, K. (2001). Amino acid substitution analyses of the DNA contact region, two amphipathic alpha-helices and a recognition-helix-like helix outside the dimeric beta-barrel of Epstein–Barr virus nuclear antigen 1. Intervirology *44*, 271–282.

Gadi, V.K., Zou, N., Liu, J.S., Cheng, S., Broker, T.R., Sorscher, E.J., and Chow, L.T. (1999). Components of human papillomavirus that activate transcription and support plasmid replication in human airway cells. Am. J. Respir. Cell Mol. Biol. *20*, 1001–1006.

Gilbert, D.M., and Cohen, S.N. (1987). Bovine papilloma virus plasmids replicate randomly in mouse fibroblasts throughout S phase of the cell cycle. Cell *50*, 59–68.

Griffiths, R., Harrison, S.M., Macnab, S., and Whitehouse, A. (2008). Mapping the minimal regions within the ORF73 protein required for herpesvirus saimiri episomal persistence. J. Gen. Virol. *89*, 2843–2850.

Grundhoff, A., and Ganem, D. (2003). The latency-associated nuclear antigen of Kaposi's sarcoma-associated herpesvirus permits replication of terminal repeat-containing plasmids. J. Virol. *77*, 2779–2783.

Han, S.J., Hu, J., Pierce, B., Weng, Z., and Renne, R. (2010). Mutational analysis of the latency-associated nuclear antigen DNA-binding domain of Kaposi's sarcoma-associated herpesvirus reveals structural conservation among gammaherpesvirus origin-binding proteins. J. Gen. Virol. *91*, 2203–2215.

Harrison, S., Fisenne, K., and Hearing, J. (1994). Sequence requirements of the Epstein–Barr virus latent origin of DNA replication. J. Virol. *68*, 1913–1925.

Hebner, C., Lasanen, J., Battle, S., and Aiyar, A. (2003). The spacing between adjacent binding sites in the family of repeats affects the functions of Epstein–Barr nuclear antigen 1 in transcription activation and stable plasmid maintenance. Virology *311*, 263–274.

Hegde, R.S. (2002). The papillomavirus E2 proteins: structure, function, and biology. Annu. Rev. Biophys. Biomol. Struct. *31*, 343–360.

Hegde, R.S., Grossman, S.R., Laimins, L.A., and Sigler, P.B. (1992). Crystal structure at 1.7 A of the bovine papillomavirus-1 E2 DNA- binding domain bound to its DNA target. Nature *359*, 505–512.

Hines, C.S., Meghoo, C., Shetty, S., Biburger, M., Brenowitz, M., and Hegde, R.S. (1998). DNA structure and flexibility in the sequence-specific binding of papillomavirus E2 proteins. J. Mol. Biol. *276*, 809–818.

Hoffmann, R., Hirt, B., Bechtold, V., Beard, P., and Raj, K. (2006). Different modes of human papillomavirus DNA replication during maintenance. J. Virol. *80*, 4431–4439.

Hung, S.C., Kang, M.S., and Kieff, E. (2001). Maintenance of Epstein–Barr virus (EBV) oriP-based episomes requires EBV-encoded nuclear antigen-1 chromosome-binding domains, which can be replaced by high-mobility group-I or histone H1. Proc. Natl. Acad. Sci. U.S.A. *98*, 1865–1870.

Ilves, I., Kivi, S., and Ustav, M. (1999). Long-term episomal maintenance of bovine papillomavirus type 1 plasmids is determined by attachment to host chromosomes, which Is mediated by the viral E2 protein and its binding sites. J. Virol. *73*, 4404–4412.

Ilves, I., Maemets, K., Silla, T., Janikson, K., and Ustav, M. (2006). Brd4 is involved in multiple processes of the bovine papillomavirus type 1 life cycle. J. Virol. *80*, 3660–3665.

Jang, M.K., Kwon, D., and McBride, A.A. (2009). Papillomavirus E2 proteins and the host BRD4 protein associate with transcriptionally active cellular chromatin. J. Virol. *83*, 2592–2600.

Jantzen, H.M., Admon, A., Bell, S.P., and Tjian, R. (1990). Nucleolar transcription factor hUBF contains a DNA-binding motif with homology to HMG proteins. Nature *344*, 830–836.

Jenke, A.C., Stehle, I.M., Herrmann, F., Eisenberger, T., Baiker, A., Bode, J., Fackelmayer, F.O., and Lipps, H.J. (2004). Nuclear scaffold/matrix attached region modules linked to a transcription unit are sufficient for replication and maintenance of a mammalian episome. Proc Natl. Acad. Sci. U.S.A. *101*, 11322–11327.

Kamper, N., Day, P.M., Nowak, T., Selinka, H.C., Florin, L., Bolscher, J., Hilbig, L., Schiller, J.T., and Sapp, M. (2006). A membrane-destabilizing peptide in capsid protein L2 is required for egress of papillomavirus genomes from endosomes. J. Virol. *80*, 759–768.

Kapoor, P., Shire, K., and Frappier, L. (2001). Reconstitution of Epstein–Barr virus-based plasmid partitioning in budding yeast. EMBO J. *20*, 222–230.

Kapoor, P., Lavoie, B.D., and Frappier, L. (2005). EBP2 plays a key role in Epstein–Barr virus mitotic segregation and is regulated by aurora family kinases. Mol. Cell Biol. *25*, 4934–4945.

Kelley-Clarke, B., Ballestas, M.E., Komatsu, T., and Kaye, K.M. (2007a). Kaposi's sarcoma herpesvirus C-terminal LANA concentrates at pericentromeric and peri-telomeric regions of a subset of mitotic chromosomes. Virology *357*, 149–157.

Kelley-Clarke, B., Ballestas, M.E., Srinivasan, V., Barbera, A.J., Komatsu, T., Harris, T.A., Kazanjian, M., and Kaye, K.M. (2007b). Determination of Kaposi's sarcoma-associated herpesvirus C-terminal latency-associated nuclear antigen residues mediating chromosome association and DNA binding. J. Virol. *81*, 4348–4356.

Kelley-Clarke, B., De Leon-Vazquez, E., Slain, K., Barbera, A.J., and Kaye, K.M. (2009). Role of Kaposi's sarcoma-associated herpesvirus C-terminal LANA chromosome binding in episome persistence. J. Virol. *83*, 4326–4337.

Kim, K., and Lambert, P.F. (2002). E1 protein of bovine papillomavirus 1 is not required for the maintenance of viral plasmid DNA replication. Virology *293*, 10–14.

Kim, K., Garner-Hamrick, P.A., Fisher, C., Lee, D., and Lambert, P.F. (2003). Methylation patterns of papillomavirus DNA, its influence on E2 function, and implications in viral infection. J. Virol. *77*, 12450–12459.

Kim, S.S., Tam, J.K., Wang, A.F., and Hegde, R.S. (2000). The structural basis of DNA target discrimination by papillomavirus E2 proteins. J. Biol. Chem. *275*, 31245–31254.

Komatsu, T., Barbera, A.J., Ballestas, M.E., and Kaye, K.M. (2001). The Kaposi' s sarcoma-associated herpesvirus latency-associated nuclear antigen. Viral Immunol. *14*, 311–317.

Komatsu, T., Ballestas, M.E., Barbera, A.J., Kelley-Clarke, B., and Kaye, K.M. (2004). KSHV LANA1 binds DNA as an oligomer and residues N-terminal to the oligomerization domain are essential for DNA binding, replication, and episome persistence. Virology *319*, 225–236.

Krithivas, A., Fujimuro, M., Weidner, M., Young, D.B., and Hayward, S.D. (2002). Protein interactions targeting the latency-associated nuclear antigen of Kaposi's sarcoma-associated herpesvirus to cell chromosomes. J. Virol. *76*, 11596–11604.

Lai, M.C., Teh, B.H., and Tarn, W.Y. (1999). A human papillomavirus E2 transcriptional activator. The interactions with cellular splicing factors and potential function in pre-mRNA processing. J. Biol. Chem. *274*, 11832–11841.

Lambert, P.F., Monk, B.C., and Howley, P.M. (1990). Phenotypic analysis of bovine papillomavirus type 1 E2 repressor mutants. J. Virol. *64*, 950–956.

Lehman, C.W., and Botchan, M.R. (1998). Segregation of viral plasmids depends on tethering to chromosomes and is regulated by phosphorylation. Proc Natl. Acad. Sci. U.S.A. *95*, 4338–4343.

Li, R., Knight, J., Bream, G., Stenlund, A., and Botchan, M. (1989). Specific recognition nucleotides and their DNA context determine the affinity of E2 protein for 17 binding sites in the BPV-1 genome. Genes Dev. *3*, 510–526.

Lim, D.A., Gossen, M., Lehman, C.W., and Botchan, M.R. (1998). Competition for DNA binding sites between the short and long forms of E2 dimers underlies repression in bovine papillomavirus type 1 DNA replication control. J. Virol. *72*, 1931–1940.

Lin, A., Wang, S., Nguyen, T., Shire, K., and Frappier, L. (2008). The EBNA1 protein of Epstein–Barr virus functionally interacts with Brd4. J. Virol. *82*, 12009–12019.

Lu, F., Wikramasinghe, P., Norseen, J., Tsai, K., Wang, P., Showe, L., Davuluri, R.V., and Lieberman, P.M. (2010). Genome-wide analysis of host-chromosome binding sites for Epstein–Barr Virus Nuclear Antigen 1 (EBNA1). Virol. J. *7*, 262.

Lupton, S., and Levine, A.J. (1985). Mapping genetic elements of Epstein–Barr virus that facilitate extrachromosomal persistence of Epstein–Barr virus-derived plasmids in human cells. Mol. Cell Biol. *5*, 2533–2542.

McBride, A.A. (2008). Replication and partitioning of papillomavirus genomes. Adv. Virus Res. *72*, 155–205.

McBride, A.A., and Howley, P.M. (1991). Bovine papillomavirus with a mutation in the E2 serine 301 phosphorylation site replicates at a high copy number. J. Virol. *65*, 6528–6534.

McPhillips, M.G., Ozato, K., and McBride, A.A. (2005). Interaction of bovine papillomavirus E2 protein with Brd4 stabilizes its association with chromatin. J. Virol. *79*, 8920–8932.

McPhillips, M.G., Oliveira, J.G., Spindler, J.E., Mitra, R., and McBride, A.A. (2006). Brd4 is required for e2-mediated transcriptional activation but not genome partitioning of all papillomaviruses. J. Virol. *80*, 9530–9543.

Mannik, A., Piirsoo, M., Nordstrom, K., Ustav, E., Vennstrom, B., and Ustav, M. (2003). Effective generation of transgenic mice by Bovine papillomavirus type 1 based self-replicating plasmid that is maintained as extrachromosomal genetic element in three generations of animals. Plasmid *49*, 193–204.

Marechal, V., Dehee, A., Chikhi-Brachet, R., Piolot, T., Coppey-Moisan, M., and Nicolas, J.C. (1999). Mapping EBNA-1 domains involved in binding to metaphase chromosomes. J. Virol. *73*, 4385–4392.

Martinon, F., Kaldma, K., Sikut, R., Culina, S., Romain, G., Tuomela, M., Adojaan, M., Mannik, A., Toots, U., Kivisild, T., *et al.* (2009). Persistent immune responses induced by a human immunodeficiency virus DNA vaccine delivered in association with electroporation in the skin of nonhuman primates. Hum. Gene Ther. *20*, 1291–1307.

Matsumura, S., Persson, L.M., Wong, L., and Wilson, A.C. (2010). The latency-associated nuclear antigen interacts with MeCP2 and nucleosomes through separate domains. J. Virol. *84*, 2318–2330.

Mattsson, K., Kiss, C., Platt, G.M., Simpson, G.R., Kashuba, E., Klein, G., Schulz, T.F., and Szekely, L. (2002). Latent nuclear antigen of Kaposi's sarcoma herpesvirus/human herpesvirus-8 induces and relocates RING3 to nuclear heterochromatin regions. J. Gen. Virol. *83*, 179–188.

Mazda, O., and Kishida, T. (2009). Molecular therapeutics of cancer by means of electroporation-based transfer of siRNAs and EBV-based expression vectors. Front Biosci. *1*, 316–331.

Mirkovitch, J., Mirault, M.E., and Laemmli, U.K. (1984). Organization of the higher-order chromatin loop: specific DNA attachment sites on nuclear scaffold. Cell *39*, 223–232.

Mohr, I.J., Clark, R., Sun, S., Androphy, E.J., MacPherson, P., and Botchan, M.R. (1990). Targeting the E1 replication protein to the papillomavirus origin of replication by complex formation with the E2 transactivator. Science *250*, 1694–1699.

Mok, Y.K., Bycroft, M., and de Prat-Gay, G. (1996). The dimeric DNA-binding domain of the human papillomavirus E2 protein folds through a monomeric intermediate which cannot be native-like. Nat. Struct. Biol. 3, 711–717.

Molder, T., Adojaan, M., Kaldma, K., Ustav, M., and Sikut, R. (2009). Elicitation of broad CTL response against HIV-1 by the DNA vaccine encoding artificial multi-component fusion protein MultiHIV-study in domestic pigs. vaccine *28*, 293–298.

Nayyar, V.K., Shire, K., and Frappier, L. (2009). Mitotic chromosome interactions of Epstein–Barr nuclear antigen 1 (EBNA1) and human EBNA1-binding protein 2 (EBP2). J. Cell Sci. *122*, 4341–4350.

Ohsaki, E., Suzuki, T., Karayama, M., and Ueda, K. (2009). Accumulation of LANA at nuclear matrix fraction is important for Kaposi's sarcoma-associated herpesvirus replication in latency. Virus Res. *139*, 74–84.

Oliveira, J.G., Colf, L.A., and McBride, A.A. (2006). Variations in the association of papillomavirus E2 proteins with mitotic chromosomes. Proc. Natl. Acad. Sci. U S.A. *103*, 1047–1052.

Ottinger, M., Christalla, T., Nathan, K., Brinkmann, M.M., Viejo-Borbolla, A., and Schulz, T.F. (2006). The Kaposi's Sarcoma-Associated Herpesvirus LANA-1 interacts with the short variant of BRD4 and releases cells from a. J. Virol. 10772–10786.

Parish, J.L., Bean, A.M., Park, R.B., and Androphy, E.J. (2006). ChlR1 is required for loading papillomavirus E2 onto mitotic chromosomes and viral genome maintenance. Mol. Cell *24*, 867–876.

Penrose, K.J., and McBride, A.A. (2000). Proteasome-mediated degradation of the papillomavirus E2-TA protein is regulated by phosphorylation and can modulate viral genome copy number. J. Virol. *74*, 6031–6038.

Penrose, K.J., Garcia-Alai, M., Prat-Gay, G., and McBride, A.A. (2004). CK2 phosphorylation-induced conformational switch triggers degradation of the papillomavirus E2 protein. J. Biol. Chem. *279*, 22430–22439.

Piirsoo, M., Ustav, E., Mandel, T., Stenlund, A., and Ustav, M. (1996). Cis and trans requirements for stable episomal maintenance of the BPV-1 replicator. EMBO J. *15*, 1–11.

Pittayakhajonwut, D., and Angeletti, P.C. (2008). Analysis of cis-elements that facilitate extrachromosomal persistence of human papillomavirus genomes. Virology *374*, 304–314.

Platt, G.M., Simpson, G.R., Mittnacht, S., and Schulz, T.F. (1999). Latent nuclear antigen of Kaposi's sarcoma-associated herpesvirus interacts with RING3, a homolog of the Drosophila female sterile homeotic (fsh) gene. J. Virol. 73, 9789–9795.

Poddar, A., Reed, S.C., McPhillips, M.G., and McBride A.A. (2009a). The human papillomavirus type 8 E2 tethering protein targets the ribosomal DNA loci of host mitotic chromosomes. J. Virol. *83*, 640–650.

Poddar, A., Reed, S.C., McPhillips, M.G., Spindler, J.E., and McBride, A.A. (2009b). The human papillomavirus type 8 E2 tethering protein targets the ribosomal DNA loci of host mitotic chromosomes. J. Virol. *83*, 640–650.

Powell, M.L., Smith, J.A., Sowa, M.E., Harper, J.W., Iftner, T., Stubenrauch, F., and Howley, P.M. (2010). NCoR1 mediates papillomavirus E8;E2C transcriptional repression. J. Virol. *84*, 4451–4460.

Pyeon, D., Pearce, S.M., Lank, S.M., Ahlquist, P., and Lambert, P.F. (2009). Establishment of human papillomavirus infection requires cell cycle progression. PLoS. Pathog. *5*, e1000318.

Ravnan, J.-B., Gilbert, G.M., Ten Hagen, K.G., and Cohen, S.N. (1992). Random-choice replication of extrachromosomal bovine papillomavirus (BPV) molecules in heterogeneous clonally derived BPV-infected cell lines. J. Virol. *66*, 6946–6952.

Razin, S.V., Vassetzky, Y.S., and Hancock, R. (1991). Nuclear matrix attachment regions and topoisomerase II binding and reaction sites in the vicinity of a chicken DNA replication origin. Biochem. Biophys. Res. Commun. *177*, 265–270.

Reisman, D., Yates, J., and Sugden, B. (1985). A putative origin of replication of plasmids derived from Epstein–Barr virus is composed of two cis-acting components. Mol. Cell Biol. 5, 1822–1832.

Roberts, J.M., and Weintraub, H. (1988). Cis-acting negative control of DNA replication in eukaryotic cells. Cell *52*, 397–404.

Roussel, P., Andre, C., Masson, C., Geraud, G., and Hernandez-Verdun, D. (1993). Localization of the RNA polymerase I transcription factor hUBF during the cell cycle. J. Cell Sci. *104 (Pt 2)*, 327–337.

Sakakibara, N, Mitra, R., and McBride A.A. (2011). The Papillomavirus E1 helicase activates a cellular DNA damage response in viral replication foci. J. Virol. *85*, 8981–8985.

Sanders, C.M., and Stenlund, A. (1998). Recruitment and loading of the E1 initiator protein: an ATP-dependent process catalysed by a transcription factor. EMBO J. *17*, 7044–7055.

Sarafi, T.R., and McBride, A.A. (1995). Domains of the BPV-1 E1 replication protein required for origin-specific DNA binding and interaction with the E2 transactivator. Virology *211*, 385–396.

Schaal, T.D., Mallet, W.G., McMinn, D.L., Nguyen, N.V., Sopko, M.M., John, S., and Parekh, B.S. (2003). Inhibition of human papilloma virus E2 DNA binding protein by covalently linked polyamides. Nucleic Acids Res. *31*, 1282–1291.

Schepers, A., Ritzi, M., Bousset, K., Kremmer, E., Yates, J.L., Harwood, J., Diffley, J.F., and Hammerschmidt, W. (2001). Human origin recognition complex binds to the region of the latent origin of DNA replication of Epstein–Barr virus. EMBO J. *20*, 4588–4602.

Schiller, J.T., Day, P.M., and Kines, R.C. (2010). Current understanding of the mechanism of HPV infection. Gynecol. Oncol. *118*, S12-S17.

Schuck, S., and Stenlund, A. (2006). Surface mutagenesis of the bovine papillomavirus E1 DNA-binding domain reveals residues required for multiple functions related to DNA replication. J. Virol. *80*, 7491–7499.

Sclimenti, C.R., Neviaser, A.S., Baba, E.J., Meuse, L., Kay, M.A., and Calos, M.P. (2003). Epstein–Barr virus vectors provide prolonged robust factor IX expression in mice. Biotechnol. Prog. *19*, 144–151.

Sears, J., Kolman, J., Wahl, G.M., and Aiyar, A. (2003). Metaphase chromosome tethering is necessary for the DNA synthesis and maintenance of oriP plasmids but is insufficient for transcription activation by Epstein–Barr nuclear antigen 1. J. Virol. *77*, 11767–11780.

Sears, J., Ujihara, M., Wong, S., Ott, C., Middeldorp, J., and Aiyar, A. (2004). The amino terminus of Epstein–Barr virus (EBV) nuclear antigen 1 contains AT hooks that facilitate the replication and partitioning of latent EBV genomes by tethering them to cellular chromosomes. J. Virol. *78*, 11487–11505.

Sedman, J., and Stenlund, A. (1998). The papillomavirus E1 protein forms a DNA-dependent hexameric complex with ATPase and DNA helicase activities. J. Virol. *72*, 6893–6897.

Sedman, T., Sedman, J., and Stenlund, A. (1997). Binding of the E1 and E2 proteins to the origin of replication of bovine papillomavirus. J. Virol. *71*, 2887–2896.

Sekhar, V., Reed, S.C., and McBride, A.A. (2010). Interaction of the betapapillomavirus E2 tethering protein with mitotic chromosomes. J. Virol. *84*, 543–557.

Senechal, H., Poirier, G.G., Coulombe, B., Laimins, L.A., and Archambault, J. (2007). Amino acid substitutions that specifically impair the transcriptional activity of papillomavirus E2 affect binding to the long isoform of Brd4. Virology *358*, 10–17.

Shire, K., Ceccarelli, D.F., Avolio-Hunter, T.M., and Frappier, L. (1999). EBP2, a human protein that interacts with sequences of the Epstein–Barr virus nuclear antigen 1 important for plasmid maintenance. J. Virol. *73*, 2587–2595.

Shire, K., Kapoor, P., Jiang, K., Hing, M.N., Sivachandran, N., Nguyen, T., and Frappier, L. (2006). Regulation of the EBNA1 Epstein–Barr virus protein by serine phosphorylation and arginine methylation. J. Virol. *80*, 5261–5272.

Si, H., and Robertson, E.S. (2006). Kaposi's sarcoma-associated herpesvirus-encoded latency-associated nuclear antigen induces chromosomal instability through inhibition of p53 function. J. Virol. *80*, 697–709.

Silla, T., Mannik, A., and Ustav, M. (2010). Effective formation of the segregation-competent complex determines successful partitioning of the bovine papillomavirus genome during cell division. J. Virol. *84*, 11175–11188.

Skiadopoulos, M.H., and McBride, A.A. (1998). Bovine papillomavirus type 1 genomes and the E2 transactivator protein are closely associated with mitotic chromatin. J. Virol. *72*, 2079–2088.

Smith, J.A., White, E.A., Sowa, M.E., Powell, M.L., Ottinger, M., Harper, J.W., and Howley, P.M. (2010). Genome-wide siRNA screen identifies SMCX, EP400, and Brd4 as E2-dependent regulators of human papillomavirus oncogene expression. Proc. Natl. Acad. Sci. U.S.A. *107*, 3752–3757.

Stedman, W., Deng, Z., Lu, F., and Lieberman, P.M. (2004). ORC, MCM, and histone hyperacetylation at the Kaposi's sarcoma-associated herpesvirus latent replication origin. J. Virol. *78*, 12566–12575.

Steger, G., Schnabel, C., and Schmidt, H.M. (2002). The hinge region of the human papillomavirus type 8 E2 protein activates the human p21(WAF1/CIP1) promoter via interaction with Sp1. J. Gen. Virol. *83*, 503–510.

Stubenrauch, F., Colbert, A.M., and Laimins, L.A. (1998a). Transactivation by the E2 protein of oncogenic human papillomavirus type 31 is not essential for early and late viral functions. J. Virol. *72*, 8115–8123.

Stubenrauch, F., Lim, H.B., and Laimins, L.A. (1998b). Differential requirements for conserved E2 binding sites in the life cycle of oncogenic human papillomavirus type 31. J. Virol. *72*, 1071–1077.

Stunkel, W., Huang, Z., Tan, S.H., O'Connor, M.J., and Bernard, H.U. (2000). Nuclear matrix attachment regions of human papillomavirus type 16 repress or activate the E6 promoter, depending on the physical state of the viral DNA. J. Virol. *74*, 2489–2501.

Sverdrup, F., Sheahan, L., and Khan, S. (1999). Development of human papillomavirus plasmids capable of episomal replication in human cell lines. Gene Ther. *6*, 1317–1321.

Szekely, L., Kiss, C., Mattsson, K., Kashuba, E., Pokrovskaja, K., Juhasz, A., Holmvall, P., and Klein, G. (1999). Human herpesvirus-8-encoded LNA-1 accumulates in heterochromatin- associated nuclear bodies. J. Gen. Virol. *80 (Pt 11)*, 2889–2900.

Tan, S.H., Bartsch, D., Schwarz, E., and Bernard, H.U. (1998). Nuclear matrix attachment regions of human papillomavirus type 16 point toward conservation of these genomic elements in all genital papillomaviruses. J. Virol. *72*, 3610–3622.

Thain, A., Jenkins, O., Clarke, A.R., and Gaston, K. (1996). CpG methylation directly inhibits binding of the human papillomavirus type 16 E2 protein to specific DNA sequences. J. Virol. *70*, 7233–7235.

Ustav, M., and Stenlund, A. (1991). Transient replication of BPV-1 requires two viral polypeptides encoded by the E1 and E2 open reading frames. EMBO J. *10*, 449–457.

Ustav, M., Ustav, E., Szymanski, P., and Stenlund, A. (1991). Identification of the origin of replication of bovine papillomavirus and characterization of the viral origin recognition factor E1. EMBO J. *10*, 4321–4329.

Van Tine, B.A., Dao, L.D., Wu, S.Y., Sonbuchner, T.M., Lin, B.Y., Zou, N., Chiang, C.M., Broker, T.R., and Chow, L.T. (2004). Human papillomavirus (HPV) origin-binding protein associates with mitotic spindles to enable viral DNA partitioning. Proc Natl. Acad. Sci. U.S.A. *101*, 4030–4035.

Verma, S.C., Lan, K., Choudhuri, T., Cotter, M.A., and Robertson, E.S. (2007a). An autonomous replicating element within the KSHV genome. Cell Host. Microbe *2*, 106–118.

Verma, S.C., Lan, K., and Robertson, E. (2007b). Structure and function of latency-associated nuclear antigen. Curr. Top. Microbiol. Immunol. *312*, 101–136.

Viejo-Borbolla, A., Ottinger, M., Bruning, E., Burger, A., Konig, R., Kati, E., Sheldon, J.A., and Schulz, T.F.

(2005). Brd2/RING3 interacts with a chromatin-binding domain in the Kaposi's Sarcoma-associated herpesvirus latency-associated nuclear antigen 1 (LANA-1) that is required for multiple functions of LANA-1. J. Virol. *79*, 13618–13629.

Voitenleitner, C., and Botchan, M. (2002). E1 protein of bovine papillomavirus type 1 interferes with E2 protein- mediated tethering of the viral DNA to mitotic chromosomes. J. Virol. *76*, 3440–3451.

Walboomers, J.M., Jacobs, M.V., Manos, M.M., Bosch, F.X., Kummer, J.A., Shah, K.V., Snijders, P.J., Peto, J., Meijer, C.J., and Oz, N. (1999). Human papillomavirus is a necessary cause of invasive cervical cancer worldwide. J. Pathol. *189*, 12–19.

White, R.E., Wade-Martins, R., and James, M.R. (2001). Sequences adjacent to oriP improve the persistence of Epstein–Barr virus-based episomes in B cells. J. Virol. *75*, 11249–11252.

Wu, H., Ceccarelli, D.F., and Frappier, L. (2000). The DNA segregation mechanism of Epstein–Barr virus nuclear antigen 1. EMBO Rep. *1*, 140–144.

Wu, S.Y., and Chiang, C.M. (2007). The double bromodomain-containing chromatin adaptor Brd4 and transcriptional regulation. J. Biol. Chem. *282*, 13141–13145.

Wu, S.Y., Lee, A.Y., Hou, S.Y., Kemper, J.K., Erdjument-Bromage, H., Tempst, P., and Chiang, C.M. (2006). Brd4 links chromatin targeting to HPV transcriptional silencing. Genes Dev. *20*, 2383–2396.

Wysokenski, D.A., and Yates, J.L. (1989). Multiple EBNA1-binding sites are required to form an EBNA1-dependent enhancer and to activate a minimal replicative origin within oriP of Epstein–Barr virus. J. Virol. *63*, 2657–2666.

Xiao, B., Verma, S.C., Cai, Q., Kaul, R., Lu, J., Saha, A., and Robertson, E.S. (2010). Bub1 and CENP-F can contribute to Kaposi's sarcoma-associated herpesvirus genome persistence by targeting LANA to kinetochores. J. Virol. *84*, 9718–9732.

Yang, L., Mohr, I., Fouts, E., Lim, D.A., Nohaile, M., and Botchan, M. (1993). The E1 protein of bovine papillomavirus 1 is an ATP-dependent DNA helicase. Proc. Natl. Acad. Sci. U.S.A. *90*, 5086–5090.

Yates, J.L., and Guan, N. (1991). Epstein–Barr virus-derived plasmids replicate only once per cell cycle and are not amplified after entry into cells. J. Virol. *65*, 483–488.

You, J., Croyle, J.L., Nishimura, A., Ozato, K., and Howley, P.M. (2004). Interaction of the bovine papillomavirus E2 protein with Brd4 tethers the viral DNA to host mitotic chromosomes. Cell *117*, 349–360.

You, J., Schweiger, M.R., and Howley, P.M. (2005). Inhibition of E2 binding to Brd4 enhances viral genome loss and phenotypic reversion of bovine papillomavirus-transformed cells. J. Virol. *79*, 14956–14961.

You, J., Srinivasan, V., Denis, G.V., Harrington, W.J., Jr., Ballestas, M.E., Kaye, K.M., and Howley, P.M. (2006). Kaposi's sarcoma-associated herpesvirus latency-associated nuclear antigen interacts with bromodomain protein Brd4 on host mitotic chromosomes. J. Virol. *80*, 8909–8919.

Yu, T., Peng, Y.C., and Androphy, E.J. (2007). Mitotic kinesin-like protein 2 binds and colocalizes with papillomavirus E2 during mitosis. J. Virol. *81*, 1736–1745.

Zimmerman, J.M., and Maher, L.J., III (2003). Solution measurement of DNA curvature in papillomavirus E2 binding sites. Nucleic Acids Res. *31*, 5134–5139.

Zobel, T., Iftner, T., and Stubenrauch, F. (2003). The papillomavirus E8-E2C protein represses DNA replication from extrachromosomal origins. Mol. Cell Biol. *23*, 8352–8362.

Zou, N., Lin, B.Y., Duan, F., Lee, K.Y., Jin, G., Guan, R., Yao, G., Lefkowitz, E.J., Broker, T.R., and Chow, L.T. (2000). The hinge of the human papillomavirus type 11 E2 protein contains major determinants for nuclear localization and nuclear matrix association. J. Virol. *74*, 3761–3770.

Alterations in Cellular miRNAs Induced by Human Papillomaviruses

Amy S. Gardiner, Abigail I. Wald and Saleem A. Khan

Abstract

In recent years, microRNAs (miRNAs) have been found to play important roles in the regulation of gene expression in mammalian cells. MiRNAs regulate many processes, including cell cycle progression, cell differentiation and organogenesis. Human cells encode approximately 1000 miRNAs, and their expression has been shown to be altered in a variety of human cancers. Human papillomaviruses (HPVs) are DNA tumour viruses that are associated with cancers, especially cancers of the cervix and oropharynx. Recently, several studies have shown altered expression of miRNAs in HPV-associated cervical and oral cancers. In this article, we discuss the role of HPVs and their oncogenes in altering cellular miRNA expression, possible targets of such miRNAs, and how miRNA changes may contribute to the pathogenesis of HPV-associated cancers.

Human papillomaviruses

HPVs are common papilloma (wart)-causing viruses that are associated with a variety of cutaneous and mucosal lesions, particularly cancer of the cervix (Walboomers *et al.*, 1999; Greenlee *et al.*, 2001). Recent molecular epidemiological studies have also shown a strong correlation between oncogenic HPV infections and a subset of oropharyngeal cancers (Gillison *et al.*, 2000; Wong and Munger, 2000). HPV DNA can be found in approximately 99.7% of cervical cancers and greater than 25% of oral cancers (Gillison *et al.*, 2000; Greenlee *et al.*, 2001). More than 150 HPV types exist, and they can be grouped as low-risk (LR) or high-risk (HR) based on the risk of malignant transformation (de Villiers *et al.*, 2004). LR-HPVs, such as types 6 and 11, are associated with epithelial lesions, while HR-HPVs, such as types 16 and 18, are associated with cancer (de Villiers *et al.*, 2004).

HR-HPVs encode two oncoproteins, E6 and E7, and their expression is essential for cellular transformation (Munger *et al.*, 2004). The E6 protein (~150 amino acids) interacts with the tumour suppressor protein p53 (Werness *et al.*, 1990) as a tripartite complex with the E6-associated protein (E6AP) (Huibregtse *et al.*, 1991). This interaction results in ubiquitination and degradation of p53 (Scheffner *et al.*, 1990). The E6 protein also interacts with proteins containing PDZ-binding domains (Kiyono *et al.*, 1997; Nakagawa and Huibregtse, 2000), activates telomerase (Klingelhutz *et al.*, 1996), and interferes with the function of transcription factors (TFs) involved in cellular processes such as growth, differentiation, and cell cycle progression by binding to regions of the coactivators ADA3, p300 and CBP (Patel *et al.*, 1999; Zimmermann *et al.*, 1999). The high-risk HPV E7 protein is ~100 amino acids long, interacts with the retinoblastoma family of pocket proteins, and induces the ubiquitin-mediated degradation of pRb, p107, and p130 (Dyson *et al.*, 1989; Munger *et al.*, 1989). This helps push the cell into S-phase by releasing the E2F TF from pRB–E2F complexes (Weintraub *et al.*, 1992, 1995). E7 also binds to cyclin dependent kinase (cdk) inhibitors $p21^{CIP1}$ and $p27^{KIP1}$ to further disrupt cell cycle control (Zerfass-Thome *et al.*, 1996; Funk *et al.*, 1997; Jones *et al.*, 1997). HR-HPV E6 and E7 can cooperate to induce centrosome abnormalities and genomic instability (Duensing *et al.*, 2000). Persistent infections with HR-HPVs, such as HPV16, are often characterized by integration of the viral genome into the host chromosomes, which results in increased expression of the E6 and E7 oncogenes owing to interruption of the E2 regulatory gene (Schwarz *et al.*, 1985). Integration is thought to occur at fragile sites in chromosomes (Thorland *et al.*, 2003) and may be a consequence of HR-HPV E6/E7-induced chromosomal instability.

MicroRNAs

MicroRNAs are a recently characterized class of gene regulators. Mature miRNAs are ~22 nt long single-stranded, non-protein-coding RNAs that negatively regulate their targets. MiRNAs function by binding to the 3′ untranslated regions (UTRs) of target messenger RNAs (mRNAs), thereby mediating either translational repression or mRNA degradation. MiRNAs were first discovered in *Caenorhabditis elegans* and have since been found to be conserved across many species, and may regulate thousands of targets via the RNAi pathway (Lee *et al.*, 1993; Lewis *et al.*, 2005). Most miRNAs are transcribed by RNA polymerase II, have a cap, and are polyadenylated (Cai *et al.*, 2004; Lee *et al.*, 2004). MiRNAs are either transcribed from their own promoters or are processed from the introns of protein-encoding genes (Lee *et al.*, 2002). Following transcription, the large primary miRNA transcripts are processed into ~80 nt long precursor miRNAs by the RNase III Drosha (Lee *et al.*, 2003). Pre-miRNAs are hairpin-like structures with characteristic 2 nt 3′ overhangs. They are exported to the cytoplasm by Exportin 5 (Lund *et al.*, 2004), where further processing into miRNA duplexes by the RNase Dicer occurs (Hutvagner *et al.*, 2001). In addition, non-canonical pathways have been identified for the production of small miRNA-like regulators (Miyoshi *et al.*, 2010). MiRNA duplexes associate with the RISC complex, but only one strand, the mature miRNA, remains associated with it and is delivered to its target (Khvorova *et al.*, 2003; Schwarz *et al.*, 2003). The fate of the target mRNAs depends on the degree of complementarity with the miRNA. Commonly, the 5′ 2–8 nt of the miRNA (called the seed sequence) are complementary to the target, and the remaining miRNA contains many mismatches. A low degree of complementarity results in translational repression, whereas a high degree of complementarity results in cleavage of the mRNA followed by its eventual destruction (Kim, 2005). A single miRNA can have multiple mRNA targets, while a single mRNA may also be regulated by several miRNAs (Bartel, 2009). Other classes of small RNAs that function to repress mRNA have recently been described (Kim *et al.*, 2009a; Li *et al.*, 2009). Also, some miRNAs have also been shown to promote transcriptional or translational activation of their targets (Vasudevan *et al.*, 2007; Place *et al.*, 2008).

MicroRNA expression in cancer

Distinct miRNA expression profiles have been linked to many cancers (Volinia *et al.*, 2006). MiRNA encoding sequences are often found near fragile sites in the chromosomes, as well as near integration sites of HR-HPVs (Zhang *et al.*, 2006). MiRNAs can serve as oncogenes or tumour suppressors. For example, overexpression of miRNAs that target oncogenes can result in tumour suppression, while overexpression of miRNAs that target tumour suppressors can result in oncogenic activity and tumour formation. MiRNAs have been found to regulate the oncogenes *Bcl* and *Ras* and the tumour suppressor pRb (Scarola *et al.*, 2010; Park *et al.*, 2011). The first miRNA genes shown to be associated with cancer were miR-15 and miR-16 (Calin *et al.*, 2002). These two miRNAs are usually down-regulated in chronic lymphocytic leukaemia (CLL) (Calin *et al.*, 2002). They have also been shown to negatively regulate *Bcl-2*, an oncogene that is commonly overexpressed in many cancers (Cimmino *et al.*, 2005). These miRNAs are transcribed as a cluster within the intron of the non-protein coding gene *Leu2* that is often deleted in CLL. The *let-7* family of miRNAs regulates the *Ras* oncogenes (Johnson *et al.*, 2005). *Ras* genes contain activating mutations in approximately 15%–30% of human cancers. The miRNAs controlling the expression of these oncogenes therefore serve as important regulators of cellular proliferation. Some cancers show down-regulation of the *let-7* family of miRNAs resulting in reciprocal upregulation of *Ras*. This down-regulation is most pronounced in lung cancers (Takamizawa *et al.*, 2004), but is sporadically regulated in other tissues (Johnson *et al.*, 2005).

Oncogenes and tumour suppressor genes can also regulate the expression of miRNAs. C-myc regulates the miRNA cluster miR-17~92 that in turn targets E2F1 (O'Donnell *et al.*, 2005). Upregulation of this cluster was demonstrated in B cell lymphomas (He *et al.*, 2005). C-myc is often upregulated in cancers, and it serves as a transcription factor for the miR-17~92 cluster (O'Donnell *et al.*, 2005). Because C-myc directly targets E2F1, the C-myc-mediated repression of E2F1 via the miR-17~92 cluster serves to fine-tune its regulation (O'Donnell *et al.*, 2005). Several studies have shown that p53, a major target of E6, regulates the expression of many miRNAs (Xi *et al.*, 2006). Epigenetic mechanisms such as DNA methylation also play important roles in regulating miRNA expression (Saito *et al.*, 2006; Brueckner *et al.*, 2007; Lujambio *et al.*, 2007). Changes in the expression of Dicer and other proteins associated with the processing and functions of the miRNAs have also been reported (Chiosea *et al.*, 2006, 2007). Many of the cell growth and proliferation pathways affected by HPV16 contain probable targets or mediators of cellular miRNAs.

MiRNA expression in HPV-positive cervical cancers

Recently, several laboratories have investigated changes in cellular miRNA expression in HPV-positive cervical cancer (CaCx) cell lines and tissues. Our miRNA microarray studies using HPV16-positive (CaSki and SiHa) and HPV-negative (C-33A) CaCx cell lines, HPV16 cervical cell line W2 (20863, episomal HPV16; 20861 and 201402, integrated HPV16), and normal cervix showed that 24 miRNAs (miRs 126, 145, 451, 195, 143, 199b, 133a, 368, 1, 495, 497, 133b, 223, 146a, 218, 126_AS, 150, 376a, 214, 487b, 10b, and the predicted ambi-miRs-7029, 5021, and 7070) were underexpressed and three miRNAs (miRs-210, 182, and 183) were overexpressed in all the integrated HPV16 cervical cell lines compared with the normal cervix (Table 8.1) (Martinez *et al.*, 2008). Furthermore, based on microarray, real-time qRT-PCR and Northern blot analysis, we found that miR-218 was specifically reduced in the HPV-positive cell lines compared with both the HPV-negative C-33A cell line and the normal

Table 8.1 MiRNAs differentially expressed in HPV16 integrated cell lines compared with the normal cervix

MiRNA	HPV16 integrated fold[a]
Overexpressed	
hsa_miR_210	7.3
hsa_miR_182	6.4
hsa_miR_183	5.1
Underexpressed	
hsa_miR_126	–14.5
hsa_miR_145	–14.1
hsa_miR_451	–11.3
ambi_miR_7029	–9.5
hsa_miR_195	–8.3
hsa_miR_143	–8.0
hsa_miR_199b	–7.9
hsa_miR_133a	–7.6
hsa_miR_368	–7.6
hsa_miR_1	–7.2
hsa_miR_495	–6.3
hsa_miR_497	–6.2
hsa_miR_133b	–6.0
hsa_miR_223	–5.4
hsa_miR_146a	–5.1
hsa_miR_218	–4.8
hsa_miR_126_AS	–4.7
hsa_miR_150	–4.4
hsa_miR_376a	–4.2
hsa_miR_214	–4.1
hsa_miR_487b	–4.1
hsa_miR_10b	–3.9
ambi_miR_5021	–3.6
ambi_miR_7070	–3.5

[a]Mean fold-changes in HPV16-integrated cell lines CaSki, SiHa, 20861 and 201402.

The *q*-value of all miRNAs was 0.

cervix (Fig. 8.1) (Martinez *et al.*, 2008). MiR-218 is known to be down-regulated in several other human cancers as well (Volinia *et al.*, 2006; Zhang *et al.*, 2006).

We have also analysed changes in the expression of 667 human miRNAs in additional clinical samples, including four normal cervical tissues, three cervical lesions, and six cervical cancer tissues by TaqMan® MicroRNA Array analysis (McBee *et al.*, 2011). These studies showed significant overexpression of 18 miRNAs (miRs 124, 449a, 449b, 301b, 517c, 545, 223, 135b, 21, 512_3p, 542_3p, 181c, 517a, 518f, 106b, 192, 16, and 141) and underexpression of two miRNAs (miRs 218 and 433) in cervical cancer tissues compared with normal cervical tissues (Table 8.2) (McBee *et al.*, 2011). By individual qRT-PCR assays, we further demonstrated significant overexpression of miR-16, miR-21, miR-106b, miR-135b, miR-141, miR-223, miR-301b, and miR-449a, and underexpression of miR-218 and miR-433 in cervical cancer and dysplasia tissues compared with the normal cervical tissues. Thus, these 10 miRNAs may be useful in delineating cervical cancer and dysplasia from normal cervical tissue. Importantly, we also found differential expression of some miRNAs between the cervical dysplasia and cancer tissues. MiR-21, miR-135b, miR-223, and miR-301b were found to be overexpressed in the cervical cancer tissues compared with both the normal cervical and cervical dysplasia tissues. Based on our studies, such miRNAs may be good candidates for markers of progression from normal cervix to cervical dysplasia to cervical cancer. Furthermore, loss of miR-218 expression may also be a good marker of cervical cancer progression.

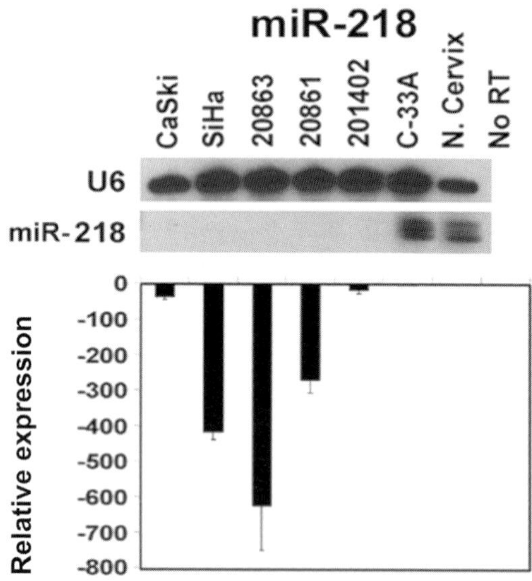

Figure 8.1 Confirmation of miR-218 microarray expression data in various cervical cell lines and normal cervical tissue. (Top panel) Northern blot analysis. The housekeeping splicing-related small U6 RNA was used as a loading control. (Bottom panel) Real-time qRT-PCR analysis. RNU43 served as the endogenous control for miRNAs.

Other studies on miRNA expression in cervical cancer have shown many similarities as well as some differences compared with our studies. Wang and coworkers profiled 455 miRNAs via miRNA microarray in four cervical cancer tissues and four normal cervical tissues (Wang *et al.*, 2008). They found significant overexpression of 18 miRNAs (miRs 223, 16, 15a, 20a, 17–5p, 106a, 20b, 224, 185, 146a, 15b, 155, 93, 21, 181c, 148a, 182, and 183) and underexpression of 15 miRNAs (miRs- 424, 125a, 30b, 450, 133b, 133a, 455, 324–5p, 126, 218, 574, 127, 191*, 422b, and 422a) in cervical cancer tissues compared with normal cervical tissues. The array results were validated by Northern blotting for several of the above miRNAs. In common with our study (McBee *et al.*, 2011), they found that miR-16, miR-21, miR-181c, and miR-223 were overexpressed, while miR-218 was underexpressed, in cervical cancer compared with normal cervical tissue.

Using a cloning strategy to identify differentially expressed miRNAs, Lui *et al.* analysed the respective cloning frequency of miRNAs in cervical carcinoma cell lines versus normal cervical tissue (Lui *et al.*, 2007). MiR-218 clones were isolated from the HPV-negative cell line C-33A but not from HPV-positive cell lines. They further compared the expression of miR-21 and miR-143 by Northern blotting in 29 cervical cancer tissues versus adjacent normal cervical tissue. Their studies demonstrated significant overexpression of miR-21 and underexpression of miR-143 in cervical cancer compared with normal cervical tissue (Lui *et al.*, 2007).

Table 8.2 MiRNAs differentially expressed in cervical cancer tissue compared with normal cervical tissue ($P<0.05$)

MiRNA	Fold change
Overexpressed	
hsa_miR_124	58.2
hsa_miR_449a	35.3
hsa_miR_449b	30.5
hsa_miR_301b	24.3
hsa_miR_517c	22.4
hsa_miR_545	17.4
hsa_miR_223	14.2
hsa_miR_135b	11.1
hsa_miR_21	9.6
hsa_miR_512_3p	9.4
hsa_miR_542_3p	7.8
hsa_miR_181c	7.6
hsa_miR_517a	6.9
hsa_miR_518f	5.0
hsa_miR_106b	4.3
hsa_miR_192	3.8
hsa_miR_16	3.8
hsa_miR_141	2.8
Underexpressed	
hsa_miR_433	−18.6
hsa_miR_218	−5.6

[a]Mean fold-changes in HPV16-integrated cell lines CaSki, SiHa, 20861 and 201402.

The *q*-value of all miRNAs was 0.

Lee *et al.* analysed the expression of 157 cellular miRNAs in ten samples each of early stage squamous cell carcinomas of the cervix and normal cervical specimens (Lee *et al.*, 2008). They found overexpression of 68 miRNAs and underexpression of two miRNAs in cervical carcinomas versus normal cervical tissues. They reported that the miRNAs most overexpressed were miRs-9, 199-s, 199a*, 199a, 199b, 145, 133a, 133b, 214, and 127. Two miRNAs, miRs-149 and 203, were underexpressed. Also, they reported that overexpression of miR-127 correlated with lymph node metastasis, and miR-199a suppressed cervical carcinoma cell growth. In common with our study (McBee *et al.*, 2011), they reported that miR-16, miR-21, miR-135b, and miR-181c were overexpressed in cervical cancer compared with the normal cervical tissue.

Hu and coworkers analysed 96 miRNAs in 59 cervical cancer tissues via real-time qRT-PCR (Hu *et al.*, 2010). They found that expression of miR-9, miR-21, miR-200a, miR-218, and miR-203 was associated with cancer survival (Hu *et al.*, 2010). They further evaluated miR-200a and miR-9 in an additional 42 cervical cancer tissues and found them to be predictive of survival.

Pereira *et al.* (2010) analysed 377 known and predicted miRNAs using miRNA microarray in four pooled normal cervical samples, 14 cervical dysplasia samples, and four squamous cell carcinomas. They found a great deal of variability in the tissues, but by pooling the normal samples they were able to identify differences in miRNA expression between the various groups. Five miRNAs (miRs-10a, 132, 148a, 196, and 302b) increased in expression from normal tissue to dysplasia to cervical cancer. Eight miRNAs (miRs-26a, 29a, 99a, 143, 145, 99a, 203, and 513) had decreased expression from normal tissue to dysplasia to cervical cancer. Surprisingly, the expression of two miRNAs (miRs-512_3p and 522) increased from normal tissue to dysplasia but decreased from dysplasia to cancer. Finally, six miRNAs (miRs-16, 27a, 106a, 142–5p, 197, and 205) showed decreased expression from normal tissue to dysplasia but increased from dysplasia to cancer.

In a recent study, Rao *et al.* investigated 802 known and 122 predicted human miRNAs using CaptialBio mammalian miRNA arrays V3.0 in 13 HPV16 and HPV18-positive cervical cancers and their adjacent normal tissues (Rao *et al.*, 2011). They found that 16 human miRNAs (miRs- 7, 429, 141, 142–5p, 31, 200a, 224, 20b, 18a, 200b, 93, 146b, 200c, 210, 20a, and the predicted 189) were upregulated, and another 16 human miRNAs (miRs-127, 376a, 214, 218, 1, 368, 145, 100, 99a, 195, 320, 152, 497, 143, 99b and 10b) were downregulated in cervical cancer tissues vs. normal tissues. In common with our study (McBee *et al.*,

2011), they found that miR-141 is overexpressed and miR-218 underexpressed in cervical cancer compared with normal cervical tissue.

Yang *et al.* analysed the expression of miR-214 via qRT-PCR and found that it was underexpressed in one cervical cancer of unknown HPV type, four HPV16-positive cervical cancers, and two HPV18-positive cervical cancers versus matched controls (Yang *et al.*, 2009).

Wang *et al.* showed by Northern blotting that miR-34a, which is regulated by the E6 proteins of HPVs 16/18 E6 via p53 destabilization, was decreased in three cervical cancer specimens with high-risk HPVs (types 16, 33, and 82) compared with age-matched controls (Wang *et al.*, 2009). Li *et al.* (2010) also analysed the expression of primary miR-34a by semi-quantitative qRT-PCR in 32 histopathologically normal cervical samples with high-risk HPV infection, 32 normal samples without HPV infection, 32 CIN lesions with high-risk HPV infection, 12 CIN lesions without HPV infection, and 32 cervical cancers. They found that pri-miR-34a was reduced in cervical cancer and dysplasia compared with the normal cervical tissue. It was also reduced in CIN 2 and CIN 3 compared with CIN 1, and in tissues with high-risk HPV infection compared with those that did not appear to contain any HPV.

As discussed above, there are some differences in the results of miRNA expression analysis in various studies. In studying the expression of genes using high-throughput methods such as microarrays, it is important to validate the results with another method. Both hybridization-based and qRT-PCR-based microarrays have inherent problems. Discrepancies in the labelling and hybridization (and possible cross-hybridization) of individual miRNAs to microarrays must be considered. In addition, there may be errors related to background signal, variation between spots on a slide, and variation between slides. While statistical analysis aims to reduce these problems and provide statistically significant information, it is best to validate at least some of the results by a second method to increase confidence in the results. Validation by Northern blotting, when possible, and/or individual qRT-PCR assays is essential. Although qRT-PCR arrays are more sensitive than hybridization arrays, they can generate false positive or negative signals. Another consideration for subsequent analysis and focusing on those miRNAs with the highest fold changes may mask smaller changes that are biologically relevant. Discrepancies between the above studies may also be due to factors such as tumour heterogeneity, small number of samples, etc. Taken together, the miRNA expression patterns in cervical cancers and dysplasias that are in common between these studies are overexpression of miR-16, miR-21, miR-135b, and miR-141, and underexpression of miR-218, miR-143, miR-145, and miR-34a. The expression levels of these miRNAs may be useful markers of disease progression and these for therapeutic targets. The mRNA targets of these miRNAs may also be important, and validation of miRNA-mRNA target interaction is the subject of ongoing studies in our laboratory.

HPVs and the squamous cell carcinoma of the head and neck (SCCHN)

SCCHN ranks sixth among cancers worldwide (Tran *et al.*, 2007). It is the most common head and neck malignancy, with more than 35,000 new cases in the USA and 500,000 new cases worldwide every year (Greenlee *et al.*, 2001). Furthermore, the prognosis still remains poor (Greenlee *et al.*, 2001). Most SCCHN cases are associated with heavy alcohol consumption and/or tobacco use, which over time induce mutations in essential genetic pathways that regulate the cell cycle (Tran *et al.*, 2007). However, HPV16 DNA has been found in approximately 30% of SCCHN, most often in the oropharynx (OP) region, where it is found in 60–70% of oropharyngeal squamous cell carcinoma (OPSCC) (D'Souza *et al.*, 2007) and often occurs in individuals without the major risk factors of alcohol and tobacco (Tran *et al.*, 2007). The tumour-suppressor protein p53 is mutated in up to half of oral cancers, but is rarely mutated in HPV-positive SCCHN. Patients with HPV-positive oral tumours have much better prognosis, with better response to radiation, chemotherapy, and surgery (Vidal and Gillison, 2008), have improved immune system activation to viral antigens (Albers *et al.*, 2005), and have a lower likelihood of metastasis than patients with HPV-negative oral tumours (Fakhry and Gillison, 2006). A recent study reported the association of HPV and OP cancers with and without the established risk factors of tobacco and alcohol use, suggesting that HPV-positive and HPV-negative SCCHN are discrete molecular, clinical and pathological diseases, but little is known about the molecular basis for these differences (Gillison, 2008).

HPV16 alters cellular miRNA profiles in SCCHN cell lines and tissues

Altered miRNA profiles have earlier been reported in head and neck cancers but these studies did not address the role of HPVs (Wong *et al.*, 2008; Childs *et al.*, 2009; Ramdas *et al.*, 2009; Hui *et al.*, 2010). Since viruses may promote pathogenesis by altering the expression of cellular miRNAs, we tested miRNA expression in HPV-positive OPSCC. Studies in our laboratory using microRNA microarrays (Ambion mirVana V2) with two HPV16 positive OPSCC cell lines, SCC2 and SCC90 and two HPV-negative OPSCC cell lines, PCI13 and PCI30 showed that three human miRNAs (miR-363, miR-33, and miR-497) were upregulated, and eight miRNAs (miRs- 155, 181a, 181b, 29a, 218, 222, 221, and 142_3p) were down-regulated in HPV-positive cell lines (Table 8.3) (Wald *et al.*, 2011). The most overexpressed miRNA in the HPV-positive cells was miR-363. qRT-PCR validation studies using four HPV16 positive OPSCC cell lines and six HPV-negative OPSCC cell lines showed higher expression of miR-363 in HPV-positive cell lines SCC2 (310-fold), SCC90 (748-fold), SCC47 (43-fold), and 93-VU-147T (140-fold) compared with the HPV-negative JHU-012 cells (Fig. 8.2A) (Wald *et al.*, 2011). The expression of two other miRNAs in the miR-17~92 family of clusters, miR-106a and miR-92a, was not altered in the HPV-positive SCCHN cell lines (data not shown). We also confirmed the underexpression of several other miRNAs, including miR-218, in HPV-positive OPSCC cell lines compared with the HPV-negative OPSCC cell lines and NOK cells by qRT-PCR (Fig. 8.2B–F) (Wald *et al.*, 2011). MiRNA microarray analysis of 11 HPV-positive and 11 HPV-negative OP tumours, and five normal oral mucosal tissue samples showed that 20 miRNAs were upregulated and one miRNA down-regulated in both HPV-positive and HPV-negative OPSCC compared with the normal oral mucosa (data not shown). In particular, miR-363 was overexpressed in the HPV-positive tumour tissues compared with the HPV-negative tissues, indicating a possible role for miR-363 in HPV-positive OPSCC.

Table 8.3 MiRNAs differentially expressed in HPV16-positive SCCHN cell lines compared with HPV-negative SCCHN cell lines

MiRNA	Fold change[a]
Overexpressed	
hsa_miR_363	5.2
rno_miR_497	3.2
hsa_miR_33	2.0
Underexpressed	
hsa_miR_155	–7.6
hsa_miR_181a	–7.3
hsa_miR_181b	–6.8
hsa_miR_29a	–4.7
hsa_miR_218	–4.2
hsa_miR_222	–3.7
hsa_miR_221	–3.4
hsa_miR_142_5p	–3.2

[a]Mean fold changes in HPV16 positive SCCHN cell lines UD-SCC-2 and UPCI:SCC90 compared with HPV-negative SCCHN cell lines PCI-13 and PCI-30.

The *q*-values of all miRNAs were 0.

miR, microRNA; hsa, human; rno, rat.

Possible cellular targets of miRNAs differentially expressed in the presence of HPVs

Many of the miRNAs found to be differentially expressed in our studies and those of other laboratories, as well as their potential targets, are also known to be differentially expressed in other cancers. Identification of the cellular targets of differentially expressed miRNAs, the functions of the target genes, and the cellular pathways in

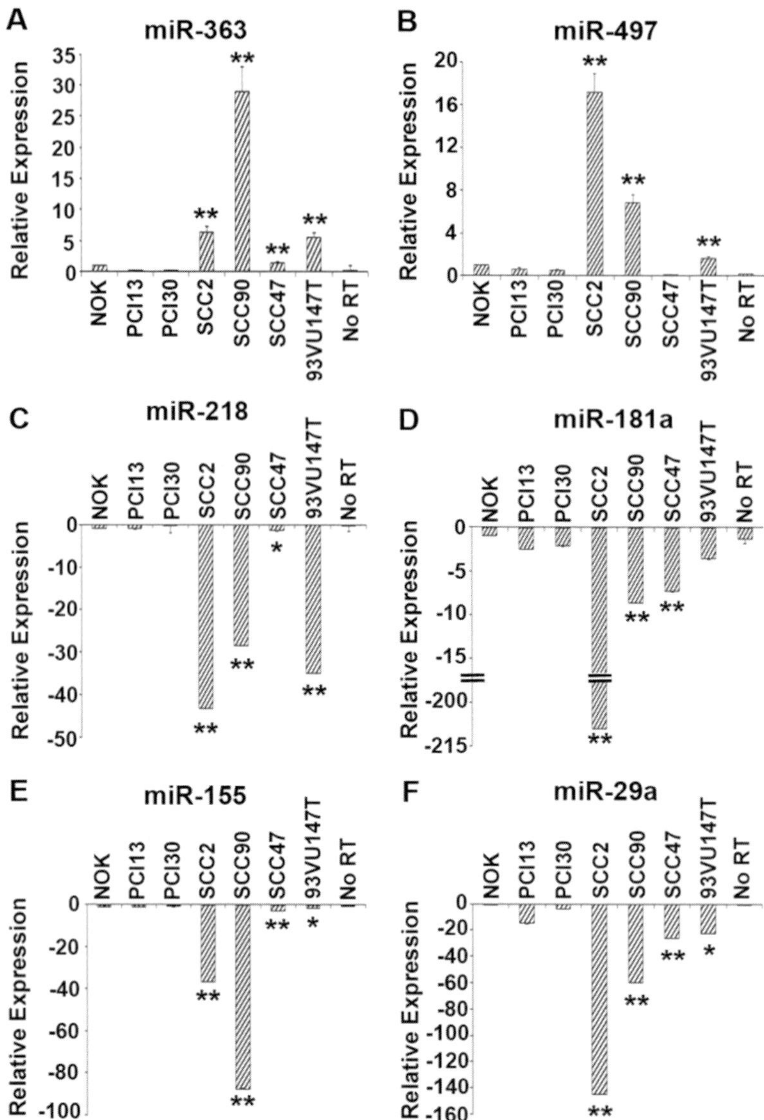

Figure 8.2 Quantitative real-time RT-PCR validation of miRNA expression data in four HPV-positive and two HPV-negative SCCHN cell lines, and transformed normal oral keratinocyte (NOK) cells. (A) miR-363. (B) miR-497. (C) miR-218. (D) miR-181a. (E) miR-155. (F) miR-29a. No RT, no reverse transcriptase added. Intensity values are relative to the NOK cells, which were arbitrarily assigned a value of 1 or –1. The *P*-values for the HPV-positive cell lines compared with the two HPV-negative SCCHN cell lines are indicated by $^{**}P<0.01$ and $^{*}P<0.05$.

which they participate could shed new light into the role of miRNAs in the pathogenesis of HPV-associated cancers. The targets of a few important miRNAs, differentially expressed in HPV-positive cells, and the function of such target genes are discussed below.

MiR-16

MiR-16 is overexpressed in cervical cancer. It is also overexpressed in ovarian cancer (Nam *et al.*, 2008) and the cancer of the head and neck (Hui *et al.*, 2010), but it is underexpressed in CLL (Calin *et al.*, 2002). Calin *et al.* (2008)

recently showed that miR-16 could regulate both oncogenes and tumour suppressors. Kaddar *et al.* (2009) observed that miR-16 can negatively regulate genes that are involved in cell proliferation, such as HMGA1 and caprin-1 in MCF-7 and HeLa cells. Interestingly, miR-16 induced G1 accumulation occurs in MCF-7 cells, but not in HeLa cells, which contain HPV18 DNA integrated into the chromosome. The authors speculate that this is most probably due to the status of p53 and Rb, which function normally in MCF-7 cells but are inactivated in HeLa cells owing to the expression of the HPV18 E6 and E7 proteins. Lack of miR-16 induced G0/G1 accumulation in HeLa cells was also observed by Linsley *et al.* (2007). It is possible that the presence of high-risk HPVs may directly or indirectly affect the expression of miR-16 and its regulatory pathways to promote cell proliferation. WNT7A, an important growth and differentiation factor in the female reproductive tract, is a potential target of miR-16 (McBee *et al.*, 2011), and it is down-regulated in cervical cancer (Gius *et al.*, 2007). WNT7A was shown to have an anti-tumorigenic effect in non-small cell lung cancer (Winn *et al.*, 2006). Downstream effectors of WNT7A include Sprouty-4, which inhibits transformed cell growth, migration and invasion, and the epithelial to mesenchymal transition in non-small cell lung cancer (Tennis *et al.*, 2010). Down-regulation of WNT7A may therefore promote these processes and could contribute to cervical tumorigenesis.

MiR-21

MiR-21, which is overexpressed in cervical cancer, is the most highly overexpressed miRNA in numerous cancers including ovarian, head and neck, lung, stomach, liver, and pancreas (Lee *et al.*, 2007; Lui *et al.*, 2007; Hui *et al.*, 2010; Guo *et al.*, 2008; Jiang *et al.*, 2008; Zhang *et al.*, 2010a). MiR-21 down-regulates the tumour suppressor PTEN in non-small cell lung cancer (Zhang *et al.*, 2010a) and in hepatocellular cancer, in which reduction in PTEN levels leads to the overexpression of matrix metalloproteinases MMP2 and MMP9 (Meng *et al.*, 2007). These proteins promote cellular migration and invasion. MiR-21 also targets the tumour suppressor genes tropomyosin 1 (TPM1), programmed cell death 4 (PDCD4), and maspin. The latter two are tumour suppressors whose loss of expression has been implicated in tumour invasion and metastasis (Zhu *et al.*, 2007, 2008). A recent paper demonstrated that miR-21 can regulate PDCD4 in HeLa cells, increasing cellular proliferation (Yao *et al.*, 2009), which may have implications for the progression of cervical cancer.

MiR-21 may target annexin 1 (ANXA1) and the eukaryotic initiation factor EIF4G2 (McBee *et al.*, 2011), which are down-regulated in cervical cancer (Gius *et al.*, 2007). Reduced expression of ANXA1 is associated with advanced stage breast cancer (Wang *et al.*, 2010a), and down-regulation of EIF4G2 correlates with invasive transitional cell carcinoma of the bladder (Buim *et al.*, 2005). ANXA1, a calcium and phospholipid binding protein that may be secreted, is involved in various cellular pathways such as inflammation, migration, and proliferation. It is a substrate for EGF receptor tyrosine kinase and inhibits EGF-mediated proliferation (Lim and Pervaiz, 2007). Thus, reduction of ANXA1 by miR-21 is likely to promote cellular proliferation. These data are consistent with the function of miR-21 as an oncogene in many cancers, including cervical cancer.

MiR-34a

Recently, Wang *et al.* showed that infection of human keratinocytes with high-risk HPVs results in decreased expression of miR-34a, and this reduced expression is dependent upon functional p53 (Wang *et al.*, 2009). They also showed that miR-34a reduces cell proliferation and has a moderate effect on apoptosis. Targets of miR-34a include proteins important in cell cycle regulation such as cyclins and cyclin-dependent kinases (Bommer *et al.*, 2007; He *et al.*, 2007; Tazawa *et al.*, 2007; Welch *et al.*, 2007).

MiR-106b

MiR-106b, which is overexpressed in cervical cancer, is also overexpressed in gastric and colorectal cancer, as well as in SCCHN (Guo *et al.*, 2008; Wang *et al.*, 2010b; Hui *et al.*, 2010). It is part of a cluster of miRNAs, along with miR-93 and miR-25, which is encoded within an intron of mini chromosome maintenance 7 (MCM7). MCM7, a target of the E2F1 transcription factor, is a marker for cervical cancer progression and may be induced by both HPV16 E6 and E7 proteins (Balsitis *et al.*, 2006; Shai *et al.*, 2007). In gastric cancer, miR-106b targets p21, antagonizing the TGF-beta tumour suppressor pathway (Petrocca *et al.*, 2008). Additionally, miR-106b may target ninjurin 2 (NINJ2) (McBee *et al.*, 2011), a cell surface adhesion molecule first identified on neural cells (Araki and Milbrandt, 2000), which is also down-regulated in cervical cancer (Gius *et al.*, 2007). Reduction of NINJ2 through overexpression of miR-106b could contribute to the loss of adhesion and increased migration of cervical cancer cells.

MiR-135b

In addition to cervical cancer, miR-135b is overexpressed in colorectal and prostate cancers (Bandres *et al.*, 2006; Sarver *et al.*, 2009; Tong *et al.*, 2009; Wang *et al.*, 2010b). A potential target of miR-135b is the serine protease inhibitor SPINK5 (McBee *et al.*, 2011), which is down-regulated in cervical cancer (Gius *et al.*, 2007). SPINK5 deficiency causes unregulated epidermal protease activity and degradation of desmoglein 1, which leads to inefficient stratum corneum adhesion and the resultant loss of skin barrier function (Descargues *et al.*, 2005). Overexpression of miR-135b and loss of SPINK5 could alter the epithelial architecture of the cervix and promote a tumorigenic phenotype.

MiR-141

MiR-141 is overexpressed in cervical, ovarian, prostate, and endometrial endometroid carcinomas (Iorio *et al.*, 2007; Porkka *et al.*, 2007; Lee *et al.*, 2011; Rao *et al.*, 2011). Also, Nam *et al.* (2008) reported that overexpression of miR-141 is associated with a poor prognosis in patients with ovarian carcinoma. The miR-141 gene is located on chromosome 12 in a cluster with miR-200c. As part of the miR-200 family of miRNAs, miR-141 is regulated by ZEB1 and ZEB2, which in a feedback loop are also regulated by the miR-200 family, mediating the epithelial to mesenchymal transition (Bracken *et al.*, 2008; Burk *et al.*, 2008). Overexpression of miR-141 in the cervical epithelium could therefore contribute to cervical tumorigenesis. Other targets of miR-141 include the tumour suppressors BRD3, UBAP1, and PTEN, which are down-regulated by miR-141 in nasopharyngeal carcinoma (Zhang *et al.*, 2010b).

MiR-223

Overexpression of miR-223 is found in cervical cancer, adult T-cell leukaemia and adenocarcinomas of the oesophagus (Wang *et al.*, 2008; Bellon *et al.*, 2009; Mathe *et al.*, 2009). MiR-223 may target desmoglein 3 (DSG3) (McBee *et al.*, 2011), which is down-regulated in cervical cancer (Gius *et al.*, 2007). DSG3, a desmosomal adhesion molecule, participates in intercellular links via desmosome–intermediate filament complexes and helps to maintain tissue integrity. Reduced expression of DSG3 increases the colony-formation efficiency and proliferative potential of primary keratinocytes (Wan *et al.*, 2007). The overexpression of miR-223 in HPV-positive cervical cancer may therefore reduce cellular adhesion and promote proliferation of cancer cells.

MiR-301b

MiR-301b is overexpressed in cervical and colon cancer tissues (Wang *et al.*, 2010b). This miRNA may target the RasGAP RASAL1 (McBee *et al.*, 2011), which is down-regulated in cervical cancer (Gius *et al.*, 2007). RASAL1 expression is lost in many colorectal cancers, which promotes activation of the *Ras* oncogene (Bernards and Settleman, 2009). *Ras* activation in HPV18 E7-expressing organoptic raft tissues increases the invasive potential of the cells via MT1-MMP and MMP9 (Yoshida *et al.*, 2008). Overexpression of miR-301b could potentially increase *Ras* activation and promote cervical cancer through the down-regulation of RASAL1.

MiR-363

MiR-363 is upregulated in HPV-positive SCCHN cell lines and tissues. MiR-363 is part of the oncogenic miR-17~92 family of clusters, which is composed of three clusters of miRNAs, miR-17~92, miR-106a~363, and miR-106b~25 and thought to have evolved through a series of deletions and duplications (Ventura *et al.*, 2008). Other members of this family of miRNA clusters have been implicated in cancers, including small cell lung cancer (Hayashita *et al.*, 2005), B-cell lymphoma (Ota *et al.*, 2004), and T-cell leukaemia (Landais *et al.*, 2007). MiRNAs in this family have similar or identical seed sequences (Ventura *et al.*, 2008), and since the seed sequence of a mature miRNA contributes significantly to its specificity for its target mRNA (Bartel, 2004), it has been hypothesized that miRNAs in the miR-17~92 family may have similar functions (Ventura *et al.*, 2008). MiR-363 has identical seed sequences to miR-92–1, miR-92–2, and miR-25 (Griffiths-Jones *et al.*, 2008). MiR-92–2 and miR-25 are also overexpressed in pancreatic, prostate, and stomach cancers (Volinia *et al.*, 2006). Recently, miR-25 has been shown to be upregulated in gastric cancers where it targets p57, an essential tumour suppressor (Kim *et al.*, 2009b). Since miR-363 and miR-25 have the same seed sequence, and miR-25 is involved in cell cycle disruption (Kim *et al.*, 2009b), it is possible that miR-363 may also be involved in the dysregulation of cell cycle in HPV-associated SCCHN.

MiR-124

A recent study found that miR-124 is underexpressed in cervical cancer cell lines and tissues (Wilting *et al.*, 2010). This underexpression is attributed to methylation at the promoter regions of miR-124 (miR-124–1, miR-124–2, and miR-124–3). High levels of methylation were seen in the HPV16-positive cell lines SiHa and CaSki and to a lesser extent in the HPV18-positive cell line HeLa. Immortalized keratinocytes (FK16A, FK16B, FK18A, and FK18B) at high passages contain greater levels of methylation at the miR-124-1 and miR-124-2 promoters than at lower passages. Greater methylation was observed in the miR-124 promoters in cervical SCCs compared with cervical neoplasias and also in neoplasias compared with normal cervical tissue. The methylation status was significantly correlated with decreased expression of miR-124. Also, in cervical scrapes, higher methylation was observed in CIN3 specimens than in disease-free specimens. The authors proposed that methylation of miR-124 promoters could be a useful marker for cervical cancer progression.

MiR-218

In addition to HPV-associated cervical cancer, miR-218 is underexpressed in a variety of cancers including, HPV-positive SCCHN, melanomas, breast, ovarian, lung, and gastric cancers (Volinia *et al.*, 2006; Zhang *et al.*, 2006; Davidson *et al.*, 2010; Gao *et al.*, 2010; Wu *et al.*, 2010; Wald *et al.*, 2011). Two identical copies of miR-218 (miR-218–1 and miR-218–2) are encoded within the introns of the Slit2 and Slit3 genes, and miR-218 is probably transcribed from the promoters of these two genes (Small *et al.*, 2010; Tie *et al.*, 2010). Polymorphisms in pri-miR-218 sequences are associated with a decreased risk for cervical cancer (Zhou *et al.*, 2010). MiR-218 can target Robo1 and Robo2, which are receptors for the Slit ligands. Slit-Robo signalling provides guidance cues for axonal development and angiogenesis (Brose *et al.*, 1999; Legg *et al.*, 2008; Ypsilanti *et al.*, 2010). Reduction of Robo via miR-218 targeting serves as a negative feedback loop for the Slit–Robo interaction, and thereby regulates vascular patterning (Small *et al.*, 2010). Targeting of Robo1 also serves to inhibit metastasis in gastric cancer (Tie *et al.*, 2010). MiR-218 can also inhibit the invasiveness of glioma cells by targeting IKK-beta (Song *et al.*, 2010).

We found that transfection of the SiHa and 20861 cell lines with pre-miR-218 significantly reduces the *LAMB3* transcript (Fig. 8.3A) (Martinez *et al.*, 2008). Furthermore, Western blot analysis showed that miR-218 expression also decreases the LAMB3 protein in SiHa cells (Fig. 8.3B) (Martinez *et al.*, 2008). These results indicate that miR-218 reduces *LAMB3* expression at the transcriptional level. LAMB3 protein is part of the polymeric cell surface receptor laminin 5 that is expressed in the basal lamina of the epithelium and is overexpressed in cervical cancers (Skyldberg *et al.*, 1999; Kohlberger *et al.*, 2003). LAMB3 increases cell migration and tumorigenicity in SCID mice, and in collaboration with its ligand α6β4 integrin, promotes tumorigenesis in human keratinocytes (Dajee *et al.*, 2003; Calaluce *et al.*, 2004). One study suggests that secreted laminin 5 can be used by HPV as a transient receptor to aid

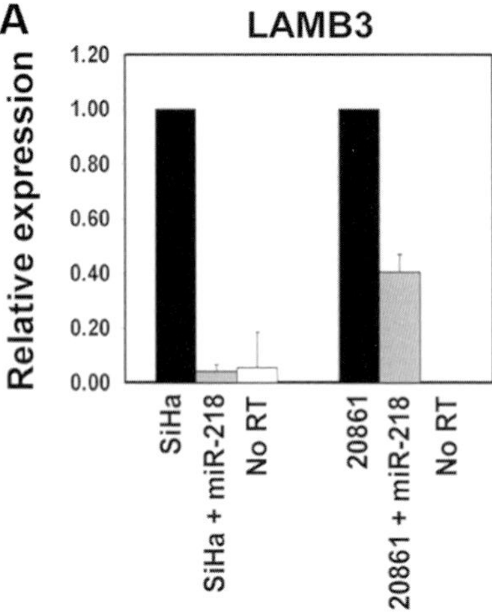

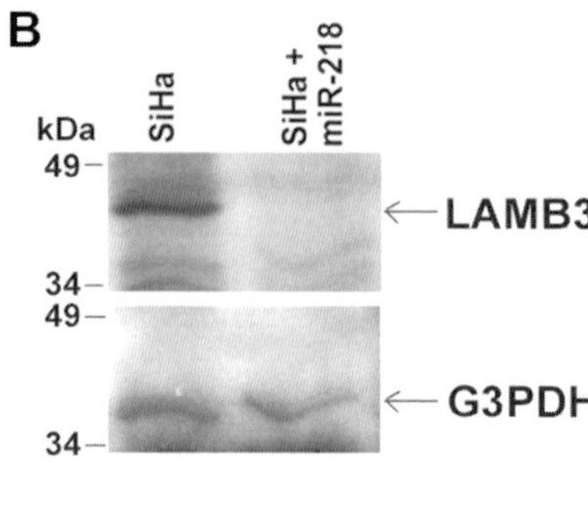

Figure 8.3 Expression of *LAMB3* is reduced in the presence of miR-218. (A) qRT-PCR analysis of *LAMB3* in SiHa and 20861 cell lines transfected with a pre-miR-218. (B) Western blot analysis of LAMB3 protein in SiHa cells transfected with pre-miR-218. Location of the 40-kDa LAMB3 protein is indicated. G3PDH was used as a control.

the virus in the infection of basal cells that express α6β4 integrin (Culp *et al.*, 2006). Thus, down-regulation of miR-218 in the presence of HPVs and the consequent overexpression of *LAMB3* may promote viral infection of the surrounding tissue and contribute to eventual tumorigenesis. We also identified CCN1 and MMP3 as potential targets of miR-218 (McBee *et al.*, 2011). MMP3 and CCN1 are extracellular matrix proteins that are overexpressed in cervical cancer (Gius *et al.*, 2007; Hagemann *et al.*, 2007), and they are important for the epithelial–mesenchymal transition and regulating tumour cell motility and metastasis (Kessenbrock *et al.*, 2010). CCN1 is overexpressed in several other cancers as well and is known to increase tumorigenicity in mice (Gery *et al.*, 2005; Nguyen *et al.*, 2006; Chai *et al.*, 2010). Thus, reduced expression of miR-218 in cervical cancer may result in the increased expression of several growth-promoting genes and contribute to tumorigenesis. We have recently found that miR-218 reduces the migration and invasion of cervical carcinoma cells. In addition, miR-218 reduces the *in vitro* wound-healing capacity of cervical carcinoma cells (A. S. Gardiner *et al.*, unpublished). Others have reported similar findings (Song *et al.*, 2010; Tie *et al.*, 2010). Based on the above results, our work supports the mounting evidence that miR-218 is a tumour suppressor miRNA in many cancers.

MiR-433

In addition to cervical cancer, miR-433 is also down-regulated in gastric cancer (Ueda *et al.*, 2010) (Luo *et al.*, 2009). Interestingly, miR-433 can potentially target cytoplasmic poly(A)-binding protein 4 (PABPC4) (McBee *et al.*, 2011), which interacts with HPV16 E6 (Katzenellenbogen *et al.*, 2010) and is upregulated in cervical cancer (Gius *et al.*, 2007). PABPC4 binding of NFX-123 is required for post-transcriptional regulation of hTERT by HPV16 E6, increasing telomerase expression and cell growth (Katzenellenbogen *et al.*, 2010). Underexpression of miR-433 could increase the expression of PABPC4 and augment the growth of E6-expressing cervical cancer cells.

Regulation of cellular miRNAs by the HPV E6 and E7 oncoproteins

To test whether E6 and/or E7 expression is directly correlated with reduced expression of miR-218, we utilized the osteosarcoma cell line

U2OS either expressing the HPV16 E6 or E7 gene, or the control neomycin resistance gene. QRT-PCR results showed that both miR-218 and the Slit2 gene are underexpressed in the U2OS-E6 cell line compared with both U2OS-E7 and the control U2OS-NEO cell line (Fig. 8.4A) (Martinez *et al.*, 2008). In another approach, the 20861 cell line containing integrated HPV16 was transfected with HPV16 E6/E7 siRNAs. The E6/E7 siRNAs reduced expression of these genes, while increasing the expression of both miR-218 and Slit2 in the 20861 cells (Fig. 8.4B) (Martinez *et al.*, 2008). These results indicate that the HPV16 E6 gene is involved in the down-regulation of miR-218 and Slit2 in HPV16 positive cell lines. Since a U2OS derivative expressing the E6 gene of a low-risk HPV was not available, we utilized normal oral keratinocytes (NOK) expressing the HPV6 E6 gene to study whether the E6 gene of a low-risk HPV also affects miR-218 expression. NOKs expressing the HPV16 E6, but not the E6 of the low-risk HPV6, display reduced levels of miR-218 (Fig. 8.4C) (Martinez *et al.*, 2008). Finally, using normal human foreskin keratinocytes (HFKs) transduced either with the retroviral LXSN vector or LXSN expressing E6 or E7, we showed that miR-218 expression is greatly reduced in the presence of E6, while E7 has a more modest effect (Fig. 8.5) (Wald *et al.*, 2011). These results demonstrate that E6 of the high-risk HPV16, but not the low-risk HPV6, contributes to the down-regulation of miR-218.

QRT-PCR analysis showed that miR-363 (upregulated in HPV-positive SCCHN cell lines compared with both HPV-negative SCCHN cell lines and NOK cells) is upregulated in the HFK-16E6 cell line, but not in the HFK-16E7 cell line (Fig. 8.5) (Wald *et al.*, 2011). Similarly, miR-181a, and miR-29a (down-regulated in HPV-positive SCCHN cell lines compared with both HPV-negative SCCHN cell lines and NOK cells) are down-regulated in the HFK-16E6 cell line (Fig. 8.5). These miRNAs are either not affected or affected to a smaller extent in the HFK-16E7 cell line (Fig. 8.5). These results show that expression of the E6 oncogene of HPV16 is associated with upregulation of miR-363 and down-regulation of miR-181a and miR-29a. Since the expression of the HPV16 E6 oncogene alters miRNA expression in the HFK cells, siRNA knockdown of HPV16 E6 was performed in the HPV-positive SCCHN cell line SCC2. The qRT-PCR results showed that reduction in the levels of E6 (Fig. 8.6, left) were accompanied by a reduction in miR-363

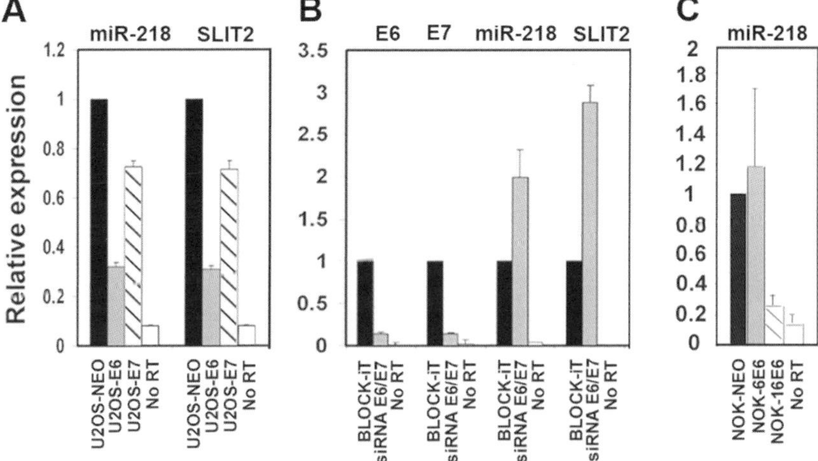

Figure 8.4 HPV16 E6 oncogene reduces the expression of miR-218. (A) qRT-PCR analysis of miR-218 and SLIT2 in U2OS-NEO, U2OS-16E6, and U2OS-16E7. (B) Expression of HPV16 E6 and E7, miR-218 and *SLIT2* in the 20861 cell line with or without RNAi against HPV16 E6/E7. (C) qRT-PCR analysis of miR-218 in NOK-NEO, NOK cell line expressing the E6 gene of either the low-risk HPV6 (NOK-6E6) or high-risk HPV16 (NOK-16E6). RNU43 served as the endogenous control for miRNAs, while G3PDH served as the endogenous control for E6, E7 and *SLIT2*.

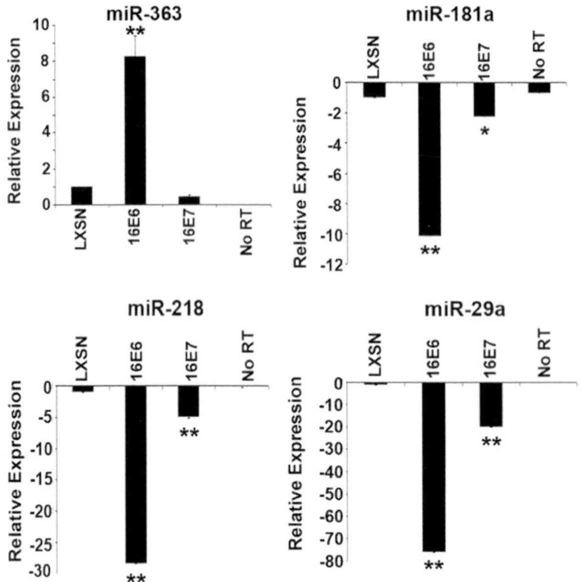

Figure 8.5 QRT-PCR analysis of miR-363, miR-181a, miR-218, and miR-29a in HFK cell lines. No RT, no reverse transcriptase added. Values are relative to the HFK-LXSN cells, which were arbitrarily assigned a value of 1 or −1. The *p* values for the HPV16 E6- and E7-expressing HFK cell lines compared the HFK-LXSN cell line are indicated by $^{**}P<0.01$ and $^{*}P<0.05$.

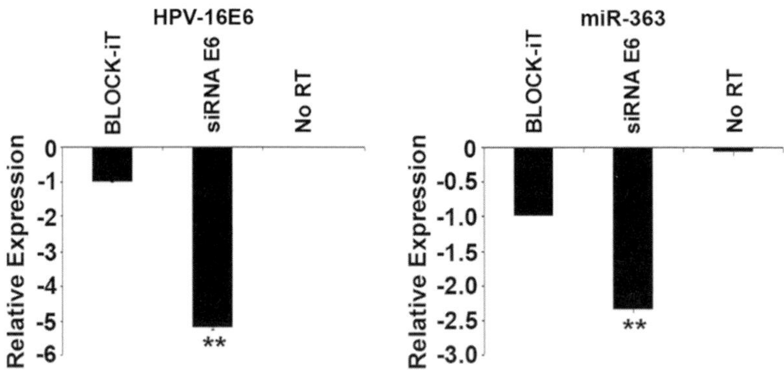

Figure 8.6 QRT-PCR analysis of HPV-positive SCCHN cell line SCC2 transfected with siRNA against HPV16 E6. Left panel, qRT-PCR analysis of HPV16 E6 oncogene expression in the HPV-positive cell line SCC2 upon transfection with a negative control BLOCK-iT siRNA or with siRNA against E6. Right panel, qRT-PCR analysis of miR-363 expression in cells transfected with the negative control siRNA or with siRNA against E6. No RT, no reverse transcriptase added. Intensity values are relative to the SCC2 cells transfected with the negative control siRNA, which were arbitrarily assigned a value of −1. The *P*-values for the SCC2+siRNA E6 cells compared with the SCC2 with control siRNA are indicated by $^{**}P<0.01$.

levels (Fig. 8.6, right) (Wald *et al.*, 2011). These results suggest that E6 is involved in the upregulation of miR-363 in HPV-positive cell lines.

Several studies have shown that p53 transcriptionally regulates the miR-34 family of miRNAs (Chang *et al.*, 2007; He *et al.*, 2007; Raver-Shapira *et al.*, 2007). The E6 oncogene binds p53 and targets p53 for degradation. This presumably results in the decreased expression of miR-34a that has been observed in some cervical cancers. It was also shown that high-risk HPVs reduce the expression of miR-34a in human keratinocytes, and this reduced expression is dependent on functional p53 (Wang *et al.*, 2009).

MiR-203 is an important regulator of epithelial cell differentiation (Yi *et al.*, 2008). Melar-New and Laimins found that keratinocytes induced to differentiate via methylcellulose exhibit increased expression of miR-203 (Melar-New and Laimins, 2010). In the presence of HPV31 E7, the cells have reduced levels of miR-203. This reduction in miR-203 was not observed in the presence of HPV31 E6. These authors reported similar observations with other HPV types. Because the putative miR-203 promoter contains binding sites for transcription factors that are induced by the MAP kinase pathway, they treated cells with PMA, which induces the MAPK pathway through activation of PKC. They found that PMA-treated HFKs exhibited increased levels of miR-203, but PMA-treated NHKs expressing HPV16 E7 did not. They hypothesized that E7-mediated interactions with the MAPK pathway contribute to the reduction of miR-203. In another recent study, McKenna *et al.* demonstrated that miR-203 is induced in HFKs in differentiated organotypic rafts (McKenna *et al.*, 2010). In contrast to Melar-New and Laimins, they found that keratinocytes expressing HPV16 E6, but not HPV16 E7, exhibited reduced levels of miR-203. E6 mutants deficient in p53 degradation did not have this effect, indicating that the reduction in miR-203 is p53 dependent. However, ChIP analysis did not reveal direct binding of p53 to the two predicted p53-binding sites in the putative miR-203 promoter. McKenna *et al.* next analysed miR-203 levels during DNA damage and found that in doxorubicin-treated cells, E6 and E7, alone or in combination, reduced the levels of miR-203. Furthermore, when they induced the keratinocytes to differentiate by Ca^{2+} treatment, both oncogenes reduced miR-203 levels. They concluded that both the E6 and E7 oncoproteins probably regulate miR-203, although they observed a greater effect with HPV16 E6. Both studies highlighted the importance of the miR-203 target, p63, a crucial regulator of epithelial cell proliferation and differentiation (Truong and Khavari, 2007). Thus, reduction of miR-203 by high-risk HPVs serves to regulate important epithelial processes by increasing expression of p63.

A recent paper has also shown that HPV16 E6 down-regulates miR-23b in cervical cancer cells in a p53-dependent manner (Au Yeung *et al.*, 2011). MiR-23b targets the urokinase-type plasminogen activator (uPA), which is implicated in cancer invasiveness and metastasis (Dass *et al.*, 2008). Thus, expression of HPV16 E6 results in increased expression of uPA and increased migration of cervical cancer cells.

Mechanisms by which the HPV oncogenes may affect cellular miRNA expression

P300 is a histone acetyltransferase (HAT) and competes with the histone methyltransferase EZH2 for modification of H3K27 (Pasini *et al.*, 2010). EZH2 binds to the SLIT2 promoter and inhibits its expression in prostate cancer cells (Yu *et al.*, 2010). We treated HPV-positive SiHa cells, which do not express miR-218, with a siRNA against EZH2, and found increased expression of miR-218 (A. S. Gardiner *et al.*, unpublished). We also found that treatment of SiHa cells with the DNA methyltransferase inhibitor 5-azacytidine (5-aza C), the histone deacetylase (HDAC) inhibitor trichostatin A (TSA), or the histone methyltransferase (HMT) inhibitor adenosine dialdehyde (AdOx) individually did not significantly affect miR-218 expression. However, treatment with all the three inhibitors together or with TSA and AdOX in combination resulted in greatly increased miR-218 levels (not shown). We also found that treatment of C-33A cells (which express high levels of miR-218) with anacardic acid (AA), which inhibits the HAT activity of p300 and reduces H3K27 acetylation, reduced miR-218 expression (not shown). We also transfected the HPV-negative cervical carcinoma C-33A cell line, which expresses high levels of miR-218, with siRNA molecules against p300 and found that miR-218 expression was greatly reduced (not shown). EZH2 is a polycomb group protein and functions in the polycomb repressive complex 2 (PRC2). PRC2 proteins, EZH2 and SUZ12, and PRC1 proteins, BMI1, RING1, and RING2, can bind to the SLIT2 promoter (Yu *et al.*, 2010). PRC1 and PRC2 are transcriptional repressors that antagonize developmental regulators in embryonic stem cells. EZH2 catalyses the trimethylation of H3K27, while p300 catalyses the acetylation of H3K27 (Pasini *et al.*, 2010). EZH2 has been reported to be overexpressed in various cancers (Varambally *et al.*, 2002; Wagener *et al.*, 2008). Interestingly, HPV16/18 E7 can increase the expression of EZH2 in cervical cancer cells (Holland *et al.*, 2008), thus providing an additional mechanism by which high-risk HPVs may target miR-218. The above results indicate that H3K27 acetylation/methylation at the SLIT2/3 promoters is important for the regulation of miR-218 expression and suggest a role for E6-mediated changes in local chromatin structure in cervical epithelial cells.

Do HPVs encode miRNAs?

Several animal viruses are known to encode miRNAs, and such miRNAs can affect both cellular and viral gene expression (Grundhoff and Sullivan, 2011). One study showed that HPV31 does not encode miRNAs during latent or productive replication in a suspension culture model that can be induced to differentiate by methylcellulose (Cai *et al.*, 2006). This study utilized a cloning-based strategy to identify cellular and viral miRNAs and resulted in the identification of over 500 miRNAs, but no HPV-encoded miRNAs were identified. Additional technical approaches as well as the use of a variety of cell lines, growth conditions, tissue samples from HPV lesions and cancers, etc. are currently being utilized to identify miRNAs that could potentially be encoded by various HPV types.

Concluding remarks

It is now well established that the presence of high-risk HPVs is associated with alterations in miRNA expression in HPV-positive cervical and oropharyngeal cancers. Such miRNAs may be useful as biomarkers of disease progression. Furthermore, the HPV E6 and E7 oncogenes have been shown to be involved in the regulation of several cellular miRNAs. While some of the miRNAs altered by the E6 protein involve the p53 pathway, others appear to be p53-independent. Some of the alterations in cellular miRNAs by the E6 and E7 oncoproteins may involve epigenetic mechanisms such as DNA methylation and/or histone modifications. The cellular targets of several differentially expressed miRNAs have been identified and our knowledge of miRNA-mediated alterations in cellular pathways and their role in the progression of HPV-promoted lesions and cancer an area of active investigation. Finally, while there is no current published evidence that HPVs encode miRNAs, this possibility is the subject of active investigations.

Acknowledgements

We thank members of our laboratory for useful discussions. Work in the authors' laboratory was supported by an NIH grant to SAK (DE016406). ASG was supported by NIH training grant 5T32 GM065100 (Biotechnology Training Grant) and by the Ruth Kirschstein predoctoral fellowship DE019028. AIW was supported by NIH training grant 2T32 AI049820 (Molecular Microbial Persistence and Pathogenesis) and by the Ruth Kirschstein predoctoral fellowship DE019749.

References

Albers, A., Abe, K., Hunt, J., Wang, J., Lopez-Albaitero, A., Schaefer, C., Gooding, W., Whiteside, T.L., Ferrone, S., DeLeo, A., *et al.* (2005). Anti-tumor activity of human papillomavirus type 16 E7-specific T cells against virally infected squamous cell carcinoma of the head and neck. Cancer Res. *65*, 11146–11155.

Araki, T., and Milbrandt, J. (2000). Ninjurin2, a novel homophilic adhesion molecule, is expressed in mature sensory and enteric neurons and promotes neurite outgrowth. J. Neurosci. *20*, 187–195.

Au Yeung, C.L., Tsang, T.Y., Yau, P.L., and Kwok, T.T. (2011). Human papillomavirus type 16 E6 induces cervical cancer cell migration through the p53/microRNA-23b/urokinase-type plasminogen activator pathway. Oncogene *30*, 2401–2430.

Balsitis, S., Dick, F., Dyson, N., and Lambert, P.F. (2006). Critical roles for non-pRb targets of human papillomavirus type 16 E7 in cervical carcinogenesis. Cancer Res. *66*, 9393–9400.

Bandres, E., Cubedo, E., Agirre, X., Malumbres, R., Zarate, R., Ramirez, N., Abajo, A., Navarro, A., Moreno, I., Monzo, M., *et al.* (2006). Identification by Real-time PCR of 13 mature microRNAs differentially expressed in colorectal cancer and non-tumoral tissues. Mol. Cancer 5, 29.

Bartel, D.P. (2004). MicroRNAs: genomics, biogenesis, mechanism, and function. Cell *116*, 281–297.

Bartel, D.P. (2009). MicroRNAs: target recognition and regulatory functions. Cell *136*, 215–233.

Bellon, M., Lepelletier, Y., Hermine, O., and Nicot, C. (2009). Deregulation of microRNA involved in hematopoiesis and the immune response in HTLV-I adult T-cell leukemia. Blood *113*, 4914–4917.

Bernards, A., and Settleman, J. (2009). Loss of the Ras regulator RASAL1: another route to Ras activation in colorectal cancer. Gastroenterology *136*, 46–48.

Bommer, G.T., Gerin, I., Feng, Y., Kaczorowski, A.J., Kuick, R., Love, R.E., Zhai, Y., Giordano, T.J., Qin, Z.S., Moore, B.B., *et al.* (2007). p53-mediated activation of miRNA34 candidate tumor-suppressor genes. Curr. Biol. *17*, 1298–1307.

Bracken, C.P., Gregory, P.A., Kolesnikoff, N., Bert, A.G., Wang, J., Shannon, M.F., and Goodall, G.J. (2008). A double-negative feedback loop between ZEB1-SIP1 and the microRNA-200 family regulates epithelial-mesenchymal transition. Cancer Res. *68*, 7846–7854.

Brose, K., Bland, K.S., Wang, K.H., Arnott, D., Henzel, W., Goodman, C.S., Tessier-Lavigne, M., and Kidd, T. (1999). Slit proteins bind Robo receptors and have an evolutionarily conserved role in repulsive axon guidance. Cell *96*, 795–806.

Brueckner, B., Stresemann, C., Kuner, R., Mund, C., Musch, T., Meister, M., Sultmann, H., and Lyko, F. (2007). The human let-7a-3 locus contains an epigenetically regulated microRNA gene with oncogenic function. Cancer Res. *67*, 1419–1423.

Buim, M.E., Soares, F.A., Sarkis, A.S., and Nagai, M.A. (2005). The transcripts of SFRP1, CEP63 and EIF4G2 genes are frequently down-regulated in transitional cell carcinomas of the bladder. Oncology *69*, 445–454.

Burk, U., Schubert, J., Wellner, U., Schmalhofer, O., Vincan, E., Spaderna, S., and Brabletz, T. (2008). A reciprocal repression between ZEB1 and members of the miR-200 family promotes EMT and invasion in cancer cells. EMBO Rep. *9*, 582–589.

Cai, X., Hagedorn, C.H., and Cullen, B.R. (2004). Human microRNAs are processed from capped, polyadenylated transcripts that can also function as mRNAs. RNA *10*, 1957–1966.

Cai, X., Li, G., Laimins, L.A., and Cullen, B.R. (2006). Human papillomavirus genotype 31 does not express detectable microRNA levels during latent or productive virus replication. J. Virol. *80*, 10890–10893.

Calaluce, R., Bearss, D.J., Barrera, J., Zhao, Y., Han, H., Beck, S.K., McDaniel, K., and Nagle, R.B. (2004). Laminin-5 beta3A expression in LNCaP human prostate carcinoma cells increases cell migration and tumorigenicity. Neoplasia *6*, 468–479.

Calin, G.A., Cimmino, A., Fabbri, M., Ferracin, M., Wojcik, S.E., Shimizu, M., Taccioli, C., Zanesi, N., Garzon, R., Aqeilan, R.I., *et al.* (2008). MiR-15a and miR-16-1 cluster functions in human leukemia. Proc. Natl. Acad. Sci. U.S.A. *105*, 5166–5171.

Calin, G.A., Dumitru, C.D., Shimizu, M., Bichi, R., Zupo, S., Noch, E., Aldler, H., Rattan, S., Keating, M., Rai, K., *et al.* (2002). Frequent deletions and down-regulation of micro RNA genes miR15 and miR16 at 13q14 in chronic lymphocytic leukemia. Proc. Natl. Acad. Sci. U.S.A. *99*, 15524–15529.

Chai, J., Norng, M., Modak, C., Reavis, K.M., Mouazzen, W., and Pham, J. (2010). CCN1 induces a reversible epithelial-mesenchymal transition in gastric epithelial cells. Lab. Invest. *90*, 1140–1151.

Chang, T.C., Wentzel, E.A., Kent, O.A., Ramachandran, K., Mullendore, M., Lee, K.H., Feldmann, G., Yamakuchi, M., Ferlito, M., Lowenstein, C.J., *et al.* (2007). Transactivation of miR-34a by p53 broadly influences gene expression and promotes apoptosis. Mol. Cell *26*, 745–752.

Childs, G., Fazzari, M., Kung, G., Kawachi, N., Brandwein-Gensler, M., McLemore, M., Chen, Q., Burk, R.D., Smith, R.V., Prystowsky, M.B., *et al.* (2009). Low-level expression of microRNAs let-7d and miR-205 are prognostic markers of head and neck squamous cell carcinoma. Am J. Pathol. *174*, 736–745.

Chiosea, S., Jelezcova, E., Chandran, U., Acquafondata, M., McHale, T., Sobol, R.W., and Dhir, R. (2006). Up-regulation of dicer, a component of the MicroRNA machinery, in prostate adenocarcinoma. Am. J. Pathol. *169*, 1812–1820.

Chiosea, S., Jelezcova, E., Chandran, U., Luo, J., Mantha, G., Sobol, R.W., and Dacic, S. (2007). Overexpression of Dicer in precursor lesions of lung adenocarcinoma. Cancer Res. *67*, 2345–2350.

Cimmino, A., Calin, G.A., Fabbri, M., Iorio, M.V., Ferracin, M., Shimizu, M., Wojcik, S.E., Aqeilan, R.I., Zupo, S., Dono, M., *et al.* (2005). miR-15 and miR-16 induce apoptosis by targeting BCL2. Proc. Natl. Acad. Sci. U.S.A. *102*, 13944–13949.

Culp, T.D., Budgeon, L.R., Marinkovich, M.P., Meneguzzi, G., and Christensen, N.D. (2006). Keratinocyte-secreted laminin 5 can function as a transient receptor for human papillomaviruses by binding virions and transferring them to adjacent cells. J. Virol. *80*, 8940–8950.

D'Souza, G., Kreimer, A.R., Viscidi, R., Pawlita, M., Fakhry, C., Koch, W.M., Westra, W.H., and Gillison, M.L. (2007). Case–control study of human papillomavirus and oropharyngeal cancer. N Engl. J. Med. *356*, 1944–1956

Dajee, M., Lazarov, M., Zhang, J.Y., Cai, T., Green, C.L., Russell, A.J., Marinkovich, M.P., Tao, S., Lin, Q., Kubo, Y., *et al.* (2003). NF-kappaB blockade and oncogenic Ras trigger invasive human epidermal neoplasia. Nature *421*, 639–643.

Dass, K., Ahmad, A., Azmi, A.S., Sarkar, S.H., and Sarkar, F.H. (2008). Evolving role of uPA/uPAR system in human cancers. Cancer Treat. Rev. *34*, 122–136.

Davidson, M.R., Larsen, J.E., Yang, I.A., Hayward, N.K., Clarke, B.E., Duhig, E.E., Passmore, L.H., Bowman, R.V., and Fong, K.M. (2010). MicroRNA-218 is deleted and down-regulated in lung squamous cell carcinoma. PLoS ONE *5*, e12560.

Descargues, P., Deraison, C., Bonnart, C., Kreft, M., Kishibe, M., Ishida-Yamamoto, A., Elias, P., Barrandon, Y., Zambruno, G., Sonnenberg, A., *et al.* (2005). Spink5-deficient mice mimic Netherton syndrome through degradation of desmoglein 1 by epidermal protease hyperactivity. Nat. Genet. *37*, 56–65.

Duensing, S., Lee, L.Y., Duensing, A., Basile, J., Piboonniyom, S., Gonzalez, S., Crum, C.P., and Munger, K. (2000). The human papillomavirus type 16 E6 and E7 oncoproteins cooperate to induce mitotic defects and genomic instability by uncoupling centrosome duplication from the cell division cycle. Proc. Natl. Acad. Sci. U.S.A. *97*, 10002–10007.

Dyson, N., Howley, P.M., Munger, K., and Harlow, E. (1989). The human papilloma virus-16 E7 oncoprotein is able to bind to the retinoblastoma gene product. Science *243*, 934–937.

Fakhry, C., and Gillison, M.L. (2006). Clinical implications of human papillomavirus in head and neck cancers. J. Clin. Oncol. *24*, 2606–2611.

Funk, J.O., Waga, S., Harry, J.B., Espling, E., Stillman, B., and Galloway, D.A. (1997). Inhibition of CDK activity and PCNA-dependent DNA replication by p21 is blocked by interaction with the HPV16 E7 oncoprotein. Genes Dev. *11*, 2090–2100.

Gao, C., Zhang, Z., Liu, W., Xiao, S., Gu, W., and Lu, H. (2010). Reduced microRNA-218 expression is associated with high nuclear factor kappa B activation in gastric cancer. Cancer *116*, 41–49.

Gery, S., Xie, D., Yin, D., Gabra, H., Miller, C., Wang, H., Scott, D., Yi, W.S., Popoviciu, M.L., Said, J.W., *et al.* (2005). Ovarian carcinomas: CCN genes are aberrantly expressed and CCN1 promotes proliferation of these cells. Clin. Cancer Res. *11*, 7243–7254.

Gillison, M.L. (2008). Human papillomavirus-related diseases: oropharynx cancers and potential implications for adolescent HPV vaccination. J. Adolesc. Health *43*, S52–60.

Gillison, M.L., Koch, W.M., Capone, R.B., Spafford, M., Westra, W.H., Wu, L., Zahurak, M.L., Daniel, R.W., Viglione, M., Symer, D.E., *et al.* (2000). Evidence for a causal association between human papillomavirus and a subset of head and neck cancers. J. Natl. Cancer Inst. *92*, 709–720.

Gius, D., Funk, M.C., Chuang, E.Y., Feng, S., Huettner, P.C., Nguyen, L., Bradbury, C.M., Mishra, M., Gao, S., Buttin, B.M., *et al.* (2007). Profiling microdissected epithelium and stroma to model genomic signatures for cervical carcinogenesis accommodating for covariates. Cancer Res. *67*, 7113–7123.

Greenlee, R.T., Hill-Harmon, M.B., Murray, T., and Thun, M. (2001). Cancer statistics, 2001. CA Cancer J. Clin. *51*, 15–36.

Griffiths-Jones, S., Saini, H.K., van Dongen, S., and Enright, A.J. (2008). miRBase: tools for microRNA genomics. Nucleic Acids Res. *36*, D154–158.

Grundhoff, A., and Sullivan, C.S. (2011). Virus-encoded microRNAs. Virology *411*, 325–343.

Guo, J., Miao, Y., Xiao, B., Huan, R., Jiang, Z., Meng, D., and Wang, Y. (2008). Differential expression of microRNA species in human gastric cancer versus non-tumourous tissues. J. Gastroenterol. Hepatol. *24*, 652–657.

Hagemann, T., Bozanovic, T., Hooper, S., Ljubic, A., Slettenaar, V.I., Wilson, J.L., Singh, N., Gayther, S.A., Shepherd, J.H., and Van Trappen, P.O. (2007). Molecular profiling of cervical cancer progression. Br. J. Cancer *96*, 321–328.

Hayashita, Y., Osada, H., Tatematsu, Y., Yamada, H., Yanagisawa, K., Tomida, S., Yatabe, Y., Kawahara, K., Sekido, Y., and Takahashi, T. (2005). A polycistronic microRNA cluster, miR-17–92, is overexpressed in human lung cancers and enhances cell proliferation. Cancer Res. *65*, 9628–9632.

He, L., He, X., Lim, L.P., de Stanchina, E., Xuan, Z., Liang, Y., Xue, W., Zender, L., Magnus, J., Ridzon, D., *et al.* (2007). A microRNA component of the p53 tumour suppressor network. Nature *447*, 1130–1134.

He, L., Thomson, J.M., Hemann, M.T., Hernando-Monge, E., Mu, D., Goodson, S., Powers, S., Cordon-Cardo, C., Lowe, S.W., Hannon, G.J., *et al.* (2005). A microRNA polycistron as a potential human oncogene. Nature *435*, 828–833.

Holland, D., Hoppe-Seyler, K., Schuller, B., Lohrey, C., Maroldt, J., Durst, M., and Hoppe-Seyler, F. (2008). Activation of the enhancer of zeste homologue 2 gene by the human papillomavirus E7 oncoprotein. Cancer Res. *68*, 9964–9972.

Hu, X., Schwarz, J.K., Lewis, J.S., Jr., Huettner, P.C., Rader, J.S., Deasy, J.O., Grigsby, P.W., and Wang, X. (2010). A microRNA expression signature for cervical cancer prognosis. Cancer Res. *70*, 1441–1448.

Hui, A.B., Lenarduzzi, M., Krushel, T., Waldron, L., Pintilie, M., Shi, W., Perez-Ordonez, B., Jurisica, I., O'Sullivan, B., Waldron, J., *et al.* (2010). Comprehensive MicroRNA profiling for head and neck squamous cell carcinomas. Clin. Cancer Res. *16*, 1129–1139.

Huibregtse, J.M., Scheffner, M., and Howley, P.M. (1991). A cellular protein mediates association of p53 with the E6 oncoprotein of human papillomavirus types 16 or 18. EMBO J. *10*, 4129–4135.

Hutvagner, G., McLachlan, J., Pasquinelli, A.E., Balint, E., Tuschl, T., and Zamore, P.D. (2001). A cellular function for the RNA-interference enzyme Dicer in the maturation of the let-7 small temporal RNA. Science *293*, 834–838.

Iorio, M.V., Visone, R., Di Leva, G., Donati, V., Petrocca, F., Casalini, P., Taccioli, C., Volinia, S., Liu, C.G., Alder, H., *et al.* (2007). MicroRNA signatures in human ovarian cancer. Cancer Res. *67*, 8699–8707.

Jiang, J., Gusev, Y., Aderca, I., Mettler, T.A., Nagorney, D.M., Brackett, D.J., Roberts, L.R., and Schmittgen, T.D. (2008). Association of MicroRNA expression in hepatocellular carcinomas with hepatitis infection, cirrhosis, and patient survival. Clin. Cancer Res. *14*, 419–427.

Johnson, S.M., Grosshans, H., Shingara, J., Byrom, M., Jarvis, R., Cheng, A., Labourier, E., Reinert, K.L., Brown, D., and Slack, F.J. (2005). RAS is regulated by the let-7 microRNA family. Cell *120*, 635–647.

Jones, D.L., Alani, R.M., and Munger, K. (1997). The human papillomavirus E7 oncoprotein can uncouple cellular differentiation and proliferation in human keratinocytes by abrogating p21Cip1-mediated inhibition of cdk2. Genes Dev. *11*, 2101–2111.

Kaddar, T., Rouault, J.P., Chien, W.W., Chebel, A., Gadoux, M., Salles, G., Ffrench, M., and Magaud, J.P. (2009). Two new miR-16 targets: caprin-1 and HMGA1, proteins implicated in cell proliferation. Biol. Cell *101*, 511–524.

Katzenellenbogen, R.A., Vliet-Gregg, P., Xu, M., and Galloway, D.A. (2010). Cytoplasmic poly (A) binding proteins regulate telomerase activity and cell growth in human papillomavirus type 16 E6-expressing keratinocytes. J. Virol. *84*, 12934–12944.

Kessenbrock, K., Plaks, V., and Werb, Z. (2010). Matrix metalloproteinases: regulators of the tumour microenvironment. Cell *141*, 52–67.

Khvorova, A., Reynolds, A., and Jayasena, S.D. (2003). Functional siRNAs and miRNAs exhibit strand bias. Cell *115*, 209–216.

Kim, V.N. (2005). MicroRNA biogenesis: coordinated cropping and dicing. Nat. Rev. Mol. Cell. Biol. *6*, 376–385.

Kim, V.N., Han, J., and Siomi, M.C. (2009a). Biogenesis of small RNAs in animals. Nat. Rev. Mol. Cell. Biol. *10*, 126–139.

Kim, Y.K., Yu, J., Han, T.S., Park, S.Y., Namkoong, B., Kim, D.H., Hur, K., Yoo, M.W., Lee, H.J., Yang, H.K., *et al.* (2009b). Functional links between clustered microRNAs: suppression of cell-cycle inhibitors by microRNA clusters in gastric cancer. Nucleic Acids Res. *37*, 1672–1681.

Kiyono, T., Hiraiwa, A., Fujita, M., Hayashi, Y., Akiyama, T., and Ishibashi, M. (1997). Binding of high-risk human papillomavirus E6 oncoproteins to the human homologue of the Drosophila discs large tumour suppressor protein. Proc. Natl. Acad. Sci. U.S.A. *94*, 11612–11616.

Klingelhutz, A.J., Foster, S.A., and McDougall, J.K. (1996). Telomerase activation by the E6 gene product of human papillomavirus type 16. Nature *380*, 79–82.

Kohlberger, P., Beneder, C., Horvat, R., Leodolter, S., and Breitenecker, G. (2003). Immunohistochemical expression of laminin-5 in cervical intraepithelial neoplasia. Gynecol. Oncol. *89*, 391–394.

Landais, S., Landry, S., Legault, P., and Rassart, E. (2007). Oncogenic potential of the miR–106–363 cluster and its implication in human T-cell leukemia. Cancer Res. *67*, 5699–5707.

Lee, E.J., Gusev, Y., Jiang, J., Nuovo, G.J., Lerner, M.R., Frankel, W.L., Morgan, D.L., Postier, R.G., Brackett, D.J., and Schmittgen, T.D. (2007). Expression profiling identifies microRNA signature in pancreatic cancer. Int. J. Cancer *120*, 1046–1054.

Lee, J.W., Choi, C.H., Choi, J.J., Park, Y.A., Kim, S.J., Hwang, S.Y., Kim, W.Y., Kim, T.J., Lee, J.H., Kim, B.G., *et al.* (2008). Altered MicroRNA expression in cervical carcinomas. Clin. Cancer Res. *14*, 2535–2542.

Lee, J.W., Park, Y.A., Choi, J.J., Lee, Y.Y., Kim, C.J., Choi, C., Kim, T.J., Lee, N.W., Kim, B.G., and Bae, D.S. (2011). The expression of the miRNA-200 family in endometrial endometrioid carcinoma. Gynecol. Oncol. *120*, 56–62.

Lee, R.C., Feinbaum, R.L., and Ambros, V. (1993). The *C. elegans* heterochronic gene lin-4 encodes small RNAs with antisense complementarity to lin-14. Cell *75*, 843–854.

Lee, Y., Jeon, K., Lee, J.T., Kim, S., and Kim, V.N. (2002). MicroRNA maturation: stepwise processing and subcellular localization. EMBO J. *21*, 4663–4670.

Lee, Y., Ahn, C., Han, J., Choi, H., Kim, J., Yim, J., Lee, J., Provost, P., Radmark, O., Kim, S., *et al.* (2003). The nuclear RNase III Drosha initiates microRNA processing. Nature *425*, 415–419.

Lee, Y., Kim, M., Han, J., Yeom, K.H., Lee, S., Baek, S.H., and Kim, V.N. (2004). MicroRNA genes are transcribed by RNA polymerase II. EMBO J. *23*, 4051–4060.

Legg, J.A., Herbert, J.M., Clissold, P., and Bicknell, R. (2008). Slits and Roundabouts in cancer, tumour angiogenesis and endothelial cell migration. Angiogenesis *11*, 13–21.

Lewis, B.P., Burge, C.B., and Bartel, D.P. (2005). Conserved seed pairing, often flanked by adenosines, indicates that thousands of human genes are microRNA targets. Cell *120*, 15–20.

Li, B., Hu, Y., Ye, F., Li, Y., Lv, W., and Xie, X. (2010). Reduced miR-34a expression in normal cervical tissues and cervical lesions with high-risk human papillomavirus infection. Int. J. Gynecol. Cancer *20*, 597–604.

Li, Z., Kim, S.W., Lin, Y., Moore, P.S., Chang, Y., and John, B. (2009). Characterization of viral and human RNAs smaller than canonical MicroRNAs. J. Virol. *83*, 12751–12758.

Lim, L.H., and Pervaiz, S. (2007). Annexin 1: the new face of an old molecule. FASEB J. *21*, 968–975.

Linsley, P.S., Schelter, J., Burchard, J., Kibukawa, M., Martin, M.M., Bartz, S.R., Johnson, J.M., Cummins, J.M., Raymond, C.K., Dai, H., *et al.* (2007). Transcripts targeted by the microRNA-16 family cooperatively regulate cell cycle progression. Mol. Cell. Biol. *27*, 2240–2252.

Lui, W.O., Pourmand, N., Patterson, B.K., and Fire, A. (2007). Patterns of known and novel small RNAs in human cervical cancer. Cancer Res. *67*, 6031–6043.

Lujambio, A., Ropero, S., Ballestar, E., Fraga, M.F., Cerrato, C., Setien, F., Casado, S., Suarez-Gauthier, A., Sanchez-Cespedes, M., Git, A., *et al.* (2007). Genetic unmasking of an epigenetically silenced microRNA in human cancer cells. Cancer Res. *67*, 1424–1429.

Lund, E., Guttinger, S., Calado, A., Dahlberg, J.E., and Kutay, U. (2004). Nuclear export of microRNA precursors. Science *303*, 95–98.

Luo, H., Zhang, H., Zhang, Z., Zhang, X., Ning, B., Guo, J., Nie, N., Liu, B., and Wu, X. (2009). Down-regulated miR-9 and miR-433 in human gastric carcinoma. J. Exp. Clin. Cancer Res. *28*, 82.

McBee, W.C., Gardiner, A.S., Edwards, R.P., Lesnock, J.L., Bhargava, R., Austin, R.M., Guido, R.S., and Khan, S.A. (2011). MicroRNA Analysis in Human

Papillomavirus (HPV)- Associated Cervical Neoplasia and Cancer. J. Carcinog. Mutag. *2*, 114.

McKenna, D.J., McDade, S.S., Patel, D., and McCance, D.J. (2010). MicroRNA 203 expression in keratinocytes is dependent on regulation of p53 levels by E6. J. Virol. *84*, 10644–10652.

Martinez, I., Gardiner, A.S., Board, K.F., Monzon, F.A., Edwards, R.P., and Khan, S.A. (2008). Human papillomavirus type 16 reduces the expression of microRNA-218 in cervical carcinoma cells. Oncogene *27*, 2575–2582.

Mathe, E.A., Nguyen, G.H., Bowman, E.D., Zhao, Y., Budhu, A., Schetter, A.J., Braun, R., Reimers, M., Kumamoto, K., Hughes, D., *et al.* (2009). MicroRNA expression in squamous cell carcinoma and adenocarcinoma of the esophagus: associations with survival. Clin. Cancer Res. *15*, 6192–6200.

Melar-New, M., and Laimins, L.A. (2010). Human papillomaviruses modulate expression of microRNA 203 upon epithelial differentiation to control levels of p63 proteins. J. Virol. *84*, 5212–5221.

Meng, F., Henson, R., Wehbe-Janek, H., Ghoshal, K., Jacob, S.T., and Patel, T. (2007). MicroRNA-21 regulates expression of the PTEN tumour suppressor gene in human hepatocellular cancer. Gastroenterology *133*, 647–658.

Miyoshi, K., Miyoshi, T., and Siomi, H. (2010). Many ways to generate microRNA-like small RNAs: non-canonical pathways for microRNA production. Mol. Genet. Genomics *284*, 95–103.

Munger, K., Werness, B.A., Dyson, N., Phelps, W.C., Harlow, E., and Howley, P.M. (1989). Complex formation of human papillomavirus E7 proteins with the retinoblastoma tumour suppressor gene product. EMBO J. *8*, 4099–4105.

Munger, K., Baldwin, A., Edwards, K.M., Hayakawa, H., Nguyen, C.L., Owens, M., Grace, M., and Huh, K. (2004). Mechanisms of human papillomavirus-induced oncogenesis. J. Virol. *78*, 11451–11460.

Nakagawa, S., and Huibregtse, J.M. (2000). Human scribble (Vartul) is targeted for ubiquitin-mediated degradation by the high-risk papillomavirus E6 proteins and the E6AP ubiquitin-protein ligase. Mol. Cell. Biol. *20*, 8244–8253.

Nam, E.J., Yoon, H., Kim, S.W., Kim, H., Kim, Y.T., Kim, J.H., Kim, J.W., and Kim, S. (2008). MicroRNA expression profiles in serous ovarian carcinoma. Clin. Cancer Res. *14*, 2690–2695.

Nguyen, N., Kuliopulos, A., Graham, R.A., and Covic, L. (2006). Tumour-derived Cyr61(CCN1) promotes stromal matrix metalloproteinase-1 production and protease-activated receptor 1-dependent migration of breast cancer cells. Cancer Res. *66*, 2658–2665.

O'Donnell, K.A., Wentzel, E.A., Zeller, K.I., Dang, C.V., and Mendell, J.T. (2005). c-Myc-regulated microRNAs modulate E2F1 expression. Nature *435*, 839–843.

Ota, A., Tagawa, H., Karnan, S., Tsuzuki, S., Karpas, A., Kira, S., Yoshida, Y., and Seto, M. (2004). Identification and characterization of a novel gene, C13orf25, as a target for 13q31-q32 amplification in malignant lymphoma. Cancer Res. *64*, 3087–3095.

Park, J.K., Henry, J.C., Jiang, J., Esau, C., Gusev, Y., Lerner, M.R., Postier, R.G., Brackett, D.J., and Schmittgen, T.D. (2011). miR-132 and miR-212 are increased in pancreatic cancer and target the retinoblastoma tumour suppressor. Biochem. Biophys. Res. Commun. *406*, 518–523.

Pasini, D., Malatesta, M., Jung, H.R., Walfridsson, J., Willer, A., Olsson, L., Skotte, J., Wutz, A., Porse, B., Jensen, O.N., *et al.* (2010). Characterization of an antagonistic switch between histone H3 lysine 27 methylation and acetylation in the transcriptional regulation of Polycomb group target genes. Nucleic Acids Res. *38*, 4958–4969.

Patel, D., Huang, S.M., Baglia, L.A., and McCance, D.J. (1999). The E6 protein of human papillomavirus type 16 binds to and inhibits co-activation by CBP and p300. EMBO J. *18*, 5061–5072.

Pereira, P.M., Marques, J.P., Soares, A.R., Carreto, L., and Santos, M.A. (2010). MicroRNA expression variability in human cervical tissues. PLoS ONE *5*, e11780.

Petrocca, F., Visone, R., Onelli, M.R., Shah, M.H., Nicoloso, M.S., de Martino, I., Iliopoulos, D., Pilozzi, E., Liu, C.G., Negrini, M., *et al.* (2008). E2F1-regulated microRNAs impair TGFbeta-dependent cell-cycle arrest and apoptosis in gastric cancer. Cancer Cell *13*, 272–286.

Place, R.F., Li, L.C., Pookot, D., Noonan, E.J., and Dahiya, R. (2008). MicroRNA-373 induces expression of genes with complementary promoter sequences. Proc. Natl. Acad. Sci. U.S.A. *105*, 1608–1613.

Porkka, K.P., Pfeiffer, M.J., Waltering, K.K., Vessella, R.L., Tammela, T.L., and Visakorpi, T. (2007). MicroRNA expression profiling in prostate cancer. Cancer Res. *67*, 6130–6135.

Ramdas, L., Giri, U., Ashorn, C.L., Coombes, K.R., El-Naggar, A., Ang, K.K., and Story, M.D. (2009). miRNA expression profiles in head and neck squamous cell carcinoma and adjacent normal tissue. Head Neck *31*, 642–654.

Rao, Q., Shen, Q., Zhou, H., Peng, Y., Li, J., and Lin, Z. (2011). Aberrant microRNA expression in human cervical carcinomas. Med. Oncol. doi: 10.1007/s12032-011-9830-2.

Raver-Shapira, N., Marciano, E., Meiri, E., Spector, Y., Rosenfeld, N., Moskovits, N., Bentwich, Z., and Oren, M. (2007). Transcriptional activation of miR-34a contributes to p53-mediated apoptosis. Mol. Cell *26*, 731–743.

Saito, Y., Liang, G., Egger, G., Friedman, J.M., Chuang, J.C., Coetzee, G.A., and Jones, P.A. (2006). Specific activation of microRNA-127 with down-regulation of the proto-oncogene BCL6 by chromatin-modifying drugs in human cancer cells. Cancer Cell *9*, 435–443.

Sarver, A.L., French, A.J., Borralho, P.M., Thayanithy, V., Oberg, A.L., Silverstein, K.A., Morlan, B.W., Riska, S.M., Boardman, L.A., Cunningham, J.M., *et al.* (2009). Human colon cancer profiles show differential microRNA expression depending on mismatch repair status and are characteristic of undifferentiated proliferative states. BMC Cancer *9*, 401.

Scarola, M., Schoeftner, S., Schneider, C., and Benetti, R. (2010). miR-335 directly targets Rb1 (pRb/p105) in a proximal connection to p53-dependent stress response. Cancer Res. *70*, 6925–6933.

Scheffner, M., Werness, B.A., Huibregtse, J.M., Levine, A.J., and Howley, P.M. (1990). The E6 oncoprotein encoded by human papillomavirus types 16 and 18 promotes the degradation of p53. Cell *63*, 1129–1136.

Schwarz, E., Freese, U.K., Gissmann, L., Mayer, W., Roggenbuck, B., Stremlau, A., and zur Hausen, H. (1985). Structure and transcription of human papillomavirus sequences in cervical carcinoma cells. Nature *314*, 111–114.

Schwarz, D.S., Hutvagner, G., Du, T., Xu, Z., Aronin, N., and Zamore, P.D. (2003). Asymmetry in the assembly of the RNAi enzyme complex. Cell *115*, 199–208.

Shai, A., Brake, T., Somoza, C., and Lambert, P.F. (2007). The human papillomavirus E6 oncogene dysregulates the cell cycle and contributes to cervical carcinogenesis through two independent activities. Cancer Res. *67*, 1626–1635.

Simion, A., Laudadio, I., Prevot, P.P., Raynaud, P., Lemaigre, F.P., and Jacquemin, P. (2010). MiR-495 and miR-218 regulate the expression of the Onecut transcription factors HNF-6 and OC-2. Biochem. Biophys. Res. Commun. *391*, 293–298.

Skyldberg, B., Salo, S., Eriksson, E., Aspenblad, U., Moberger, B., Tryggvason, K., and Auer, G. (1999). Laminin-5 as a marker of invasiveness in cervical lesions. J. Natl. Cancer Inst. *91*, 1882–1887.

Small, E.M., Sutherland, L.B., Rajagopalan, K.N., Wang, S., and Olson, E.N. (2010). MicroRNA-218 regulates vascular patterning by modulation of Slit-Robo signalling. Circ. Res. *107*, 1336–1344.

Song, L.B., Huang, Q., Chen, K., Liu, L., Lin, C., Dai, T., Yu, C., Wu, Z., and Li, J. (2010). miR-218 inhibits the invasive ability of glioma cells by direct down-regulation of IKK-beta. Biochem. Biophys. Res. Commun. *402*, 135–140.

Takamizawa, J., Konishi, H., Yanagisawa, K., Tomida, S., Osada, H., Endoh, H., Harano, T., Yatabe, Y., Nagino, M., Nimura, Y., *et al.* (2004). Reduced expression of the let-7 microRNAs in human lung cancers in association with shortened postoperative survival. Cancer Res. *64*, 3753–3756.

Tazawa, H., Tsuchiya, N., Izumiya, M., and Nakagama, H. (2007). Tumor-suppressive miR-34a induces senescence-like growth arrest through modulation of the E2F pathway in human colon cancer cells. Proc. Natl. Acad. Sci. U.S.A. *104*, 15472–15477.

Tennis, M.A., Van Scoyk, M.M., Freeman, S.V., Vandervest, K.M., Nemenoff, R.A., and Winn, R.A. (2010). Sprouty-4 inhibits transformed cell growth, migration and invasion, and epithelial-mesenchymal transition, and is regulated by Wnt7A through PPARgamma in non-small cell lung cancer. Mol. Cancer Res. *8*, 833–843.

Thorland, E.C., Myers, S.L., Gostout, B.S., and Smith, D.I. (2003). Common fragile sites are preferential targets for HPV16 integrations in cervical tumors. Oncogene *22*, 1225–1237.

Tie, J., Pan, Y., Zhao, L., Wu, K., Liu, J., Sun, S., Guo, X., Wang, B., Gang, Y., Zhang, Y., *et al.* (2010). MiR-218 inhibits invasion and metastasis of gastric cancer by targeting the Robo1 receptor. PLoS Genet. *6*, e1000879.

Tong, A.W., Fulgham, P., Jay, C., Chen, P., Khalil, I., Liu, S., Senzer, N., Eklund, A.C., Han, J., and Nemunaitis, J. (2009). MicroRNA profile analysis of human prostate cancers. Cancer Gene Ther. *16*, 206–216.

Tran, N., Rose, B.R., and O'Brien, C.J. (2007). Role of human papillomavirus in the etiology of head and neck cancer. Head Neck *29*, 64–70.

Truong, A.B., and Khavari, P.A. (2007). Control of keratinocyte proliferation and differentiation by p63. Cell Cycle *6*, 295–299.

Ueda, T., Volinia, S., Okumura, H., Shimizu, M., Taccioli, C., Rossi, S., Alder, H., Liu, C.G., Oue, N., Yasui, W., *et al.* (2010). Relation between microRNA expression and progression and prognosis of gastric cancer: a microRNA expression analysis. Lancet Oncol. *11*, 136–146.

Varambally, S., Dhanasekaran, S.M., Zhou, M., Barrette, T.R., Kumar-Sinha, C., Sanda, M.G., Ghosh, D., Pienta, K.J., Sewalt, R.G., Otte, A.P., *et al.* (2002). The polycomb group protein EZH2 is involved in progression of prostate cancer. Nature *419*, 624–629.

Vasudevan, S., Tong, Y., and Steitz, J.A. (2007). Switching from repression to activation: microRNAs can up-regulate translation. Science *318*, 1931–1934.

Ventura, A., Young, A.G., Winslow, M.M., Lintault, L., Meissner, A., Erkeland, S.J., Newman, J., Bronson, R.T., Crowley, D., Stone, J.R., *et al.* (2008). Targeted deletion reveals essential and overlapping functions of the miR-17 through 92 family of miRNA clusters. Cell *132*, 875–886.

Vidal, L., and Gillison, M.L. (2008). Human papillomavirus in HNSCC: recognition of a distinct disease type. Hematol. Oncol. Clin. North Am. *22*, 1125–1142, vii.

de Villiers, E.M., Fauquet, C., Broker, T.R., Bernard, H.U., and zur Hausen, H. (2004). Classification of papillomaviruses. Virology *324*, 17–27.

Volinia, S., Calin, G.A., Liu, C.G., Ambs, S., Cimmino, A., Petrocca, F., Visone, R., Iorio, M., Roldo, C., Ferracin, M., *et al.* (2006). A microRNA expression signature of human solid tumors defines cancer gene targets. Proc. Natl. Acad. Sci. U.S.A. *103*, 2257–2261.

Wagener, N., Holland, D., Bulkescher, J., Crnkovic-Mertens, I., Hoppe-Seyler, K., Zentgraf, H., Pritsch, M., Buse, S., Pfitzenmaier, J., Haferkamp, A., *et al.* (2008). The enhancer of zeste homolog 2 gene contributes to cell proliferation and apoptosis resistance in renal cell carcinoma cells. Int. J. Cancer *123*, 1545–1550.

Walboomers, J.M., Jacobs, M.V., Manos, M.M., Bosch, F.X., Kummer, J.A., Shah, K.V., Snijders, P.J., Peto, J., Meijer, C.J., and Munoz, N. (1999). Human papillomavirus is a necessary cause of invasive cervical cancer worldwide. J. Pathol. *189*, 12–19.

Wald, A.I., Hoskins, E.E., Wells, S.I., Ferris, R.L., and Khan, S.A. (2011). Alteration of microRNA profiles in squamous cell carcinoma of the head and neck cell lines by human papillomavirus. Head Neck *33*, 504–512.

Wan, H., Yuan, M., Simpson, C., Allen, K., Gavins, F.N., Ikram, M.S., Basu, S., Baksh, N., O'Toole, E.A., and Hart, I.R. (2007). Stem/progenitor cell-like properties of desmoglein 3dim cells in primary and immortalized keratinocyte lines. Stem Cells *25*, 1286–1297.

Wang, L.P., Bi, J., Yao, C., Xu, X.D., Li, X.X., Wang, S.M., Li, Z.L., Zhang, D.Y., Wang, M., and Chang, G.Q. (2010a). Annexin A1 expression and its prognostic significance in human breast cancer. Neoplasma *57*, 253–259.

Wang, X., Tang, S., Le, S.Y., Lu, R., Rader, J.S., Meyers, C., and Zheng, Z.M. (2008). Aberrant expression of oncogenic and tumour-suppressive microRNAs in cervical cancer is required for cancer cell growth. PLoS ONE *3*, e2557.

Wang, X., Wang, H.K., McCoy, J.P., Banerjee, N.S., Rader, J.S., Broker, T.R., Meyers, C., Chow, L.T., and Zheng, Z.M. (2009). Oncogenic HPV infection interrupts the expression of tumor-suppressive miR-34a through viral oncoprotein E6. Rna *15*, 637–647.

Wang, Y.X., Zhang, X.Y., Zhang, B.F., Yang, C.Q., Chen, X.M., and Gao, H.J. (2010b). Initial study of microRNA expression profiles of colonic cancer without lymph node metastasis. J. Dig. Dis. *11*, 50–54.

Weintraub, S.J., Prater, C.A., and Dean, D.C. (1992). Retinoblastoma protein switches the E2F site from positive to negative element. Nature *358*, 259–261.

Weintraub, S.J., Chow, K.N., Luo, R.X., Zhang, S.H., He, S., and Dean, D.C. (1995). Mechanism of active transcriptional repression by the retinoblastoma protein. Nature *375*, 812–815.

Welch, C., Chen, Y., and Stallings, R.L. (2007). MicroRNA-34a functions as a potential tumour suppressor by inducing apoptosis in neuroblastoma cells. Oncogene *26*, 5017–5022.

Werness, B.A., Levine, A.J., and Howley, P.M. (1990). Association of human papillomavirus types 16 and 18 E6 proteins with p53. Science *248*, 76–79.

Wilting, S.M., van Boerdonk, R.A., Henken, F.E., Meijer, C.J., Diosdado, B., Meijer, G.A., le Sage, C., Agami, R., Snijders, P.J., and Steenbergen, R.D. (2010). Methylation-mediated silencing and tumour suppressive function of hsa-miR-124 in cervical cancer. Mol. Cancer *9*, 167.

Winn, R.A., Van Scoyk, M., Hammond, M., Rodriguez, K., Crossno, J.T., Jr., Heasley, L.E., and Nemenoff, R.A. (2006). Anti-tumorigenic effect of Wnt 7a and Fzd 9 in non-small cell lung cancer cells is mediated through ERK-5-dependent activation of peroxisome proliferator-activated receptor gamma. J. Biol. Chem. *281*, 26943–26950.

Wong, D.T., and Munger, K. (2000). Association of human papillomaviruses with a subgroup of head and neck squamous cell carcinomas. J. Natl. Cancer Inst. *92*, 675–677.

Wong, T.S., Liu, X.B., Chung-Wai Ho, A., Po-Wing Yuen, A., Wai-Man Ng, R., and Ignace Wei, W. (2008). Identification of pyruvate kinase type M2 as potential oncoprotein in squamous cell carcinoma of tongue through microRNA profiling. Int. J. Cancer *123*, 251–257.

Wu, D.W., Cheng, Y.W., Wang, J., Chen, C.Y., and Lee, H. (2010). Paxillin predicts survival and relapse in non-small cell lung cancer by microRNA-218 targeting. Cancer Res. *70*, 10392–10401.

Xi, Y., Shalgi, R., Fodstad, O., Pilpel, Y., and Ju, J. (2006). Differentially regulated microRNAs and actively translated messenger RNA transcripts by tumour suppressor p53 in colon cancer. Clin. Cancer Res. *12*, 2014–2024.

Yang, Z., Chen, S., Luan, X., Li, Y., Liu, M., Li, X., Liu, T., and Tang, H. (2009). MicroRNA-214 is aberrantly expressed in cervical cancers and inhibits the growth of HeLa cells. IUBMB Life *61*, 1075–1082.

Yao, Q., Xu, H., Zhang, Q.Q., Zhou, H., and Qu, L.H. (2009). MicroRNA-21 promotes cell proliferation and down-regulates the expression of programmed cell death 4 (PDCD4) in HeLa cervical carcinoma cells. Biochem. Biophys. Res. Commun. *388*, 539–542.

Yi, R., Poy, M.N., Stoffel, M., and Fuchs, E. (2008). A skin microRNA promotes differentiation by repressing 'stemness'. Nature *452*, 225–229.

Yoshida, S., Kajitani, N., Satsuka, A., Nakamura, H., and Sakai, H. (2008). Ras modifies proliferation and invasiveness of cells expressing human papillomavirus oncoproteins. J. Virol. *82*, 8820–8827.

Ypsilanti, A.R., Zagar, Y., and Chedotal, A. (2010). Moving away from the midline: new developments for Slit and Robo. Development *137*, 1939–1952.

Yu, J., Cao, Q., Wu, L., Dallol, A., Li, J., Chen, G., Grasso, C., Cao, X., Lonigro, R.J., Varambally, S., *et al.* (2010). The neuronal repellent SLIT2 is a target for repression by EZH2 in prostate cancer. Oncogene *29*, 5370–5380.

Zerfass-Thome, K., Zwerschke, W., Mannhardt, B., Tindle, R., Botz, J.W., and Jansen-Durr, P. (1996). Inactivation of the cdk inhibitor p27KIP1 by the human papillomavirus type 16 E7 oncoprotein. Oncogene *13*, 2323–2330.

Zhang, J.G., Wang, J.J., Zhao, F., Liu, Q., Jiang, K., and Yang, G.H. (2010a). MicroRNA-21 (miR-21) represses tumor suppressor PTEN and promotes growth and invasion in non-small cell lung cancer (NSCLC). Clin. Chim. Acta *411*, 846–852.

Zhang, L., Deng, T., Li, X., Liu, H., Zhou, H., Ma, J., Wu, M., Zhou, M., Shen, S., Niu, Z., *et al.* (2010b). microRNA-141 is involved in a nasopharyngeal carcinoma-related genes network. Carcinogenesis *31*, 559–566.

Zhang, L., Huang, J., Yang, N., Greshock, J., Megraw, M.S., Giannakakis, A., Liang, S., Naylor, T.L., Barchetti, A., Ward, M.R., *et al.* (2006). microRNAs exhibit high frequency genomic alterations in human cancer. Proc. Natl. Acad. Sci. U.S.A. *103*, 9136–9141.

Zhou, X., Chen, X., Hu, L., Han, S., Qiang, F., Wu, Y., Pan, L., Shen, H., Li, Y., and Hu, Z. (2010). Polymorphisms involved in the miR-218-LAMB3 pathway and susceptibility of cervical cancer, a case–control study in Chinese women. Gynecol. Oncol. *117*, 287–290.

Zhu, S., Si, M.L., Wu, H., and Mo, Y.Y. (2007). MicroRNA-21 targets the tumor suppressor gene tropomyosin 1 (TPM1). J. Biol. Chem. *282*, 14328–14336.

Zhu, S., Wu, H., Wu, F., Nie, D., Sheng, S., and Mo, Y.Y. (2008). MicroRNA-21 targets tumor suppressor genes in invasion and metastasis. Cell Res. *18*, 350–359.

Zimmermann, H., Degenkolbe, R., Bernard, H.U., and O'Connor, M.J. (1999). The human papillomavirus type 16 E6 oncoprotein can down-regulate p53 activity by targeting the transcriptional coactivator CBP/p300. J. Virol. *73*, 6209–6219.

Viral Deregulation of DNA Damage Responses

Sergei Boichuk and Ole Gjoerup

Abstract

Incoming viral genomes, aberrant viral replication structures or individual viral proteins are potential triggers of DNA damage responses (DDRs). In an emerging theme, viruses interfere with, and frequently commandeer, DDR and repair signalling pathways to promote the viral life cycle. Here we review the diverse mechanisms that small DNA tumour viruses utilize to deregulate DDR pathways. Adenoviruses (Ad) encode gene products that specifically degrade the MRN (Mre11, Rad50, Nbs1) damage sensor or sequester it in nuclear tracks. This causes an inhibition of both ATM and ATR responses. Failure to inactivate MRN leads to attenuation of viral replication and concatemerization of the viral genome, thus preventing efficient packaging. Conversely, polyomaviruses, like SV40, as well as human papillomaviruses (HPVs) appear to exploit the DDR, since they use components of it to positively regulate their life cycle, while generally inhibiting downstream checkpoint responses. SV40, mouse polyomavirus and HPV activate ATM signalling and benefit from it. Complexities of the interplay between small DNA tumour viruses and the DDR are continuously evolving and illuminate both critical aspects of the viral life cycle as well as basic cellular mechanisms operating in a non-viral setting.

Introduction

Viruses are well-known obligatory parasites and therefore require components of the host cell to replicate their genomes. Despite the distinct replication strategies used by different viruses, many of them are now known to exploit specific cellular mechanisms involved in protecting host cell genome integrity. These mechanisms include a complex network of DNA damage signalling and repair pathways, collectively termed the DNA damage response (DDR). Here we briefly review the current knowledge about the main branches of DDR pathways implicated in repair of damaged DNA in the absence of virus infection. Several excellent reviews have been previously published on the DDR (Zhou and Elledge, 2000; Petrini and Stracker, 2003; Huen and Chen, 2009; Jackson and Bartek, 2009; Ciccia and Elledge, 2010). We will discuss the role of DDR mechanisms in the maintenance of genome integrity and cancer prevention, as well as provide evidence of the relationship between DDR abnormalities and cancer predisposition. Next, we will focus on DNA damage signalling and repair mechanisms triggered by infection with a small DNA tumour virus or owing to the expression of different viral proteins. Along the way, we will discuss potential links between the DDR, replication of DNA tumour viruses and virally induced oncogenic transformation. Since defects in many components of the DDR are strongly associated with multiple human diseases, including cancer predisposition, knowledge about the mechanisms of how DNA tumour viruses can deregulate multiple DDR pathways, and therefore interfere with the DNA repair machinery, is important and can provide opportunities for improving cancer prevention, disease detection and treatment.

A plethora of DNA damage

Each cell in the human body experiences a tremendous number of DNA lesions every day, arising from exogenous and endogenous attacks on the DNA backbone and base pairs. Exogenous DNA damage, originating from the environment, can be produced by physical or chemical sources. Physical factors include the exposure to ultraviolet (UV) light in the form of sunlight, ionizing radiation (IR) or X-ray treatment. All of these factors are able to induce the oxidation of DNA bases and formation of single-strand and double-strand DNA breaks (SSBs and DSBs, respectively). Common examples of the chemical factors inducing DNA damage include chemicals found in cigarette smoke and chemical agents used for cancer therapy such as alkylating or crosslinking agents or topoisomerase inhibitors. DNA lesions, particularly DSBs, can also arise from agents inside the cell, owing to exposure to metabolic products such as reactive oxygen species (ROS), known byproducts of normal cellular respiration. DSBs can also be the result of programmed rearrangements of the genome, for example V(D)J recombination during the maturation of the immune system genes and meiotic exchange. DNA lesions are also generated spontaneously during DNA replication owing to nucleotide misincorporation or collapsed replication forks. Most DNA lesions referred to above are hazardous and therefore trigger a DDR in order to recruit cellular factors to repair the DNA and restore genome integrity. DNA damage triggers both cell cycle checkpoints that arrest the cell cycle at critical transition points and repair. Alternatively, if repair is not possible, programmed cell death (apoptosis) ensues. Major checkpoints operate at G1/S, to control entry into cellular DNA synthesis, and at G2/M, to control entry into mitosis. It is imperative that the cell does not progress with major tasks like replication or cell division with a damaged genome, as this may allow further propagation of mutant cells. A checkpoint also operates within S-phase, the intra S-phase checkpoint, to slow down firing of late origins in case of DNA damage. DSBs are considered to be among the most dangerous cellular lesions. If repaired incorrectly, DSBs can lead to genome rearrangements, translocations, large- or small-scale deletions, or insertions, which can be potentially carcinogenic.

A multiplicity of DNA damage response pathways

To preserve the genetic information encoded by DNA and prevent the transmission of genetic mutations, eukaryotic cells have evolved multiple DDR pathways, which include mechanisms to detect DNA damage via sensors, transduce the signal from DNA lesions to inhibit cell cycle progression via effectors and finally recruit repair factors to ensure faithful DNA repair (Fig. 9.1). Various types of DNA lesions activate different DDR pathways that can be distinguished from each other because of the ser/thr kinase involved in transduction of the DNA damage signal. These kinases are members of the phosphatidylinositol 3-kinase-like protein kinases (PIKKs), which include ATM (ataxia-telangiectasia mutated), ATR (ATM and Rad3-related) and DNA-dependent protein kinase (DNA-PK). These kinases are involved in multiple DDR pathways and become recruited to the sites of DNA lesions using various sensors. Once activated, the kinases transduce a signal from the site of DNA damage, phosphorylate downstream target proteins (mediators and effectors), activate cell cycle checkpoints and recruit repair factors to heal DNA defects. Importantly, despite the fact that the various PIKKs respond to distinct types of DNA lesions, they are implicated in multiple, but partially overlapping, DDR pathways and DNA repair mechanisms. Indeed, ATM and DNA-PK are commonly activated after DNA damage, typically IR, and respond primarily to DSBs. In contrast, ATR primarily responds to replication-associated lesions generating single-stranded DNA (ssDNA) and plays a critical role in the cellular response triggered by stalled or collapsed DNA replication forks, which arise during normal DNA replication. ATR also responds to UV-induced DNA damage. Perhaps it is not surprising that these PIKKs are recruited to sites of DNA damage via different molecular mechanisms, exhibit differences in target proteins specificity in response to the various forms of DNA damage and accordingly, activate different DNA repair pathways.

Figure 9.1 Classes of players in the DNA damage response. DNA lesions are sensed via multiple different sensor proteins that signal to apical PIKKs. The signal is amplified by ATM/ATR phosphorylation of mediators, large scaffolds upon which signalling complexes are assembled, additional ATM/ATR substrates are phosphorylated and the Chk1/Chk2 transducer kinases activated. Finally, effectors are phosphorylated, either directly by ATM/ATR or by the Chk1/2 kinases, resulting in checkpoint arrest, repair and other cellular outcomes. Note that some components, for example Nbs1, play a role at multiple steps.

Molecular sensors and transducers of the DNA damage response

Detection of DNA lesions like strand breaks or crosslinks occurs using molecular sensors, which are often specific for each type of DNA damage, but also exhibit partial redundancy. To detect DNA strand breaks, eukaryotic cells use four partially independent sensors: Mre11-Rad50-Nbs1 (MRN) complex, Ku heterodimer (Ku70/80), PARP 1/2 and replication protein A (RPA). These four types of DNA damage sensors activate the 3 major PIKKs involved in DDR and repair: ATM, ATR and DNA-PK. RPA stabilizes ssDNA regions and plays a critical role in activation of the ATR pathway. Loading of Ku heterodimer onto DSBs induces the activation of the catalytic subunit of DNA-PK (DNA-PKcs) and initiates DNA repair via the non-homologous end-joining (NHEJ) pathway. PARP1/2 competes with Ku binding to DSBs and promotes DNA repair via mechanisms that are dependent on nucleolytic processing of the DSB (referred to as resection): Homologous recombination (HR) and alternative NHEJ (alt-NHEJ). PARP also facilitates the initial accumulation of the MRN complex at DSBs, which is known as the 'classic sensor' of DSBs that triggers the ATM-dependent downstream signalling cascade from the sites of DNA damage, chromatin alterations and recruitment of additional signalling factors to sites of DNA lesions. Once activated, the kinases transduce a signal from DNA lesions, phosphorylate downstream target proteins, activate cell cycle checkpoints and recruit repair factors to orchestrate DNA repair.

The ATR pathway

In contrast to ATM, mouse knockouts of ATR are early embryonic lethal, making an important distinction between the two pathways. Even in somatic cells, ATR is an essential gene, presumably because it is involved in surveillance of normal DNA replication and repair of aberrant replication structures. In order to activate the ATR pathway, ssDNA regions must be recognized and coated by RPA. RPA stabilizes ssDNA ends, thus providing the persistent signal required for recruitment. ATR is recruited by its interacting partner ATR-interacting protein (ATRIP) to RPA-coated ssDNA that accumulates at stalled replication forks or is generated as a result of DSB resection after initial DNA damage. Thus, ATM activation after DSB generation is followed by resection of the DSB creating regions of ssDNA, recruitment of RPA and secondarily ATR activation. Although the ATR and ATM signalling branches are independent, they show partial overlap and the pathways resemble an interconnected network (Fig. 9.2). After localizing to RPA-ssDNA, ATR/ATRIP can be directly activated by TopBP1, which is recruited to ATR owing to activity of the 9-1-1 (Rad9-Hus1-Rad1) PCNA-like sliding clamp complex, in turn loaded

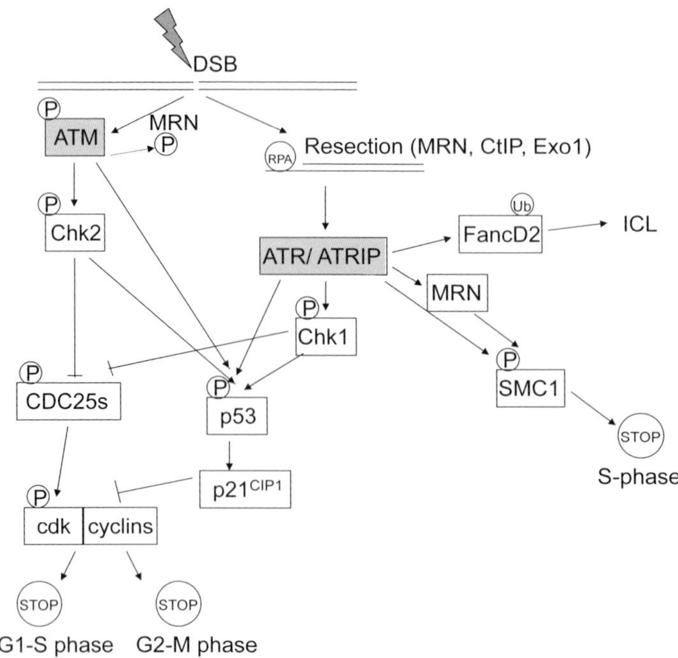

Figure 9.2 Overview of signal transduction events following a DSB. A DSB is sensed by the MRN complex, leading to ATM activation and autophosphorylation on S1981. ATM phosphorylates numerous targets, including Chk2, which in turn phosphorylates and inactivates the Cdc25 family of phosphatases. Cdc25 is required to dephosphorylate CDKs, thus activating cyclin/CDK complexes and triggering cell cycle progression. Conversely, the DNA damage-induced inhibition of Cdc25 results in checkpoint arrest at the G1/S or G2/M transition. Another important target of ATM/ATR and Chk1/2 is p53, which is activated by phosphorylation, causing transcriptional induction of the CDK inhibitor $p21^{CIP1}$. The DSB is also processed by resection, thus creating ssDNA regions with the help of MRN, CtIP and Exo1. These ssDNA ends are coated by RPA, and subsequently ATR/ATRIP are recruited and activated. The ssDNA can also arise from collapse of replication forks, known to cause ATR activation. Once activated, ATR triggers phosphorylation of Chk1 and checkpoint responses similarly to the ATM/Chk2 branch. ATR, via SMC1 phosphorylation, is involved in regulation of the intra-S phase checkpoint. ATR also regulates FancD2 monoubiquitination, which is important in interstrand crosslink repair (ICL).

to ssDNA via a Rad17 clamp loader mechanism. The downstream steps of the ATR-mediated cascade involve phosphorylation of a number of substrates, including Chk1 kinase, which targets downstream proteins to activate cell cycle checkpoints and facilitate repair. Claspin acts as a mediator protein that specifically stimulates ATR phosphorylation of Chk1. After Chk1 kinase becomes phosphorylated on S317/S345 and activated, it dissociates from chromatin and spreads rapidly throughout the nucleus, where it phosphorylates and inactivates cell cycle phosphatases of the Cdc25 family. Cdc25 inactivation prevents Cdk1/2 activation and thereby halts cell cycle progression. Besides Chk1 kinase, other DDR proteins also become phosphorylated in an ATR-dependent manner. These include BRCA1 and BRCA2, known breast cancer tumour suppressors, p53, FancD2 (a component of the Fanconi anaemia (FA) complex), histone H2AX, Claspin and MCMs. ATR is required for FancD2 monoubiquitination, thus controlling the FA pathway implicated in repair of interstrand crosslinks (ICLs). Several ATR substrates contribute to cell cycle arrest or DNA repair. For example, phosphorylation of p53 at N-terminal sites, like Ser15, promotes transcriptional upregulation of the cdk inhibitor $p21^{CIP1}$, thus blocking cell cycle progression at the G1/S transition. BRCA2 plays an essential role in homology-directed DNA repair, in part by loading of the Rad51 recombinase at sites of DSBs.

The ATM pathway

Ataxia telangiectasia, caused by mutations in the ATM gene, is a rare human disorder characterized by extreme radiation sensitivity and predisposition to cancer. ATM knockout mice are viable but hypersensitive to DNA damage and various malignancies. Activation of the ATM-dependent pathway is probably initiated by conformational changes of DNA owing to chromatin alterations or decondensation, primarily following a DSB. These changes in DNA structure enhance the accessibility of sites of DNA damage to repair factors. Chromatin remodelling facilitates the recruitment of the MRN complex, as well as the multifunctional tumour suppressor BRCA1 to DSBs. The MRN complex plays a critical role in DNA end resection owing to the exo- and endonuclease activity of the Mre11 component, and therefore promotes subsequent DNA repair via mechanisms dependent on DSB resection: HR, alt-NHEJ or single-strand annealing (SSA). The MRN complex also plays an important role in the activation of ATM kinase and its recruitment to damaged sites via direct interaction between ATM and a C-terminal motif on Nbs1. In unstressed cells, ATM kinase is present as an inactive dimer owing to intermolecular interactions hiding the major damage-induced phosphorylation site of ATM (S1981). After DSB generation, ATM undergoes a conformational change, which causes dissociation of the dimer and conversion to the highly active monomer possessing kinase activity. The earliest outcome of ATM activation is its intermolecular autophosphorylation on S1981, detectable within few minutes after DNA damage. After activation, ATM phosphorylates a broad spectrum of proteins involved in DNA repair, apoptosis signalling and cell cycle checkpoints. Recent studies indicate that ATM/ATR phosphorylate >700 substrates with functions in not just repair and checkpoint signalling but also cellular assembly/organization, metabolic signalling, gene expression and RNA splicing. Histone H2AX, one of the earliest targets, undergoes phosphorylation on S139, yielding the so-called γ-H2AX. Regions of γ-H2AX spread for distances up to 1–2 megabases around DSBs and play an important role for the maintenance and stabilization of DDR signalling and repair factors at sites of DNA damage. Indeed, γ-H2AX is a commonly used marker for DSBs. Phosphorylation of H2AX sets in motion a cascade of sequential recruitments of different DDR factors, which is regulated by post-translational modifications. The phosphorylated site on H2AX creates a phophoepitope, which is subsequently recognized by the Mdc1 mediator protein. Once recruited to DNA damage sites via interaction with γ-H2AX, Mdc1 is phosphorylated by ATM at multiple sites, in turn allowing Mdc1 to recruit ubiquitin ligases like RNF8 and RNF168. These ubiquitination events serve to amplify and maintain DDR signalling. Several proteins involved in the DDR undergo mono- and polyubiquitination, which is required for their recruitment and maintenance at DNA damage sites. Modifications also include histone methylation, which is required for recruitment of tumour suppressor p53-binding protein 1 (53BP1) via its tudor domain. Altogether, these proteins assemble into a conglomerate of recruited damage response factors known as irradiation-induced foci (IRIF). These cytologically discernible foci accumulate in a highly regulated spatiotemporal manner after DNA damage, thus representing surrogate markers of DNA damage or repair. Besides H2AX, major downstream targets that undergo ATM-dependent phosphorylation include p53 (S15 and S20), Chk2 (T68) and Smc1 (S966). The major outcomes of these events are cell cycle arrest and repair or apoptosis.

The DNA-PK pathway

DNA-PKcs has a vital role in repair of DSBs associated with for example the programmed endonucleolytic DNA cleavage in immunoglobulin V(D) J gene rearrangements. DNA-PK is also implicated in telomere maintenance, transcriptional control, post-translational modifications and plays an important role in the DDR by initiating DNA repair via NHEJ. Similar to the ATM pathway, activation of DNA-PK is triggered by DSBs. Therefore, cells lacking DNA-PKcs are radiosensitive and have defective DSB repair, while mice lacking DNA-PKcs remain viable but are immunodeficient and suffer from significant telomere fusion events. Activation of DNA-PK-mediated pathway begins from the loading of the Ku (Ku70/80) regulatory subunits onto DSBs.

The Ku70/80 heterodimer then recruits the catalytic subunit of DNA-PK (DNA-PKcs) to DNA ends, induces kinase activation and stabilizes its DNA binding. Activated DNA-PKcs undergoes autophoshorylation (S2056 and T2609), and also phosphorylates a number of substrates such as Ku70, Ku80, Artemis, XRCC4 and DNA ligase IV. Phosphorylation of DNA-PKcs is required for recruitment of the end-processing factors, such as the Artemis protein, which possess both 5′–3′ exonuclease and endonuclease activities. The final step consists of the joining of DNA ends, which is carried out by XRCC4, DNA ligase IV and XLF (also known as X4–L4 complex).

While it is generally accepted that DNA-PKcs is not directly involved in activation of cellular checkpoints, this kinase can trigger a p53-dependent apoptotic pathway in case of severe, non-repairable DNA damage.

Double-strand break repair mechanisms

DSBs are generally repaired via HR or NHEJ pathways, which are often characterized as 'error-free' and 'error-prone', respectively. The choice of DSB repair pathway is regulated by the extent of DNA end processing: HR, SSA and alt-NHEJ are dependent on DSB resection, whereas classical NHEJ does not require this step, since DSBs can be sealed without a requisite end processing step. The choice of repair pathway also depends on the cell cycle phase. Since HR utilizes the sister chromatid as template for DNA repair, it is generally restricted to late S and G2 phases of the cell cycle, where a sister chromatid is available. In contrast, the NHEJ pathway is particularly important in G0/G1 and early S phases, where a sister chromatid is unavailable and DSBs cannot be repaired via HR. HR and NHEJ pathways become activated via distinct PIKKs. Activation of the ATM pathway is known to be essential for HR repair, whereas DNA-PK is primarily implicated in NHEJ.

Homologous recombination

DNA repair via HR is initiated by resection of DSBs, which is mainly orchestrated by the MRN complex in conjunction with CtIP, BRCA1, Exo1 and ATM. The result of resection is generation of 3′ ssDNA ends, which become rapidly coated with RPA. RPA binding to ssDNA regions facilitates the recruitment of HR proteins from the Rad52 epistasis group (Rad51, Rad52, Rad54, Rad55, RAd57 and RPA) to sites of DNA damage. With the help of BRCA2, RPA is displaced by Rad51, which is the main RecA-like recombinase essential for HR repair. Owing to the activity of Rad51, the ssDNA end invades the undamaged homologous sequence, which is used as a template for repair. After strand invasion, the 3′ end can be displaced and after pairing with the other 3′ tail, DNA synthesis can complete repair in what is known as synthesis-dependent strand annealing (SDSA). Alternatively, beginning with second end capture, cross-shaped structures known as Holliday junctions (HJs) can form that by resolution yield crossover or non-crossover products.

Non-homologous end-joining

In contrast to HR, DNA repair via NHEJ has no requirements for sequence homology and therefore is not restricted to a specific phase of the cell cycle. NHEJ becomes particularly important in G1 and early S phase of the cell cycle, and this type of repair involves only limited end processing. NHEJ mechanisms usually involve alignment of one or a few complementary bases to direct repair, therefore leading to loss of a small amount of genetic material. Therefore, NHEJ has been considered as 'less accurate' when compared with HR. NHEJ pathway is generally mediated by DNA-PKcs, which is activated and loaded to DSBs via the Ku70/80-dependent mechanism. After being activated, DNA-PK undergoes autophoshorylation and phosphorylates a number of substrates involved in end processing and joining.

DNA repair and cancer susceptibility

The important role of DDR pathways in tumour suppression has been well documented. For instance, human syndromes such as ataxia-telangiectasia (AT) and Nijmegen breakage syndrome (NBS) resulting from mutations of ATM and Nbs1 genes, respectively, are characterized by defective responses to DSBs, chromosomal instability, immunodeficiency and cancer predisposition, primarily lymphomas and to a lesser extent other types of malignancies. Further examples of human cancer predisposition linkage with aberrant DNA repair include FA (defects in ICL repair), Bloom and Werner syndromes (defects in RecQ DNA helicases), Li-Fraumeni syndrome (defects in p53 or Chk2 genes), Cockayne syndrome and xeroderma pigmentosum (defects in nucleotide excision repair). Mutations in BRCA1 and BRCA2 tumour suppressor genes predispose to the development of breast, ovarian and other types of cancers.

Human syndromes associated with DDR defects clearly illustrate that integrity of DDR pathways plays an important role in tumour prevention. Since DDR defects can result in accumulation of genetic mutations, chromosomal alterations, and therefore can contribute to malignant transformation and tumorigenesis, knowledge about how small DNA tumour viruses induce DNA damage and deregulate DDR pathways can provide novel insights on the molecular mechanisms of tumorigenesis.

Polyomaviruses

SV40 permissive infection induces G2 arrest and cellular re-replication

Since SV40 interactions with the DDR are the most studied among the polyomaviruses, we will start out by reviewing the important lessons that have been learned. Early work demonstrated that SV40 infection of permissive African green monkey kidney cells causes arrest at the G2 phase concomitant with accumulation of activated Chk1 and a failure to enter mitosis (Okubo *et al.*, 2003). Cyclin B/Cdk1 is present in its inhibited form owing to Y15 phosphorylation of Cdk1 (Scarano *et al.*, 1994). Interestingly, SV40 was also found to induce accumulation of >G2 DNA content without intervening mitosis, with most of the viral replication probably occurring in this phase (Lehman *et al.*, 2000). Mitosis could be artificially induced by treatment with Chk1 inhibitor (UCN01) or caffeine, which inhibits ATM/ATR kinases (Okubo *et al.*, 2003). Collectively, these observations demonstrated relatively early on that DDR and checkpoint signalling were activated, but it was unclear if viral replication complexes were triggering the G2 checkpoint.

SV40 infection activates ATM signalling

Subsequent studies demonstrated that SV40 permissive infection activates ATM signalling, since ATM is autophosphorylated on S1981 (Shi *et al.*, 2005). Attention centred quickly on the viral initiator of replication, large T antigen (LT). ATM causes phosphorylation *in vitro* and in cells of LT on S120, and additional ATM targets like Chk2 are activated. Notably, S120 phosphorylation of LT is reported to be inhibitory for SV40 replication *in vitro*, but paradoxically is essential for efficient replication *in vivo* (Schneider and Fanning, 1988; Shi *et al.*, 2005). ATM signalling was found to facilitate viral replication, potentially via the regulation of LT through S120 phosphorylation. While it was not known whether viral replication intermediates or host cell chromosome damage are responsible for the ATM response, it was concluded that checkpoint responses did not arrest viral replication, unlike the cellular counterpart.

Additional detail was revealed in a subsequent study, where it was demonstrated that during a productive SV40 infection, pATM S1981, γ-H2AX and the MRN complex colocalize with LT in large foci representing viral replication centres (Zhao *et al.*, 2008). Furthermore, p53 is stabilized, consistent with DDR activation. Intriguingly, Rad50 and Nbs1 are proteasomally degraded late in infection by LT interaction with the Cul7 ubiquitin ligase complex (Zhao *et al.*, 2008). This degradation event and the ability to form distinct replication compartments are dependent on the activated ATM signalling pathway. Thus, SV40 replication occurs through reprogramming of cell cycle and DDR pathways. Consistent with the earlier report, ATM was found to be required for optimal viral replication. It was proposed that SV40 replication may in fact proceed through a DNA repair pathway or via recovery from host DNA damage (Zhao *et al.*, 2008).

SV40 LT induces multiple DNA damage and repair signalling pathways

Later, it was demonstrated that LT is intrinsically capable of inducing DDR signalling, even in the absence of a viral origin of replication (Hein *et al.*, 2009). It was shown that stable expression of LT in semipermissive normal human fibroblasts (BJ/tert) induces large γ-H2AX foci through binding of LT to the Bub1 mitotic checkpoint kinase (Fig. 9.3). Binding of pRB family members did not play any major role in γ-H2AX foci formation. Expression of the SV40 17k gene product, which is equivalent to an N-terminal LT fragment (identical to LT 1–131 but ends with 4 unique amino acids), even when mutated to not bind pRB family members, induces γ-H2AX foci and a premature cellular senescence programme in a Bub1 binding dependent manner (Gjoerup *et al.*, 2007; Hein *et al.*, 2009). LT expression induces ATR/ATM signalling, since Chk1 and Chk2 were activated, and p53 was phosphorylated/stabilized, in large part through Bub1 binding (Hein *et al.*, 2009). Phosphorylation of p53 on S15 was found to be instrumental for its stabilization by LT. Interestingly, LT also induces tetraploid DNA content via Bub1 binding, establishing a possible connection between the DDR and genomic instability.

Recently, LT was shown to induce overt DNA damage in BJ/tert cells using the comet assay (single cell gel electrophoresis) (Boichuk *et al.*, 2010). Comet tails under neutral conditions indicate formation of DSBs, among other lesions. Strikingly, LT was found to induce multiple, additional DDR and repair signalling pathways, which were surprisingly activated independent of Bub1 binding. The FA pathway, which is involved in ICL repair, was activated by LT as evidenced by FancD2 accumulation in its monoubiquitinated (activated) form in chromatin-localized foci (Boichuk *et al.*, 2010). These observations suggest that LT somehow deregulates the cellular replication machinery, causing replication stress, possibly in the form of stalled or collapsed replication forks. Furthermore, LT causes foci formation of the Rad51 protein critical for HR repair. The FancD2 and Rad51 foci were found to be spatially distinct and dependent on different domains of LT. LT was localized in chromatin-bound foci in close proximity with Rad51, PML and the MRN complex, whereas the dl89–97 mutant was mislocalized. This apparent upregulation of the HR machinery

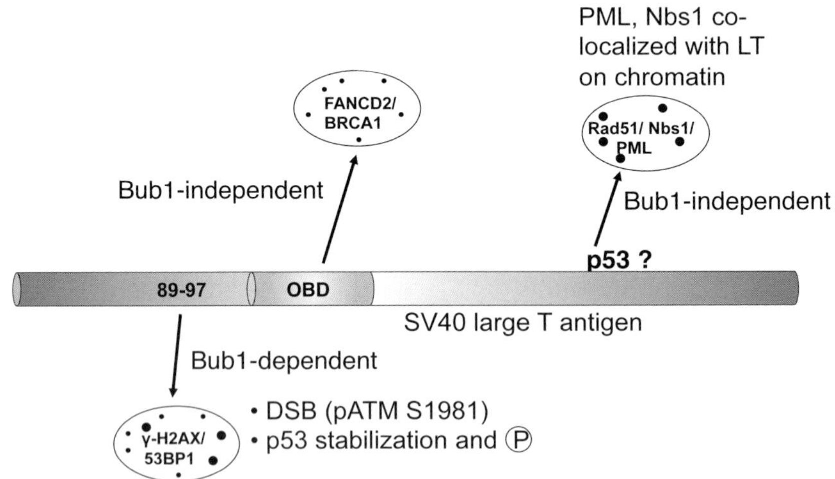

Figure 9.3 Complex interplay of SV40 LT with DDR and repair pathways. LT induces formation of γ-H2AX/53BP1 foci dependent on the Bub1 binding site contained within residues 89–97 (Hein *et al.*, 2009). Induction of DSBs, as determined by pATM S1981 accumulation, and p53 stabilization/phosphorylation are also largely dependent on the Bub1 binding site. Additional modulations of the DDR and repair machinery were found to be Bub1 binding-independent (Boichuk *et al.*, 2010). The origin-binding domain (OBD) of LT is required for induction of FancD2/BRCA1 foci (S. Boichuk and O. Gjoerup, unpublished observations). LT also induces distinct foci of the HR repair factor Rad51. It is suspected, but not yet proven that these foci depend on p53 binding by LT. LT itself was found to be colocalized in foci with Rad51, PML and Nbs1.

suggests the need for ongoing repair to preserve viability. Interestingly, several of the DDR and repair factors, including FancD2 and Rad51, were recruited into replication centres during a productive infection. Importantly, Rad51 was found to be critical for efficient origin-dependent replication *in vivo*, perhaps because high-level extrachromosomal replication causes frequent collapsed replication forks that must be repaired by HR (Boichuk *et al.*, 2010).

PML nuclear bodies, potential sites of DNA repair, are preferred sites for SV40 replication

Evidence indicates that SV40 replication occurs in discrete subnuclear sites that are at, or in close proximity with, promyelocytic leukaemia protein (PML) nuclear bodies (NBs) (Ishov and Maul, 1996; Tang *et al.*, 2000; Zhao *et al.*, 2008). The PML tumour suppressor is critical for the formation of PML NBs. Their exact function is unknown, but they play fundamental roles in key cellular processes like transcription, apoptosis, DNA repair and antiviral responses (Bernardi and Pandolfi, 2007). Many viruses deposit their incoming genomes adjacent to PML NBs (Tavalai and Stamminger, 2008). While most viruses subsequently destroy the PML NBs by degrading PML (herpes simplex virus ICP0 protein; Everett and Maul, 1994) or mislocalizing it (Ad E4orf3; Carvalho *et al.*, 1995), SV40 markedly differs in its strategy by initiating replication without disrupting PML NB function (Ishov and Maul, 1996; Tang *et al.*, 2000). Intriguingly, there are parallels between DDR and PML NB regulation. PML and DNA repair proteins are recruited to both viral replication compartments and sites of DNA damage, thus making it possible that viral genomes, and LT itself, are recruited to PML NBs as part of a DDR (Everett, 2006). Various reports indicate that PML NBs may be involved in DDR sensing, processing of DNA lesions or repair (Dellaire and Bazett-Jones, 2004; Boe *et al.*, 2006; Dellaire *et al.*, 2006). It remains to be shown if LT targets PML NBs to gain access to repair factors needed for viral replication. The targeting of PML protein and PML nuclear bodies by DNA tumour viruses is reviewed in this book by Leppard, K.N. and Wright, J.

SV40 infection induces ATR-Δp53 signalling

During a productive infection, LT also activates ATR signalling and specifically triggers an intra-S phase checkpoint associated with the Δp53 isoform, which LT does not bind to (Rohaly *et al.*, 2010). The Δp53 isoform is generated by differential splicing and lacks part of the DNA binding domain, thus believed to result in selective induction of cell cycle arrest, but not apoptosis-related, genes. This induction of the p21^{cip1} cdk inhibitor causes S phase arrest, which is probably beneficial for SV40 replication. Furthermore, the ensuing down-regulation of cyclin A/cdk activity causes hypophosphorylation of the DNA polymerase α-primase complex, which is the form that LT interacts productively with to drive replication. In support of this model, inactivation of ATR-Δp53 signalling can attenuate viral replication. However, it remains controversial if ATR contributes to SV40 replication proficiency (Boichuk *et al.*, 2010).

SV40 LT binds Nbs1 to enhance viral replication

LT specifically interacts with the Nbs1 protein through the LT origin-binding domain (OBD), and this was demonstrated to be important for LT-induced chromosome re-replication following IR (Wu *et al.*, 2004). While cellular origins fire only once per cell cycle, the viral origin is known to fire multiple times. SV40 appears to overcome this obstacle to efficient replication by LT inhibiting the function of Nbs1 that suppresses rereplication (Wu *et al.*, 2004). As predicted, Nbs1 expression suppresses LT-induced rereplication and amplification of SV40 DNA from an integrated origin. Other functions of Nbs1, like nuclear foci formation and intra-S phase checkpoint control appear unperturbed by LT.

Mouse polyomavirus infection activates ATM signalling

Early studies with mouse polyomavirus (MPyV) also found evidence of DDR activation. Productively infected cells arrest at both S and G2 phase and activate p53 (Dey *et al.*, 2002). Since, unlike SV40 LT, p53 is not inactivated by MPyV LT, target genes like p21^{cip1} are induced, and p53 phosphorylated on S18 (the equivalent of human p53 S15) accumulates. Induction of the p53 response is significantly impaired with a virus expressing an pRB binding mutant of LT (Dey *et al.*, 2002). Similarly to SV40, ATM is activated, since pATM S1981 accumulates (Dahl *et al.*, 2005). Experiments with caffeine and ATM deficient cells demonstrate that replication is enhanced up to 10 fold by ATM signalling (Dahl *et al.*, 2005). This is a p53-independent effect, because p53 deficient cells were compared. MPyV replication was also inhibited in SMC1^{S957A,S966A} knockin cells (Dahl *et al.*, 2005). The SMC1 cohesin is an ATM/ATR target. The molecular mechanisms whereby SMC1 phosphorylation by ATM contributes to viral replication proficiency have not been resolved.

JC polyomavirus induces DNA damage response signalling

Finally, infection with the human JC polyomavirus (JCV) also induces ATM/ATR signalling, which is responsible for G2 checkpoint arrest (Orba *et al.*, 2010). This arrest appears to promote viral replication, since abrogation of the G2 arrest by inhibiting ATM, ATR, Chk1 or Wee1 results in impaired genome replication. Interestingly, JCV LT was capable of causing G2 arrest in the absence of a functional origin, although the latter exacerbated the response. Notably, the G2 arrest depends on the OBD and binding to cellular DNA, possible causing interference with host replication (Orba *et al.*, 2010). In the more physiological context of oligodendrocytes from a progressive multifocal leukoencephalopathy brain, G2 arrest was also observed. Another report indicated that JCV induces genomic instability, γ-H2AX and Rad51 accumulation (Darbinyan *et al.*, 2007). Rad51 appears in nuclear foci following infection of human astrocytes. Homology-directed repair assays suggest that JCV LT inhibits repair, possibly by forcing complex formation between insulin receptor substrate 1 (IRS1) and Rad51 in the nucleus (Trojanek *et al.*, 2006). However, this inhibitory effect of JCV LT is at present difficult to reconcile with SV40 and MPyV LTs that are generally believed to promote recombination (Jasin *et al.*, 1985; St-Onge *et al.*, 1990; Xia *et al.*, 1997). Finally, the JCV agnoprotein was found to sensitize cells to cisplatin and promote aneuploidy, possibly by inhibiting DNA repair via binding and relocalization of Ku70 (Darbinyan *et al.*, 2004).

How does DNA damage response activation modulate oncogenic transformation by polyomaviruses?

SV40 LT is known to induce both structural and numerical chromosome instability, including tetraploidy/aneuploidy, chromosome breaks, fragments, dicentrics and double minutes in human fibroblasts (Stewart and Bacchetti, 1991; Ray *et al.*, 1992; Woods *et al.*, 1994; Chang *et al.*, 1997). Furthermore, LT has been shown to be mutagenic (Theile and Grabowski, 1990) and promote sister chromatid exchanges (Nichols *et al.*, 1978) as well as micronuclei formation (Hein *et al.*, 2009). These effects are likely to be at least in part mediated through disruption of DDR and checkpoint mechanisms. However, a significant unanswered question remains: Does the genomic instability contribute to, or even drive, long-term tumorigenesis?

We do not yet have a clear picture of whether DDR interference plays a positive or negative role in malignant transformation by SV40. Consistent with the conventional view that the DDR serves as an early barrier to tumorigenesis (Bartkova *et al.*, 2005), fibroblasts from FA patients are hypersensitive to SV40 transformation (Todaro *et al.*, 1966; Liu *et al.*, 1996). Nonetheless, p53 is stabilized as a consequence of DDR activation by LT, and evidence indicates that p53 stabilization is conferring a gain of function in transformation (Deppert *et al.*, 1989; Bocchetta *et al.*, 2008; Hermannstadter *et al.*, 2009). This occurs in part by LT altering the p53 transcriptional programme to not just block conventional p53 target induction but also turn on novel genes with a growth-promoting function. Taken together, this seems to portray a complex picture of DDR contributions to SV40 malignant transformation mechanisms.

Adenovirus

In contrast to the polyomaviruses, which exploit DDR mechanisms to achieve optimal viral replication, adenovirus (Ad) inhibits multiple steps of the DDR machinery to facilitate replication of its own linear genome. It has been established that DDR signalling is detrimental for Ad replication and progeny production, since cellular DNA repair factors can process the viral genome ends, which results in joining of the genomes into concatemers through NHEJ, thus preventing their packaging due the large size of these structures (Boyer *et al.*, 1999; Stracker *et al.*, 2002). The concatemers may also block viral replication, since Ad viral terminal repeats containing the replication origin become buried within genome concatemers and loss of viral sequences occurs (Karen *et al.*, 2009). This, in part, explains why Ad mutants lacking the proteins involved in degradation and/or mislocalization of DDR proteins are defective for growth and viral DNA replication. Conversely, effective Ad replication is associated with minimal activation of DDR pathways. Besides the slight activation of ATM kinase and phosphorylation of 53BP1, the broad spectrum of proteins known as a DNA damage sensors or substrates for PIKKs remain inactive. The levels of ATM, ATR and DNA-PKcs kinases also remain unchanged, confirming the lack of, or minimal, activation of the 3 major DDR pathways regulated by PIKKs (Carson *et al.*, 2003). This is owing to the functions of several Ad proteins, which interfere with the initial steps of DDR signalling pathways by mislocalizing or degrading the factors involved in sensing of DNA damage.

Adenovirus proteins target the MRN sensor of DNA damage

During infection with wild-type Ad, the MRN complex becomes degraded or mislocalized. Perturbation of the MRN complex is mediated by E1B55k and E4 gene products: E4orf3 and E4orf6. E1B55k forms a complex with E4orf6 capable of assembling a Cullin5 ubiquitin ligase complex targeting MRN for proteasomal degradation, whereas E4orf3 immobilizes the MRN complex and induces its mislocalization into track-like structures and cytoplasmic aggresomes (Stracker *et al.*, 2002, 2005; Carson *et al.*, 2003; Evans and Hearing, 2005). Degradation of the MRN complex via the E1B55k/E4orf6-dependent mechanism prevents ATM and ATR-signalling in response to Ad infection (Carson *et al.*, 2003, 2009). In contrast, Ad mutants lacking both E4 proteins (*dl1004*) or E4orf3 and E1B55K (*dl1016*) have no effects on MRN function, induce a robust DDR, form concatemers and are defective in viral replication, as well as late gene expression. Time point analysis of wild-type Ad infection indicates that the MRN complex is inactivated and redistributed via E4orf3 at an early time post-infection (6h) and before the initiation of viral DNA replication (Evans and Hearing, 2005; Karen *et al.*, 2009). E4orf6/E1B-55k-induced degradation of the MRN complex occurs later, but prior to viral DNA accumulation. Recent data indicates that the incoming viral genome is not recognized by the MRN complex owing to the Ad core protein VII (Karen and Hearing, 2011).

Although E4-deleted Ad mutants exhibit a defect in viral replication, which can be suppressed by inactivation of the MRN complex, this defect surprisingly does not appear to be caused by the genome concatemerization or even the DDR activation (Lakdawala *et al.*, 2008). Mutant analysis of Nbs1 indicates that there is a C-terminal sequence required for suppression of E4-mutant Ad replication. How this C-terminal Nbs1 sequence, known to bind ATM and Mre11, works mechanistically to inhibit viral replication has not been deduced. It may involve the Nbs1-dependent physical association of Mre11 with the E4 mutant viral genome, which has been demonstrated by chromatin immunoprecipitation (Mathew and Bridge, 2008). How the MRN complex recognizes the viral genome also remains uncertain, but it appears to be distinct from sensing of a cellular DSB (Lakdawala *et al.*, 2008).

While deregulation of the MRN complex offers significant insight to the Ad infection strategy and general principles of tumour virus targeting, it also yields fundamental insight to cellular DDR and checkpoint mechanisms operating in a non-viral setting. Adenovirus was used as a tool to inactivate the MRN complex and analyse the cellular consequences (Carson *et al.*, 2003). Thus, it was found that E1B55k/E4orf6-mediated degradation or the Mre11 complex caused a failure to mount a DDR and G2 checkpoint responses after IR (Carson *et al.*, 2003). Furthermore, restoration of Mre11 in A-TLD1 cells, an Mre11 hypomorphic cell line, allowed normal DDR signalling in either E4 mutant Ad infected, or IR-treated, cells. Consequently, Mre11 was demonstrated to be required for activation of ATM and the G2 checkpoint.

Although the regulatory and sensory role of MRN was originally considered to be limited to ATM signalling, more recent observations suggest that ATR signalling is also impacted (Lee and Paull, 2005; Jazayeri *et al.*, 2006; Stiff *et al.*, 2006; Olson *et al.*, 2007). Accordingly, deregulation of the MRN complex by Ad infection also affects activation of the ATR signalling cascade (Carson *et al.*, 2009). Infection with an E4 mutant Ad5 induces ATR signalling in a manner that depends on the MRN complex. Conversely, expression of Ad5 E4orf3 reduced ATR, but not ATM, signalling by mislocalization and immobilization of the MRN complex. It remains unclear why Ad bothers to disable ATR signalling. For some Ad serotypes, ATR is activated late in infection via E1B55k association with the E1B-AP5 host protein (Blackford *et al.*, 2008). While E4orf3 from all serotypes formed tracks and relocalized the PML protein into these, only subgroup C viruses (Ad1, Ad2, Ad5) mislocalized MRN (Carson *et al.*, 2009). This might be because of redundancy with E1B55k/E4orf6, which degrades MRN. Strikingly, Ad5 E4orf3 also prevents ATR signalling induced by external sources of damage such as laser microirradiation or hydroxyurea treatment, which causes stalling of replication forks.

Interestingly, Ad12 uses a different strategy to eliminate ATR signalling (Blackford *et al.*, 2010). Whereas E1B55k normally acts as a putative substrate receptor and E4orf6 as the adaptor that recruits the cullin complex, Ad12 E4orf6 both recruits a cullin 2-based ubiquitin ligase complex and the TOPBP1 activator of ATR. This results in targeted proteasomal degradation of TOPBP1 and concurrent inhibition of ATR signalling following Ad12 infection or replication stress.

Adenovirus proteins target p53

To prevent cell cycle arrest, promote G1/S transition and block premature apoptosis, Ad proteins inactivate p53. This is accomplished at multiple levels. E1B55k binds and inhibits the transcriptional activity of p53 (Yew and Berk, 1992) and translocates it into cytoplasmic aggresomes (Liu *et al.*, 2005). E1B55k together with E4orf6 induce p53 ubiquitination and proteasome-dependent degradation by recruiting a cullin 5-containing E3 ubiquitin ligase complex (Nevels *et al.*, 1997; Querido *et al.*, 1997, 2001). E1B55k, in the absence of E4orf6, also acts as an E3 SUMO ligase, which tethers the sumyolated form of p53 in PML NBs, thus rendering it non-functional (Pennella *et al.*, 2010). Finally, it was recently discovered that E4orf3 selectively turns off activation of p53 target genes via epigenetic silencing owing to H3K9me3 heterochromatin formation (Soria *et al.*, 2010).

Adenoviruses target NHEJ on multiple levels to inhibit its genome concatenation

Because the ends of the Ad genome are recognized by the host cell repair machinery as a DSB, and considering that one of the major outcomes is concatenation of the viral genomes through NHEJ, it is not surprising, that adenoviruses use various strategies to inactivate this type of repair. Initial reports indicated that both E4orf3 and E4orf6 interact with DNA-PKcs, block V(D) J recombination, which requires DNA-PKcs, and potentially prevent Ad genome concatenation (Boyer *et al.*, 1999). Expression of E4orf6 alone radiosensitizes human tumour cells, blocks DSB repair and prolongs the autophosphorylation of DNA-PKcs on T2609 (Hart *et al.*, 2005). A complementary strategy used by Ad to inhibit NHEJ is the E1B55k/E4orf6-dependent proteasomal degradation of DNA ligase IV, known to be involved in the final joining step of NHEJ repair (Baker *et al.*, 2007). The final step of NHEJ can be additionally prevented by E4orf6 through dissociation of the DNA ligase IV/XCCR4 complex (Jayaram *et al.*, 2008). The relative contributions of these various mechanisms, versus that of MRN targeting, to prevent genome concatenation, is not clear.

Is interference with DNA repair by adenovirus linked to transformation?

It can be envisaged that inhibition of DNA repair mechanisms by Ad would promote genomic instability, which in turn could drive malignant transformation. There is not yet a significant body of literature to definitively validate this concept. However, one report concludes that Ad E1A can cooperate with E4orf6 or E4orf3 to transform primary rat cells by a 'hit and run' mechanism (Nevels *et al.*, 2001). In this scenario, transformation is initiated by a viral oncoprotein through an initial 'hit', but maintenance of the transformed state is associated with its loss ('run'), probably owing to cell deleterious effects. Unlike cells transformed by E1A together with E1B55k, which maintain expression of both oncoproteins, transformants from E1A combinations with either E4orf3 or E4orf6 often fail to express both proteins. Coexpression of E1A with E4orf3 or E4orf6 was correlated with an increased mutation frequency in the HPRT assay. However, at this point, it has not been established if the observed effects in transformation are really due to inhibition of repair mechanisms, and even if so, what the relevant cellular targets may be.

Human papillomavirus

HPV life cycle, cancer and links to genome instability

Human papillomaviruses (HPV) are small DNA tumour viruses with a pronounced tropism for the squamous epithelium. HPVs induce chromosomal instability, deregulate the DDR and alter cell cycle control mechanisms to achieve their own genome replication and induce hyperproliferative lesions in mucosa and skin. Since HPV productive infection occurs in epithelial cells from suprabasal layers, which are composed of terminally differentiated cells that have exited the cell cycle, HPV has evolved mechanisms to induce re-entry of host cells into S-phase and therefore gain access to cellular replication factors. To achieve this goal, HPV primarily utilizes the E6 and E7 oncoproteins, which have been shown to play an essential role in HPV-induced carcinogenesis (McLaughlin-Drubin and Munger, 2009a,b). High-risk HPVs are associated with cervical carcinoma, while low-risk viruses are causally linked with genital warts (reviewed in Moody and Laimins, 2010). E6 and E7 can cooperate to immortalize a variety of cell types including human foreskin keratinocytes and baby rat kidney cells (Hawley-Nelson *et al.*, 1989; Munger *et al.*, 1989a). Expression of E7 alone is sufficient to induce cervical dysplasia and invasive malignancies in transgenic mouse models (Riley *et al.*, 2003). E6, while not sufficient, can synergize with the E7 protein to transform cells and induce malignancies in transgenic mice. However, immortalized cells were found to be non-tumorigenic in nude mouse models, indicating that E6 and E7 proteins are not sufficient to fully transform cells (Kaur and McDougall, 1988). To achieve this, additional factors are required, for example prolonged passaging of cells in culture or expression of oncogenes such as v-Ras or v-Fos (Durst *et al.*, 1989; Pei *et al.*, 1993). This is consistent with a prolonged period of latency between HPV infection and development of malignancies (Schiffman *et al.*, 2007). During this period, HPV-infected cells can gradually gain additional mutations and acquire genomic instability, which is characterized by chromosomal rearrangements that include gains and losses of whole chromosomes, leading to aneuploidy. However, initial signs of genomic abnormalities can be detected in precancerous lesions, even before integration of HPV into the host genome (Duensing *et al.*, 2001). This indicates that HPV-infected cells develop chromosome instabilities long before acquiring the tumour phenotype. Signs of DDR activation are also detectable at the earliest steps of tumorigenesis (Bartkova *et al.*, 2005) and may reflect the activation of a tumour suppressor mechanism to protect cells from acquisition and maintenance of genomic instability, which is a common feature of various types of malignancies.

The current view of genomic instability induction in HPV-associated malignancies posits that HPV oncoproteins alone or in cooperation with other oncogenes induce DNA damage and therefore activate a DDR to ensure DNA repair. However, DDR pathways become deregulated on multiple levels owing to HPV oncoprotein interactions with a broad spectrum of cellular factors involved in sensing and transducing signals from DNA lesions, cell cycle checkpoint control and DNA repair. The common outcome of these complex interactions between viral and host proteins is an attenuation of DNA repair mechanisms, followed by acquisition and maintenance of genome instability, which can be a driving force in HPV-associated tumorigenesis.

Here we will focus on mechanisms used by HPV oncoproteins to deregulate the DDR and induce genome instability. We will also discuss possible consequences of DDR activation for HPV amplification and HPV-associated transformation.

HPV E6 and E7 induce mitotic aberrations, DNA damage and DNA damage response activation

As previously underscored, HPV-associated genomic instability results from expression of the E6 and E7 oncoproteins, both of which can induce PARP and DNA breaks in keratinocytes as determined by a comet assay (Duensing and Munger, 2002). Signs of host genome destabilization include abnormal centrosome numbers, multipolar mitotic spindles, anaphase bridges, lagging chromosomes and aneuploidy. These chromosomal abnormalities may in part arise because of abnormal centrosome numbers, due to centriole overduplication caused by E7 (Duensing *et al.*, 2007). However, E6 can also interfere with single-strand break repair by binding to the XRCC1 protein, thus exerting mutagenic activity (Iftner *et al.*, 2002). E7 expression induces γ-H2AX foci, modest increases in p53 levels and S15 phosphorylation, as well as an increase in Cdk1 Y15 phosphorylation, effects that are consistent with checkpoint activation (Duensing and Munger, 2002). It was also demonstrated that adenoviral expression of E7 induces the ATM/Chk2 pathway via pRB binding and release of free E2F1 (Rogoff *et al.*, 2004).

DNA damage responses are important for genome amplification in differentiating cells

Exactly how ATM is activated by E7 remains unclear, but recent findings have shed more light on this important aspect. Episomal expression of HPV31 activates an ATM response, leading to phosphorylation of CHK2, BRCA1 and NBS1 in differentiating keratinocytes (Moody and Laimins, 2009). Importantly, ATM inhibition specifically decreased viral genome amplification in differentiated keratinocytes and blocked formation of HPV replication foci, while episomal maintenance did not require ATM signalling. E7 was found to induce pATM S1981 and associated with this species. Interestingly, CHK2 was also activated and was required for caspase activation, which is known to promote genome amplification through cleavage of the E1 protein (Moody *et al.*, 2007). Taken together, HPV exploits the ATM branch of the DDR to drive productive replication in differentiating cells.

Currently, there is limited mechanistic insight into how DDR activation specifically affects HPV replication. It has been reported that HPV replication, mediated by the E1 and E2 proteins, is not arrested by DNA damage signalling *in vitro* or *in vivo*. This is in contrast to SV40 replication, which, it was found, can be inhibited owing to ATR phosphorylation of LT (King *et al.*, 2010). Interestingly, the E2 protein binds TOPBP1, an activator of the ATR pathway, and this enhances E2's ability to activate transcription and replication (Boner *et al.*, 2002). Notably, HPV replication is believed to occur in a prolonged G2 phase, which results from E7 induction of cytoplasmic cyclin B/cdk1 and inactivating phosphorylations at T14 and Y15, consistent with checkpoint arrest (Wang *et al.*, 2009; Banerjee *et al.*, 2011). The rationale for the G2 arrest, a phenomenon observed by several viruses following infection, may be that it creates an 'uncontested window of opportunity', where viral replication needs not compete with cellular replication forks (Wang *et al.*, 2009).

Genomic instability driven by integrated HPV sequences

Intriguingly, expression of the HPV replication factors E1 and E2 induced overamplification of an integrated viral origin through 'onion skin' type replication (Kadaja *et al.*, 2007). This leads to rearrangements of the integrated HPV sequence as well as the flanking cellular sequences, thus driving a form of genomic instability that is probably relevant for HPV carcinogenesis, where an integrated viral genome coexists with episomal copies of the genome. Replication of the integrated sequences occurs in nuclear foci positive for E1, E2, the MRN complex, Ku70/80, ATM, Chk2 and ATRIP, consistent with recruitment of the repair and recombination machinery (Kadaja *et al.*, 2009).

Mechanisms used by HPV to deregulate cell cycle, DNA damage response and checkpoint mechanisms

Despite that a DDR becomes activated upon HPV infection, and this can be beneficial for viral propagation, HPV oncoproteins use multiple strategies to deregulate the DDR, mainly through the E6 and E7 oncoproteins.

To promote cell cycle progression in the presence of DNA damage and subvert G1/S cell cycle control mechanisms, E7 interacts with the pRB family via an LXCXE motif (Dyson *et al.*, 1989; Munger *et al.*, 1989b). This disrupts pRB/E2F complexes and derepresses E2F target genes required for the G1/S cell cycle transition. E7 associates with the cullin 2 ubiquitin ligase complex, providing a probable mechanism for ubiquitin-mediated proteasomal degradation of pRB (Huh *et al.*, 2007). Other 'pocket proteins', p107 and p130, are also targeted by E7 via the LXCXE motif. High-risk E7 proteins bind to Rb family members with increased affinity compared with low-risk E7 (Gage *et al.*, 1990). The major outcome of HPV16 E7 interactions with pocket proteins is deregulation of cell cycle, leading to S phase re-entry, thus providing a favourable cellular milieu, where HPV genomes can be amplified.

E7 furthermore attenuates G1/S checkpoint control by promoting Cdk2 activity. This may be due to the up-regulation of the Cdc25A phosphatase, which is a known Cdk2 activator required for G1/S transition (Nguyen *et al.*, 2002). Moreover, to overcome cell cycle inhibition signals mediated by the CDK inhibitors $p21^{CIP1}$ and $p27^{KIP1}$, E7 binds and neutralizes these (Zerfass-Thome *et al.*, 1996; Jones *etal.*, 1997; Zehbe *et al.*, 1999).

To attenuate cell cycle checkpoints and allow mitotic progression under DNA damage condition, E7 also induces accelerated degradation of claspin (Spardy *et al.*, 2009). Claspin is normally involved in recovery from the G2 checkpoint (Mailand *et al.*, 2006), thus suggesting that E7 exploits this mechanism to overcome checkpoint restrictions. The effect on claspin may involve β-TrCP/SCF, the ubiquitin ligase complex associated with claspin degradation (Peschiaroli *et al.*, 2006), and kinases PLK1 and aurora that are also implicated in checkpoint recovery.

As emphasized, E7 initiates unscheduled DNA synthesis in terminally differentiated cells and simultaneously attenuates cell cycle checkpoints. Together, this contributes to the maintenance of DNA damage, which cannot be repaired properly, and therefore an apoptotic programme is triggered (Eichten *et al.*, 2004). To prevent this undesirable outcome of apoptotic cell death, E6 inactivates several key regulators of the apoptotic programme such as p53, caspase 8, Bak, receptor for tumour necrosis factor (TNFR) and Fas-associated death domain protein (FADD) (reviewed in Moody and Laimins, 2010).

E6 inactivates the p53 tumour suppressor via multiple different mechanisms. High-risk E6 proteins, by co-opting the cellular E6AP ligase, induce p53 degradation via ubiquitin-mediated proteasomal degradation (Scheffner *et al.*, 1990, 1993). E6 can also suppress p53 activity via interaction with histone acetyltransferases CBP, p300 and hADA3. E6 binding to p300 inhibits p53 acetylation at p300-dependent sites and significantly reduces p53 transcriptional activity (Zimmermann *et al.*, 1999). Similarly, the interaction of E6 with hADA3 induces its degradation and blocks p53 activation (Kumar *et al.*, 2002).

HPV links with the Fanconi anaemia machinery

FA patients may be more susceptible to head and neck squamous cell carcinomas (HNSCCs) caused by HPV (Kutler *et al.*, 2003). Interestingly, E7 activates the FA replication stress pathway, inducing FancD2 foci formation and its monoubiquitination on chromatin (Spardy *et al.*, 2007). This induction is in part dependent on the pRB-binding motif. When E7 was expressed in FancD2-deficient human fibroblasts, it accelerated chromosomal instability, suggesting that a compromised FA genetic background can promote genetic instability and perhaps cancer incidence. Indeed, later studies demonstrated that FA deficiency is promoting hyperplastic growth in organotypic raft cultures of HPV E6/E7 immortalized keratinocytes (Hoskins *et al.*, 2009). Furthermore, a study with E7 transgenic mice underscores that knockout of FancD2 increases the incidence of HNSCC after treatment with a chemical carcinogen (Park *et al.*, 2010).

Conclusion

Studies of the interplay between viruses and the DDR have yielded tremendous insight to both viral and cellular regulatory mechanisms. Viruses have illuminated critical components of both DDR and checkpoint signalling. As discussed, viruses either inactivate the DDR or exploit it for their own benefit. Inactivation is often associated with targeted degradation or mislocalization of specific DDR factors. In some instances upstream responses may have been co-opted to promote the viral life cycle, whereas the 'end points', for example checkpoint responses and apoptosis, often are blocked. It remains unclear what specific aspects of the viral life cycle dictate whether the DDR is exerting a positive or negative effect.

Frequently, viral infection is a trigger of the DDR owing to the incoming genomes or aberrant replication structures. However, expression of individual viral proteins is also capable of inducing the DDR, for example in the case of SV40 LT and HPV E7. The exact lesions that are being sensed have not been defined but may involve cellular replication stress, owing to altered firing of cellular replication origins. While some of the DDR induction by viruses may be an inevitable consequence of stimulating cell cycle entry and replicating viral genomes, it also appears that viruses deliberately trigger activation of certain pathways that can be used to facilitate virus propagation. Generation of DSBs may also occur as a mechanism to promote integration of viral genomes.

Various molecular mechanisms have been uncovered that facilitate viral replication through DDR pathways, frequently via stimulated ATM signalling. These include delay or arrest at G2 or S phase to exploit the abundance of replication factors in those cell cycle phases, intranuclear localization at DDR foci, often near PML NBs, which may allow recruitment of needed repair factors, and regulatory phosphorylation of the viral initiator protein or other components of the replication machinery. It is probable that additional mechanisms will emerge as our knowledge becomes more refined. Insight to these important regulatory mechanisms will probably aid in the future rational design of antiviral therapeutics.

Several viruses induce both mitotic aberrations (micronuclei, anaphase bridges, lagging chromosomes, numerical and structural chromosome changes) and the DDR, but the exact relationship between these, if any, has not been resolved. Thus, it is unclear if one comes before the other, and if there is a causal relationship. Mis-segregation of chromosomes could conceivably trigger a DDR, but it is also probable that DDR induction causes aberrant mitosis and cytokinesis. Genomic instability can be a driving force in tumorigenesis. However, the DDR is generally viewed as an early barrier to carcinogenic progression. Exactly how the DDR modulates the outcome of virally induced transformation is likely to be complex and perhaps depend on specific branches within the DDR, for example the FA pathway.

References

Baker, A., Rohleder, K.J., Hanakahi, L.A., and Ketner, G. (2007). Adenovirus E4 34k and E1b 55k oncoproteins target host DNA ligase IV for proteasomal degradation. J. Virol. *81*, 7034–7040.

Banerjee, N.S., Wang, H.K., Broker, T.R., and Chow, L.T. (2011). Human papillomavirus (HPV) E7 induces prolonged G2 following S-Phase reentry in differentiated human keratinocytes. J. Biol. Chem. *286*, 15473–15482

Bartkova, J., Horejsi, Z., Koed, K., Kramer, A., Tort, F., Zieger, K., Guldberg, P., Sehested, M., Nesland, J.M., Lukas, C., *et al.* (2005). DNA damage response as a candidate anti-cancer barrier in early human tumorigenesis. Nature *434*, 864–870.

Bernardi, R., and Pandolfi, P.P. (2007). Structure, dynamics and functions of promyelocytic leukaemia nuclear bodies. Nat. Rev. Mol. Cell. Biol. *8*, 1006–1016.

Blackford, A.N., Bruton, R.K., Dirlik, O., Stewart, G.S., Taylor, A.M., Dobner, T., Grand, R.J., and Turnell, A.S. (2008). A role for E1B-AP5 in ATR signalling pathways during adenovirus infection. J. Virol. *82*, 7640–7652.

Blackford, A.N., Patel, R.N., Forrester, N.A., Theil, K., Groitl, P., Stewart, G.S., Taylor, A.M., Morgan, I.M., Dobner, T., Grand, R.J., *et al.* (2010). Adenovirus 12 E4orf6 inhibits ATR activation by promoting TOPBP1 degradation. Proc. Natl. Acad. Sci. U.S.A. *107*, 12251–12256.

Bocchetta, M., Eliasz, S., De Marco, M.A., Rudzinski, J., Zhang, L., and Carbone, M. (2008). The SV40 large T antigen–p53 complexes bind and activate the insulin-like growth factor-I promoter stimulating cell growth. Cancer Res. *68*, 1022–1029.

Boe, S.O., Haave, M., Jul-Larsen, A., Grudic, A., Bjerkvig, R., and Lonning, P.E. (2006). Promyelocytic leukemia nuclear bodies are predetermined processing sites for damaged DNA. J. Cell Sci. *119*, 3284–3295.

Boichuk, S., Hu, L., Hein, J., and Gjoerup, O.V. (2010). Multiple DNA damage signalling and repair pathways deregulated by simian virus 40 large T antigen. J. Virol. *84*, 8007–8020.

Boner, W., Taylor, E.R., Tsirimonaki, E., Yamane, K., Campo, M.S., and Morgan, I.M. (2002). A Functional interaction between the human papillomavirus 16 transcription/replication factor E2 and the DNA damage response protein TopBP1. J. Biol. Chem. *277*, 22297–22303.

Boyer, J., Rohleder, K., and Ketner, G. (1999). Adenovirus E4 34k and E4 11k inhibit double strand break repair and are physically associated with the cellular DNA-dependent protein kinase. Virology *263*, 307–312.

Carson, C.T., Schwartz, R.A., Stracker, T.H., Lilley, C.E., Lee, D.V., and Weitzman, M.D. (2003). The Mre11 complex is required for ATM activation and the G2/M checkpoint. EMBO J. *22*, 6610–6620.

Carson, C.T., Orazio, N.I., Lee, D.V., Suh, J., Bekker-Jensen, S., Araujo, F.D., Lakdawala, S.S., Lilley, C.E., Bartek, J., Lukas, J., *et al.* (2009). Mislocalization of the MRN complex prevents ATR signalling during adenovirus infection. EMBO J. *28*, 652–662.

Carvalho, T., Seeler, J.S., Ohman, K., Jordan, P., Pettersson, U., Akusjarvi, G., Carmo-Fonseca, M., and Dejean, A. (1995). Targeting of adenovirus E1A and E4-ORF3 proteins to nuclear matrix-associated PML bodies. J. Cell. Biol. *131*, 45–56.

Chang, T.H., Ray, F.A., Thompson, D.A., and Schlegel, R. (1997). Disregulation of mitotic checkpoints and regulatory proteins following acute expression of SV40 large T antigen in diploid human cells. Oncogene *14*, 2383–2393.

Ciccia, A., and Elledge, S.J. (2010). The DNA damage response: making it safe to play with knives. Mol. Cell *40*, 179–204.

Dahl, J., You, J., and Benjamin, T.L. (2005). Induction and utilization of an ATM signalling pathway by polyomavirus. J. Virol. *79*, 13007–13017.

Darbinyan, A., Siddiqui, K.M., Slonina, D., Darbinian, N., Amini, S., White, M.K., and Khalili, K. (2004). Role of JC virus agnoprotein in DNA repair. J. Virol. *78*, 8593–8600.

Darbinyan, A., White, M.K., Akan, S., Radhakrishnan, S., Del Valle, L., Amini, S., and Khalili, K. (2007). Alterations of DNA damage repair pathways resulting from JCV infection. Virology *364*, 73–86.

Dellaire, G., and Bazett-Jones, D.P. (2004). PML nuclear bodies: dynamic sensors of DNA damage and cellular stress. Bioessays *26*, 963–977.

Dellaire, G., Ching, R.W., Ahmed, K., Jalali, F., Tse, K.C., Bristow, R.G., and Bazett-Jones, D.P. (2006). Promyelocytic leukemia nuclear bodies behave as DNA damage sensors whose response to DNA double-strand breaks is regulated by NBS1 and the kinases ATM, Chk2, and ATR. J. Cell. Biol. *175*, 55–66.

Deppert, W., Steinmayer, T., and Richter, W. (1989). Cooperation of SV40 large T antigen and the cellular protein p53 in maintenance of cell transformation. Oncogene *4*, 1103–1110.

Dey, D., Dahl, J., Cho, S., and Benjamin, T.L. (2002). Induction and bypass of p53 during productive infection by polyomavirus. J. Virol. *76*, 9526–9532.

Duensing, S., Duensing, A., Crum, C.P., and Munger, K. (2001). Human papillomavirus type 16 E7 oncoprotein-induced abnormal centrosome synthesis is an early event in the evolving malignant phenotype. Cancer Res. *61*, 2356–2360.

Duensing, S., and Munger, K. (2002). The human papillomavirus type 16 E6 and E7 oncoproteins independently induce numerical and structural chromosome instability. Cancer Res. *62*, 7075–7082.

Duensing, A., Liu, Y., Perdreau, S.A., Kleylein-Sohn, J., Nigg, E.A., and Duensing, S. (2007). Centriole overduplication through the concurrent formation of multiple daughter centrioles at single maternal templates. Oncogene *26*, 6280–6288.

Durst, M., Gallahan, D., Jay, G., and Rhim, J.S. (1989). Glucocorticoid-enhanced neoplastic transformation of human keratinocytes by human papillomavirus type 16 and an activated ras oncogene. Virology *173*, 767–771.

Dyson, N., Howley, P.M., Munger, K., and Harlow, E. (1989). The human papilloma virus-16 E7 oncoprotein is able to bind to the retinoblastoma gene product. Science *243*, 934–937.

Eichten, A., Rud, D.S., Grace, M., Piboonniyom, S.O., Zacny, V., and Munger, K. (2004). Molecular pathways executing the 'trophic sentinel' response in HPV16 E7-expressing normal human diploid fibroblasts upon growth factor deprivation. Virology *319*, 81–93.

Evans, J.D., and Hearing, P. (2005). Relocalization of the Mre11–Rad50-Nbs1 complex by the adenovirus E4 ORF3 protein is required for viral replication. J. Virol. *79*, 6207–6215.

Everett, R.D. (2006). Interactions between DNA viruses, ND10 and the DNA damage response. Cell. Microbiol. *8*, 365–374.

Everett, R.D., and Maul, G.G. (1994). HSV-1 IE protein Vmw110 causes redistribution of PML. EMBO J. *13*, 5062–5069.

Gage, J.R., Meyers, C., and Wettstein, F.O. (1990). The E7 proteins of the nononcogenic human papillomavirus type 6b (HPV-6b) and of the oncogenic HPV16 differ in retinoblastoma protein binding and other properties. J. Virol. *64*, 723–730.

Gjoerup, O.V., Wu, J., Chandler-Militello, D., Williams, G.L., Zhao, J., Schaffhausen, B., Jat, P.S., and Roberts, T.M. (2007). Surveillance mechanism linking Bub1 loss to the p53 pathway. Proc. Natl. Acad. Sci. U.S.A. *104*, 8334–8339.

Hart, L.S., Yannone, S.M., Naczki, C., Orlando, J.S., Waters, S.B., Akman, S.A., Chen, D.J., Ornelles, D., and Koumenis, C. (2005). The adenovirus E4orf6 protein inhibits DNA double strand break repair and radiosensitizes human tumour cells in an E1B-55K-independent manner. J. Biol. Chem. *280*, 1474–1481.

Hawley-Nelson, P., Vousden, K.H., Hubbert, N.L., Lowy, D.R., and Schiller, J.T. (1989). HPV16 E6 and E7 proteins cooperate to immortalize human foreskin keratinocytes. EMBO J. *8*, 3905–3910.

Hein, J., Boichuk, S., Wu, J., Cheng, Y., Freire, R., Jat, P.S., Roberts, T.M., and Gjoerup, O.V. (2009). Simian virus 40 large T antigen disrupts genome integrity and activates a DNA damage response via Bub1 binding. J. Virol. *83*, 117–127.

Hermannstadter, A., Ziegler, C., Kuhl, M., Deppert, W., and Tolstonog, G.V. (2009). Wild-type p53 enhances efficiency of simian virus 40 large-T-antigen-induced cellular transformation. J. Virol. *83*, 10106–10118.

Hoskins, E.E., Morris, T.A., Higginbotham, J.M., Spardy, N., Cha, E., Kelly, P., Williams, D.A., Wikenheiser-Brokamp, K.A., Duensing, S., and Wells, S.I. (2009). Fanconi anemia deficiency stimulates HPV-associated hyperplastic growth in organotypic epithelial raft culture. Oncogene *28*, 674–685.

Huen, M.S., and Chen, J. (2009). Assembly of checkpoint and repair machineries at DNA damage sites. Trends Biochem. Sci. *35*, 101–108.

Huh, K., Zhou, X., Hayakawa, H., Cho, J.Y., Libermann, T.A., Jin, J., Harper, J.W., and Munger, K. (2007). Human papillomavirus type 16 E7 oncoprotein associates with the cullin 2 ubiquitin ligase complex, which contributes to degradation of the retinoblastoma tumour suppressor. J. Virol. *81*, 9737–9747.

Iftner, T., Elbel, M., Schopp, B., Hiller, T., Loizou, J.I., Caldecott, K.W., and Stubenrauch, F. (2002). Interference of papillomavirus E6 protein with single-strand break repair by interaction with XRCC1. EMBO J. *21*, 4741–4748.

Ishov, A.M., and Maul, G.G. (1996). The periphery of nuclear domain 10 (ND10) as site of DNA virus deposition. J. Cell. Biol. *134*, 815–826.

Jackson, S.P., and Bartek, J. (2009). The DNA-damage response in human biology and disease. Nature *461*, 1071–1078.

Jasin, M., de Villiers, J., Weber, F., and Schaffner, W. (1985). High frequency of homologous recombination in mammalian cells between endogenous and introduced SV40 genomes. Cell *43*, 695–703.

Jayaram, S., Gilson, T., Ehrlich, E.S., Yu, X.F., Ketner, G., and Hanakahi, L. (2008). E1B 55k-independent dissociation of the DNA ligase IV/XRCC4 complex by E4 34k during adenovirus infection. Virology *382*, 163–170.

Jazayeri, A., Falck, J., Lukas, C., Bartek, J., Smith, G.C., Lukas, J., and Jackson, S.P. (2006). ATM- and cell cycle-dependent regulation of ATR in response to DNA double-strand breaks. Nat. Cell. Biol. *8*, 37–45.

Jones, D.L., Alani, R.M., and Munger, K. (1997). The human papillomavirus E7 oncoprotein can uncouple cellular differentiation and proliferation in human keratinocytes by abrogating p21Cip1-mediated inhibition of cdk2. Genes Dev. *11*, 2101–2111.

Kadaja, M., Isok-Paas, H., Laos, T., Ustav, E., and Ustav, M. (2009). Mechanism of genomic instability in cells infected with the high-risk human papillomaviruses. PLoS Pathog. *5*, e1000397.

Kadaja, M., Sumerina, A., Verst, T., Ojarand, M., Ustav, E., and Ustav, M. (2007). Genomic instability of the host cell induced by the human papillomavirus replication machinery. EMBO J. *26*, 2180–2191.

Karen, K.A., and Hearing, P. (2011). Adenovirus Core Protein VII Protects the Viral Genome from a DNA Damage Response at Early Times after Infection. J. Virol. *85*, 4135–4142

Karen, K.A., Hoey, P.J., Young, C.S., and Hearing, P. (2009). Temporal regulation of the Mre11–Rad50–Nbs1 complex during adenovirus infection. J. Virol. *83*, 4565–4573.

Kaur, P., and McDougall, J.K. (1988). Characterization of primary human keratinocytes transformed by human papillomavirus type 18. J. Virol. *62*, 1917–1924.

King, L.E., Fisk, J.C., Dornan, E.S., Donaldson, M.M., Melendy, T., and Morgan, I.M. (2010). Human papillomavirus E1 and E2 mediated DNA replication is not arrested by DNA damage signalling. Virology *406*, 95–102.

Kumar, A., Zhao, Y., Meng, G., Zeng, M., Srinivasan, S., Delmolino, L.M., Gao, Q., Dimri, G., Weber, G.F., Wazer, D.E., *et al.* (2002). Human papillomavirus oncoprotein E6 inactivates the transcriptional coactivator human ADA3. Mol. Cell. Biol. *22*, 5801–5812.

Kutler, D.I., Wreesmann, V.B., Goberdhan, A., Ben-Porat, L., Satagopan, J., Ngai, I., Huvos, A.G., Giampietro, P., Levran, O., Pujara, K., *et al.* (2003). Human papillomavirus DNA and p53 polymorphisms in squamous cell carcinomas from Fanconi anemia patients. J. Natl. Cancer Inst. *95*, 1718–1721.

Lakdawala, S.S., Schwartz, R.A., Ferenchak, K., Carson, C.T., McSharry, B.P., Wilkinson, G.W., and Weitzman, M.D. (2008). Differential requirements of the C terminus of Nbs1 in suppressing adenovirus DNA replication and promoting concatemer formation. J. Virol. *82*, 8362–8372.

Lee, J.H., and Paull, T.T. (2005). ATM activation by DNA double-strand breaks through the Mre11–Rad50–Nbs1 complex. Science *308*, 551–554.

Lehman, J.M., Laffin, J., and Friedrich, T.D. (2000). Simian virus 40 induces multiple S phases with the majority of viral DNA replication in the G2 and second S phase in CV-1 cells. Exp. Cell Res. *258*, 215–222.

Liu, J.M., Poiley, J., Devetten, M., Kajigaya, S., and Walsh, C.E. (1996). The Fanconi anemia complementation group C gene (FAC) suppresses transformation of mutant fibroblasts by the SV40 virus. Biochem. Biophys. Res. Commun. *223*, 685–690.

Liu, Y., Shevchenko, A., and Berk, A.J. (2005). Adenovirus exploits the cellular aggresome response to accelerate inactivation of the MRN complex. J. Virol. *79*, 14004–14016.

Mailand, N., Bekker-Jensen, S., Bartek, J., and Lukas, J. (2006). Destruction of Claspin by SCFbetaTrCP restrains Chk1 activation and facilitates recovery from genotoxic stress. Mol Cell *23*, 307–318.

Mathew, S.S., and Bridge, E. (2008). Nbs1-dependent binding of Mre11 to adenovirus E4 mutant viral DNA is important for inhibiting DNA replication. Virology *374*, 11–22.

McLaughlin-Drubin, M.E., and Munger, K. (2009a). Oncogenic activities of human papillomaviruses. Virus Res. *143*, 195–208.

McLaughlin-Drubin, M.E., and Munger, K. (2009b). The human papillomavirus E7 oncoprotein. Virology *384*, 335–344.

Moody, C.A., and Laimins, L.A. (2009). Human papillomaviruses activate the ATM DNA damage pathway for viral genome amplification upon differentiation. PLoS Pathog. *5*, e1000605.

Moody, C.A., and Laimins, L.A. (2010). Human papillomavirus oncoproteins: pathways to transformation. Nat. Rev. Cancer *10*, 550–560.

Moody, C.A., Fradet-Turcotte, A., Archambault, J., and Laimins, L.A. (2007). Human papillomaviruses activate caspases upon epithelial differentiation to induce viral genome amplification. Proc. Natl. Acad. Sci. U.S.A. *104*, 19541–19546.

Munger, K., Phelps, W.C., Bubb, V., Howley, P.M., and Schlegel, R. (1989a). The E6 and E7 genes of the human papillomavirus type 16 together are necessary and sufficient for transformation of primary human keratinocytes. J. Virol. *63*, 4417–4421.

Munger, K., Werness, B.A., Dyson, N., Phelps, W.C., Harlow, E., and Howley, P.M. (1989b). Complex formation of human papillomavirus E7 proteins with the retinoblastoma tumour suppressor gene product. EMBO J. *8*, 4099–4105.

Nevels, M., Rubenwolf, S., Spruss, T., Wolf, H., and Dobner, T. (1997). The adenovirus E4orf6 protein can promote E1A/E1B-induced focus formation by interfering with p53 tumour suppressor function. Proc. Natl. Acad. Sci. U.S.A. *94*, 1206–1211.

Nevels, M., Tauber, B., Spruss, T., Wolf, H., and Dobner, T. (2001). 'Hit-and-run' transformation by adenovirus oncogenes. J. Virol. *75*, 3089–3094.

Nguyen, D.X., Westbrook, T.F., and McCance, D.J. (2002). Human papillomavirus type 16 E7 maintains elevated levels of the cdc25A tyrosine phosphatase during deregulation of cell cycle arrest. J. Virol. *76*, 619–632.

Nichols, W.W., Bradt, C.I., Toji, L.H., Godley, M., and Segawa, M. (1978). Induction of sister chromatid exchanges by transformation with simian virus 40. Cancer Res. *38*, 960–964.

Okubo, E., Lehman, J.M., and Friedrich, T.D. (2003). Negative regulation of mitotic promoting factor by the checkpoint kinase chk1 in simian virus 40 lytic infection. J. Virol. *77*, 1257–1267.

Olson, E., Nievera, C.J., Lee, A.Y., Chen, L., and Wu, X. (2007). The Mre11–Rad50-Nbs1 complex acts both upstream and downstream of ataxia telangiectasia mutated and Rad3-related protein (ATR) to regulate the S-phase checkpoint following UV treatment. J. Biol. Chem. *282*, 22939–22952.

Orba, Y., Suzuki, T., Makino, Y., Kubota, K., Tanaka, S., Kimura, T., and Sawa, H. (2010). Large T antigen promotes JC virus replication in G2-arrested cells by inducing ATM- and ATR-mediated G2 checkpoint signalling. J. Biol. Chem. *285*, 1544–1554.

Park, J.W., Pitot, H.C., Strati, K., Spardy, N., Duensing, S., Grompe, M., and Lambert, P.F. (2010). Deficiencies in the Fanconi anemia DNA damage response pathway increase sensitivity to HPV-associated head and neck cancer. Cancer Res. *70*, 9959–9968.

Pei, X.F., Meck, J.M., Greenhalgh, D., and Schlegel, R. (1993). Cotransfection of HPV-18 and v-fos DNA induces tumorigenicity of primary human keratinocytes. Virology *196*, 855–860.

Pennella, M.A., Liu, Y., Woo, J.L., Kim, C.A., and Berk, A.J. (2010). Adenovirus E1B 55-kilodalton protein is a p53-SUMO1 E3 ligase that represses p53 and stimulates its nuclear export through interactions with promyelocytic leukemia nuclear bodies. J. Virol. *84*, 12210–12225.

Peschiaroli, A., Dorrello, N.V., Guardavaccaro, D., Venere, M., Halazonetis, T., Sherman, N.E., and Pagano, M. (2006). SCFbetaTrCP-mediated degradation of Claspin regulates recovery from the DNA replication checkpoint response. Mol. Cell *23*, 319–329.

Petrini, J.H., and Stracker, T.H. (2003). The cellular response to DNA double-strand breaks: defining the sensors and mediators. Trends Cell. Biol. *13*, 458–462.

Querido, E., Blanchette, P., Yan, Q., Kamura, T., Morrison, M., Boivin, D., Kaelin, W.G., Conaway, R.C., Conaway, J.W., and Branton, P.E. (2001). Degradation of p53 by adenovirus E4orf6 and E1B55K proteins occurs via a novel mechanism involving a Cullin–containing complex. Genes Dev. *15*, 3104–3117.

Querido, E., Marcellus, R.C., Lai, A., Charbonneau, R., Teodoro, J.G., Ketner, G., and Branton, P.E. (1997). Regulation of p53 levels by the E1B 55-kilodalton protein and E4orf6 in adenovirus-infected cells. J. Virol. *71*, 3788–3798.

Ray, F.A., Meyne, J., and Kraemer, P.M. (1992). SV40 T antigen induced chromosomal changes reflect a process that is both clastogenic and aneuploidogenic and is ongoing throughout neoplastic progression of human fibroblasts. Mutat. Res. *284*, 265–273.

Riley, R.R., Duensing, S., Brake, T., Munger, K., Lambert, P.F., and Arbeit, J.M. (2003). Dissection of human papillomavirus E6 and E7 function in transgenic mouse models of cervical carcinogenesis. Cancer Res. *63*, 4862–4871.

Rogoff, H.A., Pickering, M.T., Frame, F.M., Debatis, M.E., Sanchez, Y., Jones, S., and Kowalik, T.F. (2004). Apoptosis associated with deregulated E2F activity is dependent on E2F1 and Atm/Nbs1/Chk2. Mol. Cell. Biol. *24*, 2968–2977.

Rohaly, G., Korf, K., Dehde, S., and Dornreiter, I. (2010). Simian virus 40 activates ATR-Delta p53 signalling to override cell cycle and DNA replication control. J. Virol. *84*, 10727–10747.

Scarano, F.J., Laffin, J.A., Lehman, J.M., and Friedrich, T.D. (1994). Simian virus 40 prevents activation of M-phase-promoting factor during lytic infection. J. Virol. *68*, 2355–2361.

Scheffner, M., Werness, B.A., Huibregtse, J.M., Levine, A.J., and Howley, P.M. (1990). The E6 oncoprotein encoded by human papillomavirus types 16 and 18 promotes the degradation of p53. Cell *63*, 1129–1136.

Scheffner, M., Huibregtse, J.M., Vierstra, R.D., and Howley, P.M. (1993). The HPV16 E6 and E6-AP complex functions as a ubiquitin-protein ligase in the ubiquitination of p53. Cell *75*, 495–505.

Schiffman, M., Castle, P.E., Jeronimo, J., Rodriguez, A.C., and Wacholder, S. (2007). Human papillomavirus and cervical cancer. Lancet *370*, 890–907.

Schneider, J., and Fanning, E. (1988). Mutations in the phosphorylation sites of simian virus 40 (SV40) T antigen alter its origin DNA-binding specificity for sites I or II and affect SV40 DNA replication activity. J. Virol. *62*, 1598–1605.

Shi, Y., Dodson, G.E., Shaikh, S., Rundell, K., and Tibbetts, R.S. (2005). Ataxia-telangiectasia-mutated (ATM) is a T-antigen kinase that controls SV40 viral replication *in vivo*. J. Biol. Chem. *280*, 40195–40200.

Soria, C., Estermann, F.E., Espantman, K.C., and O'Shea, C.C. (2010). Heterochromatin silencing of p53 target genes by a small viral protein. Nature *466*, 1076–1081.

Spardy, N., Duensing, A., Charles, D., Haines, N., Nakahara, T., Lambert, P.F., and Duensing, S. (2007). The human papillomavirus type 16 E7 oncoprotein activates the Fanconi anemia (FA) pathway and causes accelerated chromosomal instability in FA cells. J. Virol. *81*, 13265–13270.

Spardy, N., Covella, K., Cha, E., Hoskins, E.E., Wells, S.I., Duensing, A., and Duensing, S. (2009). Human papillomavirus 16 E7 oncoprotein attenuates DNA damage checkpoint control by increasing the proteolytic turnover of claspin. Cancer Res. *69*, 7022–7029.

St-Onge, L., Bouchard, L., Laurent, S., and Bastin, M. (1990). Intrachromosomal recombination mediated by papovavirus large T antigens. J. Virol. *64*, 2958–2966.

Stewart, N., and Bacchetti, S. (1991). Expression of SV40 large T antigen, but not small t antigen, is required for the induction of chromosomal aberrations in transformed human cells. Virology *180*, 49–57.

Stiff, T., Walker, S.A., Cerosaletti, K., Goodarzi, A.A., Petermann, E., Concannon, P., O'Driscoll, M., and Jeggo, P.A. (2006). ATR-dependent phosphorylation and activation of ATM in response to UV treatment or replication fork stalling. EMBO J. *25*, 5775–5782.

Stracker, T.H., Carson, C.T., and Weitzman, M.D. (2002). Adenovirus oncoproteins inactivate the Mre11-Rad50-NBS1 DNA repair complex. Nature *418*, 348–352.

Stracker, T.H., Lee, D.V., Carson, C.T., Araujo, F.D., Ornelles, D.A., and Weitzman, M.D. (2005). Serotype-specific reorganization of the Mre11 complex by adenoviral E4orf3 proteins. J. Virol. *79*, 6664–6673.

Tang, Q., Bell, P., Tegtmeyer, P., and Maul, G.G. (2000). Replication but not transcription of simian virus 40 DNA is dependent on nuclear domain 10. J. Virol. *74*, 9694–9700.

Tavalai, N., and Stamminger, T. (2008). New insights into the role of the subnuclear structure ND10 for viral infection. Biochim. Biophys. Acta *1783*, 2207–2221.

Theile, M., and Grabowski, G. (1990). Mutagenic activity of BKV and JCV in human and other mammalian cells. Arch. Virol. *113*, 221–233.

Todaro, G.J., Green, H., and Swift, M.R. (1966). Susceptibility of human diploid fibroblast strains to transformation by SV40 virus. Science *153*, 1252–1254.

Trojanek, J., Croul, S., Ho, T., Wang, J.Y., Darbinyan, A., Nowicki, M., Del Valle, L., Skorski, T., Khalili, K., and Reiss, K. (2006). T-antigen of the human polyomavirus JC attenuates faithful DNA repair by forcing nuclear interaction between IRS-1 and Rad51. J. Cell Physiol. *206*, 35–46.

Wang, H.K., Duffy, A.A., Broker, T.R., and Chow, L.T. (2009). Robust production and passaging of infectious HPV in squamous epithelium of primary human keratinocytes. Genes Dev. *23*, 181–194.

Woods, C., LeFeuvre, C., Stewart, N., and Bacchetti, S. (1994). Induction of genomic instability in SV40 transformed human cells: sufficiency of the N-terminal 147 amino acids of large T antigen and role of pRB and p53. Oncogene *9*, 2943–2950.

Wu, X., Avni, D., Chiba, T., Yan, F., Zhao, Q., Lin, Y., Heng, H., and Livingston, D. (2004). SV40 T antigen interacts with Nbs1 to disrupt DNA replication control. Genes Dev. *18*, 1305–1316.

Xia, S.J., Shammas, M.A., and Shmookler Reis, R.J. (1997). Elevated recombination in immortal human cells is mediated by HsRAD51 recombinase. Mol. Cell. Biol. *17*, 7151–7158.

Yew, P.R., and Berk, A.J. (1992). Inhibition of p53 transactivation required for transformation by adenovirus early 1B protein. Nature *357*, 82–85.

Zehbe, I., Ratsch, A., Alunni-Fabbroni, M., Burzlaff, A., Bakos, E., Durst, M., Wilander, E., and Tommasino, M. (1999). Overriding of cyclin-dependent kinase inhibitors by high and low risk human papillomavirus types: evidence for an *in vivo* role in cervical lesions. Oncogene *18*, 2201–2211.

Zerfass-Thome, K., Zwerschke, W., Mannhardt, B., Tindle, R., Botz, J.W., and Jansen-Durr, P. (1996). Inactivation of the cdk inhibitor p27KIP1 by the human papillomavirus type 16 E7 oncoprotein. Oncogene *13*, 2323–2330.

Zhao, X., Madden-Fuentes, R.J., Lou, B.X., Pipas, J.M., Gerhardt, J., Rigell, C.J., and Fanning, E. (2008). Ataxia telangiectasia-mutated damage-signalling kinase- and proteasome-dependent destruction of Mre11-Rad50-Nbs1 subunits in Simian virus 40-infected primate cells. J. Virol. *82*, 5316–5328.

Zhou, B.B., and Elledge, S.J. (2000). The DNA damage response: putting checkpoints in perspective. Nature *408*, 433–439.

Zimmermann, H., Degenkolbe, R., Bernard, H.U., and O'Connor, M.J. (1999). The human papillomavirus type 16 E6 oncoprotein can down-regulate p53 activity by targeting the transcriptional coactivator CBP/p300. J. Virol. *73*, 6209–6219.

Structural 'Snap-shots' of the Initiation of SV40 Replication

10

Gretchen Meinke and Peter A. Bullock

Abstract

The initiation of DNA replication, which is one of the fundamental processes in eukaryotic cells, is not understood at the molecular level. Therefore, several laboratories are attempting to understand this process using the simian virus 40 replication system; a relatively simple and well-characterized model for studies of eukaryotic DNA replication. The replication initiator encoded by simian virus 40 is termed T-antigen. Molecular structures have been obtained for the three major domains of T-antigen. Moreover, several structures are now available of the central 'origin binding domain' in complex with various DNA substrates. Collectively these structures have provided a series of 'snap-shots' of the initiation process. They have also significantly improved our understanding of such diverse processes as site specific binding during origin recognition, melting of origin sequences, oligomerization of T-antigen (to form hexamers and double hexamers) and the molecular basis for helicase activity.

Introduction

The initiation of DNA replication in higher eukaryotic organisms is catalysed via mechanisms that are poorly understood. This reflects the complexity of the process (greater than 40 cellular factors are involved; Bell and Dutta, 2002; Sclafani *et al.*, 2004; Masai *et al.*, 2010) and uncertainty regarding the sequences that serve as origins of DNA replication (Gilbert, 2001). The failure to understand the initiation of DNA replication represents a significant gap in our knowledge of an essential biological process. There are additional reasons for wanting to understand the initiation of eukaryotic DNA replication; for example, to establish how it is coupled to cell cycle regulatory events that are disrupted in many types of neoplasia (e.g. Burhans *et al.*, 2006).

The simian virus 40 (SV40) replication system has been used to overcome certain of these limitations and to establish basic mechanisms required for the initiation of eukaryotic DNA replication. SV40 contains a well-defined origin of replication and initiation of nascent DNA strands requires only the virally encoded T-antigen (T-Ag) and the cellular factors replication protein A (RPA), topoisomerase I and the polymerase α-primase (pol α-primase) complex (Fanning and Zhao, 2009, and references therein). In view of this relative simplicity, it is anticipated that a molecular understanding of the initiation of DNA replication will result from the characterization of these proteins, their interactions with each other and the viral origin. As an initial step towards this goal, several laboratories have sought to provide high-resolution molecular descriptions of T-Ag and its interactions with the SV40 origin of replication. The results of these studies are the focus of this review.

The SV40 core origin and its binding partner T-antigen

Essential for the initiation of DNA replication are the SV40 replication origin and T-Ag, the viral factor that interacts with the origin *in trans*. To facilitate an appreciation of the 'structural snapshots' that will be presented, the following section provides a brief overview of these two elements.

The SV40 core origin

The 64-bp SV40 core origin is necessary and sufficient for T-Ag catalysed initiation of SV40 replication (Stillman *et al.*, 1985; DeLucia *et al.*, 1986; Li *et al.*, 1986; and references therein). The sequence of this well characterized region of DNA is presented in Fig. 10.1. Prominent features of the core origin include the four GAGGC pentanucleotides (P) located in the central or 'site II' region of the origin. Additional features of interest include the AT rich and EP 'flanking regions'. The EP region includes the origin sequences that are melted upon T-Ag assembly (Borowiec *et al.*, 1991); however, the AT region is also known to undergo structural changes upon T-Ag binding (Borowiec *et al.*, 1991; Chen *et al.*, 1997).

T-antigen

The SV40 core origin is site-specifically bound by T-Ag, a 708 amino acid phospho-protein that has many roles during initiation of DNA replication (for reviews see Fanning and Knippers, 1992; Pipas, 1992; Bullock, 1997; Simmons, 2000). A schematic of this multifunctional and multidomain protein is presented in Fig. 10.2 and an overview of the functions of the domains described below.

The J-domain

The N′-terminus of the molecule contains the J-domain (amino acids 1–76; Kelley and Landry, 1994; Kelley and Georgopoulos, 1997). As with other J-domains, the J-domain in T-Ag recruits chaperones like Hsc 70 (Campbell *et al.*, 1997; Srinivasan *et al.*, 1997) and therefore it is thought to play important role(s) in regulating the rearrangement of multiprotein complexes. The T-Ag J-domain is needed for viral replication *in vivo* (Spence and Pipas, 1994; Campbell *et al.*, 1997), but not *in vitro* (Maulbecker *et al.*, 1992; Weisshart *et al.*, 1996); a finding consistent with the hypothesis that its mitogenic activity is important for promoting S-phase entry (e.g. Dickmanns *et al.*, 1994). The J domain is followed by a flexible linker (amino acids 77–130) that contains important regulatory sequences including the binding site for the retinoblastoma protein (Rb) (LXCXE; amino acids 103–107), the nuclear localization signal (amino acids 125–132) as well as important phosphorylation sites (see below).

The origin-binding domain

T-Ag residues 131–260 constitute the origin-binding domain (OBD), a region having multiple functions during the initiation of DNA replication (Wun-Kim *et al.*, 1993). One critical function of

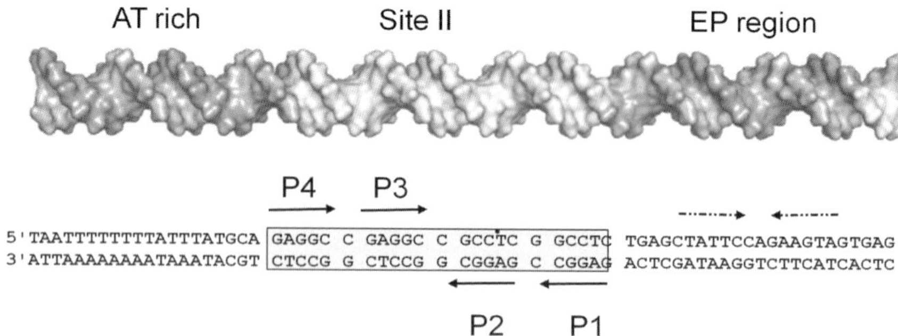

Figure 10.1 Depictions of the SV40 core origin. Top: Surface rendering of idealized DNA structure of the origin region showing the relative locations of elements within the 64-bp core origin [i.e. AT-rich, Site II and early palindrome (EP) regions]. All molecular graphics figures were made using the programme PyMOL (DeLano, 2002). Bottom: the DNA sequence of these regions including the four high affinity GAGGC binding sites for T-Ag comprising Site II (i.e. pentanucleotides P4–P1). The pentanucleotides in site II are arranged in two pairs that are inverted relative to each other; thus it contains a large inverted repeat. The sequences of additional regions of interest, including the AT and EP regions, are also presented. The imperfect inverted repeat in the EP is shown as a pair of dashed arrows.

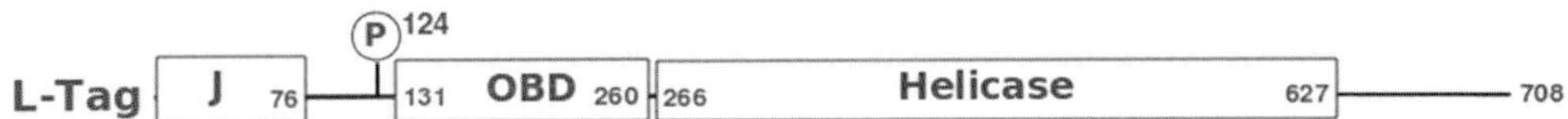

Figure 10.2 A simple rendering of T-Ag highlighting the relative locations of its three major domains; the J domain, the origin binding domain (OBD) and helicase domain (the relative positions of their boundaries are indicated). Cell-cycle dependent phosphorylation of T-Ag on residue Thr 124 is critical for initiation of viral DNA replication (Fanning, 1994).

this domain is that it is responsible for site-specific binding to the GAGGC sequences situated in site II (reviewed in Fanning and Knippers, 1992; Bullock, 1997; Simmons, 2000). Mutagenesis studies of the T-Ag-OBD identified the A1 (amino acids 147–155) and B2 motifs (amino acids 203–207) as critical for site-specific binding (Simmons *et al.*, 1990; Wun-Kim *et al.*, 1993). Residues in the A1 and B2 regions are also involved in many of the other interactions that the T-Ag-OBD makes with DNA [e.g. binding to duplex DNA in a non-sequence-specific manner (Simmons *et al.*, 1990; Luo *et al.*, 1996), to single-stranded DNA (ssDNA) (Reese *et al.*, 2006; Meinke *et al.*, 2011) and to DNA containing fork-like structures (Meinke *et al.*, 2011)]. The relative affinities of the T-Ag-OBDs for DNA are sequence specific duplex > ssDNA > non-sequence-specific duplex (Fradet-Turcotte *et al.*, 2007). An additional critical function of the T-Ag-OBD is that it interacts with many cellular proteins (e.g. Wu *et al.*, 2004; Jiang *et al.*, 2006; reviewed in Bullock, 1997).

The helicase domain

The large helicase domain, extending between residues 266 and 627 (e.g. Li *et al.*, 2003; Gai *et al.*, 2004; references therein), provides a number of additional activities that are required for SV40 DNA replication. For example, during origin recognition it interacts with the flanking regions (Reese *et al.*, 2004; Shen *et al.*, 2005; Kumar *et al.*, 2007; references therein). Once bound, the helicase domain oligomerizes into an ATP dependent DNA helicase (the assembly of the helicase domain is stimulated by binding of ATP, but ATP hydrolysis is not required (reviewed in Borowiec *et al.*, 1990). Functions of the hexameric helicase include separation of duplex DNA into two single-stranded DNA molecules and recruitment of additional cellular replication factors (e.g. the pol α-primase complex) to the replication complex (e.g. Arunkumar *et al.*, 2005; Taneja *et al.*, 2007; Khopde and Simmons, 2008; Huang *et al.*, 2010; and references therein).

The assembly of T-antigen on the SV40 origin; a complicated *pas de deux*

EM-based observations

EM studies have generated fundamental insights into the oligomerization products formed by T-Ag on the SV40 core origin. For instance, they established that, in the presence of ATP and incubation at 37°C, T-Ag assembles on the core origin into hexamers and bilobed double hexamers (Dean *et al.*, 1987; Dodson *et al.*, 1987; Mastrangelo *et al.*, 1989; Valle *et al.*, 2000; Gomez-Lorenzo *et al.*, 2003). The EM studies also confirmed that the OBD is situated over the central GAGGC containing region of the origin and that the J-domain is probably positioned above the T-Ag-OBD. Furthermore, they helped to establish that the helicase domains are positioned over the flanking sequences. Finally, they established the dimensions of the complex (Valle *et al.*, 2000) and provided evidence for conformational rearrangements of T-Ag during initiation events (Cuesta *et al.*, 2010).

Outline of one model for T-Ag assembly on the origin

In addition to the EM studies described above, a number of biochemical and structural studies have led to models depicting T-Ag assembly; one of which is presented in Fig. 10.3. According to this model (Kumar *et al.*, 2007; Meinke *et al.*, 2007), the GAGGC sequences within the

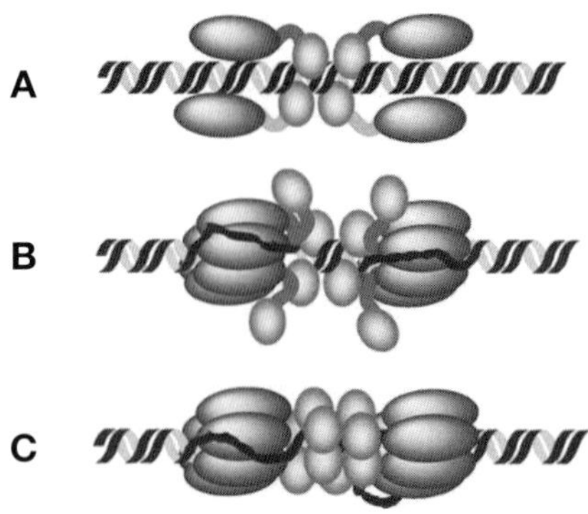

Figure 10.3 Schematic of T-Ag assembly into double hexamers on the SV40 origin. (A) The T-Ag-obds (shaded light grey) bind to the central region of the core origin and position the helicase domains over the flanking regions. (B) This model posits that upon binding to the flanking sequences, the helicase domain generates ssDNA; a prerequisite for subsequent oligomerization of this domain around one strand of DNA. (C) Once oligomerization is completed, the second strand is routed over the outer surface of the helicase domain. An additional feature of this model is that the T-Ag-obd forms a spiral once its local concentration is increased owing to oligomerization of the helicase domain. This open-ring structure contains a gap through which ssDNA may pass. Adapted from Figure 8 in Meinke *et al.* (2007).

central region of the core origin are recognized by the OBDs of individual T-Ag monomers (step A; small spheres). As a result of these site-specific binding events, the helicase domains are positioned over the proximal flanking regions. Following these initial interactions, additional T-Ag molecules are recruited to the origin; a process that is thought to reflect the high affinity of the helicase domains for each other (Li *et al.*, 2003). At this stage in the process it has been proposed that the helicase domains melt the EP (and perhaps AT-rich) regions (via a complex mechanism that is discussed later in this review). The helicase domains then assemble around one of the newly generated single strands in a manner consistent with the 3′ to 5′ polarity of this helicase (step B). Following oligomerization of the helicase domains, the T-Ag-OBDs are predicted to shift from site-specific binding of the GAGGC pentanucleotides to single-stranded DNA binding (Meinke *et al.*, 2007). This transition in DNA binding specificity of the T-Ag-OBD has been proposed to be associated with the formation of a spiral or 'lock washer' conformation (Meinke *et al.*, 2006) see below) that contains a gap through which ssDNA may pass (Kumar *et al.*, 2007; Meinke *et al.*, 2007; Step C). Once oligomerization on the origin is completed, T-Ag is assembled into the previously described double hexamer; allegedly, the active form of the helicase (Wessel *et al.*, 1992; Smelkova and Borowiec, 1997; Weisshart *et al.*, 1999; Alexandrov *et al.*, 2002). Finally, T-Ag assembled on the origin then recruits the cellular factors needed for the initiation of DNA replication (e.g. Weinberg *et al.*, 1990; Ishimi *et al.*, 1988; Tsurimoto *et al.*, 1989; Simmons *et al.*, 2004).

Structure-based advances in understanding the interactions between SV40 T-antigen and the SV40 origin

The SV40 replication system has provided many critical insights into DNA replication in eukaryotes. However, at the molecular level, there is still much about the initiation events described above that we do not understand. Nevertheless, recent biophysical studies, reviewed herein, have provided several structural 'snapshots' of this process that have greatly advanced this field. It should be noted that the structure of the J-domain and the flexible hinge bound to the 'pocket domain' of Rb was previously reported (Kim *et al.*, 2001). However, since this region is not directly required for initiation of DNA replication, it is not discussed at length in this review.

Structural studies of T-antigen's origin-binding-domain (T-Ag-OBD)

NMR-based studies

The first structure of the SV40 T-Ag OBD was solved by nuclear magnetic resonance spectroscopy (Luo *et al.*, 1996). The protein construct used in these structural studies consists of residues 131–260 and is designated as T-Ag OBD. This study revealed that the overall fold of the T-Ag OBD consists of a central five-stranded antiparallel beta-sheet flanked by two alpha-helices on one side and one alpha-helix and one $3_{(10)}$-helix on the other (reviewed in Bullock, 1997). A search of known structures established that the T-Ag OBD contained a novel fold. The NMR structure also demonstrated that the previously described A1 (147–155) and B2 (203–207) motifs form a continuous surface on the T-Ag-OBD (Fig. 10.4). As structures of the DNA binding domains of other replication initiators were solved, it became apparent that the 'T-Ag OBD fold' was evolutionarily conserved (Campos-Olivas *et al.*, 2002). Proteins having this fold act on nucleic acid sequences that undergo transitions between double and single-stranded DNA. Collectively, these structures enabled a number of additional structure–function relationships to be established (reviewed in Campos-Olivas *et al.*, 2002).

Crystal Structures of the T-Ag OBD on site II based fragments of the origin

Several high-resolution structures of T-Ag OBD were subsequently solved in the presence of site II based DNA substrates, thereby providing many fundamental insights into origin recognition. For instance, although the NMR structure of the free T-Ag OBD had been determined, it was unclear how the T-Ag OBD site-specifically bound to the four GAGGC sequences in site II. The determination of two crystal structures of the T-Ag OBD bound to subfragments of site II (Bochkareva *et al.*, 2006; Meinke *et al.*, 2007) enabled these issues to be addressed. One of these crystal structures showed four T-Ag OBDs bound to a double-stranded DNA (dsDNA) fragment containing P1-P4 (Fig. 10.5A). The second co-structure had two T-Ag OBDs bound to DNA substrates containing only P1 and P3 (Fig. 10.5B; this substrate was used because it was previously shown to support double hexamers formation (Joo *et al.*, 1998)). Some of the novel findings derived from these studies are presented below.

Sequence specificity

The T-Ag OBDs within these structures interact with the GAGGC pentanucleotides in nearly identical ways. The A1 and B2 motifs engage the double-stranded DNA in the major groove and make extensive contacts with both the bases and the phosphate backbone. Two residues (Asn153 and Arg154) in the A1 loop provide the bulk of the sequence specific contacts with the DNA, and are splayed in two directions across the major groove, not unlike the legs of a rider sitting on a saddle (Fig. 10.5C). A schematic of all of the site-specific protein–DNA interactions are shown in Fig. 10.5D. Collectively, these structures explained the sequence specificity of the T-Ag OBD and provided a structural explanation for a great body of related mutagenesis (e.g. Reese *et al.*, 2006) and biochemical data (e.g. Fradet-Turcotte *et al.*, 2007).

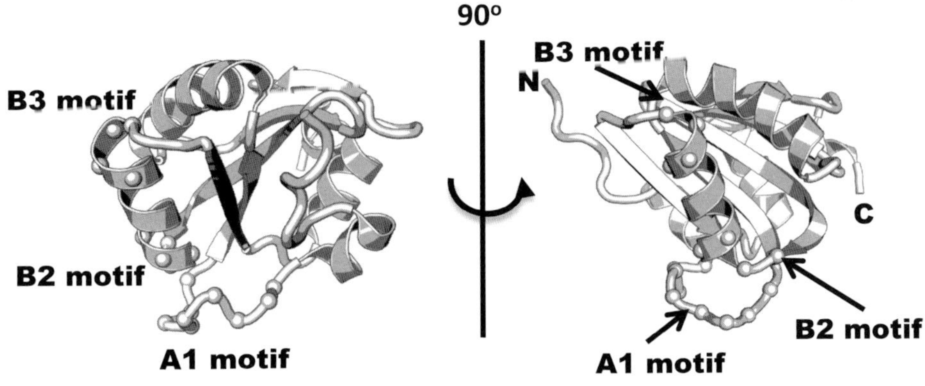

Figure 10.4 Two views of a ribbon diagram of the T-Ag OBD with the important DNA binding motifs A1 (147–155), B2 (203–207), and the B3 motif (213–220) indicated by small spheres.

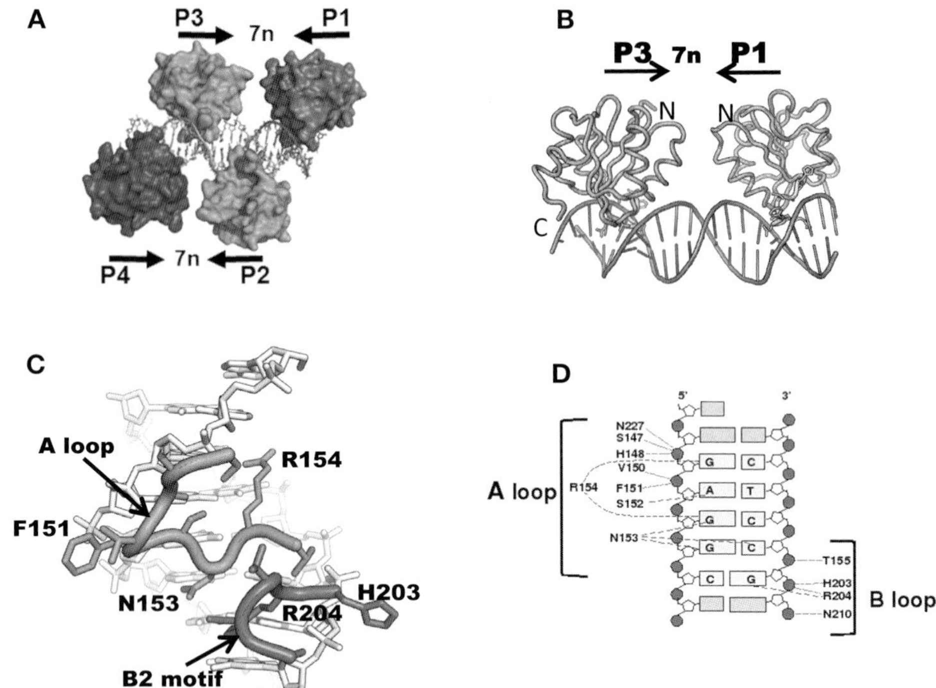

Figure 10.5 Crystal Structures of T-Ag OBD bound to GAGGC containing duplex DNA. (A) Overview of T-Ag OBD bound to all four GAGGCs. The T-Ag OBDs are depicted as surface rendering, and the DNA is shown as sticks. The number of nucleotide spacers between the head-to-head T-Ag OBDs is indicated as 7n. Arrows indicate the direction of the GAGGC sequence. Adapted from Figure 7 in Meinke *et al.* (2007). (B) Overview of T-Ag OBD bound to P1 and P3 shown as a ribbon diagram. The side chains in the DNA binding loops are shown. (C) A close-up view of the A1 and B2 loops interactions with the GAGGC sequence looking down the major groove of the DNA. Note the Asn153 and Arg 154 splayed in opposite directions along the base of the major groove. Additional residues in the A1 and B2 loops are labelled. (D) A schematic of the protein–DNA interactions. The circles represent the phosphate groups; the bases are indicated by rectangles. The dashed lines indicate sequence specific contacts and solid lines indicate phosphate interactions. Adapted from Figure 3 in Meinke *et al.* (2007).

Spacing and orientation of the GAGGC/OBD complex

A feature of the site II based co-structures is that, while close, the T-Ag OBDs bound to the pentanucleotides in site II do not interact with each other. This is in contrast to the related co-structure of the papilloma virus initiator E1 OBDs bound as a dimer to its DNA target (Enemark *et al.*, 2000); in this co-structure, interactions were noted between the OBD dimer. The T-Ag-OBDs co-structures also revealed the relative orientations of the T-Ag-OBDs while bound to the GAGGC sequences. For example, the T-Ag OBDs bound to P1 and P2 (or P3 and P4) are arranged in a head-to-tail manner on opposite faces of DNA (~160° apart). In contrast, the T-Ag OBDs on P1 and P3 (or P2 and P4) are on the same face of the DNA and arranged in a head-to-head fashion. Presumably, the T-Ag OBDs bound to P1 and P2 ultimately participate in the formation of one hexamer, while the T-Ag OBDs on P3 and P4 participate in the formation of the second hexamer. Double-hexamer formation is considered in a following section; however, it is instructive to consider the residues of the T-Ag OBDs that will ultimately reside on opposing hexamers, but are near each other while bound to P1 and P3 (or P2 and P4). Many of these residues are derived from the B3 motif (Wun-Kim *et al.*, 1993; residues 213–220), and represent a flexible region of the T-Ag OBD.

Structural distortions arising from the T-ag-OBD/pentanucleotide complex

Structures of the T-Ag OBD in both the presence and absence of DNA have been determined. Therefore, portions of the OBD that change conformation upon binding to DNA in a sequence specific manner can be identified. Superposition of the apo T-Ag OBD onto the GAGGC bound T-Ag OBD revealed that the primary backbone movement occurs in the A1 motif (Fig. 10.6), with residue Phe151 at the apex of that movement. Indeed, the structural studies suggest that at least in T-Ag monomers, the A1 loop is a switch, toggling between the 'apo' and 'bound' conformations of the T-Ag OBD. In the bound conformation the A1 loop is extended out from the rest of the molecule where it is able to interact with the GAGGC pentanucleotides. In its free or apo conformation the A1 loop is retracted and closer to the core of the T-Ag-OBD. In essence, the A1 loop serves as 'landing gear' for the T-Ag OBD, with residues such as Phe151 and Arg154 functioning as the wheels.

Changes to the origin DNA upon binding of the T-Ag-OBD are much more subtle. For instance, in the structure with 4 OBDs bound (Bochkareva *et al.*, 2006) the overall helical axis of DNA is straight. In contrast, in the structure where two OBDs are bound at P1 and P3 (Meinke *et al.*, 2007), the minor groove of the DNA compresses, inducing a 18 degree bend in the helical axis of the DNA. As a result of this bend, the OBDs on P1 and P3 are closer to each other than on straight DNA. The significance of the relatively minor differences in the DNA distortions observed in these two structures is not currently understood.

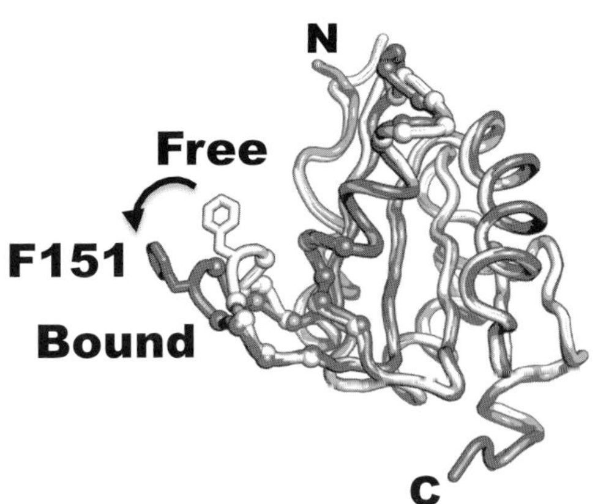

Figure 10.6 Conformational changes that occur in the T-Ag OBD upon binding in a sequence-specific manner to GAGGC sequences: Two representative structures of the T-Ag OBD in the 'free' or apo conformation and dsDNA 'bound' are superimposed and shown as a ribbon diagram. The main conformational change occurs in the A1 loop that acts as a toggle having Phe 151 at the apex of the movement. The positions of the N and C termini are indicated.

Structures of the T-Ag OBD in the absence of DNA

The spiral hexamer

In the absence of DNA, the T-Ag OBD crystallized as a helical filament with a pitch of 35.8 Å per turn (Meinke *et al.*, 2006). The helix (or spiral) is left-handed, having six OBD monomers per turn. In the spiral, the T-Ag OBD monomers are arranged head-to-tail and each monomer of the spiral is rotated 60 degrees and translated 1/6 of the pitch (~6 Å). Knowing that the functional oligomeric form of full-length T-Ag is a hexamer, it was intriguing that six T-Ag-OBDs complete one turn of the spiral. Moreover, the overall dimensions of a single turn of the T-Ag-OBD spiral were consistent with the EM studies (Valle *et al.*, 2000). Therefore, we postulated that one turn might represent a biologically relevant structure. In the following sections, we will refer to this as the 'open-ring' hexamer or 'lock-washer' structure of the T-Ag-OBD.

This open ring structure of the T-Ag-OBD has several interesting features (Fig. 10.7). The central channel of the spiral is positively charged and has a diameter of ~30 Å; large enough to accommodate double-stranded DNA (~23 Å) or even two separate single-stranded DNAs (12 Å). In addition, this open-ring structure contains a 'gap' through which DNA could transit (Meinke *et al.*, 2006). Recent EM studies of T-Ag double hexamers assembled on the SV40 origin have also reported a spiral with a gap in the central lobe (Cuesta *et al.*, 2010). Additional noteworthy features of the lock-washer structure observed in the crystallography studies are described below.

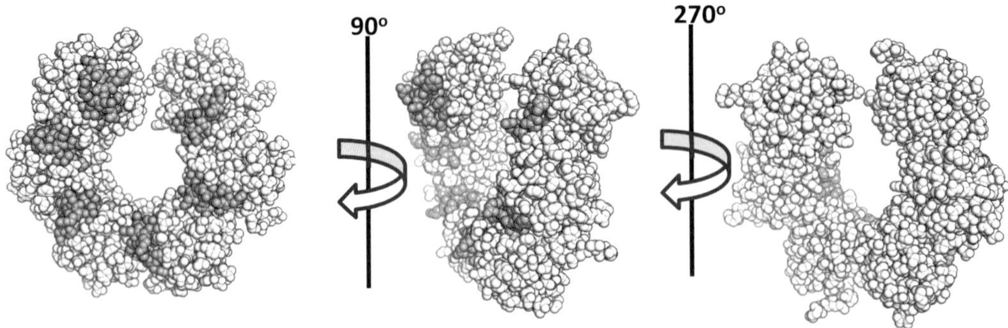

Figure 10.7 Spiral hexamer structure of the T-Ag OBD: The hexameric spiral 'lock-washer' structure is shown as spheres. Three views are shown. The left view looking down the 6-fold screw axis clearly shows that the A1 and B2 motifs (darker spheres) occur on one face of the hexamer. The central view rotated ~90° shows the 'lock-washer' nature of the open ring. The right view shows the 'back' of the hexamer that does not contain the DNA binding loops, although this face does contain the B3 motif and the N-termini.

The open-ring structure of the T-Ag-OBD provided the first views of the protein/protein interfaces that may form during hexamerization of the T-Ag OBD monomers (Meinke *et al.*, 2006). The protein–protein interface in the T-Ag-OBD spiral consists, in part, of residues from the A1 motif; in particular, Phe 151, which forms the lynchpin of the interface and makes many contacts. The relatively small buried interface (~650 Å2/monomer) is suggestive of a weak interaction and helps to explain why only monomers are seen in solution, even at high concentration (Luo *et al.*, 1996). It follows that in the context of the full-length T-Ag, the OBDs may interact with each other in a very dynamic way, sometimes as the spiral presented in Fig. 10.7.

In the spiral structure, solved in the absence of DNA, the A1 motif is in the 'apo' conformation. Furthermore, the spiral arrangement of the T-Ag OBDs positions the A1 and B2 motifs (shown as darker spheres in Fig. 10.7) on one face of the hexamer, along the inner portion of the central channel. The A1 and B2 are partially occluded and as a result, several of the residues in these motifs are predicted to be unavailable for sequence specific binding to the pentanucleotides. Thus, spiral formation would require major conformational rearrangements of the T-Ag-OBD from its site-specific binding mode to the spiral structure. This is one reason for suggesting that the T-Ag-OBD spiral represents a structural 'snapshot' that occurs after binding of the GAGGCs. Finally, although the 'lock-washer' structure is incompatible with sequence-specific binding to GAGGC sequences, it may interact with other forms of DNA (e.g. non-sequence specific duplex and single-stranded DNA; see below).

Exactly how pairs of GAGGCs in Site II (i.e. P1 and P2 or P3 and P4) contribute to the hexameric form of the T-Ag-OBD remains a puzzle. While a single GAGGC sequence can promote the formation of a hexamer of full-length T-Ag, two correctly oriented and spaced GAGGC sequences (e.g. P1 and P2) are needed for the formation of a functional hexamer (Joo *et al.*, 1998; references therein). Therefore, it is tempting to speculate that the T-Ag-OBDs could form hexamers while site-specifically bound to P1 and P2. However, molecular modelling of a hexamer nucleated by these, or any other, pentanucleotides does not seem possible owing to serious steric collisions of the T-Ag OBDs with each other and with DNA (Meinke, unpublished). In summary, T-Ag-OBDs cannot simultaneously bind the GAGGC in a site-specific manner and form a hexameric structure. For such a hexameric structure of T-Ag-OBDs to form, the site-specifically bound T-Ag-OBDs must translate radially away from the DNA such that they no longer interact with the GAGGC sequences.

In view of these considerations, we have analysed whether the spiral 'lock-washer' structure might be relevant to this discussion. Examination of the locations of the OBDs while bound to P1 and P2, and the locations of the corresponding OBDs in the open ring structure, reveals an interesting correlation between the spacing of the pentanucleotides and the pitch of the lock-washer structure. That is, when the lock-washer structure is modelled with DNA in the centre along the 6-fold screw axis, one can imagine how the OBDs bound to P1 and P2 transition to the lock-washer form. This process would involve translation of the OBDs away from the pentanucleotidies such that the OBD on P1 would become the first subunit of the hexamer while the OBD on P2 would be the fourth subunit. Formation of the spiral would also require the recruitment of additional T-Ag-OBD monomers and a change in the A1 loop from the 'bound' to 'apo' conformation. Viewed in this manner, the structure of the spiral hexamer reflects the architectural arrangement of the pentanucleotides in the SV40 core origin.

The 'lock-washer' structure also allowed a model to be proposed for a T-Ag-OBD double hexamer. Initially, it was necessary to decide how to orient the two spirals relative to each other. As noted above, in the open-ring form, the A1 and B2 motifs are partially occluded and located on the inner surface of one face of the hexamer. Of interest, a separate study of full-length T-Ag had previously established that a number of residues, mainly centred on the B3 motif (213–220), are involved in double hexamer formation (Weisshart *et al.*, 1999). When these data were considered in terms of the spiral structure, it was noted that the residues important for double hexamer formation (Weisshart *et al.*, 1999), and the N-termini, both occur on the face opposite to that containing the A1 and B2 residues. This information suggested an orientation of the spiral structure in the context of single or double hexamers; namely, the face containing the A1 and B2 motifs is oriented towards the helicase domain while the opposite face (containing the B3 region) is oriented towards the second hexamer (Fig. 10.8). Consistent with this proposal, the C-termini are also positioned on the helicase proximal side of the spiral, a spatial arrangement that would promote linkage of the two domains.

Recall that the B3 regions are initially opposite each other, but do not touch, when bound

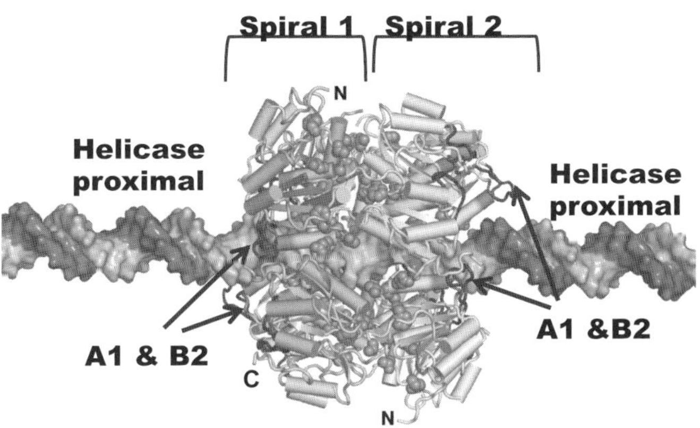

Figure 10.8 Spiral structure suggests a double hexamer model of the T-Ag OBDs. Idealized dsDNA is shown as a surface representation and is positioned going through the central channel of the spiral hexamer. The two spirals are named 'spiral 1' and 'spiral 2' to indicate the two different spiral structures. The T-Ag OBD is shown as a ribbon representation. The orientation chosen positions the two spirals back-to-back. The A1 and B2 loops are shown in darker colour and are positioned on the 'helicase proximal' side. Mutations known to impair double hexamer formation are shown as spheres and occur at the interface of the double T-Ag OBD hexamer model. The N termini of the spiral also occur at the periphery of the OBD spiral and would easily allow for the modelling of the J domain here. Adapted from Figure 4 in Meinke *et al.* (2006).

to pentanucleotides 1 and 3 (or 2 and 4). In view of the locations of the B3 motif in the double spiral (Fig. 10.8), it is possible that B3 dependent interactions take place during the 'rearrangement' events that have been proposed to take occur during formation of the double spiral. Finally, it is apparent from the positions of the A1 and B2 loops in the double spiral structure that particular monomers in both spirals (i.e. those initially bound to the pentanucleotides) must undergo 'rearrangements' prior to spiral formation. We do not know, however, if this is a concerted activity or whether the rearrangements required to form one spiral precede those needed to form the second.

A disulfide-linked dimer structure of the T-Ag-OBD

Other examples now exist of crystal structures of the T-Ag OBD in the absence of DNA. Without exception, the 'landing gear' of the A1 loop in these structures is retracted; that is, in the apo conformation. In one of these 'apo' structures, the T-Ag OBD formed a dimer through the disulfide linkage of Cys216-Cys216 (PDB code 2IF9; Meinke *et al.*, 2007). The 2-fold symmetrical dimer interface includes residues from the previously described B3 motif (i.e. those previously reported to be important for double hexamer formation). Interestingly, while the B3 motif is somewhat disordered in the previous T-Ag OBD structures, it is ordered in this structure. In light of these observations, it is tempting to postulate that the inherent flexibility of the B3 motif in full-length T-Ag plays a role in double hexamer formation. Of interest, a biological role for Cys 216 mediated dimerization has yet to be reported; nevertheless, Cys 216 is completely conserved in the large T-antigens and mutation of this residue in full-length T-Ag generates molecules that are defective for DNA unwinding (Wun-Kim *et al.*, 1993). Lastly, structure-based alignment of the disulfide linked T-Ag OBD dimer with the Bovine papilloma virus E1 OBD dimer revealed structural conservation of secondary features that participate in the dimer interface (Meinke *et al.*, 2007).

The structure of the T-Ag-OBD bound to DNA in a non-sequence-specific manner

There are now two structures of the T-Ag OBD bound to non-sequence specific DNA substrates. In the first structure (PDB code 2NL8; Bochkareva *et al.*, 2006), the OBD was crystallized in the presence of a GAGGC-containing DNA substrate. However, crystal-packing forces prevented the OBD from docking onto the GAGGC sequence in the expected manner and it was moved to a non-sequence specific portion of the DNA substrate. In this instance, the overall conformation of the T-Ag OBD/non specific dsDNA interaction is similar to that observed in the sequence specific interaction (e.g. the A1 loop landing gear is down). In the second structure, the T-Ag OBD was crystallized on an artificial 'fork' containing regions of double- and single-stranded (ds and ss, respectively) DNA (Meinke *et al.*, 2011). In this structure, the DNA formed a 'DNA lattice' (i.e. the ssDNA ends crossed over and continued the dsDNA helix of a neighbouring strand). Two T-Ag OBDs sandwiched a terminal Adenine base, using residues from the A1 loop in stacking-type interactions. Interestingly, in this structure, the A1 loop landing gear is retracted; as in the previously described apo structures. Whether this structure reflects the structure of the T-Ag-OBD while interacting with a replication fork has yet to be determined.

Interactions of the T-Ag-OBD with single-stranded DNA (ssDNA)

The forked DNA/T-Ag-OBD co-structure in Meinke *et al.* (2011) established that T-Ag OBD residues Arg 204, Asn 210, and Lys 214 are oriented to interact with the phosphate backbone of ssDNA (Fig. 10.9A). This result is consistent with previous NMR solution studies of T-Ag OBD that showed that the residues in the B2 motif (which includes Arg 204) displayed the largest chemical shift changes upon titration with ssDNA (Reese *et al.*, 2006). Related biochemical studies support the co-structure derived hypothesis that particular basic and aromatic residues, largely in the T-Ag-OBD A1 and B2 loops, play important roles in ssDNA binding (Meinke *et al.*, 2011).

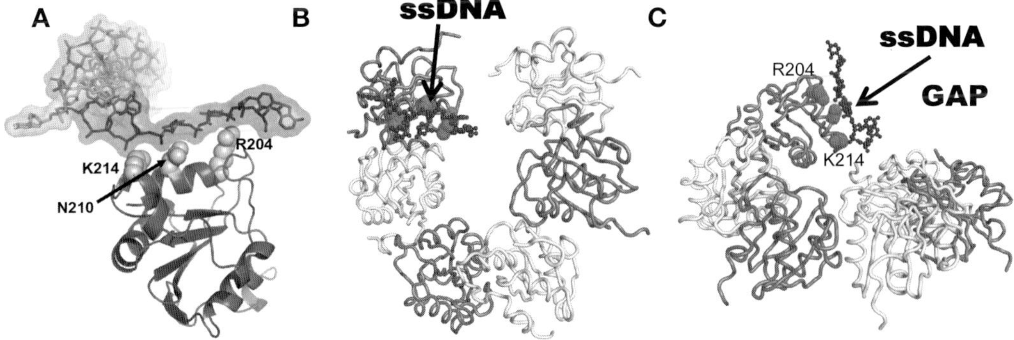

Figure 10.9 The T-Ag OBD/ssDNA interaction. (A) A view of the crystal structure of T-Ag OBD bound to an artificial replication fork that indicated how the T-Ag OBD interacts with the phosphate backbone of ssDNA. The DNA is shown as a stick model with a transparent surface. The T-Ag OBD is shown as a ribbon diagram and the interacting residues (R204, N210, K214) are represented with spheres. (B) A model of how ssDNA might interact with the T-Ag OBD spiral. The T-Ag OBD from the ssDNA-T-Ag OBD co-structure was superimposed on the terminal T-Ag OBD, which is the only subunit of the spiral to have completely accessible DNA binding residues. The T-Ag OBDs are shown as a ribbon representation and the ssDNA is shown as a ball and stick representation. (C) A view of the T-Ag OBD spiral/ssDNA model rotated ~90° relative to the view in B. The position of ssDNA is shown and is near the 'GAP'.

Modelling interactions between the T-Ag-OBD spiral and ssDNA

As previously noted, the T-Ag-OBD lock-washer contains a gap through which ssDNA could pass. Because this structure consists of a head-to-tail arrangement of T-Ag-OBDs, the Phe151 residue in the T-Ag-OBD subunit proximal to the gap is free; the five other Phe 151 residues are engaged in the previously described protein/protein interactions. Since Phe151 plays a critical role in ssDNA binding (Meinke *et al.*, 2011), and only one is available in the spiral, it is possible that the route for ssDNA through the spiral may be limited to the gap proximal surface. In view of this possibility, the T-Ag OBD-ssDNA co-structure was superimposed onto a terminal subunit of the T-Ag OBD spiral (Fig. 10.9B and C). As a result, a working model depicting how ssDNA might interact with the spiral has been generated. Verification of this model will, however, require additional experimentation. Nevertheless, the possibility that there may be limited paths for ssDNA through the spiral is consistent with recent biochemical findings suggesting that the central channel residues of only one OBD monomer are involved in the unwinding reaction (Foster and Simmons, 2010).

Concluding remarks regarding the OBD structures

The structures of the T-Ag OBD reviewed herein establish its ability to adopt multiple conformations and oligomeric states and hint at the plasticity inherent in this domain. These different states have allowed us to propose detailed molecular models that are consistent with much of the existing biochemical data. One such model involving the OBD is shown in Fig. 10.10A and B (discussed below and in the figure legend). Nevertheless, some of these structures await further experimentation to validate their biological importance in the context of full-length T-Ag (e.g. the lock-washer structure and the disulfide-linked dimer). Nonetheless, they are useful 'snap-shots' and have enabled us to conduct modelling studies of replication initiation events that will be presented after a brief review of what is known about the structure of the helicase domain.

Structural studies of the SV40 helicase domain

The SV40 T-Ag helicase domain is a member of the helicase SF3 and AAA^+ superfamilies (Neuwald *et al.*, 1999). The hexameric helicase domain from T-Ag was the first member of this family to

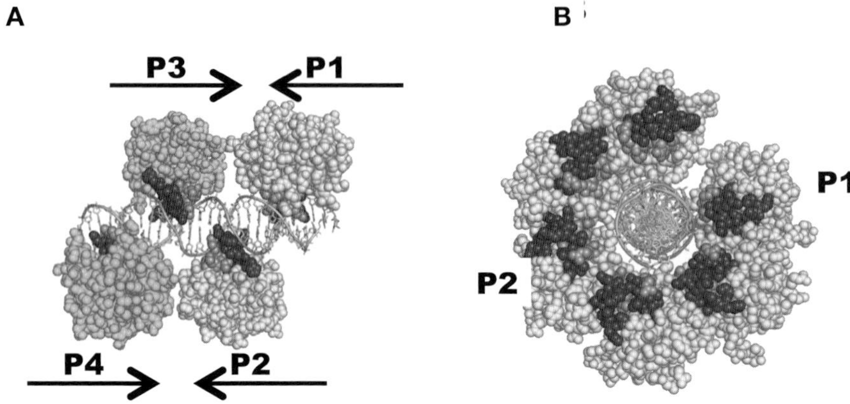

Figure 10.10 Model of transion of T-Ag OBDs from sequence specific bound monomers to non-sequence specific hexamer form: (A). This shows a model of four OBDs bound to GAGGC sequences P1-P4. The A1 and B2 motifs are shown as dark spheres and positioned deep in the major groove. (B) This is a view of the spiral structure of a single hexamer, initiated by the OBDs bound to pentanucleotides P1 and P2, rotated ~90° relative to view A. The DNA, shown in a stick representation, is modelled through the central channel, parallel to the 6-fold screw axis. The helicase proximal face of the T-Ag OBD hexamer is shown; the T-Ag OBDs that were bound at P1 and P2 now make up the subunits at position one and position four of the spiral hexamer. (A cartoon of the transition between site-specifically bound OBDs and spiral formation is shown in Fig. 10.12). This figure was adapted from Figure 7 in Meinke *et al.* (2007).

be solved crystallographically (Li *et al.*, 2003). Since the initial structure was solved, several more structures of the helicase domain have been determined in the presence of ADP, ATP (Gai *et al.*, 2004) and also in complex with p53 (Lilyestrom *et al.*, 2006). The helicase domain was shown to form flat ring structures having a positively charged central channel. A side view established that there are two tiers; at the N-terminus is the smaller zinc binding domain (D1) that is ~86 Å in diameter. The larger C-terminal tier contains the D2 and D3 domains and is ~128 Å in diameter (Fig. 10.11A). Overall, the helicase domain is ~80 Å long and is thus capable of interacting with ~23 bp of DNA. Within a given hexamer, the P-loop for ATP binding resides in the D2 region of one monomer, while residues on the opposite monomer make contact with either the ATP/ADP or bound Mg^{2+}. In the structures solved, the six ATP binding sites are either completely occupied, or completely free of nucleotide. The authors proposed that the two tiers could rotate relative to each other in an 'iris-like' motion to distort or melt double-stranded DNA (Li *et al.*, 2003). Several excellent reviews discuss a number of additional structural features of the helicase domain (e.g. Sclafani *et al.*, 2004; Fanning and Zhao, 2009; Brewster and Chen, 2010); therefore, we will focus on detailing features of this domain that pertain to origin recognition.

The role of the beta-hairpin in origin recognition

Previous studies established that origin recognition requires not only interactions between the T-Ag-OBD and site II, but additional interactions between the helicase domain and the flanking sequences (e.g. Parsons *et al.*, 1990; Kim *et al.*, 1999; Sreekumar *et al.*, 2000). The publication of the structure of the helicase domain (Li *et al.*, 2003) permitted a greatly refined analysis of this interaction. For example, a series of EMSAs (electrophoretic mobility shift assays) and modelling studies were used to predict that a region of the helicase domain, termed the beta-hairpin (shown in Fig. 10.11B), was critical for its interaction with the EP (Reese *et al.*, 2004). In subsequent studies, mutations were introduced into the tip of the beta-hairpin in either full-length T-Ag (Kumar *et al.*, 2007) or a fragment of T-Ag (i.e.

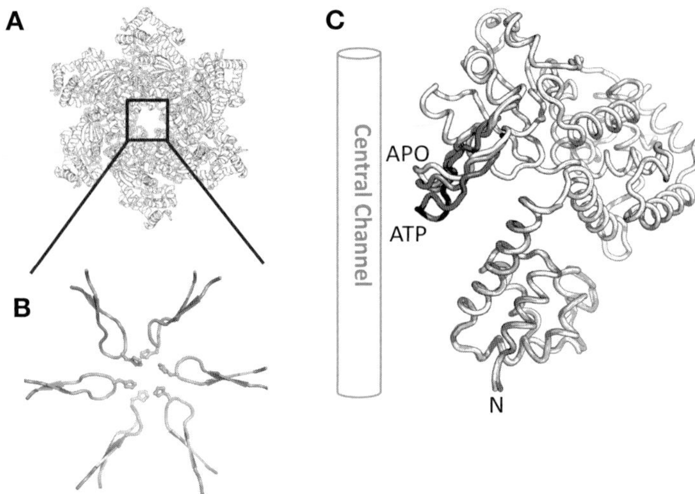

Figure 10.11 Structure of the SV40 T-Ag helicase domain: (A) A ribbon representation of the helicase domain (PDB code 1SVM) is shown as a hexamer looking down the central channel. (B) A close-up of the six beta-hairpins that protrude in the central channel is shown as a ribbon representation. The side chain of the tip of the hairpin (His 513) is shown as a stick representation. (C) A side view of a monomer of the helicase domain showing the movement the beta-hairpin undergoes from apo (PDB code 1N25), to ADP bound (PDB code 1SVL) to ATP bound (PDB code 1SVM). The relative location of the central channel is indicated by a cylinder.

T-Ag$_{131-627}$; Shen *et al.*, 2005) and additional EMSAs performed. These studies confirmed that the beta-hairpin plays a role in origin recognition (Shen *et al.*, 2005; Kumar *et al.*, 2007). Further characterization (e.g. via $KMnO_4$ assays) revealed that T-Ag molecules mutated at the tip of the beta-hairpin were incapable of melting the EP region, suggesting that the beta-hairpin plays a direct role in locally unwinding the origin (Kumar *et al.*, 2007).

Molecular modelling of T-Ag during oligomerization and DNA melting

There is still considerable uncertainty in terms of how T-Ag assembly mediates origin melting and the routes taken by DNA through the double hexamer complex. Nevertheless, the structures of the T-Ag domains and related biochemical studies have been analysed in terms of how origin recognition and oligomerization events may be coupled to origin melting. The model that we have proposed (Kumar *et al.*, 2007) is initially described and then discussed in terms of opposing models.

Structure-based modelling of the T-Ag-OBD during initial assembly events

As previously discussed, the OBD plays a fundamental role during the process whereby dsDNA is scanned for the high affinity GAGGC sequences (reviewed in Borowiec *et al.*, 1990). In addition, it has been proposed that GAGGC scanning is paired with beta-hairpin dependent monitoring for sequences that can be readily melted (Kumar *et al.*, 2007). Once a sequence with the requisite features is identified, the T-Ag OBD then 'locks' into position at the GAGGC sequences by making the interactions seen in the sequence specific co-structures (Fig. 10.5).

We have suggested that, upon helicase assembly, a transition must occur from sequence specific to non-specific OBD binding and that additional non-interacting OBDs must be recruited to form the OBD hexamer (Kumar *et al.*, 2007; Meinke *et al.*, 2007). Recall that the basis for this proposal is that in the spiral structure of the hexameric T-Ag OBD, the A1 and B2 loops cannot engage in a sequence specific manner with the GAGGC sites without serious steric collisions. Modelling of the transition to the spiral structure requires that the T-Ag OBDs must translate away from the DNA, rotate and then interact with previously unbound OBDs (Fig. 10.10A and B). In Fig. 10.10B, the T-Ag OBD initially bound to P1 is labelled as 'P1', while the T-Ag OBD bound to P2 is now 'P2'. The remaining OBDs must be recruited during the assembly process.

Modelling of the assembly of the helicase domain

Site-specific binding of the T-Ag-OBD to the pentanucleotides positions the helicase domains over the flanking sequences. Kumar *et al.* (2007) reported that full-length T-Ag molecules containing mutations at the tip of the beta-hairpin were defective for hexamer formation. However, they also observed that the block to assembly experienced by the beta-hairpin mutant could be overcome by using substrates having ssDNA in the EP region (Kumar *et al.*, 2007). These and related observations led to the proposal that a prerequisite for assembly of the helicase domain was (1) beta-hairpin-dependent melting of the EP region and (2) subsequent assembly of the helicase domains around one of the newly generated single strands of DNA (i.e. the one required for T-Ags 3′ to 5′ helicase activity) (Kumar *et al.*, 2007). It follows that upon completion of assembly, the second strand of single-stranded DNA will transit over the helicase domain. A summary of this model, that depicts the proposed functions of the beta-hairpin during origin recognition, melting to form single-stranded DNA and subsequent assembly events, is depicted in Fig. 10.12.

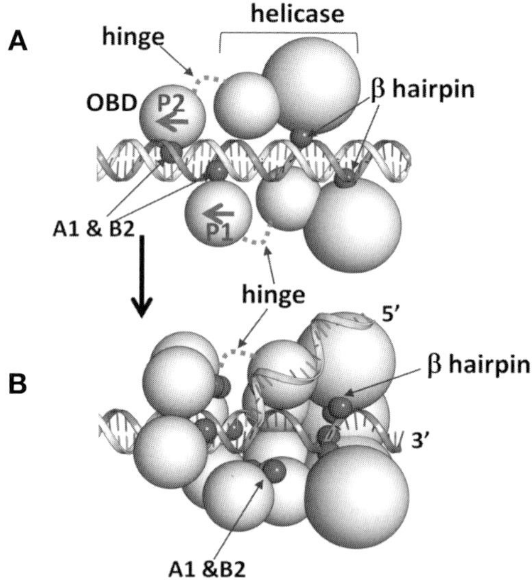

Figure 10.12 Structure-based model of T-Ag assembly upon origin containing DNA: This model shows the T-Ag OBD as light grey spheres, and the A1 and B2 motifs as a single small grey sphere. Similarly, the helicase domain is depicted by two light grey spheres (to represent the two tiers of the helicase domain); small dark spheres represent the beta hairpins. Adapted from Figure 7 in Kumar *et al.* (2007). (A). This model focuses on the formation of a single hexamer on P1 and P2. The first view (from the crystal structure) shows the OBDs engaged in a sequence specific DNA interaction with the A1 and B2 motifs deep in the major groove of the DNA at P1 and P2. It depicts the hinge that connects the OBD to the helicase domain as a dashed line. The model shows the helicase domain oriented such that the beta-hairpins interact with the dsDNA and promote the melting of this region. (B) Hexamerization of the OBDs and T-Ag helicase domain: Based on the E1 model, the beta-hairpins are depicted interacting with ssDNA such that the central channels interact with the 3′ extensions of the flanking sequences. In addition, during spiral formation, the A1 and B2 motifs have rearranged such that they now face the helicase domain.

Opposing views of origin melting and the assembly process

In contrast to the model described above, others favour models in which the helicase domain initially assembles around duplex DNA (e.g. Li *et al.*, 2003; Cuesta *et al.*, 2010). According to this model, single-stranded DNA is subsequently extruded through a positively charged side channel (e.g. Li *et al.*, 2003; Wang *et al.*, 2007). While the interior of the central channel of the helicase domain can accommodate dsDNA, the opening of the channel is too small (~12–15 Å). Therefore, if this model is correct, then some structural rearrangement must occur to widen the mouth of the channel. Furthermore, the hypothesis that duplex DNA initially transits the central channel of the helicase domain results in a topological problem; namely, how does one strand of ssDNA transit from the central channel to the outer surface of the helicase domain? This would require either a nick in the ssDNA, an activity that T-Ag does not possess, or selective removal of helicase monomers to enable passage of a strand of single-stranded DNA. Obviously, additional studies are needed to resolve issues relating to the assembly of the helicase domain on the origin and to address how the DNA is routed through the complex.

Structures of viral helicase domains have provided critical insights into how helicases function

No high-resolution structure presently exists of the SV40 helicase domain with DNA. However, the T-Ag helicase domain can oligomerize in the absence of DNA and as noted above, the structure of the DNA free SV40 helicase domain has been determined (Li *et al.*, 2003; Gai *et al.*, 2004). Therefore, while there is some controversy regarding the mechanisms associated with T-Ag assembly, there is none regarding the structure of the hexameric helicase domain. This structure (Fig. 10.11A) revealed that the previously described 'beta-hairpins' protrude into the central region of the channel (Fig. 10.11B). In a subsequent study it was shown that there is a large ~17 Å movement of this motif upon binding ATP (Gai *et al.*, 2004). Indeed, the position of this loop correlates with the nucleotide state of the helicase, with the 'apo' conformation at one extreme, ATP at the other, and ADP in the middle (Fig. 10.11C). These findings allowed the authors to propose a mechanism that correlates translocation of DNA with ATP hydrolysis (Gai *et al.*, 2004).

In addition, a crystal structure of the helicase domain from a structurally related initiator protein E1 from Bovine Papilloma virus has been solved with single-stranded DNA (Enemark and Joshua-Tor, 2006). In this structure, the E1 helicase domain encircles the ssDNA tightly, and there is no room for dsDNA. Each beta-hairpin tip interacts with the phosphate backbone of a single base and the nucleotide state varies along the hexamer from 'apo' to ATP. These authors proposed a 'spiral' stair-casing model of ssDNA translocation coupled with ATP hydrolysis. [Similar models have been published for other hexameric helicases including Rho (Thomsen and Berger, 2009) and T7gp4 (Singleton *et al.*, 2000)] Based on the high amino acid conservation of the residues involved in the stair-casing interactions with those in SV40, these authors propose that SV40 selectively encircles a single strand of DNA. Furthermore, in light of how the ssDNA is routed through the helicase domain, it was proposed that subsequent unwinding events take place via the 'steric exclusion' model (Kaplan and O'Donnell, 2004). We have also suggested that T-Ag unwinds DNA in a manner similar to DNA B (Reese *et al.*, 2006) and thus favour the steric exclusion model (according to this model, each helicase ring surrounds single-stranded DNA, while the second DNA strand is positioned outside the ring).

A working model for T-Ag-dependent initiation of replication

Recent findings obtained with the SV40 system have been considered in terms of a working model for T-Ag dependent initiation events (Fig. 10.13). The first three illustrations in Fig. 10.13 are non-molecular renderings depicting OBD and beta-hairpin-dependent origin recognition (step 1), oligomerization of the helicase domains and subsequent assembly around a single strand of DNA (step 2; singles stranded DNA is symbolized

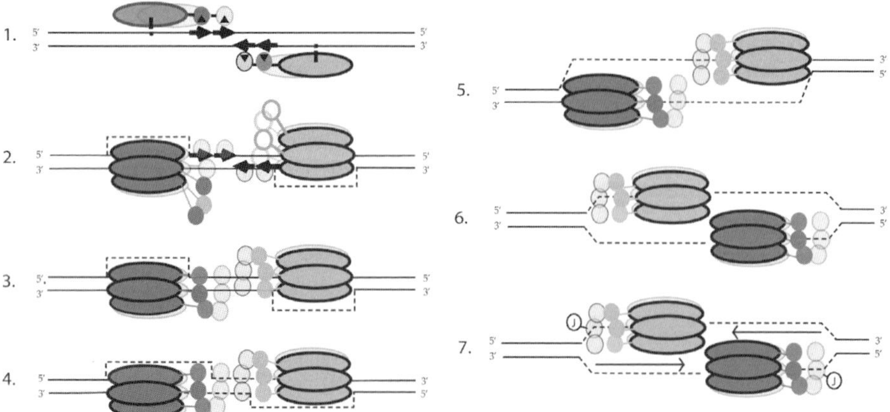

Figure 10.13 An overview depicting the functions of T-Ag during the initiation of SV40 DNA replication: (1) Four T-Ag molecules are depicted interacting with the SV40 origin; the OBDs (circles; A1 and B2 motifs shown as triangles) bind the pentanucleotides (arrows), while the helicase domains (ovals) interact with the flanking-sequences via their beta-hairpins (rectangles). (2) Additional T-Ag molecules are recruited to the origin and during assembly the flanking sequences are melted (the evidence is better for melting the EP than the AT rich region; Borowiec *et al.*, 1990). Upon completion of the oligomerization process, the helicase domains are assembled around one of the newly generated single-strands, while their complements are routed over the outer surface of the helicase domains (single-stranded DNA is symbolized by the dotted lines). (3) Following oligomerization of the helicase domain, the local concentration of the OBDs is increased; as a result, they adopt the 'lock-washer' conformation. To do this, they must release the GAGGC sequences; it is possible that Site II becomes melted at this time. (4) It is proposed that single-stranded DNA emerges through the gap in the spiral (this possibility remains to be proven). (5) Owing to the 3′ to 5′ helicase activity of T-Ag, the hexamers are poised to transit past each other. Consistent with this proposal, recent studies indicate that MCM2-7 helicases function independently (Yardimci *et al.*, 2010). (6) After translocation, the hexamers are positioned at the replication forks; the OBDs are predicted to be proximal to the forks. (7) At any given fork, the T-Ag hexamer is predicted to surround the 'leading-strand' template; therefore, the 'lagging' strand is 'free' and likely to serve as the template for initial DNA synthesis. Consistent with this proposal, it was previously reported that DNA synthesis starts outside of the origin on 'lagging-strand' templates [reviewed in Bullock (1997); the newly formed nascent strands are symbolized by the arrows]. Finally, this model also predicts that the J-domains are positioned at the replication forks where they might promote the removal of proteins that are bound in front of the fork.

by the dotted lines) and spiral formation by the OBDs (step 3). Given that the spiral structure of the T-Ag-obd contains a gap (Meinke *et al.*, 2006), it is postulated that single-strand DNA passes through the gap following the melting of site II [(step 4); there is, however, no direct evidence for this proposal]. Once site II is melted, the individual hexamers are associated with only one strand of DNA (step 5) and are free to translocate along ssDNA via the spiral staircase model (see above). An important recent observation is that sister MCM2-7 helicases normally uncouple upon activation (Yardimci *et al.*, 2010). In view of this finding, and the 3′ to 5′ polarity of T-Ag (Goetz *et al.*, 1988), the hexamers are proposed to move past each other (step 6); a possibility previously suggested for the papillomavirus E1 protein (Enemark and Joshua-Tor, 2006). One outcome of this transition is that the OBD's are predicted to be at the replication forks. Two additional features of this model are depicted in step 7. First, owing to the proximity of the J-domain to the OBDs, the J-domain will also be proximal to the replication fork; a location that suggests that it may play a role in removing proteins bound to duplex DNA in front of the replication fork. Secondly, T-Ag hexamers surround the leading strands at the forks while the lagging strands are free and therefore likely to serve as the template for the initial synthesis of nascent DNA. Consistent with this hypothesis, it was previously demonstrated that DNA synthesis starts outside of the core origin using the 'lagging-strands' templates (Bullock *et al.*, 1991; Denis and Bullock, 1993).

Regulation of the initiation of SV40 DNA replication via cell-cycle dependent phosphorylation of Thr124

Phosphorylation of T-Ag on residue Thr 124 is essential for the initiation of SV40 DNA replication (Schneider and Fanning, 1988; McVey *et al.*, 1989). Unfortunately, the precise function of this phosphorylation event is not known. Findings that hint at its function include the observation that the assembly of single hexamers on the core origin does not depend upon the phosphorylation of Thr 124 (McVey *et al.*, 1993; Moarefi *et al.*, 1993). In contrast, the assembly of the second hexamer, particularly on certain pairs of pentanucleotides, is influenced by the phosphorylation status of Thr124 (e.g. Weisshart *et al.*, 1999; Barbaro *et al.*, 2000). Therefore, since phosphorylation of Thr124 may regulate some aspect of T-Ag assembly, it is apparent that structures including Thr124 would be of tremendous interest. However, a high-resolution structure of a T-Ag domain that includes residue Thr 124 is not yet available. Nevertheless, a relatively low-resolution (~3.2 Å) co-structure of the T-Ag$_{112-260}$ bound to an origin fragment has been obtained (Meinke *et al*; unpublished). Based on the approximate location of Thr124, this structure indicates that phosphorylation of this residue is probably to regulate protein/protein interactions between hexamers and not interactions between T-Ag residues (e.g. those comprising the neighbouring NLS) and origin DNA sequences.

The SV40 replication system will continue to serve as the paradigm for the replication of polyomaviruses

There is still much that we do not understand regarding the initiation of SV40 DNA replication. For example, since the structure of full-length T-Ag is not known, we do not understand the relative orientations of the domains within T-Ag. We

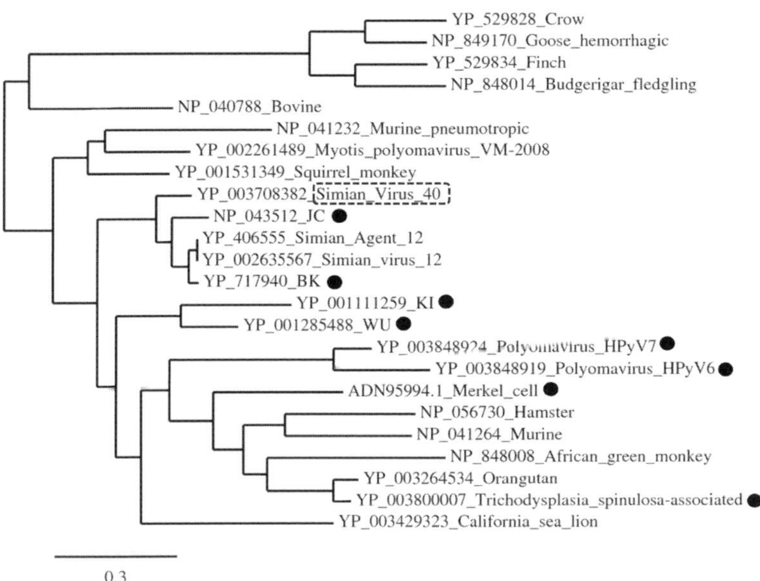

Figure 10.14 Phylogenetic relationship among 24 polyomavirus T-antigens determined via the programme Phylogeny.fr (Dereeper *et al.*, 2008). Note that SV40 T-Ag, shown in a dotted line box, is highly homologous to those encoded by human pathogens including BK and JC polyomaviruses. The accession numbers and the common name are indicated; the eight known human polyomaviruses are indicated by black circles (Gjoerup, 2011).

also do not know how site II is melted, although the T-Ag-OBD has been implicated in this process (Foster and Simmons, 2010). In addition, we do not know if the initiation assemblies described herein are maintained during elongation (Yardimci *et al.*, 2010, and references therein), nor do we understand the mechanism(s) by which the cellular factors are recruited to the origin (e.g. Simmons *et al.*, 2004; Jiang *et al.*, 2006; references therein). There are, however, ample precedents to suggest that these and related issues can be addressed in future experiments using the SV40 replication system.

It has been frequently noted that the sequence features of the SV40 origin are conserved in other members of the Polyomavirus family (e.g. Li *et al.*, 1995). Moreover, inspection of Fig. 10.14 demonstrates that there is a strong evolutionary relationship between the T-Ags encoded by this family. In view of these conserved relationships, SV40 will continue to serve as a powerful model for future studies of the propagation of polyomaviruses, including disease causing human polyomaviruses. In support of this claim, SV40 is the paradigm for those studying Merkel Cell Polyomavirus (Feng *et al.*, 2008); a deadly human pathogen that has been linked to skin cancer (e.g. Kwun *et al.*, 2009). Furthermore, SV40 is an important model for studies of the initiation of DNA replication of JC virus, the causative agent of Progressive Multifocal Leukoencephalopathy (PML) (Padgett *et al.*, 1971). Additional correlations between SV40 DNA replication and human disease, especially for immunocompromised individuals, have been reviewed (Fanning and Pipas, 2006). Finally, the rich history of the SV40 model system augurs well for it's continuing to play significant role(s) in helping to establish how the initiation of DNA replication takes place in eukaryotic organisms.

Acknowledgements

We thank Jacques Archambault, William Bachovchin, Andrew Bohm, Amélie Fradet-Turcotte, Paul Phelan and Jim Sudmeier for their contributions to these studies and Howard Robinson at Brookhaven National Laboratory for data collection. This work was funded by grants from the NIH (9R01GM55397).

References

Alexandrov, A.I., Botchan, M.R., and Cozzarelli, N.R. (2002). Characterization of simian virus 40 T-antigen double hexamers bound to a replication fork. J. Biol. Chem. *277*, 44886–44897.

Arunkumar, A.I., Klimovich, V., Jiang, X., Ott, R.D., Mizoue, L., Fanning, E., and Chazin, W.J. (2005). Insights into hRPA32 C-terminal domain-mediated assembly of the simian virus 40 replisome. Nat. Struct. Mol. Biol. *12*, 332–339.

Barbaro, B.A., Sreekumar, K.R., Winters, D.R., Prack, A.E., and Bullock, P.A. (2000). Phosphorylation of simian virus 40 T-antigen on Thr 124 selectively promotes double-hexamer formation on subfragments of the viral core origin. J. Virol. *74*, 8601–8613.

Bell, S.P., and Dutta, A. (2002). DNA replication in eukaryotic cells. Annu. Rev. Biochem. *71*, 333–374.

Bochkareva, E., Martynowski, D., Seitova, A., and Bochkarev, A. (2006). Structure of the origin-binding domain of simian virus 40 large T antigen bound to DNA. EMBO J. *25*, 5961–5969.

Borowiec, J.A., Dean, F.B., Bullock, P.A., and Hurwitz, J. (1990). Binding and unwinding -how T antigen engages the SV40 origin of DNA replication. Cell *60*, 181–184.

Borowiec, J.A., Dean, F.B., and Hurwitz, J. (1991). Differential induction of structural changes in the simian virus 40 origin of replication by T-antigen. J. Virol. *65*, 1228–1235.

Brewster, A.S., and Chen, X.S. (2010). Insights into the MCM functional mechanism: lessons learnded from the archaeal MCM complex. Crit. Rev. Biochem. Mol. Biol. *45*, 243–256.

Bullock, P.A. (1997). The initiation of simian virus 40 DNA repliation *in vitro*. Crit. Rev. Biochem. Mol. Biol. *32*, 503–568.

Bullock, P.A., Seo, Y.S., and Hurwitz, J. (1991). Initiation of simian virus 40 DNA Synthesis *in vitro*. Mol. Cell. Biol. *11*, 2350–2361.

Burhans, W.C., Carr, A.M., and Wahl, G.M. (2006). DNA replication and cancer. In DNA Replication and Human Diseases, DePamphilis, M., ed. (Cold Spring Harbor Laboratory Press, Cold Spring Harbor), pp. 481–500.

Campbell, K.S., Mullane, K.P., Aksoy, I.A., Stubdal, H., Pipas, J.M., Silver, P.A., Roberts, T.M., Schaffhausen, B.S., and DeCaprio, J.A. (1997). DnaJ/hsp40 chaperone domain of SV40 large T antigen promotes efficient viral DNA replication. Genes Dev. *11*, 1098–1110.

Campos-Olivas, R., Louis, J.M., Clerot, D., Gronenborn, B., and Gronenborn, A.M. (2002). The structure of a replication initiator unites diverse aspects of nucleic acid metabolism. Proc. Natl. Acad. Sci. U.S.A. *99*, 10310–10315.

Chen, L., Joo, W.S., Bullock, P.A., and Simmons, D.T. (1997). The N-terminal side of the origin-binding domain of simian virus 40 large T-antigen is involved in A/T untwisting. J. Virol. *71*, 8743–8749.

Cuesta, I., Nunez-Ramirez, R., Scheres, S.H.W., Gai, D., Chen, X.S., Fanning, E., and Carazo, J.M. (2010).

Conformational Rearrangements of SV40 Large T Antigen during Early Replication Events. J. Mol. Biol. *397*, 1276–1286.

Dean, F.B., Dodson, M., Echols, H., and Hurwitz, J. (1987). ATP-dependent formation of a specialized nucleoprotein structure by simian virus 40 (SV40) large tumour antigen at the SV40 replication origin. Proc. Natl. Acad. Sci. U.S.A. *84*, 8981–8985.

DeLano, W.L. (2002). The PyMOL Molecular Graphics System (Delano Scientific, Palo Alto, CA, USA).

DeLucia, A.L., Deb, S., Partin, K., and Tegtmeyer, P. (1986). Functional interactions of the simian virus 40 core origin of replication with flanking regulatory sequences. J. Virol. *57*, 138–144.

Denis, D., and Bullock, P.A. (1993). Primer-DNA formation during simian virus 40 DNA replication *in vitro*. Mol. Cell. Biol. *13*, 2882–2890.

Dereeper, A., Guignon, V., Blanc, G., Audic, S., Buffet, S., Chevenet, F., Dufayard, J.F., Guindon, S., Lefort, V., Lescot, M., *et al.* (2008). Phylogeny. fr: robust phylogenetic analysis for the non-specialist. Nucleic Acids Res. *36*, W465–469.

Dickmanns, A., Zeitvogel, A., Simmersbach, F., Weber, R., Arthur, A.K., Dehde, S., Wildeman, A.G., and Fanning, E. (1994). The kinetics of simian virus 40-induced progression of quiescent cells into S phase depend on four independent functions of large T antigen. J. Virol. *68*, 5496–5508.

Dodson, M., Dean, F.B., Bullock, P., Echols, H., and Hurwitz, J. (1987). Unwinding of duplex DNA from the SV40 origin of replication by T antigen. Science *238*, 964–967.

Enemark, E.J., Chen, G., Vaughn, D.E., Stenlund, A., and Joshua-Tor, L. (2000). Crystal structure of the DNA-binding domain of the replication initiation protein E1 from papillomavirus. Mol. Cell. *6*, 149–158.

Enemark, E.J., and Joshua-Tor, L. (2006). Mechanism of DNA translocation in a replicative hexameric helicase. Nature *442*, 270–275.

Fanning, E. (1994). Control of SV40 DNA replication by protein phosphorylation: a model for cellular DNA replication? Trends Cell Biol. *4*, 250–255.

Fanning, E., and Knippers, R. (1992). Structure and function of simian virus 40 large tumour antigen. Ann. Rev. Biochem. *61*, 55–85.

Fanning, E., and Pipas, J.M. (2006). Polyomavirus. DNA Rep 627 641.

Fanning, E., and Zhao, K. (2009). SV40 DNA replication: From the A gene to a nanomachine. Virology *384*, 352–359.

Feng, H., Shuda, M., Chang, Y., and Moore, P.S. (2008). Clonal integration of a polyomavirus in human Merkel cell carcinoma. Science *319*, 1096–1100.

Foster, E.C., and Simmons, D.T. (2010). The SV40 large T-antigen origin binding domain directly participates in DNA unwinding. Biochemistry *49*, 2087–2096.

Fradet-Turcotte, A., Vincent, C., Joubert, S., Bullock, P.A., and Archambault, J. (2007). Quantitative analysis of the binding of simian virus 40 large T antigen to DNA. J. Virol. *81*, 9162–9174.

Gai, D., Zhao, R., Li, D., Finkielstein, C.V., and Chen, X.S. (2004). Mechanisms of conformational change for a replicative hexameric helicase of SV40 large tumour antigen. Cell *119*, 47–60.

Gilbert, D.M. (2001). Making sense of eukaryotic DNA replication origins. Science *294*, 96–100.

Gjoerup, O. (2011). Polyomaviruses and cancer. In Cancer-Associated Viruses, Robertson, E., ed. (Springer).

Goetz, G.S., Dean, F.B., Hurwitz, J., and Matson, S.W. (1988). The unwinding of duplex regions in DNA by the simian virus 40 large tumour antigen-associated DNA helicase activity. J. Biol. Chem. *263*, 383–392.

Gomez-Lorenzo, M.G., Valle, M., Donate, L.E., Gruss, C., Sorzano, C.O.S., Radermacher, M., Frank, J., and Carazo, J.M. (2003). Large T Antigen on the SV40 Origin of Replication: A 3D Snapshot Prior to DNA Replication. The EMBO journal *22*, 6205–6213.

Huang, H., Zhao, K., Arnett, D.R., and Fanning, E. (2010). A Specific Docking Site for DNA Polymerase a-primase on the SV40 Helicase is Required for Viral Primosome Activity, But Helicase Activity is Dispensable. J. Biol. Chem. *285*, 33475–33484.

Ishimi, Y., Claude, A., Bullock, P., and Hurwitz, J. (1988). Complete enzymatic synthesis of DNA containing the SV40 origin of replication. J. Biol. Chem. *263*, 19723–19733.

Jiang, X., Klimovich, V., Arunkumar, A.I., Hysinger, E.B., Wang, Y., Ott, R.D., Guler, G.D., Weiner, B., Chazin, W.J., and Fanning, E. (2006). Structural mechanism of RPA loading on DNA during activation of a simple pre-replication complex. EMBO J. *25*, 5516–5526.

Joo, W.S., Kim, H.Y., Purviance, J.D., Sreekumar, K.R., and Bullock, P.A. (1998). Assembly of T-antigen double hexamers on the simian virus 40 core origin requires only a subset of the available binding sites. Mol. Cell. Biol. *18*, 2677–2687.

Kaplan, D.L., and O'Donnell, M. (2004). Twin DNA pumps of a hexameric helicase provide power to simultaneously melt two duplexes. Mol. Cell *15*, 453–465.

Kelley, W.L., and Georgopoulos, C. (1997). The T/t common exon of simian virus 40, JC, and BK polyomavirus T antigens can functionally replace the J-domain of the *Escherichia coli* Dna J molecular chaperone. Proc. Natl. Acad. Sci. U. .A. *94*, 3679–3684.

Kelley, W.L., and Landry, S.J. (1994). Chaperone power in a virus? TIBS *19*, 277–278.

Khopde, S., and Simmons, D.T. (2008). Simian virus 40 DNA replication is dependent on an interaction between topoisomerase I and the C-terminal end of T antigen. J. Virol. *82*, 1136–1145.

Kim, H.Y., Barbaro, B.A., Joo, W.S., Prack, A., Sreekumar, K.R., and Bullock, P.A. (1999). Sequence requirements for the assembly of simian virus40 T-antigen and T-antigen origin binding domain on the viral core origin of replication. J. Virol. *73*, 7543–7555.

Kim, H.-Y., Ahn, B.Y., and Cho, Y. (2001). Structural basis for the inactivation of retinoblastoma tumor suppressor by SV40 large T antigen. EMBO J. *20*, 295–304.

Kumar, A., Meinke, G., Reese, D.K., Moine, S., Phelan, P.J., Fradet-Turcotte, A., Archambault, J., Bohm, A., and Bullock, P.A. (2007). Model for T-antigen-dependent melting of the simian virus 40 core origin based on studies of the interaction of the beta-hairpin with DNA. J. Virol. *81*, 4808–4818.

Kwun, H.J., Guastafierro, A., Shuda, M., Meinke, G., Bohm, A., Moore, P.S., and Chang, Y. (2009). The minimum replication origin of Merkel cell polyomavirus has a unique large T-antigen loading architecture and requires small T-antigen expression for optimal replication. J. Virol. *83*, 12118–12128.

Li, D., Zhao, R., Lilyestrom, W., Gai, D., Zhang, R., DeCaprio, J.A., Fanning, E., Jochimiak, A., Szakonyi, G., and Chen, X.S. (2003). Structure of the replicative helicase of the oncoprotein SV40 large tumour antigen. Nature *423*, 512–518.

Li, J.J., Peden, K.W.C., Dixon, R.A.F., and Kelly, T. (1986). Functional organization of the simian virus 40 origin of DNA replication. Mol. Cell. Biol. *6*, 1117–1128.

Li, L., Li, B.L., Hock, M., Wang, E., and Folk, W.R. (1995). Sequences flanking the pentanucleotide T-antigen binding sites in the polyomavirus core origin help determine selectivity of DNA replication. J. Virol. *69*, 7570 – 7578.

Lilyestrom, W., Klein, M.G., Zhang, R., Joachimiak, A., and Chen, X.S. (2006). Crystal structure of SV40 large T-antigen bound to p53: interplay between a viral oncoprotein and a cellular tumor suppressor. Genes Dev. *20*, 2373–2382.

Luo, X., Sanford, D.G., Bullock, P.A., and Bachovchin, W.W. (1996). Structure of the origin specific DNA-binding domain from simian virus 40 T-antigen. Nature struct. Biol. 3, 1034–1039.

McVey, D., Brizuela, L., Mohr, I., Marshak, D.R., Gluzman, Y., and Beach, D. (1989). Phosphorylation of large tumour antigen by cdc2 stimulates SV40 DNA replication. Nature *341*, 503–507.

McVey, D., Ray, S., Gluzman, Y., Berger, L., Wildeman, A.G., Marshak, D.R., and Tegtmeyer, P. (1993). cdc2 phosphorylation of threonine 124 activates the origin-unwinding functions of simian virus 40 T antigen. J. Virol. *67*, 5206–5215.

Masai, H., Matsumoto, S., You, Z., Yoshizawa-Sugata, N., and Oda, M. (2010). Eukaryotic chromosome DNA replication: where, when, and how? Annu. Rev. Biochem. *79*, 89–130.

Mastrangelo, I.A., Hough, P.V.C., Wall, J.S., Dodson, M., Dean, F.B., and Hurwitz, J. (1989). ATP-dependent assembly of double hexamers of SV40 T antigen at the viral origin of DNA replication. Nature 338, 658–662.

Maulbecker, C., Mohr, I., Gluzman, Y., Bartholomew, J., and Botchan, M. (1992). A deletion in the simian virus 40 large T antigen impairs lytic replication in monkey cells *in vivo* but enhances DNA replication *in vitro*: new complementation function of T antigen. J. Virol. *66*, 2195–2207.

Meinke, G., Bullock, P.A., and Bohm, A. (2006). Crystal structure of the simian virus 40 large T-antigen origin-binding domain. J. Virol. *80*, 4304–4312.

Meinke, G., Phelan, P.J., Moine, S., Bochkareva, E., Bochkarev, A., Bullock, P.A., and Bohm, A. (2007). The crystal structure of the SV40 T-antigen origin binding domain in complex with DNA. PLoS Biol. *5*, e23.

Meinke, G., Phelan, P.J., Fradet-Turcotte, A., Bohm, A., Archambault, J., and Bullock, P.A. (2011). Structure-Based Analysis of the Interaction between the SV40 T-Antigen Origin Binding Domain and ssDNA. J. Virol. *85*, 818–827.

Moarefi, I.F., Small, D., Gilbert, I., Hopfner, M., Randall, S.K., Schneider, C., Russo, A.A.R., Ramsperger, U., Arthur, A.K., Stahl, H., *et al.* (1993). Mutation of the cyclin-dependent kinase phosphorylation site in simian virus 40 (SV40) large T antigen specifically blocks SV40 origin DNA unwinding. J. Virol. *67*, 4992–5002.

Neuwald, A.F., Aravind, L., Spouge, J.L., and Koonin, E.V. (1999). AAA^+: A class of chaperone-like ATPases associated with the assembly, operation, and disassembly of protein complexes. Genome Res. *9*, 27–43.

Padgett, B.L., Zurhein, G.M., Walker, D.L., Eckroade, R.J., and Dessel, B.H. (1971). Cultivation of papova-like virus from human brain wiht progressive multifocal leukoencephalopathy. Lancet *297*, 1257–1260.

Parsons, R., Anderson, M.E., and Tegtmeyer, P. (1990). Three domains in the simian virus 40 core origin orchestrate the binding, melting, and DNA helicase activities of T antigen. J. Virol. *64*, 509–518.

Pipas, J.M. (1992). Common and unique features of T antigens encoded by the polyomavirus group. J Virol. *66*, 3979–3985.

Reese, D.K., Sreekumar, K.R., and Bullock, P.A. (2004). Interactions required for binding of simian virus40 T antigen to the viral origin and molecular modeling of initial assembly events. J. Virol. *78*, 2921–2934.

Reese, D.K., Meinke, G., Kumar, A., Moine, S., Chen, K., Sudmeier, J.L., Bachovchin, W., Bohm, A., and Bullock, P.A. (2006). Analyses of the interaction between the origin binding domain from simian virus40 T-antigen and single stranded DNA provides insights into DNA unwinding and initiation of DNA replication. J. Virol. *80*, 12248–12259.

Schneider, J., and Fanning, E. (1988). Mutations in the phosphorylation sites of simian virus 40 (SV40) T antigen alter its origin DNA-binding specificity for sites I or II and affect SV40 DNA replication activity. J. Virol. *62*, 1598–1605.

Sclafani, R.A., Fletcher, R.J., and Chen, X.S. (2004). Two heads are better than one: regulation of DNA replication by hexameric helicases. Genes Dev. *18*, 2039–2045.

Shen, J., Gai, D., Patrick, A., Greenleaf, W.B., and Chen, X.S. (2005). The roles of the residues on the channel β-hairpin and loop structures of simian virus 40 hexameric helicase. Proc. Natl. Acad. Sci. U.S.A. *102*, 11248–11253.

Simmons, D.T. (2000). SV40 large T antigen functions in DNA replication and transformation. Adv. Virus Res. *55*, 75–134.

Simmons, D.T., Loeber, G., and Tegtmeyer, P. (1990). Four major sequence elements of simian virus 40 large T antigen coordinate its specific and nonspecific DNA binding. J. Virol. *64*, 1973–1983.

Simmons, D.T., Gai, D., Parsons, R., Debes, A., and Roy, R. (2004). Assembly of the replication initiation complex on SV40 origin DNA. Nucleic Acids Res. *32*, 1103–1112.

Singleton, M.R., Sawaya, M.R., Ellenberger, T., and Wigley, D.B. (2000). Crystal structure of T7 gene 4 ring helicase indicates a mechanism for sequential hydrolysis of nucleotides. Cell *101*, 589–600.

Smelkova, N.V., and Borowiec, J.A. (1997). Dimerization of simian virus40 T-antigen hexamers activates T-antigen DNA helicase activity. J. Virol. *71*, 8766–8773.

Spence, S.L., and Pipas, J.M. (1994). SV40 Large T antigen functions at two distinct steps in virion assembly. Virology *204*, 200–209.

Sreekumar, K.R., Prack, A.E., Winters, D.R., Barbaro, B.A., and Bullock, P.A. (2000). The simian virus 40 core origin contains two separate sequence modules that support T-antigen double-hexamer assembly. J. Virol. *74*, 8589–8600.

Srinivasan, A., McClellan, A.J., Vartikar, J., Marks, I., Cantalupo, P., Li, Y., Whyte, P., Rundell, K., Brodsky, J.L., and Pipas, J.M. (1997). The amino-terminal transforming region of simian virus 40 large T and small T antigens functions as a J domain. Mol. Cell Biol. *17*, 4761–4773.

Stillman, B., Gerard, R.D., Guggenheimer, R.A., and Gluzman, Y. (1985). T antigen and template requirement for SV40 DNA replication *in vitro*. EMBO J. *4*, 2933–2939.

Taneja, P., Nasheuer, H.P., Hartmann, H., Grosse, F., Fanning, E., and Weisshart, K. (2007). Timed interactions between viral and cellular replication factors during the initiation of SV40 *in vitro* DNA replication. Biochem. J. *407*, 313–320.

Thomsen, N.D., and Berger, J.M. (2009). Running in reverse: the structural basis for translocation polarity in hexameric helicases. Cell *139*, 523–535.

Tsurimoto, T., Fairman, M.P., and Stillman, B. (1989). Simian virus 40 DNA replication *in vitro*: identification of multiple stages of initiation. Mol. Cell. Biol. *9*, 3839–3849.

Valle, M., Gruss, C., Halmer, L., Carazo, J.M., and Donate, L.E. (2000). Large T-antigen double hexamers imaged at the simian virus40 origin of replication. Mol. Cell. Biol. *20*, 34–41.

Wang, W., Manna, D., and Simmons, D.T. (2007). Role of the hydrophilic channels of simian virus 40 T-antigen helicase in DNA replication. J. Virol. *81*, 4510–4519.

Weinberg, D.H., Collins, K.L., Simancek, P., Russo, A., Wold, M.S., Virshup, D.M., and Kelly, T.J. (1990). Reconstitution of simian virus 40 DNA replication with purified proteins. Proc. Natl. Acad. Sci. U.S.A. *87*, 8692–8696.

Weisshart, K., Bradley, M.K., Weiner, B.M., Schneider, C., Moarefi, I., Fanning, E., and Arthur, A.K. (1996). An N-terminal deletion mutant of simian virus 40 (SV40) large T antigen oligomerizes incorrectly on SV40 DNA but retains the ability to bind to DNA polymerase a and replicate SV40 DNA *in vitro*. J. Virol. *70*, 3509–3516.

Weisshart, K., Taneja, P., Jenne, A., Herbig, U., Simmons, D.T., and Fanning, E. (1999). Two regions of simian virus40 T antigen determine cooperativity of double-hexamer assembly on the viral origin of DNA replication and promote hexamer interactions during bidirectional origin DNA unwinding. J. Virol. *73*, 2201–2211.

Wessel, R., Schweizer, J., and Stahl, H. (1992). Simian virus 40 T-antigen DNA helicase is a hexamer which forms a binary complex during bidirectional unwinding from the viral origin of DNA replication. J. Virol. *66*, 804 – 815.

Wu, X., Avni, D., Chiba, T., Yan, F., Zhao, Q., Lin, Y., Heng, H., and Livingston, D. (2004). SV40 T antigen interacts with Nbs1 to disrupt DNA replication control. Genes Dev. *18*, 1305–1316.

Wun-Kim, K., Upson, R., Young, W., Melendy, T., Stillman, B., and Simmons, D.T. (1993). The DNA-binding domain of simian virus 40 tumour antigen has multiple functions. J. Virol. *67*, 7608–7611.

Yardimci, H., Loveland, A.B., Habuchi, S., van Oijen, A.M., and Walter, J.C. (2010). Uncoupling of Sister Replisomes during Eukaryotic DNA Replication. Mol. Cell *40*, 834–840.

Human Papillomavirus DNA Replication: Insights into the Structure and Regulation of a Eukaryotic DNA Replisome

11

Claudia M. D'Abramo, Amélie Fradet-Turcotte and Jacques Archambault

Abstract

Human papillomaviruses (HPV) replicate their double-stranded, circular DNA genome using two virally encoded proteins, E1 and E2, and several components of the host DNA replication machinery. Viral DNA replication is initiated by the recruitment of the E1 helicase to the viral origin of replication, through its interaction with E2, where it assembles into double hexamers and interacts with cellular factors to form an active replication complex. As such, replication of the HPV genome bears many similarities with that of the host DNA at the biochemical level, making these small DNA tumour viruses excellent models for the study of eukaryotic DNA replication. This chapter will provide a detailed view of HPV DNA replication with an emphasis on the different mechanisms utilized by these viruses to maintain and amplify their genome in undifferentiated and differentiated keratinocytes, respectively, and how these processes are regulated by phosphorylation of the E1 protein.

Introduction

DNA replication in eukaryotic cells is a highly regulated and complex process dependent on the interplay between more than 16 cellular proteins (Sclafani and Holzen, 2007). Much of what is known about the basic mechanisms of eukaryotic DNA replication has been initially revealed from extensive studies on double-stranded DNA tumour viruses including human papillomaviruses (HPV). Unlike mammalian cells, which contain several origins of replication that lack a clear consensus sequence, HPV initiate replication of their circular double-stranded DNA genome (episome) from a single, well-characterized origin of replication. Moreover, given the small number of proteins encoded in their genome, HPV rely intensively upon host DNA replication factors to replicate their DNA. Indeed, only two viral gene products participate in viral DNA synthesis. The rest of the required proteins are from the host, making replication of the HPV genome akin to that of cellular DNA at the biochemical level. As such, several studies on HPV have provided fundamental insights into the mechanism and regulation of eukaryotic DNA replication. For instance, assembly of the viral E1 helicase at the origin has provided a useful framework to understand how the MCM complex, the cellular replicative helicase, assembles during the initiation of eukaryotic replication. Additionally, the fact that E1 interacts with several cellular replication factors to regulate DNA replication has helped establish their function in the assembly and activation of the eukaryotic replication complex, as well as led to the identification of new cellular proteins involved in this process. This chapter aims to provide a detailed look at HPV DNA replication, with a particular emphasis on the role of the viral protein E1, its assembly into replication-competent double hexamers, and its interaction with viral and cellular factors that regulate DNA synthesis. In the following sections, we will review our current understanding of the mechanisms involved in the initiation of HPV DNA replication and viral DNA synthesis and how these steps are regulated by post-translational modification of E1. Finally, as the mode of DNA replication changes during the viral life cycle, a unique and characteristic feature of HPV, the differentiation-dependent amplification of the HPV genome will also be reviewed.

Initiation of viral DNA replication

Components of the HPV initiation complex

E1 and E2 are among the early proteins that are expressed upon HPV infection. These two proteins form a complex at the viral origin of replication to initiate the assembly of a bidirectional replication fork. The structure and function of E1, E2 and the origin are summarized below with an emphasis on how these factors assemble into a DNA replication initiation complex. E1 has a dual role in HPV replication in that it functions both as a DNA-binding protein that binds to the viral origin of replication, and as an ATP-dependent helicase that unwinds duplex DNA ahead of the replication fork. Structure–function studies have shown that E1, which varies in size from 600 to 650 amino acids depending on the HPV type, contains three functional domains: a C-terminal enzymatic domain with ATPase/helicase activity and capable of oligomerization into hexamers (Titolo *et al.*, 2000; White *et al.*, 2001), a central origin-binding domain (OBD) that contains an important dimerization interface (Enemark *et al.*, 2000, 2002; Titolo *et al.*, 2003b; Auster and Joshua-Tor, 2004), and an N-terminal regulatory region that is essential for replication *in vivo* but dispensable *in vitro* (Fig. 11.1A) (Sun *et al.*, 1998; Amin *et al.*, 2000; Morin *et al.*, 2011). The N-terminal region contains several conserved sequences that modulate E1 function and activity in cells. These include sequences for nuclear localization (NLS) and export (NES), a conserved cyclin-binding motif (CBM) that interacts with cyclin A/E-Cdk2, and several phosphorylation sites for Cdk and other kinases (Fig. 11.1B) (Ma *et al.*, 1999; Deng *et al.*, 2004; Yu *et al.*, 2007; Fradet-Turcotte *et al.*, 2010a).

In addition to E1, the E2 protein is required for efficient initiation of viral DNA replication *in vivo*. E2 is organized into two functional domains: an N-terminal transactivation domain (TAD) that is involved in transcriptional regulation and direct association with E1, and a C-terminal DNA-binding/dimerization domain (DBD) (Fig. 11.2A) (Harris and Botchan, 1999; Antson *et al.*, 2000; Hegde, 2002; Dell *et al.*, 2003; Wang *et al.*, 2004; de Prat-Gay *et al.*, 2008). Both these domains are separated by a hinge region that is thought to be flexible and whose function has been poorly characterized. E2 is a multifunctional protein that binds to specific sites in the regulatory region of the viral genome, known as the Long-Control Region (LCR), to promote viral DNA replication (Remm *et al.*, 1992; Lu *et al.*, 1993; Stubenrauch *et al.*, 1998). In addition, E2 regulates viral gene transcription, governs proper segregation of the viral episome to daughter cells at mitosis, and exerts pro-apoptotic activity (Ilves *et al.*, 1999; Webster *et al.*, 2000; Howley, 2001; Hegde, 2002; Blachon and Demeret, 2003; McBride *et al.*, 2004; Parish *et al.*, 2006). These latter functions of E2 have been extensively reviewed and are beyond the scope of this chapter. Rather, the role of E2 during viral replication will be discussed in greater detail below.

Viral DNA replication is initiated by the recruitment of E1 to specific DNA sequences within the viral origin of replication. The origins of replication of DNA tumour viruses are well characterized. For HPV, the origin is localized to the 3′ region of the LCR, upstream of the early open reading frames (Chiang *et al.*, 1992; Del Vecchio *et al.*, 1992; Sverdrup and Khan, 1994; Lee *et al.*, 1997). The E1-binding region spans an 18-nucleotide AT-rich imperfect palindrome that contains four E1-binding sites (E1BS) of the consensus sequence ATTGTT, arranged as two overlapping pairs of inverted repeats (E1BS 1 and 3 or E1BS 2 and 4); both E1BS within each repeat being separated by three nucleotides from each other (Fig. 11.2B) (Sun *et al.*, 1996; Titolo *et al.*, 2003b). An additional AT-rich domain lies adjacent to the E1 palindrome and is required for viral DNA replication. The HPV LCR also contains four copies of a 12 nucleotide palindromic sequence to which the E2 protein binds as a dimer (Remm *et al.*, 1992; Lu *et al.*, 1993; Stubenrauch *et al.*, 1998). Unlike the E1BS, the E2-binding sites (E2BS), with a

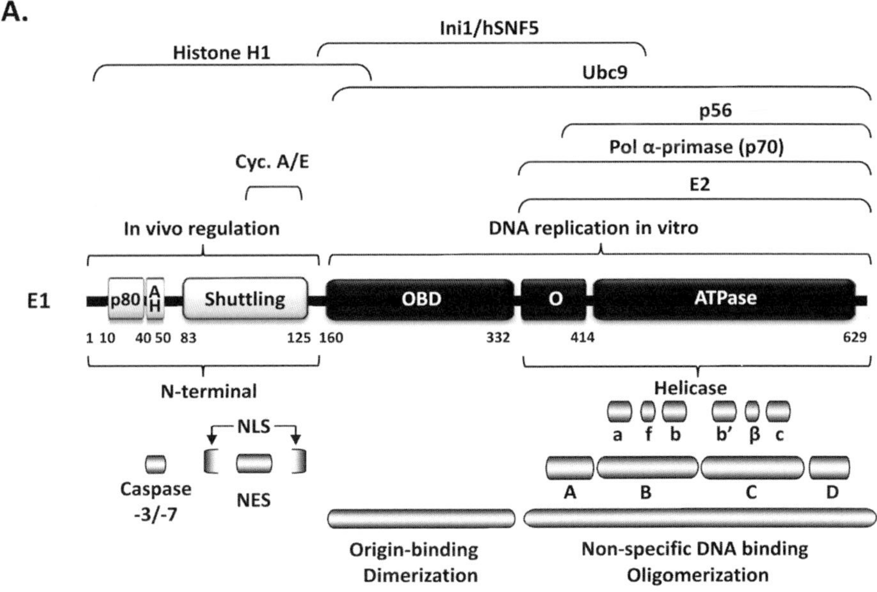

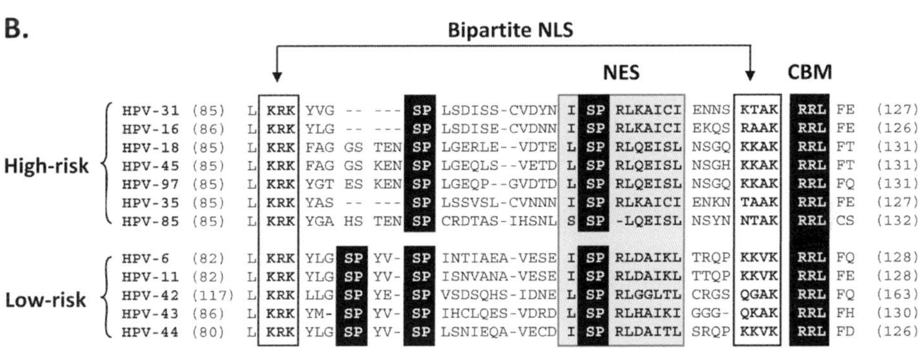

Figure 11.1 Domain structure of the HPV E1 helicase. (A) Schematic representation of the HPV31 E1 protein. Locations of the different functional domains of E1 are indicated, including the p80-binding site (p80), the amphipathic helix (AH) that contains caspase cleavage sites (caspase-3/-7), the shuttling module, and the domains required for origin binding (OBD), oligomerization (O) and ATPase activity. Also shown are regions with sequence homology to the large T-Ag of SV40 (A–D) and the motifs a, f, b, b′, β and c representing the Walker A, FWL, Walker B, B′, β-hairpin and C motifs, respectively. The domains and proteins involved in mediating important protein–protein interactions required for viral DNA synthesis are listed at the top of each domain. (B) Sequence alignment of the shuttling modules of E1 from high- and low-risk HPV types. The location of the conserved bipartite nuclear localization (NLS) and nuclear export (NES) signals are indicated. Also shown is the location of the cyclin-binding motif (CBM) and Cdk phosphorylation sites (S/T-P).

consensus sequence ACCN_6GGT, are very well conserved among different HPV types (Androphy *et al.*, 1987; Alexander and Phelps, 1996) and are considered the major *cis*-acting sequences required for viral DNA replication. In transient DNA replication assays, two of the three E2BS located close to the E1 palindrome are sufficient to achieve near wild type levels of viral DNA replication (Chiang *et al.*, 1992; Lu *et al.*, 1993; Russell and Botchan, 1995; Sverdrup and Khan, 1995; Titolo *et al.*, 1999; Dixon *et al.*, 2000). Robust transient DNA replication is typically achieved

A

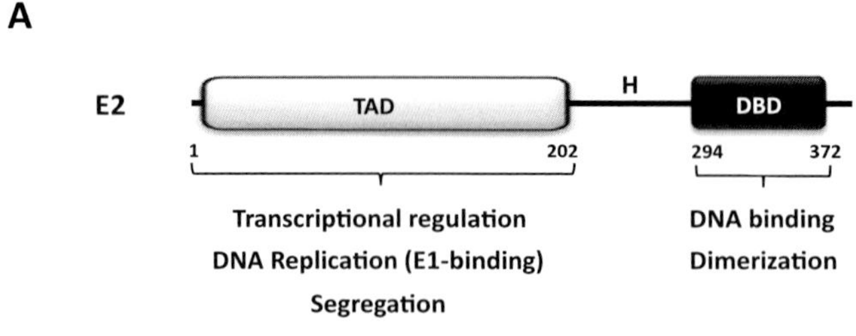

B

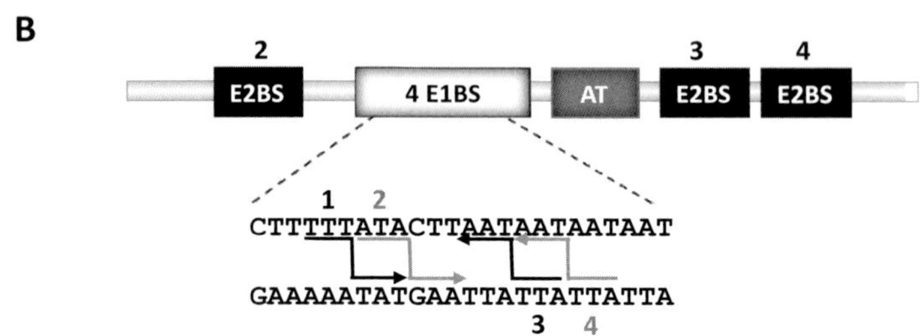

Figure 11.2 Functional regions of the E2 protein and of the origin. (A) Schematic representation of the HPV31 E2 protein. The location and functions of each of the different domains within the E2 protein are indicated: transactivation domain (TAD), hinge region (H), and the DNA-binding/dimerization domain (DBD). (B) Diagram of the minimal HPV replication origin. The E2-binding sites (E2BS) 2, 3 and 4 are depicted by black boxes, the AT rich region by a light grey box, and the region containing the 4 E1-binding sites (E1BS) is in white. The nucleotide sequence of the E1BS is enlarged and shows the 4 binding sites upon which the first and second E1 dimers are formed, as indicated by the grey and black arrows, respectively.

with a minimal origin fragment containing three E2BS, the E1 palindrome and the AT-rich region, all of which contribute to the overall efficiency of the process (Fig. 11.2B) (Sverdrup and Khan, 1994; Russell and Botchan, 1995).

Assembly of E1 at the origin

A model of the assembly of E1 at the origin is shown in Fig. 11.3. Monomeric E1 binds to DNA with little sequence specificity and weak affinity (Dixon *et al.*, 2000). Two different domains of E1 interact with DNA. The OBD, located in the central region of the protein, binds to DNA in a sequence-specific manner, while the C-terminal helicase domain binds with little to no sequence specificity (Wilson *et al.*, 2002; Titolo *et al.*, 2003a). While the non-specific DNA-binding activity of E1 is necessary for its assembly into double hexamers, it is initially inhibited by E2 to allow for the specific interaction of the OBD with the origin.

Assembly of the E1–E2-ori ternary complex is dependent upon the interaction of E1 and E2 with DNA, and upon a critical protein–protein interaction between E1 and E2 which has been characterized by X-ray crystallography (Abbate *et al.*, 2004). Specifically, the C-terminal ATPase domain of E1 interacts with the helical domain of the E2 TAD, which lies opposite to the surface of the TAD required for its transactivation activity and interaction with Brd4 (Fig. 11.4A) (Sakai *et al.*, 1996; Cooper *et al.*, 1998; Abbate *et al.*, 2006; Senechal *et al.*, 2007). Thus, E2 acts as a specificity factor by recruiting E1 directly to the origin and by suppressing its non-specific DNA-binding activity (Mohr *et al.*, 1990; Yang *et al.*, 1991; Seo *et al.*, 1993a; Berg and Stenlund, 1997; Bonne-Andrea *et al.*, 1997; Sanders and Stenlund, 1998).

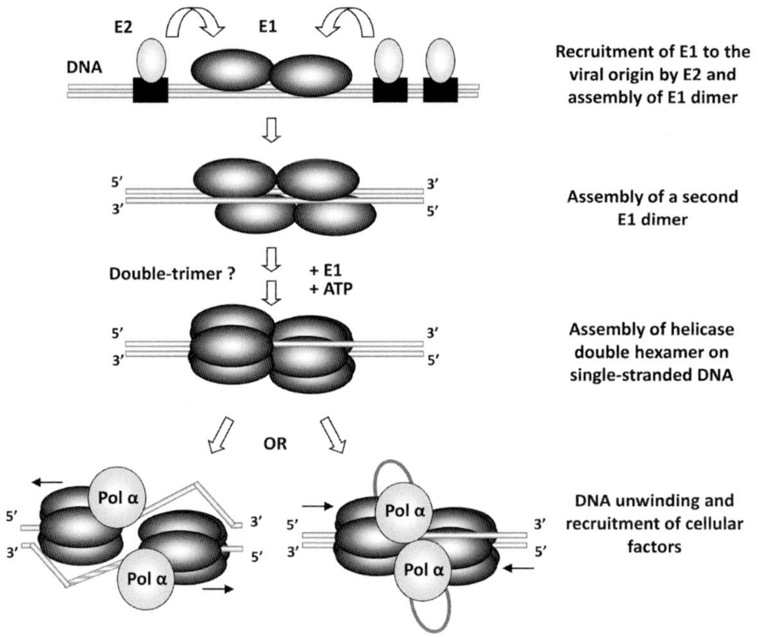

Figure 11.3 Assembly of the E1 double hexamer at the origin. Schematic diagram of the initiation of HPV DNA replication. Replication is initiated by the recruitment of E1 (dark grey), by E2 (light grey), to the viral origin for the assembly of the first E1 dimer. Additional E1 molecules are then recruited for the assembly of a second E1 dimer and/or double-trimer, which then serves as a platform for the assembly of other E1 proteins into a replication-competent double hexamer capable of bidirectional DNA unwinding. ATP stimulates the oligomerization of E1 and is further needed to power the helicase activity of E1. Finally, E1 interacts with host cell replication factors such as DNA polymerase α primase (Pol α) to promote DNA synthesis. Unwinding of DNA can proceed either with both E1 double hexamers remaining assembled at the origin (right) or with the E1 hexamers separating and migrating in opposite directions on complementary strands of DNA (left).

Mutagenesis studies have identified key amino acids in E2 that are critical for its interaction with E1, in particular a highly conserved glutamic acid (E39 in HPV31 E2) located in the central α-helix of the TAD which, in the crystal structure of the E1–E2 complex, forms a salt-bridge with a conserved arginine in the C-terminal region of E1 (R427 in HPV31 E1) (Sakai *et al.*, 1996; Abbate *et al.*, 2004; Wang *et al.*, 2004). Accordingly, mutation of E39 for either alanine or glutamine (E39A/Q) was found to drastically reduce DNA replication (Sakai *et al.*, 1996; Cooper *et al.*, 1998; Fradet-Turcotte *et al.*, 2010b). In addition, alanine substitutions of leucine 15 (L15A), tyrosine 19 (Y19A), and of serine 98 (S98A) in the TAD of HPV11 E2 also appear to specifically inhibit assembly of the E1–E2 complex without affecting the overall structure of the TAD (Wang *et al.*, 2004). For E1, substitution of R247 for alanine (R247A) completely inhibited binding to E2, consistent with R247 forming an ion pair with E39 of the TAD (Abbate *et al.*, 2004). Furthermore, while ATP itself is not required for E1 to interact with E2, the integrity of the P-loop within the ATP-binding domain appears to be essential (Titolo *et al.*, 1999).

The ternary complex formed between E1, E2, and the origin serves as a platform for the assembly of additional E1 proteins into replication-competent double hexamers capable of bidirectional DNA unwinding (Sedman and Stenlund, 1998; Fouts *et al.*, 1999; Titolo *et al.*, 2000). How the E1 double hexamer is assembled is not completely understood, but may involve a double dimer (tetramer) and/or double trimer intermediate (Fig. 11.3) (Schuck and Stenlund, 2005a). Assembly of the E1 double hexamer also requires ATP to promote E1 hexamerization and to release

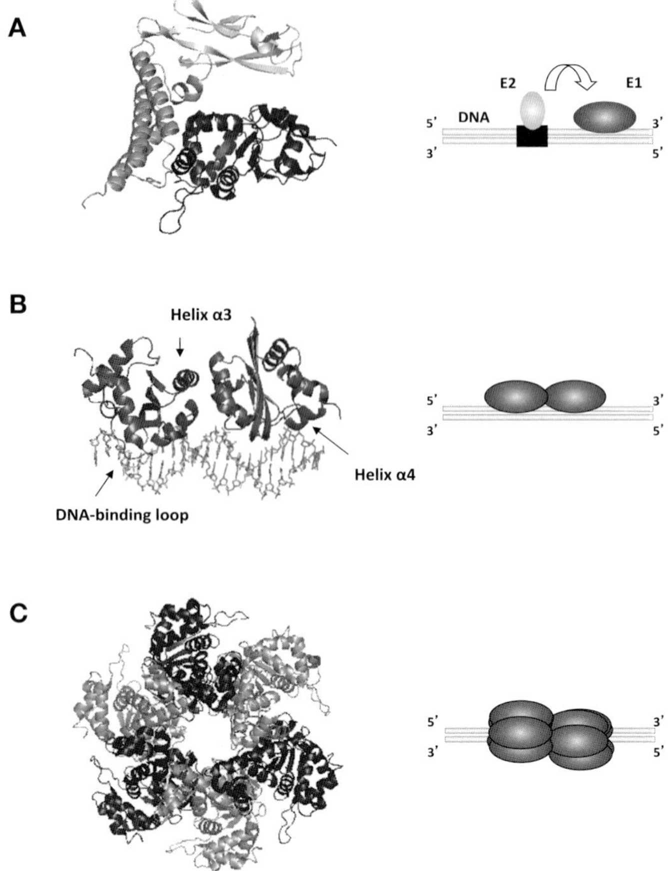

Figure 11.4 Crystal structures of key intermediates in the assembly of the E1 double hexamer. (A) Structure of the complex formed between the E2 TAD and the C-terminal domain of E1 (PDB accession number 1TUE) (Abbate *et al.*, 2004). The N-terminal helical subdomain of the HPV18 E2 TAD is coloured grey, while its C-terminal β-pleated sheet is coloured light grey. The C-terminal domain of the HPV18 E1 protein is depicted in dark grey. (B) Structure of two OBDs of the bovine papillomavirus (BPV) E1 protein (i.e. E1 dimer) bound to double-stranded DNA (PDB accession number 1KSX) (Enemark *et al.*, 2002). The locations of the DNA-binding region of E1, which includes the helix α4 and the flexible DNA-binding loop, as well as the helix α3 located at the E1 dimer interface, are represented by arrows. (C) Crystal structure of the hexameric C-terminal helicase domain of BPV E1 (PDB accession number 2GXA) (Enemark and Joshua-Tor, 2006). Each monomer is coloured differently. Schematic representations of each intermediate during the assembly of the E1 double hexamer is depicted to the left: (A) E1–E2 complex at the viral origin, (B) E1 dimerization, and (C) E1 double hexamer assembly.

E2 from the complex. In particular, the binding of ATP to the ATP-binding domain in E1 plays an important role during initiation by stabilizing the interaction of E1 with DNA (i.e. the viral origin) and by promoting the interaction between E1 monomers (i.e. oligomerization of E1 into hexamers). Studies with HPV11 E1 have shown that E2 binding hinders the ATPase activity of E1 and similarly, that ATP weakens the E1–E2 interaction and impairs the cooperative binding of E1 and E2 to the origin (White *et al.*, 2001). These and other studies have suggested that E2 release probably occurs through conformational changes in E1 that take place upon ATP binding and oligomerization (Titolo *et al.*, 1999, 2000; White *et al.*, 2001; Abbate *et al.*, 2004). For BPV, hydrolysis of ATP is required to displace E2 from the origin (Lusky *et al.*, 1994; Sanders and Stenlund, 1998, 2000,

2001). Release of E2 from the HPV initiation complex may also be dependent upon the action of molecular chaperones, including the heat shock proteins Hsp40 and Hsp70 (Liu *et al.*, 1998; Lin *et al.*, 2002). Indeed, *in vitro* studies have demonstrated that both of these proteins favour the assembly of the E1 double hexamer and help promote initiation of viral replication by facilitating the removal of E2 from the origin. Hsp40 and Hsp70 have also been shown to facilitate HPV replication by increasing the affinity of the HPV11 E1 protein for the viral origin.

Aside from the interaction of E1 with E2, the binding of E1 to DNA is also essential for the formation of an E1–E2 complex at the origin (Titolo *et al.*, 1999). The sequence specific DNA-binding activity of E1 is mediated by its OBD, which is highly conserved among different papillomaviruses (Sun *et al.*, 1998; Titolo *et al.*, 2003a). Furthermore, the OBDs of E1 and of the large T antigen (T-Ag) of SV40 display a very high degree of conservation in their secondary and tertiary structures, despite limited sequence homology (Luo *et al.*, 1996; Enemark *et al.*, 2000). The OBD of papillomaviruses is composed of five antiparallel β-sheets, surrounded by α-helices. Part of these helices lie at the interface of the E1 dimer that is formed at the origin, while the others face the C-terminal helicase domain of the protein (Fig. 11.4B) (Enemark *et al.*, 2000, 2002; Auster and Joshua-Tor, 2004). The DNA-binding region of E1 contains helix α4 and a flexible region called the DNA-binding loop, which together form a positively charged DNA–binding interface on the protein (Fig. 11.4B) (Enemark *et al.*, 2000, 2002; Auster and Joshua-Tor, 2004). These two structures form separate interactions with the individual strands of the double-stranded DNA. In particular, the DNA-binding loop contacts the first three nucleotides of the E1BS through van der Waals interactions, while helix α4 engages in non-specific hydrophilic interactions with the DNA phosphate backbone of the remaining three nucleotides, on the opposite DNA strand (Chen and Stenlund, 1998; Enemark *et al.*, 2002). Mutation of either A1, T2, or G4 within the consensus E1BS (<u>AT</u>T<u>G</u>TT) significantly affects the DNA-binding affinity of E1 (Enemark *et al.*, 2002; Titolo *et al.*, 2003a). Consistent with this observation, the crystal structure of Enemark *et al* revealed that many of the protein-DNA contacts occur between the DNA-binding loop and T2 (Enemark *et al.*, 2002).

Dimerization of E1 on specific pairs of E1BS increases its affinity for DNA. It was established that dimerization is favoured on two E1BS placed in an opposite orientation and separated by three base pairs, as is the case for E1BS1 and 3 or E1BS2 and 4 in the origin (Fig. 11.2B) (Enemark *et al.*, 2002; Titolo *et al.*, 2003a). Amino acid substitutions that abolish E1 dimerization have been identified. These mutations affect residues located in helix α3 at the dimer interface, which creates a hydrophobic surface between two monomers of E1 (Fig. 11.4B) (Enemark *et al.*, 2000, 2002; Auster and Joshua-Tor, 2004). These mutations significantly reduce transient DNA replication highlighting the importance of the OBD dimerization in this process (Titolo *et al.*, 2003a; Schuck and Stenlund, 2005b; Fradet-Turcotte *et al.*, 2010b). As a dimer, E1 has a higher affinity for specific and non-specific DNA. The C-terminal domain of E1 also contributes to the non-specific DNA-binding activity of E1, by interacting with sequences flanking the E1BS. In fact, in the absence of E2, both the OBD and the C-terminal domain of E1 can interact with an area of more than 80 base pairs (Stenlund, 2003). Competition assays have shown that residues within the highly conserved FWL and β-hairpin motifs are required for the non-specific DNA-binding activity of E1, suggesting the importance of these residues for this function (Liu *et al.*, 2010).

The N-terminal region of E1 has also been shown to affect the binding of E1 to DNA. An initial study by Titolo *et al.* revealed that deletion of the first 71 amino acids of HPV11 E1 increases its binding and oligomerization at the origin as measured in a protein-DNA co-immunoprecipitation assay (McKay DNA-binding assay) (Titolo *et al.*, 2000). Our recent data suggests that the N-terminal region of E1 neither affects the affinity of the OBD for DNA nor its dimerization, suggesting that it stimulates the assembly of E1 at the origin indirectly (AFT and JA, unpublished). Interestingly, we recently reported that this region of E1 contains a short amphipathic α-helix whose mutation inhibits the DNA replication activity of E1 by 50%, at a step following assembly of the E1-E2-origin ternary complex (Morin *et al.*, 2011). Collectively, these findings raise the possibility that the N-terminal region of E1 regulates the conversion of E1 from a sequence-specific DNA-binding protein to a non-specific hexameric helicase, although more work is needed to ascertain whether this is the case.

E1 double hexamer formation

The C-terminal domain of E1 contains four highly conserved regions termed A, B, C, and D which share significant similarities to the corresponding regions of the large T-Ag of SV40 and other polyomaviruses (Clertant and Seif, 1984; Bream *et al.*, 1993; Hughes and Romanos, 1993; Seo *et al.*, 1993b; Yang *et al.*, 1993; Sarafi and McBride, 1995; Jenkins *et al.*, 1996; Mansky *et al.*, 1997). Three of these regions have been identified as important for E1 oligomerization, namely region A and the ATP-binding site, which is located in regions B and C (Titolo *et al.*, 2000). The minimal E1–E1 interaction site within region A has been localized to residues 332 to 410 of HPV31. Mutations of some of the highly conserved residues within this region inhibits E1 oligomerization and replication activity *in vivo* (Titolo *et al.*, 2000). Likewise, the integrity of the ATP-binding site is essential for E1 oligomerization, since mutation of the conserved lysine in this region abolishes this activity. Finally, studies performed with HPV11 and BPV E1 proteins have also suggested a role for the OBD in oligomerization (Titolo *et al.*, 2000; Castella *et al.*, 2006b).

It has also been determined that the two DNA strands are separated during the assembly of active E1 double hexamers. This is believed to be initiated by distortion of the DNA induced upon formation of the E1 tetramer and/or double trimer intermediate at the origin (Enemark *et al.*, 2002; Schuck and Stenlund, 2005a). Both ATP-binding to the C-terminus of E1 and an intact β-hairpin appear to be necessary for DNA unwinding in the AT-rich region and for double hexamer formation (Gillette *et al.*, 1994; Castella *et al.*, 2006a; Liu *et al.*, 2007; Schuck and Stenlund, 2007). In support of this, mutations of residues within the β-hairpin (K530 or H531 in HPV31) were found to inhibit E1 DNA unwinding (Castella *et al.*, 2006a; Liu *et al.*, 2007; Liu and Stenlund, 2010). Likewise, the integrity of the FWL motif (F464, W465 and L466), which is located between the Walker A and Walker B of E1, was determined to be important for both DNA unwinding and double hexamer formation (Castella *et al.*, 2006a).

DNA synthesis

Unwinding activity of E1

The ATPase and 3′–5′ DNA helicase activities of E1 depend on residues located within conserved regions B and C of its C-terminus (amino acids 332–629 in HPV31) (White *et al.*, 2001). Through sequence homology, the E1 hexameric helicase is classified as a member of the superfamily III (SF3) of nucleoside triphosphate (NTP)-binding proteins, which consists of helicases encoded by small DNA and RNA viruses (Gorbalenya *et al.*, 1990). This family of helicases is characterized by four highly conserved regions, namely the Walker A, Walker B, B′, and C motifs. The Walker A (also called the phosphate-binding loop or P-loop) and Walker B motifs form the catalytic site that binds to and hydrolyses ATP as a magnesium chelate (Fig. 11.1A). Motif C, also referred to as sensor 1, is characterized by the presence of an invariant asparagine residue. This motif, along with motif B′, detect the phosphorylation state of the nucleotide and stabilize a water molecule necessary for nucleophilic attack of the γ-phosphate of ATP (Gorbalenya *et al.*, 1990; Ogura and Wilkinson, 2001; Hickman and Dyda, 2005). These motifs lie within a region comprising a five-stranded parallel β-sheet flanked on both sides by several α-helices, characteristic of AAA$^+$ proteins (ATPases associated with various cellular activities) (Hickman and Dyda, 2005). In addition to these four highly conserved regions, these ATPases contain an 'arginine finger' that detects the phosphorylation state of the nucleotide on the adjacent monomer, and links ATP hydrolysis to the conformational change needed for DNA unwinding. Amino acid substitutions of conserved residues of the Walker A (K484A/R and P479S, HPV11), Walker B (D523N, D523A/D524A, HPV11), and C motifs (T556A and N568A, HPV11) or of the 'arginine finger' (R589A, HPV18) are all associated with a loss of both the ATPase and replication activities of E1 (Yasugi *et al.*, 1997; Titolo *et al.*, 1999; Rocque *et al.*, 2000; White *et al.*, 2001; Lin *et al.*, 2002; Abbate *et al.*, 2004).

ATP hydrolysis plays an important role during elongation by powering the unwinding activity of the E1 helicase. The unwinding rate of an E1 hexamer is estimated to be approximately 0.3 pM substrate/h/nM of purified E1, when the substrate is an oligonucleotide annealed to single-stranded circular DNA (Rocque *et al.*, 2000; White *et al.*, 2001). Crystal structures of the C-terminal region of BPV E1, in the absence or presence of different nucleotides (ATP or ADP), have demonstrated that the E1 protein forms a hexamer around single-stranded DNA (Fig. 11.4C). Upon ATP binding, each of the six monomers contact the sugar moiety of the nucleic acid through the β-hairpin, which points towards the centre of the canal that measures approximately 13 Å in diameter (Enemark and Joshua-Tor, 2006). In particular, this DNA contact involves an interaction between the ammonium of the conserved lysine of the β-hairpin and the oxygen of the phosphates in the single-stranded DNA, as well as formation of a hydrogen bond between the conserved histidine of the β-hairpin and the phosphate in the adjacent nucleotide. These interactions are stabilized by van der Waals forces created by both the aliphatic group of lysine and the aromatic ring of histidine with the sugar moiety of the single-stranded DNA. Hydrolysis of ATP to ADP, involves the displacement of the β-hairpin towards the C-terminal end of the E1 protein, which then remains bound to the DNA phosphate backbone and escorts the single-stranded DNA through the central channel of the hexamer (Enemark and Joshua-Tor, 2006). Thus, each E1 monomer bound to ATP initially binds DNA, and then translocates the DNA towards the helicase domain through conformational changes in E1 that are induced by the hydrolysis of ATP to ADP and by the subsequent release of ADP (Apo form). When each monomer passes through the ATP-ADP-Apo process, DNA is translocated by a length of 6 nucleotides inside the central channel of the hexamer. It has been suggested that both E1 hexamers remain assembled during unwinding and that the DNA is pumped through the double hexamer (Liu *et al.*, 1998). In contrast, other studies have suggested that the two E1 hexamers separate and migrate in the opposite direction on complementary DNA strands to unwind DNA (Fouts *et al.*, 1999). This last model is supported by the recent observation that the double MCM heterohexamer of the eukaryotic replication complex is converted into two independent replisomes during elongation (Yardimci *et al.*, 2010).

Interaction of E1 with the cellular DNA replication machinery

In agreement with its central role in viral DNA replication, E1 has been reported to interact with several cellular replication factors. These include the DNA polymerase (Pol) α-primase complex, the single-stranded DNA-binding protein replication protein A (RPA), and topoisomerases I and II (Topo I and Topo II). *In vitro*, papillomavirus DNA replication also requires RFC and PCNA, although interaction of these factors with E1 has not been reported (Kuo *et al.*, 1994; Melendy *et al.*, 1995). The E1 protein appears to interact with the p70 and/or p180 subunit of the Pol α-primase complex through its C-terminal helicase domain (Fig. 11.1A) (Park *et al.*, 1994; Bonne-Andrea *et al.*, 1995; Masterson *et al.*, 1998; Conger *et al.*, 1999; Amin *et al.*, 2000). This interaction is mutually exclusive with the E1–E2 interaction, suggesting that the polymerase interacts with E1 following the removal of E2 from the initiation complex and the subsequent assembly of E1 into active double hexamers. E1 has also been reported to interact with RPA using ELISA (enzyme-linked immunoabsorbent assay) or GST pulldown assays with recombinant proteins (Han *et al.*, 1999; Loo and Melendy, 2004). The E2 TAD has also been shown to interact with RPA, raising the possibility that association of E1 with RPA is stabilized in the presence of E2 (Li and Botchan, 1993). Lastly, *in vitro* analyses revealed that both the OBD and C-terminal domains of E1 interact with Topo I. This interaction increases both the activity of Topo I and the binding of E1 to the viral origin (Fig. 11.1A) (Clower *et al.*, 2006a; Hu *et al.*, 2006). A similar interaction has been observed between E2 and Topo I, although further studies are needed to elucidate the role of this interaction in viral DNA synthesis (Clower *et al.*, 2006b).

Interaction of E1 with other cellular proteins

In addition to known DNA replication factors, E1 has been reported to associate with other cellular proteins to regulate viral DNA replication. Among these are p80, hSNF5, histone H1, Ubc9, E1BP, and p56. We recently showed that the N-terminal 40 amino acids of E1 from anogenital HPV types interact with the cellular WD-repeat-containing protein p80 (WDR48 or USP1-associated factor 1 (UAF1)) using tandem affinity purification of HPV11 E1 (Cote-Martin *et al.*, 2008). Co-localization studies showed that E1 from both low- and high-risk HPV types can redistribute p80 from the cytoplasm to the nucleus. Mutant E1 proteins defective in p80 binding have a reduced capacity to maintain the viral episome in immortalized keratinocytes, suggesting a key role for the E1–p80 interaction in HPV genome replication and episomal maintenance. While the cellular function of p80 remains to be established, recent studies have identified p80 in complex with deubiquitinating enzymes (DUBs) (Cohn *et al.*, 2007, 2009; Kee *et al.*, 2010). Whether these complexes are important during HPV DNA replication remains to be determined.

The HPV genome is assembled into chromatin. Interestingly, E1 was reported to interact with two proteins that could influence replication by modifying chromatin structure, namely histone H1 and Ini1/SNF5, a subunit of the SWI/SNF chromatin remodelling complex (Swindle and Engler, 1998; Lee *et al.*, 1999; Tang *et al.*, 2010). Overexpression of Ini1/hSNF5 was found to activate viral DNA replication, while repression of Ini1/hSNF5 expression by antisense RNA significantly reduced viral DNA levels. Furthermore, the binding of nucleosomes to viral DNA inhibited viral replication, suggesting that E1 interaction with the SWI/SNF complex destabilizes nucleosomes present at the origin to facilitate DNA replication (Favre *et al.*, 1977; Li and Botchan, 1994; del Mar Pena and Laimins, 2001). Critical residues that mediate this interaction were mapped to amino acids 143–437 of HPV31 E1 (Fig. 11.1A). Amino acid substitutions of highly conserved residues within this region (i.e. S225P/F226S, L305K, N395H/A396E and H418E/Y419H) were identified that abolished E1 interaction with Ini1/SNF5 (Lee *et al.*, 1999). Additionally, histone H1 has been shown to be dislodged from DNA upon its interaction with E1 (Swindle and Engler, 1998). Given that histone H1 condenses chromatin, E1 binding to histone H1 may facilitate the destabilization of chromatin and the disruption of nucleosomes at the replication fork to promote viral replication.

Viral DNA synthesis may also be regulated by the interaction of E1 with Ubc9, the E2-conjugating enzyme that mediates SUMO (small ubiquitin-like modifier) transfer to the lysine residue of target proteins (Yasugi and Howley, 1996; Yasugi *et al.*, 1997; Rangasamy and Wilson, 2000; Fradet-Turcotte *et al.*, 2009). The role of sumoylation as a post-translational modification of E1 will be discussed in greater detail below. Of relevance is the observation that several mutations in E1 that disrupt interaction with Ubc9 also abrogate its ability to support viral DNA replication. However, the subsequent finding that these mutations also affect other essential functions of E1 including DNA-binding, oligomerization and ATPase activity complicated the interpretation of these results (Yasugi *et al.*, 1997; Fradet-Turcotte *et al.*, 2009). Similarly, mutations in E1 that prevent its association with E1-BP, a putative ATPase of unknown function that was identified in a yeast-two-hybrid screen, were also found to affect the ability of E1 to support viral DNA replication *in vivo*, hydrolyse ATP or oligomerize on DNA (Yasugi *et al.*, 1997). These mutational studies suggested that the integrity of the E1 DNA-binding and helicase domains is necessary for interaction with Ubc9 and E1-BP, raising the possibility that E1 associates with these cellular proteins only after its assembly into a double hexamer.

In contrast, the association of E1 with p56, a cellular protein induced by activation of the interferon (IFN) pathway, was shown to inhibit viral DNA replication *in vitro* and *in vivo*, possibly by inhibiting the steps leading to assembly of the E1 double hexamer. Specifically, p56 directly inhibits the helicase activity of E1 and interferes with its binding to E2 and to the origin (Terenzi *et al.*, 2008; Saikia *et al.*, 2010). The amino acid phenylalanine 399 of HPV18 (F399) was identified as a critical residue required for the E1–p56 interaction. The exact role of this amino acid in HPV replication, however, is difficult to assess.

In one study, deletion of F399 did not have any major effect on DNA replication other than rendering it insensitive to inhibition by IFN (Saikia *et al.*, 2010). In contrast, mutation or deletion of this residue in the E1 protein of other HPV types has been found to completely abrogate transient DNA replication (CMD and JA, unpublished; Titolo *et al.*, 2000; White *et al.*, 2001; Fradet-Turcotte *et al.*, 2010b). Although the precise mechanism by which IFN and/or p56 inhibits the replication functions of E1 remains to be identified and confirmed in other HPV types, these findings help explain the effectiveness of treatments for HPV infections by either IFN or Imiquimod.

Post-translational modifications of E1

Nucleo-cytoplasmic shuttling of E1

Post-translational modifications of E1 have, so far, been associated mainly with the regulation of its nucleo-cytoplasmic shuttling. As mentioned above, the N-terminal region of E1 contains several regulatory motifs including a bipartite NLS and a CRM1-dependent NES (Fig. 11.1B). Together, these two motifs have been shown by mutagenesis to regulate the nucleo-cytoplasmic shuttling of E1 *in vivo*. For instance, substitutions of conserved amino acids in either of the two basic regions of the bipartite NLS (K-R-K-x_{30}-K-x-x-K) were found to significantly reduce the nuclear accumulation of E1 (Yu *et al.*, 2007; Fradet-Turcotte *et al.*, 2010a) while substitutions of conserved hydrophobic residues in the NES (L109 and I113 in HPV31), or treatment of the cells with the Crm-1 inhibitor leptomycin B (LMB), were shown to inhibit E1 nuclear export (Deng *et al.*, 2004; Fradet-Turcotte *et al.*, 2010a). Although the presence of an NLS in E1 is easily rationalized by the fact that viral DNA replication takes place in the nucleus, the need for a nuclear export sequence was far less obvious. Our studies have shown that nuclear export of E1 is essential for the maintenance of the viral episome in keratinocytes (Fradet-Turcotte *et al.*, 2010a), in part to inhibit the antiproliferative effects of E1. Indeed, we observed that when present in high amounts in the nucleus, but not in the cytoplasm, E1 induces cell cycle arrest in S-phase and triggers an ataxia-telangiectasia mutated (ATM)-dependent DNA damage response (Fradet-Turcotte *et al.*, 2010a, 2011). Overall, these studies highlighted the importance of nuclear export in controlling the nuclear accumulation of E1 in order to achieve the optimal balance between promoting viral DNA replication without interfering with cellular proliferation.

Regulation of E1 nuclear accumulation by phosphorylation

In addition to the NLS and NES, the E1 N-terminal region also contains a CBM that binds to cyclin A/E-Cdk2 (Cdk: cyclin-dependent-kinase) and Cdk2 phosphorylation sites that serve to regulate the nucleo-cytoplasmic shuttling of E1 (Fig. 11.1B). Cyclin A/E-Cdk2 can bind and phosphorylate E1 *in vitro* and *in vivo*. This interaction is mediated by the highly conserved CBM (R-x-L) in the N-terminal domain of E1 (Ma *et al.*, 1999; Fradet-Turcotte *et al.*, 2010a). A mutation in this motif abrogates cyclin binding and reduces HPV DNA replication *in vivo* (Ma *et al.*, 1999; Deng *et al.*, 2004; Fradet-Turcotte *et al.*, 2010a), highlighting the importance of this interaction for efficient viral DNA replication. For HPV31 and 11 E1, the effect of the CBM mutation was almost completely rescued by mutation of the NES, consistent with the CBM being required to prevent E1 nuclear export (Deng *et al.*, 2004; Fradet-Turcotte *et al.*, 2010a). Accordingly, E1 contains several putative Cdk2 phosphorylation sites (S/T-P) that are critical for its nuclear retention. Among the putative phosphorylation sites in HPV11 and 31 E1, those located within the N-terminal region were found to be essential for nuclear accumulation of the protein. Indeed, mutation of three (S89, S93 and S107 in HPV11) or two (S92 and S106 in HPV31) serine residues to alanine located near or within the NES was found to promote E1 nuclear export (Deng *et al.*, 2004). Interestingly, while mutation of S107 in HPV11 E1 inhibited transient DNA replication, the analogous mutation S106 in HPV31 E1 had no such effect, suggesting that this serine residue may have additional functions in low-risk HPV besides regulating nuclear export. In support of this suggestion, the replication defect imposed by mutation of S107 in

HPV11 E1 was not completely rescued by mutation of the NES (Deng *et al.*, 2004). Surprisingly, phosphorylation of E1 appears to play an opposite role in BPV. In this case, Cdk2 phosphorylation of a serine residue within the OBD was found to promote, rather than inhibit the export of BPV E1 (Hsu *et al.*, 2007).

In addition to cyclin/Cdk, the HPV11 E1 protein has also been shown to specifically interact with and to be phosphorylated by ERK1 and JNK2, two MAPK (mitogen-activated protein kinase) family members (Yu *et al.*, 2007). Accordingly, two MAPK docking motifs have been identified within the C-terminal region of E1: an L-x-L motif followed by a hydrophobic stretch of amino acids (L555, T556, L557, L563, L564, V565) and an F-x-F-P motif (F587, T588, F589, and P590). Alanine substitution of hydrophobic residues within these motifs, as well as the addition of specific MAPK inhibitors, was found to significantly impair MAPK binding and phosphorylation, resulting in cytoplasmic accumulation of E1 (Yu *et al.*, 2007). This study therefore raises the possibility that kinases other than Cdk2 may regulate the nucleo-cytoplasmic shuttling of E1.

Sumoylation and ubiquitination of E1

In addition to phosphorylation, the E1 protein from several HPV types and from BPV were reported to interact with the cellular protein Ubc9 and to be modified by sumoylation, a modification that often affects the activity or cellular localization of target proteins. Sumoylation of E1 from BPV, HPV1 and HPV18 has been demonstrated *in vitro* and *in vivo* and the site of SUMO attachment in BPV E1 was identified as lysine 514 in the C-terminal helicase domain (Rangasamy and Wilson, 2000; Rangasamy *et al.*, 2000; Rosas-Acosta *et al.*, 2005; Hsu *et al.*, 2007; Rosas-Acosta and Wilson, 2008). Substitution of this conserved lysine or of leucine residues needed for interaction with Ubc9, was shown to abrogate the ability of BPV E1 to support transient DNA replication *in vivo* (Rangasamy *et al.*, 2000). Results obtained with BPV E1 mutant proteins defective for sumoylation or interaction with Ubc9 led to the proposal that conjugation with SUMO-1 and interaction with Ubc9 are both required for intranuclear accumulation of BPV E1. Our data, however, does not support a role for the sumoylation pathway in regulating the intracellular localization of HPV E1. Indeed, we have observed that HPV11 and HPV16 E1 mutants that are defective for interaction with Ubc9 are still capable of accumulating in the nuclear compartment even when the sumoylation pathway is inhibited by the Gam1 protein of the CELO adenovirus or by a dominant negative form of Ubc9 (Fradet-Turcotte *et al.*, 2009). As mentioned previously, a difficulty in addressing the function of the E1–Ubc9 interaction in viral DNA replication is that all of the mutations known to disrupt this interaction also affect the origin-binding and/or oligomerization of E1 (Yasugi *et al.*, 1997). By extension, these observations raise the possibility that Ubc9 interacts with the oligomerized form of E1 bound at the origin. Further work will be required to determine the role of the E1–Ubc9 interaction and of sumoylation in viral DNA synthesis.

E1 has also been shown to be modified by ubiquitination at least for BPV. In this case, E1 has been found to be a substrate of the ubiquitin-dependent proteasome pathway *in vitro* and *in vivo* (Malcles *et al.*, 2002; Mechali *et al.*, 2004). Interestingly, E1 degradation has been determined to be enhanced under replication conditions, suggesting that it occurs at the end of replication perhaps as a mechanism to prevent further rounds of DNA synthesis.

E1 phosphorylation-dependent switch in viral DNA replication mode

As described above, unregulated nuclear accumulation of E1 impedes cellular proliferation by imposing an S-phase delay. Hence, controlling the amount of E1 in the nucleus is needed to prevent its deleterious effect on cellular proliferation, and possibly to control the copy number of the episome (Fradet-Turcotte *et al.*, 2010a). Several studies have described the existence of two different modes of viral DNA replication during the maintenance of the genome, namely random and ordered replication. In the random model, replication of the viral genome is uncoupled from that of cellular DNA such that viral DNA is replicated multiple times within one cell cycle (Gilbert and

Cohen, 1987; Ravnan *et al.*, 1992; Piirsoo *et al.*, 1996). In the ordered model, viral DNA is replicated once per cell cycle during the S-phase and is probably strictly controlled by host cellular factors, analogous to the replication of cellular DNA (Botchan *et al.*, 1986; Roberts and Weintraub, 1988). Along with the study of Hoffman *et al.*, our recent study supports the hypothesis that the mode of replication used during the maintenance phase of the HPV life cycle is dictated by the levels of E1. Indeed, Hoffman *et al.* showed that DNA replication switches from an ordered to a random mode of replication upon the expression of E1 from a heterologous promoter, while our data revealed that sustained accumulation of E1 in the nucleus interferes with cell cycle progression in S-phase (Fradet-Turcotte *et al.*, 2010a; Hoffmann *et al.*, 2006). The observation that an E1 mutant protein that accumulates in the cytoplasm is unable to delay cell cycle progression suggest that E1 nuclear export may be required to avoid a block in S-phase and to limit the levels of viral DNA replication during genome maintenance. In contrast, increased Cdk2 activity in differentiated cells would promote the phosphorylation of E1 leading to its increased nuclear accumulation, which in turn would cause cell cycle arrest during S-phase and allow for several rounds of viral DNA synthesis.

Amplification of the viral genome in differentiated cells

Overview of the HPV life cycle

While viral DNA is replicated similarly to cellular DNA when maintained in undifferentiated keratinocytes, the virus has evolved different mechanisms to amplify its genome in differentiated cells. The HPV life cycle is dependent on the differentiation programme that keratinocytes undergo within a stratified epithelium and can be divided into three main phases: establishment, maintenance, and amplification (see Chapter 1 of this book). Following virion entry into the basal cell layer, the viral genome is amplified and established as an episome at 50–100 copies per infected cell. This initial amplification is followed by a maintenance phase, where viral DNA is stably maintained at a constant copy number through low levels of replication by E1 and E2. Other viral proteins including E6, E7, E1^E4, E8^E2C, and E5 have also been reported to play an essential role in episomal maintenance, although additional studies will be required to establish the role of these proteins during this process (Klumpp *et al.*, 1997; Thomas *et al.*, 1999; Stubenrauch *et al.*, 2001; Park and Androphy, 2002; Zobel *et al.*, 2003; Longworth and Laimins, 2004; Longworth *et al.*, 2005; Wilson *et al.*, 2005, 2007; McLaughlin-Drubin *et al.*, 2005; Nakahara *et al.*, 2005; Lace *et al.*, 2008). Lastly, as the infected cells reach the upper layers of the epithelium, differentiation induces the productive phase of the viral life cycle, which includes genome amplification, late gene expression, and virion production (Hebner and Laimins, 2006). Here, high levels of DNA replication amplify the viral genome to greater than 1000 copies per cell. Genome amplification is dependent upon the expression of the viral oncoproteins, E6 and E7, which uncouple cell growth arrest and differentiation primarily through the inactivation of p53 and Rb, respectively (McLaughlin-Drubin and Munger, 2009; Moody and Laimins, 2010; Pim and Banks, 2010). Specifically, inactivation of Rb and the related pocket proteins p107 and p130 by E7 forces infected cells to remain in a proliferative state and escape cell cycle exit by binding and regulating the activity of the E2F family of transcription factors. Abrogation of p53 by E6, in turn, ensures cell survival by preventing apoptosis triggered by this aberrant growth signal. As such, the cooperation between these two viral oncoproteins promotes a cellular environment supporting viral DNA replication by stimulating infected differentiating keratinocytes to re-enter S-phase so as to utilize the host replication factors required for viral DNA synthesis. Interestingly, we recently discovered that the nuclear accumulation of E1 arrests cells in S-phase, revealing a potential role for E1 in the establishment of the S-phase environment in differentiated keratinocytes (Fradet-Turcotte *et al.*, 2010a). It has also been determined that E1^E4 and E5 are required for genome amplification in differentiated keratinocytes, although the exact mechanisms by which these proteins contribute to this process remains to be determined (Wilson *et al.*, 2005). Finally, it was recently shown that activation of a DNA damage response (DDR) dependent on the ATM pathway is required for amplification of the genome (Moody and Laimins, 2009). The role of a DDR in genome activation is discussed in greater detail below.

Role of the DNA damage response in genome amplification

Activation of a DDR can originate from two types of DNA breaks, double-stranded and single-stranded breaks (DSBs and SSBs, respectively) (reviewed in Su, 2006). DSBs induce the activation of the ATM-dependent pathway, which is characterized by the phosphorylation of ATM and its downstream effector kinase, Chk2. SSBs lead to the activation of the ataxia-telangiectasia and Rad3-related (ATR) pathway and result in the activation of ATR and phosphorylation of its cognate downstream effector kinase, Chk1. Both pathways transduce signals to downstream effectors by phosphorylation of H2AX (γH2AX), which in turn amplify the DNA damage signal. Phosphorylation of the effector kinases Chk2 and Chk1 also triggers activation of checkpoints that block cell cycle progression in order to facilitate DNA repair (Reinhardt and Yaffe, 2009; Freeman and Monteiro, 2010). As mentioned above, activation of a DDR and of the ATM pathway is necessary for genome amplification in differentiating keratinocyctes (Moody and Laimins, 2009). Specifically, it has been shown that inhibition of the ATM kinase activity by the chemical inhibitor KU-55933 completely blocks viral genome amplification, while having little to no effect on the stable maintenance of HPV episomes in undifferentiated cells or on transient DNA replication (Moody and Laimins, 2009; Fradet-Turcotte *et al.*, 2011). The precise mechanism by which a DDR is needed for genome amplification is still unclear, although it may be required to induce caspase-mediated cleavage of E1 (Moody *et al.*, 2007). Indeed, mutation of the conserved caspase-cleavage motif (DxxDxxD) in the N-terminal region of E1, targeted by caspase-3/-7, was found to inhibit amplification of viral genomes in immortalized keratinocytes upon differentiation while having little to no effect on transient DNA replication or on episomal maintenance in undifferentiated cells (Moody *et al.*, 2007; Morin *et al.*, 2011). Similarly, treatment of differentiated keratinocytes with the pan-caspase inhibitor Z-VAD-FMK prevented genome amplification. Moreover, Chk2 kinase activity was shown to be necessary for the activation of caspase-3/-7 and possibly for proteolytic cleavage of the N-terminal domain of E1 (Moody and Laimins, 2009). Strikingly, transient HPV DNA replication remains unaffected by the induction of a DDR and of the ATM pathway indicating that it is insensitive to the checkpoint that blocks cellular DNA synthesis (King *et al.*, 2010; Fradet-Turcotte *et al.*, 2011). Given that E1 usurps several host DNA replication factors for viral DNA synthesis, it is possible that the virus activates a DDR to inhibit host cell replication and alleviate competition for these proteins.

Signalling pathways leading to a DDR can be activated by viral oncogenes. In support of this, it was determined that expression of the E6 and E7 HPV oncogenes in keratinocytes is accompanied by an increase in DSBs and in γH2AX (Duensing and Munger, 2002). Moreover, high-risk E7 was found to induce Chk2 phosphorylation upon differentiation (Moody and Laimins, 2009). Likewise, studies from our group and the group of Alison McBride have demonstrated that E1 induces a DDR and that this induction requires both the origin-binding and ATPase activities of E1, thus suggesting that E1 may be causing DSBs by assembling onto host DNA (Fradet-Turcotte *et al.*, 2011; Sakakibara *et al.*, 2011). Our data also suggest that E1 nuclear export reduces activation of a DDR in the lower layers of the epithelium, leading to the proposal that DDR activation is restricted to the upper layers of the epithelium at least in part by controlling the nuclear accumulation of E1.

While the activation of a DDR by E6, E7, and E1 may be required for completion of the viral life cycle, it has also been reported that replication of the HPV genome after integration of viral DNA into the host genome activates a DDR (Kadaja *et al.*, 2007, 2009a). In this case, the DDR would be induced by the intermediate products formed during replication of integrated DNA (i.e. small linear double-stranded DNA fragments). Such 'onion-skin' replication of the integrated genome by E1 and E2 results in the recruitment of DNA damage signalling proteins including NBS1, MRE11, RAD50, ATM, Chk2, ATRIP, Chk1 and KU70/KU80 at sites of replication (Kadaja *et al.*, 2009a). The activation of a DDR under these conditions, although not part of a normal viral life cycle, would contribute to host cell genomic instability and favour cancer progression (Kadaja *et al.*, 2009b).

Unidirectional (rolling circle) mode of replication

DNA replication may proceed in one direction (unidirectional) or both directions (bidirectional) from the origin of replication. Several studies support bidirectional replication as the basic mode of HPV DNA replication. Visualization of DNA replication forks by electron microscopy have shown that E1 double hexamers induce replication bubbles at the viral origin that are typically associated with bidirectional replication (Fouts *et al.*, 1999; Lin *et al.*, 2002). Analyses of DNA replication in cells maintaining the viral genome generate a pattern of migration on 2D-gels also characteristic of bidirectional replication (Auborn *et al.*, 1994; Flores and Lambert, 1997; Kadaja *et al.*, 2009a; Geimanen *et al.*, 2011).

During keratinocyte differentiation, however, HPV replication may switch to a rolling circle mode of replication (i.e. unidirectional) to facilitate genome amplification (Dasgupta *et al.*, 1992; Flores and Lambert, 1997). Indeed, a recent study carried out using cellular extracts of differentiated keratinocytes revealed that replication generated multiple single-stranded linear copies of the template DNA arranged in a continuous head-to-tail orientation (i.e. concatemers), consistent with the products of rolling circle replication (Kusumoto-Matsuo *et al.*, 2011). Rolling circle replication, however, would require a single E1 hexamer, making it difficult to reconcile with the notion that a double hexamer is assembled at the viral origin. Additionally, the protein components that would be required for rolling circle replication, to either nick one strand of the double-stranded circular DNA or to join the ends of the nicked strands, are currently unknown. Further studies are therefore required to elucidate the role of rolling circle replication during genome amplification.

Conclusions and future perspectives

Considerable efforts have been undertaken in the past years to expand our current understanding of the mechanism and regulation of HPV DNA replication. In particular, significant progress have been made in determining the processes involved in the initiation of viral DNA synthesis, the assembly of a pre-replication complex, and how these steps are regulated by post-translational modifications of the E1 protein, specifically by phosphorylation. Regulation of E1 nuclear export has been shown to be essential for episomal maintenance in keratinocytes. When the availability of E1 in the nucleus is low, viral DNA replication probably occurs either once (ordered) or only a few times per cell cycle (random). This 'maintenance' mode of replication is analogous to the replication of cellular DNA, occurs in synchrony with host cell replication, and is dictated by the levels of E1 (Fig. 11.5). Nuclear accumulation of E1 in high amounts, however, impedes cellular proliferation by inducing a cell cycle arrest in S-phase and triggers an ATM-dependent DNA damage response. During this 'amplification' mode of replication, high levels of DNA replication drastically increase the copy number of the viral episome in the upper layers of the infected epithelium. Here, genome amplification is dependent upon the expression of the viral oncoproteins, E6 and E7, and upon Cdk2 activation. Essentially, E6 and E7 uncouple cell growth and differentiation, while increased Cdk2 activity promotes nuclear accumulation of E1 and delays cell cycle progression during S-phase to allow for multiple rounds of viral DNA synthesis. Thus, the accumulation of E1 in the nucleus needs to be tightly regulated by its nuclear export to prevent its deleterious effect on cellular proliferation, and possibly to control the episomal copy number. Failure to adequately control the nuclear accumulation and activity of E1 in the lower layers of the epithelium may lead to assembly of E1 on host DNA resulting in what we have termed the 'DNA damage' mode of replication. This mode would be associated with the induction of undesired double-stranded DNA breaks in the host genome that are recombinagenic and could favour viral genome integration. For high-risk HPV types, integration could result in overexpression

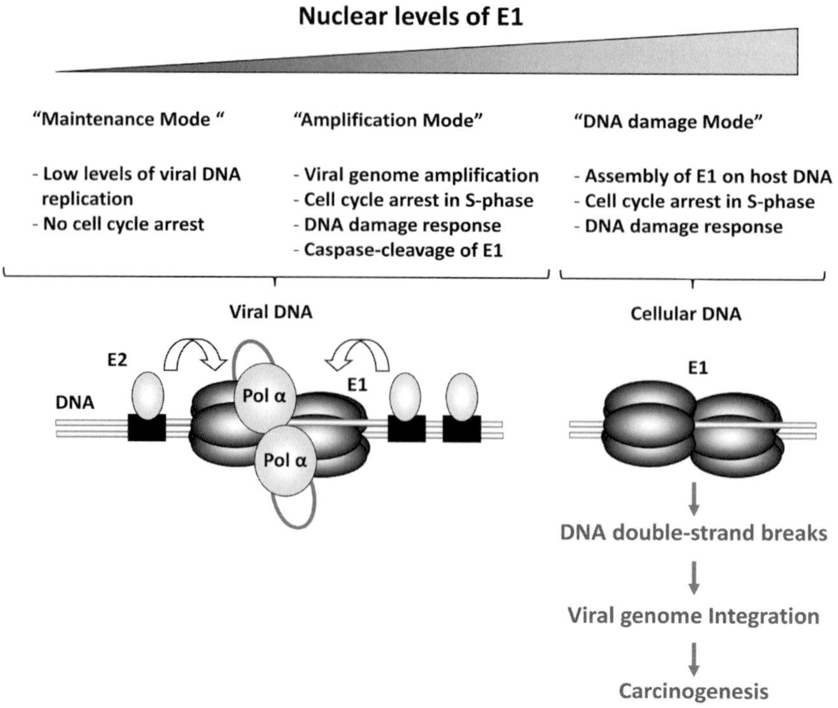

Figure 11.5 Function of E1 nuclear accumulation during viral replication. Nuclear accumulation of E1 during viral DNA synthesis is depicted as a grey triangle on the top of the figure. The 'maintenance' mode of replication occurs when E1 is present in the nucleus at low levels, whereas the 'amplification' mode takes place in S-phase arrested cells when the levels of E1 in the nucleus are high and in presence of a DNA damage response induced by E1, E6 and E7. Failure to adequately control the nuclear accumulation of E1 may lead to its assembly on host DNA and trigger the 'DNA damage' mode of replication. Under these conditions, double-strand breaks created by E1 and activation of a DNA damage response would contribute to host cell genomic instability, favour viral genome integration, and ultimately contribute to HPV-induced carcinogenesis.

of the viral oncogenes E6 and E7, thereby favouring genomic instability and cancer progression. Thus, through its ability to induce a DDR, E1 may be a more important contributor to HPV-induced carcinogenesis than previously suspected.

Despite these advances, several questions regarding the regulation of HPV DNA replication still remain. While nuclear export of the HPV31 E1 protein is inhibited by cyclin A/E-Cdk2-mediated phosphorylation of E1, all other mechanisms involved in the regulation of E1 nuclear export or import are not very well characterized, including if and how serines 92 and 106 are dephosphorylated. We have determined that nuclear export of E1 is essential for the maintenance of the viral episome in keratinocytes, in part to inhibit the antiproliferative effects of E1. Whether the nuclear export of E1 has additional functions during episomal maintenance remains to be determined. Additionally, our studies suggest a key role for the E1–p80 interaction in HPV genome replication and episomal maintenance. Very little is known, however, about the function of p80 in normal and in HPV-infected cells. Also unclear is how caspase cleavage of E1 within its amphipathic helix promotes viral genome amplification and if this process is related to the removal of the p80-binding site. Finally, the role of the E1–Ubc9 interaction and of sumoylation in viral DNA synthesis remains to be elucidated. The observations that E1 interacts with Ubc9 suggest a possible function for this interaction in the sumoylation of cellular components of the viral replication complex found at the origin. Indeed, it has been determined that PCNA is sumolyated and that this modification of the protein can inhibit recombination during replication in absence of DNA breaks (Ulrich *et al.*, 2005; Branzei *et al.*, 2006). Furthermore, when DNA breakage is induced by fork stalling, sumoylation of PCNA activates the translation DNA synthesis (TLS) repair pathway (Shaheen *et al.*, 2010; Halas *et al.*, 2011). Finally, it has also been shown that Ubc9 is recruited at sites of DNA double-strand breaks and that their repair requires the sumoylation of BRCA1 and 53BP1 (Galanty *et al.*, 2009).

Overall, the answers to these questions will continue to extend our knowledge of how both viruses and eukaryotic cells coordinate different aspects of DNA replication and repair, which will ultimately further our understanding of cancer progression.

Acknowledgements

We apologize to those whose work was not included because of space considerations or whose papers were inadvertently omitted. Work in the authors' laboratory is supported by grants from the Canadian Institutes of Health Research (CIHR), Canadian Cancer Society Research Institute, and The Cancer Research Society Inc. CMD and AFT hold a fellowship from the Fonds de la Recherche en Santé du Québec (FRSQ).

References

Abbate, E.A., Berger, J.M., and Botchan, M.R. (2004). The X-ray structure of the papillomavirus helicase in complex with its molecular matchmaker E2. Genes Dev. *18*, 1981–1996.

Abbate, E.A., Voitenleitner, C., and Botchan, M.R. (2006). Structure of the papillomavirus DNA-tethering complex E2:Brd4 and a peptide that ablates HPV chromosomal association. Mol. Cell *24*, 877–889.

Alexander, K.A., and Phelps, W.C. (1996). A fluorescence anisotropy study of DNA binding by HPV-11 E2C protein: a hierarchy of E2-binding sites. Biochemistry *35*, 9864–9872.

Amin, A.A., Titolo, S., Pelletier, A., Fink, D., Cordingley, M.G., and Archambault, J. (2000). Identification of domains of the HPV-11 E1 protein required for DNA replication *in vitro*. Virology *272*, 137–150.

Androphy, E.J., Lowy, D.R., and Schiller, J.T. (1987). Bovine papillomavirus E2 trans-activating gene product binds to specific sites in papillomavirus DNA. Nature *325*, 70–73.

Antson, A.A., Burns, J.E., Moroz, O.V., Scott, D.J., Sanders, C.M., Bronstein, I.B., Dodson, G.G., Wilson, K.S., and Maitland, N.J. (2000). Structure of the intact transactivation domain of the human papillomavirus E2 protein. Nature *403*, 805–809.

Auborn, K.J., Little, R.D., Platt, T.H., Vaccariello, M.A., and Schildkraut, C.L. (1994). Replicative intermediates of human papillomavirus type 11 in laryngeal papillomas: site of replication initiation and direction of replication. Proc. Natl. Acad. Sci. U.S.A. *91*, 7340–7344.

Auster, A.S., and Joshua-Tor, L. (2004). The DNA-binding domain of human papillomavirus type 18 E1. Crystal structure, dimerization, and DNA binding. J. Biol. Chem. *279*, 3733–3742.

Berg, M., and Stenlund, A. (1997). Functional interactions between papillomavirus E1 and E2 proteins. J. Virol. *71*, 3853–3863.

Blachon, S., and Demeret, C. (2003). The regulatory E2 proteins of human genital papillomaviruses are pro-apoptotic. Biochimie *85*, 813–819.

Bonne-Andrea, C., Santucci, S., Clertant, P., and Tillier, F. (1995). Bovine papillomavirus E1 protein binds specifically DNA polymerase alpha but not replication protein A. J. Virol. *69*, 2341–2350.

Bonne-Andrea, C., Tillier, F., McShan, G.D., Wilson, V.G., and Clertant, P. (1997). Bovine papillomavirus type 1 DNA replication: the transcriptional activator E2 acts *in vitro* as a specificity factor. J. Virol. *71*, 6805–6815.

Botchan, M., Berg, L., Reynolds, J., and Lusky, M. (1986). The bovine papillomavirus replicon. Ciba Found. Symp. *120*, 53–67.

Branzei, D., Sollier, J., Liberi, G., Zhao, X., Maeda, D., Seki, M., Enomoto, T., Ohta, K., and Foiani, M. (2006). Ubc9- and mms21-mediated sumoylation counteracts recombinogenic events at damaged replication forks. Cell *127*, 509–522.

Bream, G.L., Ohmstede, C.A., and Phelps, W.C. (1993). Characterization of human papillomavirus type 11 E1 and E2 proteins expressed in insect cells. J. Virol. *67*, 2655–2663.

Castella, S., Bingham, G., and Sanders, C.M. (2006a). Common determinants in DNA melting and helicase-catalysed DNA unwinding by papillomavirus replication protein E1. Nucleic Acids Res. *34*, 3008–3019.

Castella, S., Burgin, D., and Sanders, C.M. (2006b). Role of ATP hydrolysis in the DNA translocase activity of the bovine papillomavirus (BPV-1) E1 helicase. Nucleic Acids Res. *34*, 3731–3741.

Chen, G., and Stenlund, A. (1998). Characterization of the DNA-binding domain of the bovine papillomavirus replication initiator E1. J. Virol. *72*, 2567–2576.

Chiang, C.M., Dong, G., Broker, T.R., and Chow, L.T. (1992). Control of human papillomavirus type 11 origin of replication by the E2 family of transcription regulatory proteins. J. Virol. *66*, 5224–5231.

Clertant, P., and Seif, I. (1984). A common function for polyoma virus large-T and papillomavirus E1 proteins? Nature *311*, 276–279.

Clower, R.V., Fisk, J.C., and Melendy, T. (2006a). Papillomavirus E1 protein binds to and stimulates human topoisomerase I. J. Virol. *80*, 1584–1587.

Clower, R.V., Hu, Y., and Melendy, T. (2006b). Papillomavirus E2 protein interacts with and stimulates human topoisomerase I. Virology *348*, 13–18.

Cohn, M.A., Kowal, P., Yang, K., Haas, W., Huang, T.T., Gygi, S.P., and D'Andrea, A.D. (2007). A UAF1-containing multisubunit protein complex regulates the Fanconi anemia pathway. Mol. Cell *28*, 786–797.

Cohn, M.A., Kee, Y., Haas, W., Gygi, S.P., and D'Andrea, A.D. (2009). UAF1 is a subunit of multiple deubiquitinating enzyme complexes. J. Biol. Chem. *284*, 5343–5351.

Conger, K.L., Liu, J.S., Kuo, S.R., Chow, L.T., and Wang, T.S. (1999). Human papillomavirus DNA replication. Interactions between the viral E1 protein and two subunits of human dna polymerase alpha/primase. J. Biol. Chem. *274*, 2696–2705.

Cooper, C.S., Upmeyer, S.N., and Winokur, P.L. (1998). Identification of single amino acids in the human papillomavirus 11 E2 protein critical for the transactivation or replication functions. Virology *241*, 312–322.

Cote-Martin, A., Moody, C., Fradet-Turcotte, A., D'Abramo, C.M., Lehoux, M., Joubert, S., Poirier, G.G., Coulombe, B., Laimins, L.A., and Archambault, J. (2008). Human papillomavirus E1 helicase interacts with the WD repeat protein p80 to promote maintenance of the viral genome in keratinocytes. J. Virol. *82*, 1271–1283.

Dasgupta, S., Zabielski, J., Simonsson, M., and Burnett, S. (1992). Rolling-circle replication of a high-copy BPV-1 plasmid. J. Mol. Biol. *228*, 1–6.

Del Vecchio, A.M., Romanczuk, H., Howley, P.M., and Baker, C.C. (1992). Transient replication of human papillomavirus DNAs. J. Virol. *66*, 5949–5958.

Dell, G., Wilkinson, K.W., Tranter, R., Parish, J., Leo Brady, R., and Gaston, K. (2003). Comparison of the structure and DNA-binding properties of the E2 proteins from an oncogenic and a non-oncogenic human papillomavirus. J. Mol. Biol. *334*, 979–991.

Deng, W., Lin, B.Y., Jin, G., Wheeler, C.G., Ma, T., Harper, J.W., Broker, T.R., and Chow, L.T. (2004). Cyclin/CDK regulates the nucleocytoplasmic localization of the human papillomavirus E1 DNA helicase. J. Virol. *78*, 13954–13965.

Dixon, E.P., Pahel, G.L., Rocque, W.J., Barnes, J.A., Lobe, D.C., Hanlon, M.H., Alexander, K.A., Chao, S.F., Lindley, K., and Phelps, W.C. (2000). The E1 helicase of human papillomavirus type 11 binds to the origin of replication with low sequence specificity. Virology *270*, 345–357.

Duensing, S., and Munger, K. (2002). The human papillomavirus type 16 E6 and E7 oncoproteins independently induce numerical and structural chromosome instability. Cancer Res. *62*, 7075–7082.

Enemark, E.J., and Joshua-Tor, L. (2006). Mechanism of DNA translocation in a replicative hexameric helicase. Nature *442*, 270–275.

Enemark, E.J., Chen, G., Vaughn, D.E., Stenlund, A., and Joshua-Tor, L. (2000). Crystal structure of the DNA-binding domain of the replication initiation protein E1 from papillomavirus. Mol. Cell *6*, 149–158.

Enemark, E.J., Stenlund, A., and Joshua-Tor, L. (2002). Crystal structures of two intermediates in the assembly of the papillomavirus replication initiation complex. EMBO J. *21*, 1487–1496.

Favre, M., Breitburd, F., Croissant, O., and Orth, G. (1977). Chromatin-like structures obtained after alkaline disruption of bovine and human papillomaviruses. J. Virol. *21*, 1205–1209.

Flores, E.R., and Lambert, P.F. (1997). Evidence for a switch in the mode of human papillomavirus type 16 DNA replication during the viral life cycle. J. Virol. *71*, 7167–7179.

Fouts, E.T., Yu, X., Egelman, E.H., and Botchan, M.R. (1999). Biochemical and electron microscopic image analysis of the hexameric E1 helicase. J. Biol. Chem. *274*, 4447–4458.

Fradet-Turcotte, A., Bergeron-Labrecque, F., Moody, C., Lehoux, M., Laimins, L., and Archambault, J. (2011). Nuclear accumulation of the papillomavirus E1 helicase blocks S-phase progression and triggers an ATM-dependent DNA damage response. J. Virol. *85*, 8996–9012.

Fradet-Turcotte, A., Brault, K., Titolo, S., Howley, P.M., and Archambault, J. (2009). Characterization of papillomavirus E1 helicase mutants defective for interaction with the SUMO-conjugating enzyme Ubc9. Virology *395*, 190–201.

Fradet-Turcotte, A., Moody, C., Laimins, L.A., and Archambault, J. (2010a). Nuclear export of human papillomavirus type 31 E1 is regulated by Cdk2 phosphorylation and required for viral genome maintenance. J. Virol. *84*, 11747–11760.

Fradet-Turcotte, A., Morin, G., Lehoux, M., Bullock, P.A., and Archambault, J. (2010b). Development of quantitative and high-throughput assays of polyomavirus and papillomavirus DNA replication. Virology *399*, 65–76.

Freeman, A.K., and Monteiro, A.N. (2010). Phosphatases in the cellular response to DNA damage. Cell Commun. Signal *8*, 27.

Galanty, Y., Belotserkovskaya, R., Coates, J., Polo, S., Miller, K.M., and Jackson, S.P. (2009). Mammalian SUMO E3-ligases PIAS1 and PIAS4 promote responses to DNA double-strand breaks. Nature *462*, 935–939.

Geimanen, J., Isok-Paas, H., Pipitch, R., Salk, K., Laos, T., Orav, M., Reinson, T., Ustav, M., Jr., Ustav, M., and Ustav, E. (2011). Development of a cellular assay system to study the genome replication of high- and low-risk mucosal and cutaneous human papillomaviruses. J. Virol. *85*, 3315–3329.

Gilbert, D.M., and Cohen, S.N. (1987). Bovine papilloma virus plasmids replicate randomly in mouse fibroblasts throughout S phase of the cell cycle. Cell *50*, 59–68.

Gillette, T.G., Lusky, M., and Borowiec, J.A. (1994). Induction of structural changes in the bovine papillomavirus type 1 origin of replication by the viral E1 and E2 proteins. Proc. Natl. Acad. Sci. U.S.A. *91*, 8846–8850.

Gorbalenya, A.E., Koonin, E.V., and Wolf, Y.I. (1990). A new superfamily of putative NTP-binding domains encoded by genomes of small DNA and RNA viruses. FEBS Lett. *262*, 145–148.

Halas, A., Podlaska, A., Derkacz, J., McIntyre, J., Skoneczna, A., and Sledziewska-Gojska, E. (2011). The roles of PCNA SUMOylation, Mms2-Ubc13 and Rad5 in translesion DNA synthesis in *Saccharomyces cerevisiae*. Mol. Microbiol. *80*, 786–797.

Han, Y., Loo, Y.M., Militello, K.T., and Melendy, T. (1999). Interactions of the papovavirus DNA replication initiator proteins, bovine papillomavirus type 1 E1 and simian virus 40 large T antigen, with human replication protein A. J. Virol. *73*, 4899–4907.

Harris, S.F., and Botchan, M.R. (1999). Crystal structure of the human papillomavirus type 18 E2 activation domain. Science *284*, 1673–1677.

Hebner, C.M., and Laimins, L.A. (2006). Human papillomaviruses: basic mechanisms of pathogenesis and oncogenicity. Rev. Med. Virol. *16*, 83–97.

Hegde, R.S. (2002). The papillomavirus E2 proteins: structure, function, and biology. Annu. Rev. Biophys. Biomol. Struct. *31*, 343–360.

Hickman, A.B., and Dyda, F. (2005). Binding and unwinding: SF3 viral helicases. Curr. Opin. Struct. Biol. *15*, 77–85.

Hoffmann, R., Hirt, B., Bechtold, V., Beard, P., and Raj, K. (2006). Different modes of human papillomavirus DNA replication during maintenance. J. Virol. *80*, 4431–4439.

Howley, P.M., and Lowy, D. R (2001). Papillomaviruses and their replication. In Fields Virology. Fields, B.N., Knipe, D.M., and Howley, P.M., eds. (Lippincott-Raven: Philadelphia), pp. 2197–2229.

Hsu, C.Y., Mechali, F., and Bonne-Andrea, C. (2007). Nucleocytoplasmic shuttling of bovine papillomavirus E1 helicase down-regulates viral DNA replication in S phase. J. Virol. *81*, 384–394.

Hu, Y., Clower, R.V., and Melendy, T. (2006). Cellular topoisomerase I modulates origin binding by bovine papillomavirus type 1 E1. J. Virol. *80*, 4363–4371.

Hughes, F.J., and Romanos, M.A. (1993). E1 protein of human papillomavirus is a DNA helicase/ATPase. Nucleic Acids Res. *21*, 5817–5823.

Ilves, I., Kivi, S., and Ustav, M. (1999). Long-term episomal maintenance of bovine papillomavirus type 1 plasmids is determined by attachment to host chromosomes, which Is mediated by the viral E2 protein and its binding sites. J. Virol. *73*, 4404–4412.

Jenkins, O., Earnshaw, D., Sarginson, G., Del Vecchio, A., Tsai, J., Kallender, H., Amegadzie, B., and Browne, M. (1996). Characterization of the helicase and ATPase activity of human papillomavirus type 6b E1 protein. J. Gen. Virol. *77 (Pt 8)*, 1805–1809.

Kadaja, M., Sumerina, A., Verst, T., Ojarand, M., Ustav, E., and Ustav, M. (2007). Genomic instability of the host cell induced by the human papillomavirus replication machinery. EMBO J. *26*, 2180–2191.

Kadaja, M., Isok-Paas, H., Laos, T., Ustav, E., and Ustav, M. (2009a). Mechanism of genomic instability in cells infected with the high-risk human papillomaviruses. PLoS Pathog. *5*, e1000397.

Kadaja, M., Silla, T., Ustav, E., and Ustav, M. (2009b). Papillomavirus DNA replication – from initiation to genomic instability. Virology *384*, 360–368.

Kee, Y., Yang, K., Cohn, M.A., Haas, W., Gygi, S.P., and D'Andrea, A.D. (2010). WDR20 regulates activity of the USP12 x UAF1 deubiquitinating enzyme complex. J. Biol. Chem. *285*, 11252–11257.

King, L.E., Fisk, J.C., Dornan, E.S., Donaldson, M.M., Melendy, T., and Morgan, I.M. (2010). Human papillomavirus E1 and E2 mediated DNA replication is not arrested by DNA damage signalling. Virology *406*, 95–102.

Klumpp, D.J., Stubenrauch, F., and Laimins, L.A. (1997). Differential effects of the splice acceptor at nucleotide 3295 of human papillomavirus type 31 on stable and transient viral replication. J. Virol. *71*, 8186–8194.

Kuo, S.R., Liu, J.S., Broker, T.R., and Chow, L.T. (1994). Cell-free replication of the human papillomavirus DNA with homologous viral E1 and E2 proteins and human cell extracts. J. Biol. Chem. *269*, 24058–24065.

Kusumoto-Matsuo, R., Kanda, T., and Kukimoto, I. (2011). Rolling circle replication of human papillomavirus type 16 DNA in epithelial cell extracts. Genes Cells *16*, 23–33.

Lace, M.J., Anson, J.R., Thomas, G.S., Turek, L.P., and Haugen, T.H. (2008). The E8–E2 gene product of human papillomavirus type 16 represses early transcription and replication but is dispensable for viral plasmid persistence in keratinocytes. J. Virol. *82*, 10841–10853.

Lee, D., Kim, H., Lee, Y., and Choe, J. (1997). Identification of sequence requirement for the origin of DNA replication in human papillomavirus type 18. Virus Res. *52*, 97–108.

Lee, D., Sohn, H., Kalpana, G.V., and Choe, J. (1999). Interaction of E1 and hSNF5 proteins stimulates replication of human papillomavirus DNA. Nature *399*, 487–491.

Li, R., and Botchan, M.R. (1993). The acidic transcriptional activation domains of VP16 and p53 bind the cellular replication protein A and stimulate *in vitro* BPV-1 DNA replication. Cell *73*, 1207–1221.

Li, R., and Botchan, M.R. (1994). Acidic transcription factors alleviate nucleosome-mediated repression of DNA replication of bovine papillomavirus type 1. Proc. Natl. Acad. Sci. U.S.A. *91*, 7051–7055.

Lin, B.Y., Makhov, A.M., Griffith, J.D., Broker, T.R., and Chow, L.T. (2002). Chaperone proteins abrogate inhibition of the human papillomavirus (HPV) E1 replicative helicase by the HPV E2 protein. Mol. Cell Biol. *22*, 6592–6604.

Liu, J.S., Kuo, S.R., Makhov, A.M., Cyr, D.M., Griffith, J.D., Broker, T.R., and Chow, L.T. (1998). Human Hsp70 and Hsp40 chaperone proteins facilitate human papillomavirus-11 E1 protein binding to the origin and stimulate cell-free DNA replication. J. Biol. Chem. *273*, 30704–30712.

Liu, X., and Stenlund, A. (2010). Mutations in Sensor 1 and Walker B in the bovine papillomavirus E1 initiator protein mimic the nucleotide-bound state. J. Virol. *84*, 1912–1919.

Liu, X., Schuck, S., and Stenlund, A. (2007). Adjacent residues in the E1 initiator beta-hairpin define different roles of the beta-hairpin in Ori melting, helicase loading, and helicase activity. Mol. Cell *25*, 825–837.

Liu, X., Schuck, S., and Stenlund, A. (2010). Structure-based mutational analysis of the bovine papillomavirus E1 helicase domain identifies residues involved in the nonspecific DNA binding activity required for double trimer formation. J. Virol. *84*, 4264–4276.

Longworth, M.S., and Laimins, L.A. (2004). The binding of histone deacetylases and the integrity of zinc finger-like motifs of the E7 protein are essential for the life cycle of human papillomavirus type 31. J. Virol. *78*, 3533–3541.

Longworth, M.S., Wilson, R., and Laimins, L.A. (2005). HPV-31 E7 facilitates replication by activating E2F2 transcription through its interaction with HDACs. EMBO J. *24*, 1821–1830.

Loo, Y.M., and Melendy, T. (2004). Recruitment of replication protein A by the papillomavirus E1 protein and modulation by single-stranded DNA. J. Virol. *78*, 1605–1615.

Lu, J.Z., Sun, Y.N., Rose, R.C., Bonnez, W., and McCance, D.J. (1993). Two E2 binding sites (E2BS) alone or one E2BS plus an A/T-rich region are minimal requirements for the replication of the human papillomavirus type 11 origin. J. Virol. *67*, 7131–7139.

Luo, X., Sanford, D.G., Bullock, P.A., and Bachovchin, W.W. (1996). Solution structure of the origin DNA-binding domain of SV40 T-antigen. Nat. Struct. Biol. *3*, 1034–1039.

Lusky, M., Hurwitz, J., and Seo, Y.S. (1994). The bovine papillomavirus E2 protein modulates the assembly of but is not stably maintained in a replication-competent multimeric E1-replication origin complex. Proc. Natl. Acad. Sci. U.S.A. *91*, 8895–8899.

Ma, T., Zou, N., Lin, B.Y., Chow, L.T., and Harper, J.W. (1999). Interaction between cyclin-dependent kinases and human papillomavirus replication-initiation protein E1 is required for efficient viral replication. Proc. Natl. Acad. Sci. U.S.A. *96*, 382–387.

McBride, A.A., McPhillips, M.G., and Oliveira, J.G. (2004). Brd4: tethering, segregation and beyond. Trends Microbiol. *12*, 527–529.

McLaughlin-Drubin, M.E., and Munger, K. (2009). Oncogenic activities of human papillomaviruses. Virus Res. *143*, 195–208.

McLaughlin-Drubin, M.E., Bromberg-White, J.L., and Meyers, C. (2005). The role of the human papillomavirus type 18 E7 oncoprotein during the complete viral life cycle. Virology *338*, 61–68.

Malcles, M.H., Cueille, N., Mechali, F., Coux, O., and Bonne-Andrea, C. (2002). Regulation of bovine papillomavirus replicative helicase e1 by the ubiquitin-proteasome pathway. J. Virol. *76*, 11350–11358.

Mansky, K.C., Batiza, A., and Lambert, P.F. (1997). Bovine papillomavirus type 1 E1 and simian virus 40 large T antigen share regions of sequence similarity required for multiple functions. J. Virol. *71*, 7600–7608.

del Mar Pena, L.M., and Laimins, L.A. (2001). Differentiation-dependent chromatin rearrangement coincides with activation of human papillomavirus type 31 late gene expression. J. Virol. *75*, 10005–10013.

Masterson, P.J., Stanley, M.A., Lewis, A.P., and Romanos, M.A. (1998). A C-terminal helicase domain of the human papillomavirus E1 protein binds E2 and the DNA polymerase alpha-primase p68 subunit. J. Virol. *72*, 7407–7419.

Mechali, F., Hsu, C.Y., Castro, A., Lorca, T., and Bonne-Andrea, C. (2004). Bovine papillomavirus replicative helicase E1 is a target of the ubiquitin ligase APC. J. Virol. *78*, 2615–2619.

Melendy, T., Sedman, J., and Stenlund, A. (1995). Cellular factors required for papillomavirus DNA replication. J. Virol. *69*, 7857–7867.

Mohr, I.J., Clark, R., Sun, S., Androphy, E.J., MacPherson, P., and Botchan, M.R. (1990). Targeting the E1 replication protein to the papillomavirus origin of replication by complex formation with the E2 transactivator. Science *250*, 1694–1699.

Moody, C.A., and Laimins, L.A. (2009). Human papillomaviruses activate the ATM DNA damage pathway for viral genome amplification upon differentiation. PLoS Pathog. *5*, e1000605.

Moody, C.A., and Laimins, L.A. (2010). Human papillomavirus oncoproteins: pathways to transformation. Nat. Rev. Cancer *10*, 550–560.

Moody, C.A., Fradet-Turcotte, A., Archambault, J., and Laimins, L.A. (2007). Human papillomaviruses activate caspases upon epithelial differentiation to induce viral genome amplification. Proc. Natl. Acad. Sci. U.S.A. *104*, 19541–19546.

Morin, G., Fradet-Turcotte, A., Di Lello, P., Bergeron-Labrecque, F., Omichinski, J.G., and Archambault, J. (2011). A conserved amphipathic helix in the N-terminal regulatory region of the papillomavirus E1 helicase is required for efficient viral DNA replication. J. Virol. *85*, 5287–5300.

Nakahara, T., Peh, W.L., Doorbar, J., Lee, D., and Lambert, P.F. (2005). Human papillomavirus type 16 E1circumflexE4 contributes to multiple facets of the papillomavirus life cycle. J. Virol. *79*, 13150–13165.

Ogura, T., and Wilkinson, A.J. (2001). AAA^{+} superfamily ATPases: common structure–diverse function. Genes Cells *6*, 575–597.

Parish, J.L., Kowalczyk, A., Chen, H.T., Roeder, G.E., Sessions, R., Buckle, M., and Gaston, K. (2006). E2 proteins from high- and low-risk human papillomavirus types differ in their ability to bind p53 and induce apoptotic cell death. J. Virol. *80*, 4580–4590.

Park, P., Copeland, W., Yang, L., Wang, T., Botchan, M.R., and Mohr, I.J. (1994). The cellular DNA polymerase alpha-primase is required for papillomavirus DNA replication and associates with the viral E1 helicase. Proc. Natl. Acad. Sci. U.S.A. *91*, 8700–8704.

Park, R.B., and Androphy, E.J. (2002). Genetic analysis of high-risk e6 in episomal maintenance of human papillomavirus genomes in primary human keratinocytes. J. Virol. *76*, 11359–11364.

Piirsoo, M., Ustav, E., Mandel, T., Stenlund, A., and Ustav, M. (1996). Cis and trans requirements for stable episomal maintenance of the BPV-1 replicator. EMBO J. *15*, 1–11.

Pim, D., and Banks, L. (2010). Interaction of viral oncoproteins with cellular target molecules: infection with high-risk vs low-risk human papillomaviruses. Apmis *118*, 471–493.

de Prat-Gay, G., Gaston, K., and Cicero, D.O. (2008). The papillomavirus E2 DNA binding domain. Front. Biosci. *13*, 6006–6021.

Rangasamy, D., and Wilson, V.G. (2000). Bovine papillomavirus E1 protein is sumoylated by the host cell Ubc9 protein. J. Biol. Chem. *275*, 30487–30495.

Rangasamy, D., Woytek, K., Khan, S.A., and Wilson, V.G. (2000). SUMO-1 modification of bovine papillomavirus E1 protein is required for intranuclear accumulation. J. Biol. Chem. *275*, 37999–38004.

Ravnan, J.B., Gilbert, D.M., Ten Hagen, K.G., and Cohen, S.N. (1992). Random-choice replication of extrachromosomal bovine papillomavirus (BPV) molecules in heterogeneous, clonally derived BPV-infected cell lines. J. Virol. *66*, 6946–6952.

Reinhardt, H.C., and Yaffe, M.B. (2009). Kinases that control the cell cycle in response to DNA damage: Chk1, Chk2, and MK2. Curr. Opin. Cell Biol. *21*, 245–255.

Remm, M., Brain, R., and Jenkins, J.R. (1992). The E2 binding sites determine the efficiency of replication for the origin of human papillomavirus type 18. Nucleic Acids Res. *20*, 6015–6021.

Roberts, J.M., and Weintraub, H. (1988). Cis-acting negative control of DNA replication in eukaryotic cells. Cell *52*, 397–404.

Rocque, W.J., Porter, D.J., Barnes, J.A., Dixon, E.P., Lobe, D.C., Su, J.L., Willard, D.H., Gaillard, R., Condreay, J.P., Clay, W.C., *et al.* (2000). Replication-associated activities of purified human papillomavirus type 11 E1 helicase. Protein Expr. Purif. *18*, 148–159.

Rosas-Acosta, G., and Wilson, V.G. (2008). Identification of a nuclear export signal sequence for bovine papillomavirus E1 protein. Virology *373*, 149–162.

Rosas-Acosta, G., Langereis, M.A., Deyrieux, A., and Wilson, V.G. (2005). Proteins of the PIAS family enhance the sumoylation of the papillomavirus E1 protein. Virology *331*, 190–203.

Russell, J., and Botchan, M.R. (1995). cis-Acting components of human papillomavirus (HPV) DNA replication: linker substitution analysis of the HPV type 11 origin. J. Virol. *69*, 651–660.

Saikia, P., Fensterl, V., and Sen, G.C. (2010). The inhibitory action of P56 on select functions of E1 mediates interferon's effect on human papillomavirus DNA replication. J. Virol. *84*, 13036–13039.

Sakai, H., Yasugi, T., Benson, J.D., Dowhanick, J.J., and Howley, P.M. (1996). Targeted mutagenesis of the human papillomavirus type 16 E2 transactivation domain reveals separable transcriptional activation and DNA replication functions. J. Virol. *70*, 1602–1611.

Sakakibara, N., Mitra, R., and McBride, A.A. (2011). The papillomavirus E1 helicase activates a cellular DNA damage response in viral replication foci. J. Virol. *85*, 8981–8995.

Sanders, C.M., and Stenlund, A. (1998). Recruitment and loading of the E1 initiator protein: an ATP-dependent process catalysed by a transcription factor. EMBO J. *17*, 7044–7055.

Sanders, C.M., and Stenlund, A. (2000). Transcription factor-dependent loading of the E1 initiator reveals modular assembly of the papillomavirus origin melting complex. J. Biol. Chem. *275*, 3522–3534.

Sanders, C.M., and Stenlund, A. (2001). Mechanism and requirements for bovine papillomavirus, type 1, E1 initiator complex assembly promoted by the E2 transcription factor bound to distal sites. J. Biol. Chem. *276*, 23689–23699.

Sarafi, T.R., and McBride, A.A. (1995). Domains of the BPV-1 E1 replication protein required for origin-specific DNA binding and interaction with the E2 transactivator. Virology *211*, 385–396.

Schuck, S., and Stenlund, A. (2005a). Assembly of a double hexameric helicase. Mol. Cell *20*, 377–389.

Schuck, S., and Stenlund, A. (2005b). Role of papillomavirus E1 initiator dimerization in DNA replication. J. Virol. *79*, 8661–8664.

Schuck, S., and Stenlund, A. (2007). ATP-dependent minor groove recognition of TA base pairs is required for template melting by the E1 initiator protein. J. Virol. *81*, 3293–3302.

Sclafani, R.A., and Holzen, T.M. (2007). Cell cycle regulation of DNA replication. Ann. Rev. Genet. *41*, 237–280.

Sedman, J., and Stenlund, A. (1998). The papillomavirus E1 protein forms a DNA-dependent hexameric complex with ATPase and DNA helicase activities. J. Virol. *72*, 6893–6897.

Senechal, H., Poirier, G.G., Coulombe, B., Laimins, L.A., and Archambault, J. (2007). Amino acid substitutions that specifically impair the transcriptional activity of papillomavirus E2 affect binding to the long isoform of Brd4. Virology *358*, 10–17.

Seo, Y.S., Muller, F., Lusky, M., Gibbs, E., Kim, H.Y., Phillips, B., and Hurwitz, J. (1993a). Bovine papilloma virus (BPV)-encoded E2 protein enhances binding of E1 protein to the BPV replication origin. Proc. Natl. Acad. Sci. U.S.A. *90*, 2865–2869.

Seo, Y.S., Muller, F., Lusky, M., and Hurwitz, J. (1993b). Bovine papilloma virus (BPV)-encoded E1 protein contains multiple activities required for BPV DNA replication. Proc. Natl. Acad. Sci. U.S.A. *90*, 702–706.

Shaheen, M., Shanmugam, I., and Hromas, R. (2010). The role of PCNA posttranslational modifications in translesion synthesis. J. Nucleic Acids. pii:761217.

Stenlund, A. (2003). E1 initiator DNA binding specificity is unmasked by selective inhibition of non-specific DNA binding. EMBO J. *22*, 954–963.

Stubenrauch, F., Lim, H.B., and Laimins, L.A. (1998). Differential requirements for conserved E2 binding sites in the life cycle of oncogenic human papillomavirus type 31. J. Virol. *72*, 1071–1077.

Stubenrauch, F., Zobel, T., and Iftner, T. (2001). The E8 domain confers a novel long-distance transcriptional repression activity on the E8E2C protein of high-risk human papillomavirus type 31. J. Virol. *75*, 4139–4149.

Su, T.T. (2006). Cellular responses to DNA damage: one signal, multiple choices. Ann. Rev. Genet. *40*, 187–208.

Sun, Y., Han, H., and McCance, D.J. (1998). Active domains of human papillomavirus type 11 E1 protein for origin replication. J. Gen. Virol. *79 (Pt 7)*, 1651–1658.

Sun, Y.N., Lu, J.Z., and McCance, D.J. (1996). Mapping of HPV-11 E1 binding site and determination of other important cis elements for replication of the origin. Virology *216*, 219–222.

Sverdrup, F., and Khan, S.A. (1994). Replication of human papillomavirus (HPV) DNAs supported by the HPV type 18 E1 and E2 proteins. J. Virol. *68*, 505–509.

Sverdrup, F., and Khan, S.A. (1995). Two E2 binding sites alone are sufficient to function as the minimal origin of replication of human papillomavirus type 18 DNA. J. Virol. *69*, 1319–1323.

Swindle, C.S., and Engler, J.A. (1998). Association of the human papillomavirus type 11 E1 protein with histone H1. J. Virol. *72*, 1994–2001.

Tang, L., Nogales, E., and Ciferri, C. (2010). Structure and function of SWI/SNF chromatin remodeling complexes and mechanistic implications for transcription. Prog. Biophys. Mol. Biol. *102*, 122–128.

Terenzi, F., Saikia, P., and Sen, G.C. (2008). Interferon-inducible protein, P56, inhibits HPV DNA replication by binding to the viral protein E1. EMBO J. *27*, 3311–3321.

Thomas, J.T., Hubert, W.G., Ruesch, M.N., and Laimins, L.A. (1999). Human papillomavirus type 31 oncoproteins E6 and E7 are required for the maintenance of episomes during the viral life cycle in normal human keratinocytes. Proc. Natl. Acad. Sci. U.S.A. *96*, 8449–8454.

Titolo, S., Pelletier, A., Sauve, F., Brault, K., Wardrop, E., White, P.W., Amin, A., Cordingley, M.G., and Archambault, J. (1999). Role of the ATP-binding domain of the human papillomavirus type 11 E1 helicase in E2-dependent binding to the origin. J. Virol. *73*, 5282–5293.

Titolo, S., Pelletier, A., Pulichino, A.M., Brault, K., Wardrop, E., White, P.W., Cordingley, M.G., and Archambault, J. (2000). Identification of domains of the human papillomavirus type 11 E1 helicase involved in oligomerization and binding to the viral origin. J. Virol. *74*, 7349–7361.

Titolo, S., Brault, K., Majewski, J., White, P.W., and Archambault, J. (2003a). Characterization of the minimal DNA-binding domain of the human papillomavirus e1 helicase: fluorescence anisotropy studies and characterization of a dimerization-defective mutant protein. J. Virol. *77*, 5178–5191.

Titolo, S., Welchner, E., White, P.W., and Archambault, J. (2003b). Characterization of the DNA-binding properties of the origin-binding domain of simian virus 40 large T antigen by fluorescence anisotropy. J. Virol. *77*, 5512–5518.

Ulrich, H.D., Vogel, S., and Davies, A.A. (2005). SUMO keeps a check on recombination during DNA replication. Cell Cycle *4*, 1699–1702.

Wang, Y., Coulombe, R., Cameron, D.R., Thauvette, L., Massariol, M.J., Amon, L.M., Fink, D., Titolo, S., Welchner, E., Yoakim, C., *et al.* (2004). Crystal structure of the E2 transactivation domain of human papillomavirus type 11 bound to a protein interaction inhibitor. J. Biol. Chem. *279*, 6976–6985.

Webster, K., Parish, J., Pandya, M., Stern, P.L., Clarke, A.R., and Gaston, K. (2000). The human papillomavirus (HPV) 16 E2 protein induces apoptosis in the absence of other HPV proteins and via a p53-dependent pathway. J. Biol. Chem. *275*, 87–94.

White, P.W., Pelletier, A., Brault, K., Titolo, S., Welchner, E., Thauvette, L., Fazekas, M., Cordingley, M.G., and Archambault, J. (2001). Characterization of recombinant HPV-6 and -11 E1 helicases: effect of ATP on the interaction of E1 with E2 and mapping of a minimal helicase domain. J. Biol. Chem. *276*, 22426–22438.

Wilson, R., Fehrmann, F., and Laimins, L.A. (2005). Role of the E1–E4 protein in the differentiation-dependent life cycle of human papillomavirus type 31. J. Virol. *79*, 6732–6740.

Wilson, R., Ryan, G.B., Knight, G.L., Laimins, L.A., and Roberts, S. (2007). The full-length E1E4 protein of human papillomavirus type 18 modulates differentiation-dependent viral DNA amplification and late gene expression. Virology *362*, 453–460.

Wilson, V.G., West, M., Woytek, K., and Rangasamy, D. (2002). Papillomavirus E1 proteins: form, function, and features. Virus Genes *24*, 275–290.

Yang, L., Li, R., Mohr, I.J., Clark, R., and Botchan, M.R. (1991). Activation of BPV-1 replication *in vitro* by the transcription factor E2. Nature *353*, 628–632.

Yang, L., Mohr, I., Fouts, E., Lim, D.A., Nohaile, M., and Botchan, M. (1993). The E1 protein of bovine papilloma virus 1 is an ATP-dependent DNA helicase. Proc. Natl. Acad. Sci. U.S.A. *90*, 5086–5090.

Yardimci, H., Loveland, A.B., Habuchi, S., van Oijen, A.M., and Walter, J.C. (2010). Uncoupling of sister replisomes during eukaryotic DNA replication. Mol. Cell *40*, 834–840.

Yasugi, T., and Howley, P.M. (1996). Identification of the structural and functional human homolog of the yeast ubiquitin conjugating enzyme UBC9. Nucleic Acids Res *24*, 2005–2010.

Yasugi, T., Vidal, M., Sakai, H., Howley, P.M., and Benson, J.D. (1997). Two classes of human papillomavirus type 16 E1 mutants suggest pleiotropic conformational constraints affecting E1 multimerization, E2 interaction, and interaction with cellular proteins. J. Virol. *71*, 5942–5951.

Yu, J.H., Lin, B.Y., Deng, W., Broker, T.R., and Chow, L.T. (2007). Mitogen-activated protein kinases activate the nuclear localization sequence of human papillomavirus type 11 E1 DNA helicase to promote efficient nuclear import. J. Virol. *81*, 5066–5078.

Zobel, T., Iftner, T., and Stubenrauch, F. (2003). The papillomavirus E8-E2C protein represses DNA replication from extrachromosomal origins. Mol. Cell Biol. *23*, 8352–8362.

Induction of Genomic Instability by Human Papillomavirus Oncoproteins

12

Karl Münger and Stefan Duensing

Abstract

The high-risk human papillomavirus (HPV)-encoded oncoproteins E6 and E7 have been instrumental to dissect crucial pathways of genomic instability and carcinogenic progression. This includes the notion that cell cycle deregulation and the development of numerical and structural chromosomal instability are intricately linked. The HPV E6 and E7 oncoproteins disrupt p53 and pRB signalling, respectively, and it has become evident that these events set the stage for numerous host cellular aberrations that can promote genomic instability and thus ultimately promote malignant progression. Many if not all of these host cellular aberrations can also be detected in non-virus-associated tumours, making HPV-associated carcinogenesis an attractive model system to analyse the molecular mechanisms and functional consequences of genomic instability in cancer in general.

Introduction

High-risk HPV-associated cancer of the uterine cervix is a leading cause of cancer-related deaths in women worldwide (zur Hausen, 2002). More than 90% of cervical carcinomas harbour high-risk human papillomavirus (HPV) DNA and two viral oncoproteins, E6 and E7, are consistently expressed (Munoz *et al.*, 2003). Since the discovery of high-risk HPVs as causative agents for cervical cancer by Harald zur Hausen (Durst *et al.*, 1983), it has become clear that this link provides an exceptional model to understand the molecular mechanisms that underlie malignant progression and cancer formation.

A hallmark of high-risk HPV-associated carcinomas and cell lines derived from such lesions is the fact that they are genomically unstable (reviewed in Duensing and Munger, 2004). Cells frequently display not only numerical chromosome aberrations (known as aneuploidy) but also a high level of structural chromosomal changes (White *et al.*, 1994; Solinas-Toldo *et al.*, 1997). Although there are clearly other forms of genetic alterations that can be detected in tumour cells, this review will focus on numerical and structural aberrations on the whole chromosome level.

Genomic instability is not only detected in high-risk HPV-associated neoplasms but is a frequent finding in cancer in general. Genomic instability is an important source of tumour cell heterogeneity, it influences the clonal evolution within a tumour and is hence critical for malignant progression and the development of therapy resistance (Cahill *et al.*, 1999). A better understanding of the molecular basis of genomic instability will help to improve future preventative and therapeutic approaches and will be instrumental to either block the progressive loss of genome integrity or to exacerbate genomic instability in order to trigger apoptosis, for example in the context of synthetic lethal approaches.

Human papillomaviruses (HPVs) are small DNA tumour viruses that infect keratinocytes at various organ sites including skin, cervix, anus and oropharynx (reviewed in Longworth and Laimins, 2004b; McLaughlin-Drubin and Munger,

2009b). They contain a circular double-stranded DNA genome of approximately 8000 base pairs. The high-risk HPV open reading frames encode proteins that are important to support the viral life cycle and for production of viral progeny. However, with the exception of the E1 protein that encodes a helicase and ATPase, HPVs do not encode proteins that directly function to replicate DNA. Hence, HPVs use the host DNA synthesis machinery to replicate their genomes (Howley and Lowy, 2001). Since HPV genome amplification and viral progeny formation is confined to terminally differentiated keratinocytes within epithelial tissues, these viruses face the task of maintaining or re-activating DNA replication in cells that are otherwise undergoing differentiation and are normally permanently withdrawn from the cell division cycle (reviewed in (Moody and Laimins, 2010).

An intriguing aspect of HPV-associated carcinogenesis is the fact that malignant progression is typically preceded by years or decades of persistent infection with high-risk HPVs. This suggests that high-risk HPVs persist in cells with stem cell-like characteristics. Whether or not high-risk HPVs preferentially infect such stem cell-like cells or whether expression of HPV proteins may 'reprogramme' infected cells to a more stem-like state is an open question. The long interval between initial infection and cancer formation furthermore argues for important co-factors that influence malignant progression. There is epidemiological and experimental evidence that oestrogen and the host inflammatory response may be such co-factors (Riley *et al.*, 2003; Wei *et al.*, 2009). Moreover, accumulation of genetic and/or epigenetic changes to the host cellular genome plays a crucial role in the selection and clonal evolution of neoplastic cells that will ultimately lead to invasive and metastatic cancer (zur Hausen, 1991).

It has become evident over the past decade that high-risk HPV oncoproteins stimulate genomic instability. Exploiting the fact that the molecular activities of HPV oncoproteins have been characterized in great detail, it has been possible to identify critical pathways that contribute to chromosomal instability in cancer.

The HPV E6 and E7 oncoproteins

In order to uncouple cellular differentiation from proliferation, high-risk HPV oncoproteins inactivate crucial tumour suppressor pathways of the infected host cell.

The high-risk HPV E6 genes encode small proteins of approximately 150 amino acid that contain two metal binding motifs (reviewed in Howie *et al.*, 2009). High-risk HPV E6 proteins associate with E6-AP, the founding member of the HECT domain class of E3 ubiquitin ligases. A key function of the E6/E6–AP complex is to target the p53 tumour suppressor for degradation (Scheffner *et al.*, 1993). In addition, association of high-risk HPV E6 with p53 interferes with p53 acetylation by p300 and CBP, resulting in further impairment of p53-mediated gene activation (Thomas and Chiang, 2005; Hebner *et al.*, 2007). HPV16 E6 also abrogates p53-dependent activation of apoptotic pathway genes through destabilization of the acetyltransferase TIP60 (Jha *et al.*, 2010). High-risk HPV E6 proteins also contain a C-terminal PDZ binding site that mediates interaction with cellular PDZ motif-containing proteins including the human homologues of the *Drosophila* discs large (hDLG) and scribble (hSCRIB) tumour suppressors, MAGI1–3 as well as several others (Kiyono *et al.*, 1997; Glaunsinger *et al.*, 2000; Nakagawa and Huibregtse, 2000; Thomas *et al.*, 2002). High-risk HPV E6 oncoproteins have also been found to activate telomerase (see also below) and have been reported to interact with a variety of additional host cell proteins. Intriguingly, low-risk HPV E6 proteins, while they retain binding to E6-AP, do not cause p53 degradation and lack the C-terminal PDZ binding site. The paramount importance of p53 in cellular stress responses and the effects of impaired p53 on apoptosis induction and proliferation control make it clear that targeting p53 is a critical mechanisms for carcinogenic progression. However, there is also evidence that loss of p53 per se has no or only moderate effects on the genome destabilizing mechanisms described below. Nonetheless, there is compelling evidence that the HPV E6 oncoprotein functions as an important amplifier of many high-risk HPV E7-associated cellular alterations (Duensing and Duensing, 2005).

The high-risk HPV E7 genes encode small unstable phosphoproteins of approximately 100 amino acids that, similar to E6, lack enzymatic activities and function through association with host cellular proteins (reviewed in McLaughlin-Drubin and Munger, 2009a). The best-characterized function of high-risk HPV E7 is the binding and degradation of the retinoblastoma tumour suppressor protein (pRB) as well as the pRB-related proteins p107 and p130 (Dyson *et al.*, 1989; Gonzalez *et al.*, 2001). Low-risk HPV proteins also interact with pRB, albeit much less efficiently. Degradation of pRB is though the proteasome machinery and, in the case of HPV16 E7, involves CUL2-based ubiquitin ligase complexes (Huh *et al.*, 2007). HPV E7 mediated pRB inactivation results in the aberrant expression of S phase genes, which, in turn, allow for replication of viral DNA (Martin *et al.*, 1998). pRB represses transcription through interaction with E2Fs and recruitment of histone deacetylases (HDACs) (Brehm *et al.*, 1998) to promoter sequences and association with E7 disrupts these repressive complexes. In addition, high-risk HPV E7 proteins have also been reported to directly interact with E2F1 as well as with HDAC1 and 2 containing NURD/Mi2β complexes (Brehm *et al.*, 1999; Hwang *et al.*, 2002; Longworth and Laimins, 2004a). This is thought to lead to a profound deregulation of S phase-associated genes including cyclin E and cyclin A. These lead to deregulated CDK2 activity, which may be further enhanced by the ability of high-risk HPV E7 to interact with cyclin/CDK complexes as well as the CDK2 inhibitors $p21^{CIP1}$ and $p27^{KIP1}$ (Tommasino *et al.*, 1993; Zerfass-Thome *et al.*, 1996; Jones *et al.*, 1997; Funk *et al.*, 1997; He *et al.*, 2003; Nguyen and Munger, 2008). The HPV E7 oncoprotein has also been reported to upregulate CDC25 expression, which further activates CDK2 (Nguyen *et al.*, 2002). Why the HPV E7 oncoprotein has evolved to subvert this perplexingly large number of proteins of the pRB signalling axis remains unclear. However, it could point to a high level of control by negative feedback mechanisms.

HPV16 E7 and centrosome aberrations

Centrosomes are evolutionarily ancient cellular organelles that play important roles in mitotic spindle pole formation and assembly of cilia, among a number of other functions (Azimzadeh and Bornens, 2007). They are approximately 0.5 μm in diameter and contain a pair of centrioles, short cylinders of microtubule triplets with a fascinating 9-fold symmetry. Centrioles are embedded in pericentriolar material, an aggregation of a diverse set of proteins. Centrosomes function as major microtubule-organizing centres during interphase and mitosis in most mammalian cells. A normal, non-dividing cell contains a single centrosome. During mitosis, bipolarity as well as three-dimensional orientation of the mitotic spindle is assured by the presence of one centrosome at each of the two opposing spindle poles. This requires duplication of the single centrosome prior to mitosis (Nigg, 2007; Strnad and Gonczy, 2008). In cancer cells, including HPV-associated neoplasms, this process is frequently dysregulated leading to abnormal mitotic figures (Fig. 12.1) (Nigg, 2002; Duensing *et al.*, 2009).

During a normal centriole duplication cycle, the two pre-existing centrioles disengage in late mitosis (Tsou *et al.*, 2009) followed by recruitment and activation of Polo-like kinase 4 (PLK4) adjacent to the wall of each of the pre-existing, maternal centrioles (Kleylein-Sohn *et al.*, 2007). PLK4 subsequently mediates the recruitment of additional centriolar proteins including hSAS-6, which is found at nascent procentrioles and responsible for the establishment of the 9-fold symmetry through a self-assembly process (Kitagawa *et al.*, 2011). PLK4 is critical for daughter centriole formation and depletion of PLK4 leads to a progressive loss of centrosomes (Bettencourt-Dias *et al.*, 2005; Habedanck *et al.*, 2005).

The E7 oncoprotein from high-risk HPV type 16 (HPV16) rapidly induces centrosome overduplication in normal human cells and can thus drive genomic destabilization (Duensing *et al.*, 2000, 2001). Aberrant centriole numbers in HPV16 E7-expressing cells were found to develop rapidly and in a time frame that corresponds to approximately a single cell division cycle

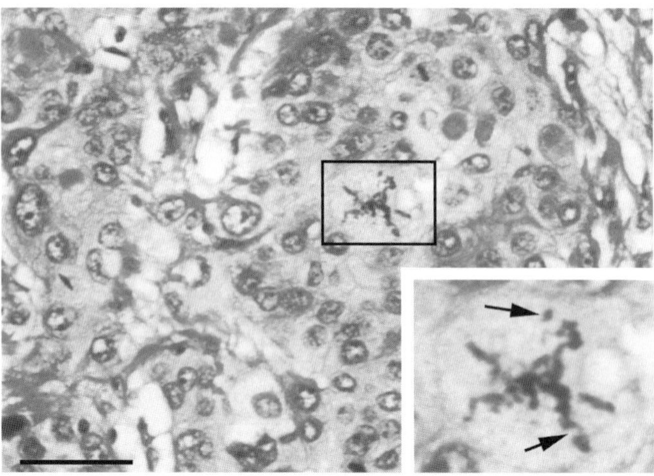

Figure 12.1 Complex mitotic defects in HPV-associated cancer. Haematoxylin and eosin (HE) staining of an high-risk HPV-associated anal squamous cell carcinoma. Inset shows a highly abnormal mitosis with severely altered polarity) as well as chromosome breakage (isolated chromosome fragment; top arrow; chromatid gap, lower arrow). Scale bar indicates 50 μm.

(Duensing *et al.*, 2004). HPV16 E7-expressing cells contained excessive numbers of immature daughter centrioles indicating that E7 induces true centriole duplication errors (Guarguaglini *et al.*, 2005). These results were initially difficult to reconcile with the prevailing model that maternal centrioles can only spawn a single daughter centriole per cell cycle. Experiments using the HPV16 E7 oncoprotein clearly demonstrated that this model was incorrect and it is now widely accepted that single maternal centrioles can serve as an assembly platform for simultaneous synthesis of multiple centrioles (centriole multiplication) (Duensing *et al.*, 2007). Mechanistically, it was found that cyclin E/CDK2 is required for HPV16 E7-induced centriole overduplication (Duensing *et al.*, 2006). Subsequent experiments showed that cyclin E/CDK2 can help to recruit PLK4 to maternal centrioles but that PLK4 levels are clearly rate-limiting in centriole multiplication (Korzeniewski *et al.*, 2009) (Fig. 12.2). One prediction from these results is that the HPV16 E7 oncoprotein not only deregulates cyclin E/CDK2 complexes but also PLK4 expression and/or activity.

A link between the degradation of pRB by HPV16 E7 has been established since low-risk HPV E7 proteins, which are much less efficient in this process, as well as mutant HPV16 E7 that cannot bind and degrade pRB, also do not cause numerical centriole aberrations (Duensing and Münger, 2003). Nonetheless, it is important to note that HPV16 E7 expression can cause centrosome abnormalities in mouse embryo fibroblasts that lack all known members of the pRB family of proteins (Duensing and Münger, 2003). HPV16 E7 has also been shown to bind to γ-tubulin and it is conceivable that this association may contribute to the pRB family member independent ability of HPV16 E7 to induce supernumerary centrosomes (Nguyen *et al.*, 2007).

In contrast to HPV16 E7, centrosome abnormalities in HPV16 E6-expressing cells develop over time and in parallel with multinucleation and other morphological alteration (Duensing *et al.*, 2001). Hence, they are likely to arise as a mere consequence of genomic instability and through defects unrelated to the centrosome duplication cycle. Interestingly, the high-risk HPV E6 oncoprotein, but to a lesser extent E7, was found to stimulate premature mitotic chromosome segregation (Plug-Demaggio and McDougall, 2002), which is likely to be one mechanisms to exacerbate the phenotypical alterations that are commonly seen when HPV E7 and E6 oncoproteins are co-expressed.

A number of oncogenic stimuli have been shown to cause aberrant centriole and centrosome

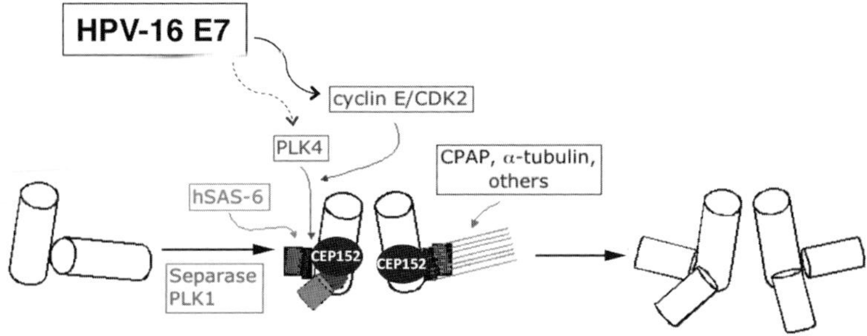

Figure 12.2 HPV16 E7 and centriole overduplication. The HPV16 E7 oncoprotein rapidly stimulates centriole overduplication through the concurrent formation of more than one daughter at single maternal centrioles (centriole multiplication). The two pre-existing centrioles first disengage, which is mediated by PLK1 and separase, followed by the recruitment of PLK4 to the proximity of maternal centrioles. PLK4 then recruits additional proteins involved in centriole duplication such as hSAS-6, CPAP, α-tubulin and others. It has recently been discovered that the centriolar protein CEP152 (*Drosophila melanogaster* Asterless) not only interacts with PLK4 but is also critical for some of the subsequent recruitment steps (Cizmecioglu *et al.* 2010; Dzhindzhev *et al.* 2010; Hatch *et al.* 2010). HPV16 E7-expressing cells show deregulated expression of cyclin E/CDK2 and this has been implicated in an aberrant recruitment of PLK4 to maternal centrioles (Korzeniewski *et al.*, 2009). Additional mechanisms of HPV16 E7 to disrupt this exquisite signalling cascade that limits centriole biogenesis remain to be discovered.

numbers (Fukasawa, 2007) but whether true centrosome overduplication occurs *in vivo* and what the functional consequences may be is still incompletely characterized. Using a marker for mature centrioles, it has been shown that genuine centrosome overduplication i.e. the overproduction of daughter centrioles, does in fact occur in HPV-associated human cancers and correlates with the presence of mitotic defects. Overall, however, the number of cells with such defects is relatively modest (Duensing *et al.*, 2008). It remains to be determined whether cells with true centrosome overduplication have certain properties that could endow them with the ability to function as 'tumour-initiating' or and/or 'tumour maintaining' cells.

HPV16 E7 and DNA damage

It has long been known that HPV16 E7-expressing cells show increased mutation rates, enhanced integration of foreign DNA, structural chromosomal alterations (Fig. 12.1) and signs of an activated DNA damage response typically associated with DNA double-strand breaks such as nuclear γ-H2AX foci (Kessis *et al.*, 1996; Liu *et al.*, 1997; Solinas-Toldo *et al.*, 1997; Duensing and Münger, 2002). In addition, high-risk HPV E6 and E7 expression causes activation of the ATM-associated DNA damage response, which plays a critical role in facilitating viral genome replication in differentiated epithelia (Moody and Laimins, 2009). In cells with integrated high-risk HPV genomes, there is evidence that replication from integrated HPV origins causes DNA damage and activation of DNA damage response pathways (Kadaja *et al.*, 2009). The mechanistic basis of enhanced DNA damage in pre-invasive lesions and in cells with episomal HPV genomes, however, are still poorly characterized.

Recent results demonstrate that HPV16 E7, but not low-risk HPV6 E7 or HPV16 E6, activates the Fanconi Anaemia (FA) DNA damage response pathway (Spardy *et al.*, 2007). In addition to FA pathway activation, it was discovered that reduction of FA protein expression leads to enhanced chromosomal breakage in the presence of HPV16 E7. The FA pathway is the major host cellular pathway that responds to DNA replication stress (as detailed below) and FA deficiency has also been shown to stimulate HPV-associated epithelial hyperplasia (Hoskins *et al.*, 2009). Importantly, Park and co-workers could show that *FancD2*-deficient mice with transgenic expression of HPV16 E7 develop head and neck squamous cell carcinomas at a significant higher incidence than mice with normal *FancD2* status (Park *et al.*, 2010).

Deficiency in one of the thirteen known FA genes leads to hypersensitivity to DNA crosslinking agents, which cause stalled replication forks, and the clinical FA cancer susceptibility syndrome (D'Andrea, 2003; Patel and Joenje, 2007; Wang and D'Andrea, 2004). The majority of the FA gene products function together as a multisubunit E3 ubiquitin ligase complex that monoubiquitinates two FA proteins, FANCD2 and FANCI. These proteins then relocate to chromatin where they recruit additional proteins involved in replication fork repair and restart such as BRCA1, BRCA2 (also known as FANCD1) and others (Garcia-Higuera *et al.*, 2001; Wang *et al.*, 2004). Stalled DNA replication forks are preferred substrates for homologous recombination (HR) repair, and it is believed that the FA pathway activation contributes importantly to protection of the genome by facilitating HR repair at stalled forks (Figs. 12.3 and 12.4). This is likely to play an important role at chromosomal regions that are intrinsically difficult to replicate because of repetitive DNA sequences such as telomeres or common fragile sites (CFSs) (Howlett *et al.*, 2005; Ohki and Ishikawa, 2004).

Patients suffering from the clinical FA syndrome have an approximately 50-fold increased risk of solid tumour formation (Alter *et al.*, 2005) including neoplasms that are commonly associated with high-risk HPV infection such

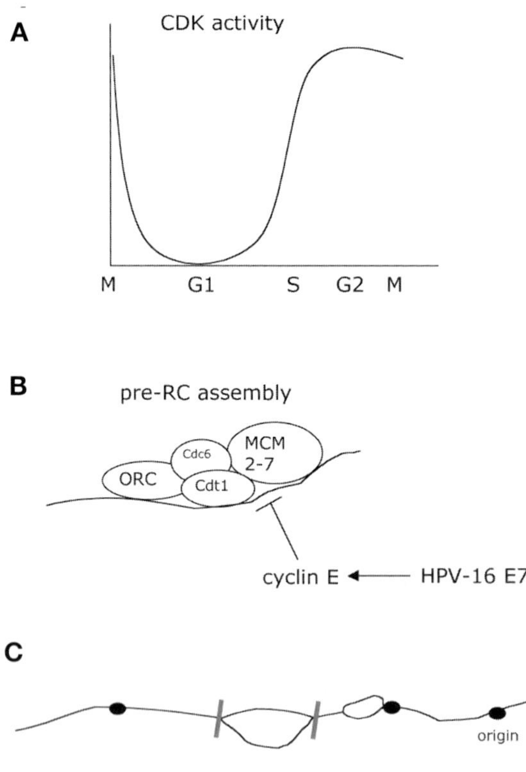

Figure 12.3 A hypothetical model for HPV16 E7-induced replication stress and DNA damage. The oscillation of CDK activity throughout the cell division cycle (A) is important to allow for assembly of pre-replicative complexes (pre-RCs) at origins of replication (licensing). Deregulation of cyclin E expression may cause impaired licensing as suggested by Ekholm-Reed *et al.* (2004) (B). Under such conditions, forks stalled at normal pause sites and or sites that contain DNA that is difficult to replicate, could not be rescued by neighbouring origins. Such forks would eventually break down and promote DNA damage as suggested by Tanaka and Diffley (2002b) (C).

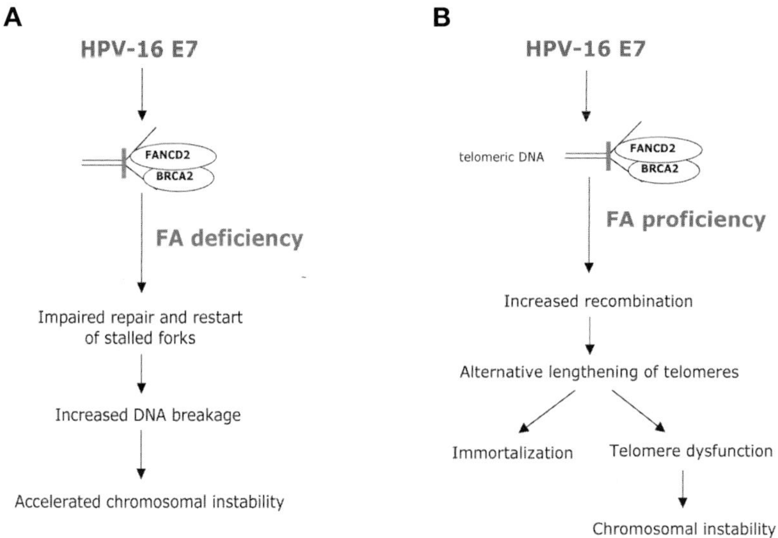

Figure 12.4 A model for different outcomes of HPV16 E7-induced DNA damage depending on the integrity of the Fanconi anaemia (FA) pathway. There is evidence that the HPV16 E7 oncoprotein induces DNA replication stress in host cells (Spardy *et al.*, 2007). The FA pathway is a major pathway that responds to stalled replication forks. However, cells in which the pathway is compromised by genetic or epigenetic events experience accelerated chromosomal breakage, which could fuel enhanced cancer formation as observed in patients suffering from the FA syndrome (A). In the context of a normal FA pathway, however, an increase of stalled forks would lead to enhanced repair by homologous recombination, which is the mechanistic basis for alternative lengthening of telomeres (ALT). ALT-associated PML bodies (APBs) were found in HPV16 E7-expressing cells that showed an extended life span *in vitro* (Spardy *et al.*, 2008). However, ALT is ultimately not efficient in maintaining an overall normal telomere length and is hence likely to also promote telomere dysfunction and chromosomal instability (B).

as squamous cell carcinomas of the oropharynx and the anogenital tract. Some studies have suggested that a substantial fraction of these cancers may harbour HPV DNA (Kutler *et al.*, 2003b) whereas other could not corroborate a strong connection (van Zeeburg *et al.*, 2008). It needs to be mentioned, however, that there are a number of differences between squamous cell carcinomas that occur in FA patients and those caused by HPV in the general population. For example tumours in FA patients occur at a significantly younger age and the tonsils are not the major anatomical site of disease (Kutler *et al.*, 2003a). Besides genetic disruption of the FA pathway, it is noteworthy that epigenetic silencing through DNA methylation of the FANCF gene has been detected in about one third of advanced stage cervical neoplasms, in particular in younger patients (Narayan *et al.*, 2004). A single nucleotide polymorphism (SNP) in the FANCA gene has recently been found to be associated with advanced stage cervical neoplasms (Wang *et al.*, 2009).

Collectively, these results show that high-risk HPV16 E7-expressing cells not only show enhanced DNA breakage and an activated DNA damage response but also experience significant DNA replication stress as evidenced by FA pathway activation and an increased cellular content of single-stranded DNA (Spardy *et al.*, 2008). Although this points to increased stalling of replication forks in the presence of high-risk HPV E7, direct fork movement is difficult to monitor and the ultimate proof that stalled DNA replication is the source of DNA breakage in HPV oncoprotein-expressing cells is still missing. Furthermore, the reason for increased DNA replication fork stalling remains to be elucidated. One possibility is that deregulation of CDK activity in HPV oncoprotein-expressing cells leads to improper licensing of replication origins and defects in pre-replicative complex (pre-RC) formation similar to what has been shown for cells with deregulated cyclin E expression (Ekholm-Reed *et al.*, 2004) (Fig. 12.3). However, other possibilities such as direct interference of HPV oncoproteins with pre-RC proteins causing problems with origin firing should not be prematurely dismissed. According to a model suggested by others, improper firing of replication origins could lead to a failure to rescue forks that stall at physiologic pause sites from neighbouring origins (Tanaka and Diffley, 2002a). In addition, pRB-deficient cells exhibit a prolonged S phase suggesting impaired origin firing (Classon *et al.*, 2000). Again, whether this scenario causes DNA damage in HPV oncoprotein-expressing cells, remains to be tested.

HPV oncoproteins and telomere maintenance

Cells normally undergo only a limited number of cell divisions before they experience a critical shortening of telomeres, the end structures of chromosomes consisting of approximately 10–12 kb of TTAGGG repeats and specialized protein complexes that prevent the recognition of a chromosome end as a DNA break (de Lange, 2005). The 3′ end consists of a single-stranded overhang that cannot be replicated by the semiconservative DNA replication machinery and instead is maintained by telomerase and its catalytic subunit telomere reverse transcriptase (hTERT) (Greider and Blackburn, 1996). Critical telomere shortening can cause apoptosis or senescence, two major barriers to oncogenic transformation, but may also spur chromosomal alterations such as DNA double-strand breaks or anaphase bridges that may increase genomic instability and in fact promote tumour growth (Hande *et al.*, 1999; Artandi *et al.*, 2000; Shay and Wright, 2005). Nonetheless, it is widely accepted that tumour cells with disrupted cell cycle checkpoints need to activate mechanisms to ensure telomere maintenance in order to sustain continued proliferation that ultimately results in immortalization (Henson *et al.*, 2002). Whereas most non-malignant cells do not express telomerase, many tumours cells show an upregulation of telomerase and maintain telomere length above the threshold that would trigger growth-suppressive responses. It is noteworthy, however, that telomerase activity is often detected only in advanced stage malignancies, for example at the transition from non-invasive to invasive cancer (Zhang *et al.*, 2004). Since pre-invasive lesions can persist for years or decades without detectable increases in telomerase activity, the question arises how critical telomere shortening is prevented at early stages of malignant progression. One mechanism that cells employ to maintain telomere length independent of telomerase activity is known as alternative lengthening of telomeres (ALT) (Dunham *et al.*, 2000). During ALT, homologous recombination, in particular between sister chromatids, is abnormally enhanced thus leading to increased telomere sister chromatid exchanges (T-SCE) but also causing a net increase of telomeric DNA through copying of a telomere template (de Lange, 2005). This leads to considerable telomere heterogeneity. It is noteworthy that some tumour types such as sarcomas have been found to rely on ALT more frequently than others (Henson and Reddel, 2010).

Remarkably, each of the two high-risk HPV oncoproteins has been linked to either one of these two principal, not mutually exclusive, telomere maintenance mechanisms in tumour cells.

There is compelling evidence that the high-risk HPV E6 oncoprotein can effectively active telomerase (Klingelhutz *et al.*, 1996). It has been suggested that the E6 binding partner E6-AP plays a critical role in hTERT transcriptional deregulation together with c-MYC, which binds to sequences in the hTERT promoter, as well as other transcriptional regulators (Gewin *et al.*, 2004; Liu *et al.*, 2005). Besides E6-AP-dependent histone acetylation at the hTERT promoter, it has been shown that E6-AP-mediated degradation of a transcriptional repressor, NFX-1-91, contributes to hTERT transcriptional upregulation and may act in concert with post-transcriptional mechanisms (Gewin *et al.*, 2004).

The common theme in HPV pathogenesis that important regulatory nodes are disrupted through several mechanisms hence also holds true for HPV E6-induced telomerase activation. There is evidence, however, that telomerase induction and telomere maintenance can become uncoupled. Primary human keratinocytes transduced with HPV16 E6 were found to exhibit telomerase activity yet failed to stabilize telomeres upon prolonged passaging. Astonishingly, HPV16 E7-expressing keratinocytes where the ones that gained telomeric DNA at later passages and in the absence of telomerase re-expression (Stoppler *et al.*, 1997). This remarkable finding may point to a role of ALT in telomere elongation in HPV16 E7-expressing cells (Stoppler *et al.*, 1997). Indeed, HPV16 E7-expressing cells show a significant increase of ALT-associated PML bodies (APBs), common hallmarks of ALT (Spardy *et al.*, 2008). This increase was associated with extension of the cellular life span whereas non-HPV16 E7-expressing keratinocytes that ultimately ceased to proliferate never displayed APBs (Spardy *et al.*, 2008). There is evidence that the APBs in HPV16 E7-expressing cells contain single-stranded DNA and single-stranded DNA binding proteins (RPA) as well as proteins that respond to stalled replication forks (ATR, FANCD2), contribute to the processing of altered replication structures (MUS81) or are involved in their repair (BRCA2) (Spardy *et al.*, 2008). Stalled replication forks are excellent substrates for HR-mediated repair, which is the basis of ALT. Hence, these results suggest a model in which increased stalling of replication forks at telomeric sequences, which are poor replication templates, trigger HR thereby promoting ALT. Again, the ultimate proof of this idea will be the direct visualization of replication dynamics at telomeric sequences and how it differs in HPV16 E7-expressing cells in comparison to normal cells. Another possibility is that HPV E7 interferes with a hypothetical repressor mechanism of ALT or can generally increase HR through other mechanisms. Since telomerase re-expression has been predominantly detected in more advanced cervical lesions, it is possible that ALT contributes to cellular immortalization at earlier stages of malignant progression (Zhang *et al.*, 2004) (Fig. 12.4).

Why telomerase tends to become detectable only late and why the HPV E6-induced telomerase upregulation does not prevail as the telomere length maintenance mechanism remains to be determined. E6-AP is involved in both, p53 degradation and telomerase activation. Whether there is competition between these two functions in HPV E6-expressing cells at early stages of infection where HPV E6 oncoprotein expression is tightly regulated is unknown. It is possible, though, that under these conditions p53 degradation is favoured in contrast to later stages where the viral genome can be found increasingly integrated leading to HPV E6 overexpression and possibly more E6/E6-AP at the telomerase promoter. These possibilities remains to be tested in future experiments.

HPV16 oncoproteins and mitotic checkpoint control

Mitotic checkpoints are vital to preserve genome integrity since they prevent cells with DNA damage entering mitosis as well as halting mitotic progression when kinetochore-spindle attachments are not complete. In addition, there appears to be a post-mitotic checkpoint that removes tetraploid cells that form after aborted mitoses from the proliferative pool (Holland and Cleveland, 2009).

As described above, high-risk HPV oncoprotein-expressing cells show increased DNA breakage and activated DNA damage response pathways even during mitosis (Spardy *et al.*, 2009). Yet, these cells are able to proliferate and there must be mechanisms to abrogate the negative, growth-suppressive consequences of DNA damage (Song *et al.*, 1998). One could argue that inhibition of p53 signalling by the HPV E6 oncoprotein is sufficient to provide this mechanism. However, cells that express the HPV E7 oncoprotein alone are able to proliferate in the presence of DNA damage suggesting that more than one mechanism exist to thwart mitotic checkpoint control (Hickman *et al.*, 1994). Although HPV E7 has been shown to functionally impair p53 (Eichten *et al.*, 2002), it is possible that there are more and other mechanisms involved in HPV16 E7-expressing cells.

It is remarkable, that cells with DNA damage as evidenced by γ-H2AX foci manage to enter mitosis and even progress to anaphase in HPV16 E7 oncoprotein-expressing populations in cultured cells as well as in clinical lesions (Spardy *et al.*, 2009). The increase of cells in mitosis with DNA damage was not simply because there were more cells with damaged chromosomes. Instead,

HPV16 E7 was found to overcome checkpoint responses triggered by exogenous stimuli such as the replication stress-inducing agent hydroxyurea (HU) (Spardy *et al.*, 2009). One mechanism for this dampened checkpoint activity is the increased proteolytic turnover of the adaptor protein claspin in HPV16 E7-expressing cells (Spardy *et al.*, 2009). Claspin functions in the ATR/CHK1 signalling cascade and its degradation has previously been implicated in the recovery from DNA damage checkpoint activation in the G2 phase of the cell division cycle (Peschiaroli *et al.*, 2006). A non-degradable mutant of claspin effectively inhibits mitotic entry in HPV16 E7-expressing cells and the E7 oncoprotein was found to upregulate components of the claspin degradation machinery. Nonetheless, baseline claspin levels are increased in the presence of HPV16 E7 suggesting that claspin is involved in facilitating DNA replication in S phase but becomes abnormally degraded as cells enter G2 to allow mitotic entry of cells with DNA damage (Spardy *et al.*, 2009). Obviously, this relaxation of checkpoint control in G2 phase can be further augmented by the HPV E6 oncoprotein-mediated impairment of p53.

Besides the G2/M checkpoint, HPV oncoproteins have also been implicated in interfering with the vigilance of the mitotic spindle checkpoint (or spindle assembly checkpoint). The main function of this checkpoint is to prevent cells with incomplete spindle-kinetochore attachments to transit from metaphase to anaphase. The HPV E6 and E7 oncoproteins have each been shown to independently compromise the spindle checkpoint leading to an attenuated arrest after microtubule-depolymerizing agents such as nocodazole and the development of polyploid cell populations (Thompson *et al.*, 1997; Thomas and Laimins, 1998). The latter is also a frequent finding *in vivo* (Southern *et al.*, 1997, 2001). How precisely the HPV oncoproteins interfere with mitotic spindle checkpoint function remains to be determined but altered expression of Ubch10 and CDC20, components of the checkpoint machinery, has been suggested (Patel and McCance, 2010). In addition, HPV oncoproteins can upregulate a number of mitotic regulators such as PLK1 and Aurora-A and this may also contribute to mitotic defects and polyploidization (Patel *et al.*, 2004).

On the other hand, HPV E7 proteins have been shown to bind to the nuclear and mitotic apparatus protein 1 (NuMA) and this has been linked to delocalization of the motor protein dynein from microtubules, which affected prometaphase to metaphase transition (Nguyen *et al.*, 2008; Nguyen and Munger, 2009). Whereas the exact mechanism of the HPV E7 induced prometaphase delay remains to be delineated it is probably caused by problems in spindle kinetochore attachments as unattached chromosomal material was observed in metaphases of HPV16 E7 expressing cells. Videomicroscopy experiments supported this model and showed that most HPV16 E7 expressing cells eventually progressed through metaphase and underwent apparently normal anaphase (Nguyen and Munger, 2009). Collectively, these experiments suggest that the mitotic spindle assembly checkpoint retains some activity on HPV16 E7 expressing cells

Checkpoints, in general, cannot remain active indefinitely and cells ultimately proceed to the next cell cycle stage (adaptation) or undergo apoptosis (van Vugt and Medema, 2004). Prolonged activation of the mitotic spindle checkpoint results in unscheduled mitotic exit with increased DNA content. A post-mitotic checkpoint has been suggested that prevents cells after such an event to re-enter the cell division cycle. HPV16 E7 has been proposed to disrupt this checkpoint but the molecular mechanisms are unclear (Heilman *et al.*, 2009). It has been suggested that the post-mitotic checkpoint relies on the p53–p21^{Cip1} axis (Lanni and Jacks, 1998), which is impaired at multiple levels in the presence of HPV oncoproteins. Whether this checkpoint does in fact monitor failed or aborted mitotic events or rather responds to other forms of cellular damage, together with the question to what extent this checkpoint is related to a normal G1/S checkpoint, is currently unclear.

Together, there is mounting evidence that HPV oncoproteins subvert multiple aspects of mitotic checkpoint control, although the molecular mechanisms are still incompletely understood. HPV-associated cervical lesions are frequently polyploid and it is possible that this ultimately contributes to aneuploidy and malignant progression. Nonetheless, there is clear evidence that mechanisms implicated in aneuploidy can arise in HPV oncoprotein expressing diploid cells without the necessity to go through a polyploid intermediate stage (Duensing and Duensing, 2010). Both pathways are not mutually exclusive and likely to occur in parallel in an emerging tumour.

Epigenetic instability

Epigenetic alterations can dramatically affect gene expression and result in expression of genes that are normally silenced in a given cell type or silence genes that are normally expressed. The concept that oncogenic pathways can be modulated through epigenetic mechanisms is well established in the literature. Epigenetic silencing of the p16^{INK4A} tumour suppressor, for example is frequently observed in human tumours (Kim and Sharpless, 2006).

Epigenetic mechanisms of gene expression include histone modifications as well as DNA methylation. The two mechanisms are intimately linked. HPV oncoproteins have been shown to modulate gene expression through alteration of DNA methylation (Rincon-Orozco *et al.*, 2009; Laurson *et al.* 2010) and the HPV E7 oncoprotein has been reported to associate with DNA (cytosine-5-)-methyltransferase 1 (DNMT-1) (Burgers *et al.*, 2007). The biological relevance of the latter interaction remains to be determined.

As mentioned earlier, HPV16 E6 and E7 are known to associate with and/or modify the activities of histone acetyl transferases (HATs) as well as histone deacetylaces (HDACs). Furthermore, the HPV E7 oncoprotein can associate with the non-canonical E2F family member, E2F6, which is a component of the polycomb repressive complexes (PRCs) (McLaughlin-Drubin *et al.*, 2008). HPV16 E7 expressing cells show a marked decrease of detectable E2F6 containing PRCs and in addition E7 also causes transcriptional activation of two histone demethylases, KDM6A and KDM6B which are responsible for removing repressive trimethyl marks on lysine 27 of Histone H3 (H3K27me3), which function as binding sites for polycomb repressive complexes (McLaughlin-Drubin *et al.*, 2011). Consequently, a number of cellular genes that are normally silenced by polycomb repressive complexes including the cervical cancer biomarker p16^{INK4A} and many homeobox genes are expressed at higher levels in E7 expressing cells than in parental cells (McLaughlin-Drubin *et al.*, 2011). In addition, HPV16 E7 expression also causes a dramatic increase in expression of the histone methyl transferase and PRC component EZH2 through an E2F dependent pathway (Holland *et al.*, 2008). It has been reported that, when overexpressed, EZH2 is incorporated into PRCs that modulate lysine 26 methylation of histone H1 (H1K26me) (Kuzmichev *et al.*, 2004, 2005) but whether HPV oncoprotein expressing cells contain increased levels of H1K26 methylation has not yet been tested experimentally.

As more and more is known regarding the enzymology and biological relevance of epigenetic modifications it will be possible to experimentally investigate the importance of epigenetic destabilization of virally infected cell for the viral life cycle and how these may contribute to the oncogenic potential of high-risk HPVs.

Summary and outlook

Human papillomavirus oncoproteins have been invaluable tools to dissect the molecular mechanisms of genomic instability in cancer. Given the 'inductive power' of this model system, it is probable that more findings that help to improve our understanding of cancer and malignant progression will be generated in the future. Areas such as the role of stem cells in cancer, the role of chromatin remodelling in genomic instability as well as the problem of viral genome integration are open for new and exciting discoveries.

Acknowledgements

Work in the authors' laboratories is supported by the NIH (R01 CA112598 to SD; CA066980, CA081135, HG004233, and CA141583 to K. M.), the American Cancer Society (RSG-07-075-01-MBC to SD) and the University of Heidelberg School of Medicine and Department of Urology (to SD). KM also gratefully acknowledges the support of Margaret A. Herradura.

References

Alter, B.P., Joenje, H., Oostra, A.B., and Pals, G. (2005). Fanconi anemia: adult head and neck cancer and hematopoietic mosaicism. Arch. Otolaryngol. Head Neck Surg. *131*, 635–639.

Artandi, S.E., Chang, S., Lee, S.L., Alson, S., Gottlieb, G.J., Chin, L., and DePinho, R.A. (2000). Telomere dysfunction promotes non-reciprocal translocations and epithelial cancers in mice. Nature *406*, 641–645.

Azimzadeh, J., and Bornens, M. (2007). Structure and duplication of the centrosome. J. Cell Sci. *120*, 2139–2142.

Bettencourt-Dias, M., Rodrigues-Martins, A., Carpenter, L., Riparbelli, M., Lehmann, L., Gatt, M.K., Carmo, N., Balloux, F., Callaini, G., and Glover, D.M. (2005). SAK/PLK4 is required for centriole duplication and flagella development. Curr. Biol. *15*, 2199–2207.

Brehm, A., Miska, E.A., McCance, D.J., Reid, J.L., Bannister, A.J., and Kouzarides, T. (1998). Retinoblastoma protein recruits histone deacetylase to repress transcription. Nature *391*, 597–601.

Brehm, A., Nielsen, S.J., Miska, E.A., McCance, D.J., Reid, J.L., Bannister, A.J., and Kouzarides, T. (1999). The E7 oncoprotein associates with Mi2 and histone deacetylase activity to promote cell growth. EMBO J. *18*, 2449–2458.

Burgers, W.A., Blanchon, L., Pradhan, S., de Launoit, Y., Kouzarides, T., and Fuks, F. (2007). Viral oncoproteins target the DNA methyltransferases. Oncogene *26*, 1650–1655.

Cahill, D.P., Kinzler, K.W., Vogelstein, B., and Lengauer, C. (1999). Genetic instability and darwinian selection in tumours. Trends Cell Biol. *9*, M57–60.

Cizmecioglu, O., Arnold, M., Bahtz, R., Settele, F., Ehret, L., Haselmann-Weiss, U., Antony, C., and Hoffmann, I. (2010). Cep152 acts as a scaffold for recruitment of Plk4 and CPAP to the centrosome. J. Cell. Biol. *191*, 731–739.

Classon, M., Salama, S., Gorka, C., Mulloy, R., Braun, P., and Harlow, E. (2000). Combinatorial roles for pRB, p107, and p130 in E2F-mediated cell cycle control. Proc. Natl. Acad. Sci. U.S.A. *97*, 10820–10825.

D'Andrea, A.D. (2003). The Fanconi road to cancer. Genes Dev. *17*, 1933–1936.

Duensing, A., and Duensing, S. (2005). Guilt by association? p53 and the development of aneuploidy in cancer. Biochem. Biophys. Res. Commun. *331*, 694–700.

Duensing, A., and Duensing, S. (2010). Centrosomes, polyploidy and cancer. Adv. Exp. Med. Biol. *676*, 93–103.

Duensing, A., Liu, Y., Tseng, M., Malumbres, M., Barbacid, M., and Duensing, S. (2006). Cyclin-dependent kinase 2 is dispensable for normal centrosome duplication but required for oncogene-induced centrosome overduplication. Oncogene *25*, 2943–2949.

Duensing, A., Liu, Y., Perdreau, S.A., Kleylein-Sohn, J., Nigg, E.A., and Duensing, S. (2007). Centriole overduplication through the concurrent formation of multiple daughter centrioles at single maternal templates. Oncogene *26*, 6280–6288.

Duensing, A., Chin, A., Wang, L., Kuan, S.F., and Duensing, S. (2008). Analysis of centrosome overduplication in correlation to cell division errors in high-risk human papillomavirus (HPV)-associated anal neoplasms. Virology *372*, 157–164.

Duensing, A., Spardy, N., Chatterjee, P., Zheng, L., Parry, J., Cuevas, R., Korzeniewski, N., and Duensing, S. (2009). Centrosome overduplication, chromosomal instability, and human papillomavirus oncoproteins. Environ Mol Mutagen. *50*, 741–747.

Duensing, S., and Münger, K. (2002). The human papillomavirus type 16 E6 and E7 oncoproteins independently induce numerical and structural chromosome instability. Cancer Res. *62*, 7075–7082.

Duensing, S., and Münger, K. (2003). Human papillomavirus type 16 E7 oncoprotein can induce abnormal centrosome duplication through a mechanism independent of inactivation of retinoblastoma protein family members. J. Virol. *77*, 12331–12335.

Duensing, S., and Munger, K. (2004). Mechanisms of genomic instability in human cancer: insights from studies with human papillomavirus oncoproteins. Int. J. Cancer *109*, 157–162.

Duensing, S., Lee, L.Y., Duensing, A., Basile, J., Piboonniyom, S., Gonzalez, S., Crum, C.P., and Munger, K. (2000). The human papillomavirus type 16 E6 and E7 oncoproteins cooperate to induce mitotic defects and genomic instability by uncoupling centrosome duplication from the cell division cycle. Proc. Natl. Acad. Sci. U.S.A. *97*, 10002–10007.

Duensing, S., Duensing, A., Crum, C.P., and Munger, K. (2001). Human papillomavirus type 16 E7 oncoprotein-induced abnormal centrosome synthesis is an early event in the evolving malignant phenotype. Cancer Res. *61*, 2356–2360.

Duensing, S., Duensing, A., Lee, D.C., Edwards, K.M., Piboonniyom, S., Manuel, E., Skaltsounis, L., Meijer, L., and Munger, K. (2004). Cyclin-dependent kinase inhibitor indirubin-3′-oxime selectively inhibits human papillomavirus type 16 E7-induced numerical centrosome anomalies. Oncogene *23*, 8206–8215.

Dunham, M.A., Neumann, A.A., Fasching, C.L., and Reddel, R.R. (2000). Telomere maintenance by recombination in human cells. Nat. Genet. *26*, 447–450.

Durst, M., Gissmann, L., Ikenberg, H., and zur Hausen, H. (1983). A papillomavirus DNA from a cervical carcinoma and its prevalence in cancer biopsy samples from different geographic regions. Proc. Natl. Acad. Sci. U.S.A. *80*, 3812–3815.

Dyson, N., Howley, P.M., Munger, K., and Harlow, E. (1989). The human papilloma virus-16 E7 oncoprotein is able to bind to the retinoblastoma gene product. Science *243*, 934–937.

Dzhindzhev, N.S., Yu, Q.D., Weiskopf, K., Tzolovsky, G., Cunha-Ferreira, I., Riparbelli, M., Rodrigues-Martins, A., Bettencourt-Dias, M., Callaini, G., and Glover, D.M. (2010). Asterless is a scaffold for the onset of centriole assembly. Nature *467*, 714–718.

Eichten, A., Westfall, M., Pietenpol, J.A., and Munger, K. (2002). Stabilization and functional impairment of the tumour suppressor p53 by the human papillomavirus type 16 E7 oncoprotein. Virology *295*, 74–85.

Ekholm-Reed, S., Mendez, J., Tedesco, D., Zetterberg, A., Stillman, B., and Reed, S.I. (2004). Deregulation of cyclin E in human cells interferes with prereplication complex assembly. J. Cell. Biol. *165*, 789–800.

Fukasawa, K. (2007). Oncogenes and tumour suppressors take on centrosomes. Nat. Rev. Cancer *7*, 911–924.

Funk, J.O., Waga, S., Harry, J.B., Espling, E., Stillman, B., and Galloway, D.A. (1997). Inhibition of CDK activity and PCNA-dependent DNA replication by p21 is blocked by interaction with the HPV16 E7 oncoprotein. Genes Dev. *11*, 2090–2100.

Garcia-Higuera, I., Taniguchi, T., Ganesan, S., Meyn, M.S., Timmers, C., Hejna, J., Grompe, M., and D'Andrea, A.D. (2001). Interaction of the Fanconi anemia proteins and BRCA1 in a common pathway. Mol. Cell *7*, 249–262.

Gewin, L., Myers, H., Kiyono, T., and Galloway, D.A. (2004). Identification of a novel telomerase repressor that interacts with the human papillomavirus type-16 E6/E6–AP complex. Genes Dev. *18*, 2269–2282.

Glaunsinger, B.A., Lee, S.S., Thomas, M., Banks, L., and Javier, R. (2000). Interactions of the PDZ-protein MAGI-1 with adenovirus E4-ORF1 and high-risk papillomavirus E6 oncoproteins. Oncogene *19*, 5270–5280.

Gonzalez, S.L., Stremlau, M., He, X., Basile, J., and Munger, K. (2001). Degradation of the retinoblastoma tumour suppressor by the human papillomavirus type 16 E7 oncoprotein is important for functional inactivation and is separable from proteasomal degradation of E7. J. Virol. *75*, 7583–7591.

Greider, C.W., and Blackburn, E.H. (1996). Telomeres, telomerase and cancer. Sci. Am. *274*, 92–97.

Guarguaglini, G., Duncan, P.I., Stierhof, Y.D., Holmstrom, T., Duensing, S., and Nigg, E.A. (2005). The Forkhead-associated Domain Protein Cep170 Interacts with Polo-like Kinase 1 and Serves as a Marker for Mature Centrioles. Mol. Biol. Cell *16*, 1095–1107.

Habedanck, R., Stierhof, Y.D., Wilkinson, C.J., and Nigg, E.A. (2005). The Polo kinase Plk4 functions in centriole duplication. Nat. Cell Biol. *7*, 1140–1146.

Hande, M.P., Samper, E., Lansdorp, P., and Blasco, M.A. (1999). Telomere length dynamics and chromosomal instability in cells derived from telomerase null mice. J. Cell. Biol. *144*, 589–601.

Hatch, E.M., Kulukian, A., Holland, A.J., Cleveland, D.W., and Stearns, T. (2010). Cep152 interacts with Plk4 and is required for centriole duplication. J. Cell. Biol. *191*, 721–729.

zur Hausen, H. (1991). Viruses in human cancers. Science *254*, 1167–1173.

zur Hausen, H. (2002). Papillomaviruses and cancer: from basic studies to clinical application. Nat. Rev. Cancer *2*, 342–350.

He, W., Staples, D., Smith, C., and Fisher, C. (2003). Direct activation of cyclin-dependent kinase 2 by human papillomavirus E7. J. Virol. *77*, 10566–10574.

Hebner, C., Beglin, M., and Laimins, L.A. (2007). Human papillomavirus E6 proteins mediate resistance to interferon-induced growth arrest through inhibition of p53 acetylation. J. Virol. *81*, 12740–12747.

Heilman, S.A., Nordberg, J.J., Liu, Y., Sluder, G., and Chen, J.J. (2009). Abrogation of the postmitotic checkpoint contributes to polyploidization in human papillomavirus E7-expressing cells. J. Virol. *83*, 2756–2764.

Henson, J.D., Neumann, A.A., Yeager, T.R., and Reddel, R.R. (2002). Alternative lengthening of telomeres in mammalian cells. Oncogene *21*, 598–610.

Henson, J.D., and Reddel, R.R. Assaying and investigating Alternative Lengthening of Telomeres activity in human cells and cancers. (2010). FEBS Lett. *584*, 3800–3811.

Hickman, E.S., Picksley, S.M., and Vousden, K. (1994). Cells expressing HPV16 E7 continue cell cycle progression following DNA damage induced p53 activation. Oncogene *9*, 2177–2181.

Holland, A.J., and Cleveland, D.W. (2009). Boveri revisited: chromosomal instability, aneuploidy and tumorigenesis. Nat. Rev. Mol. Cell Biol. *10*, 478–487.

Holland, D., Hoppe-Seyler, K., Schuller, B., Lohrey, C., Maroldt, J., Durst, M., and Hoppe-Seyler, F. (2008). Activation of the enhancer of zeste homologue 2 gene by the human papillomavirus E7 oncoprotein. Cancer Res. *68*, 9964–9972.

Hoskins, E.E., Morris, T.A., Higginbotham, J.M., Spardy, N., Cha, E., Kelly, P., Williams, D.A., Wikenheiser-Brokamp, K.A., Duensing, S., and Wells, S.I. (2009). Fanconi anemia deficiency stimulates HPV-associated hyperplastic growth in organotypic epithelial raft culture. Oncogene *28*, 674–685.

Howie, H.L., Katzenellenbogen, R.A., and Galloway, D.A. (2009). Papillomavirus E6 proteins. Virology *384*, 324–334.

Howlett, N.G., Taniguchi, T., Durkin, S.G., D'Andrea, A.D., and Glover, T.W. (2005). The Fanconi anemia pathway is required for the DNA replication stress response and for the regulation of common fragile site stability. Hum. Mol. Genet. *14*, 693–701.

Howley, P.M., and Lowy, D.R. (2001). Papillomaviruses and their replication, In Field's Virology, Knipe, D.M., and Howley, P.M., eds. (Lippincott Williams & Wilkins: Philadelphia), pp. 2197–2229.

Huh, K., Zhou, X., Hayakawa, H., Cho, J.Y., Libermann, T.A., Jin, J., Harper, J.W., and Munger, K. (2007). Human papillomavirus type 16 E7 oncoprotein associates with the cullin 2 ubiquitin ligase complex, which contributes to degradation of the retinoblastoma tumour suppressor. J. Virol. *81*, 9737–9747.

Hwang, S.G., Lee, D., Kim, J., Seo, T., and Choe, J. (2002). Human papillomavirus type 16 E7 binds to E2F1 and activates E2F1-driven transcription in a retinoblastoma protein-independent manner. J. Biol. Chem. *277*, 2923–2930.

Jha, S., Vande Pol, S., Banerjee, N.S., Dutta, A.B., Chow, L.T., and Dutta, A. (2010). Destabilization of TIP60 by human papillomavirus E6 results in attenuation of TIP60-dependent transcriptional regulation and apoptotic pathway. Mol. Cell *38*, 700–711.

Jones, D.L., Alani, R.M., and Munger, K. (1997). The human papillomavirus E7 oncoprotein can uncouple cellular differentiation and proliferation in human keratinocytes by abrogating p21Cip1-mediated inhibition of cdk2. Genes Dev. *11*, 2101–2111.

Kadaja, M., Isok-Paas, H., Laos, T., Ustav, E., and Ustav, M. (2009). Mechanism of genomic instability in cells infected with the high-risk human papillomaviruses. PLoS Pathog. *5*, e1000397.

Kessis, T.D., Connolly, D.C., Hedrick, L., and Cho, K.R. (1996). Expression of HPV16 E6 or E7 increases integration of foreign DNA. Oncogene *13*, 427–431.

Kim, W.Y., and Sharpless, N.E. (2006). The regulation of INK4/ARF in cancer and aging. Cell *127*, 265–275.

Kitagawa, D., Vakonakis, I., Olieric, N., Hilbert, M., Keller, D., Olieric, V., Bortfeld, M., Erat, M.C., Fluckiger, I., Gonczy, P., *et al.* (2011). Structural basis of the 9-fold symmetry of centrioles. Cell *144*, 364–375.

Kiyono, T., Hiraiwa, A., Fujita, M., Hayashi, Y., Akiyama, T., and Ishibashi, M. (1997). Binding of high-risk human papillomavirus E6 oncoproteins to the human homologue of the Drosophila discs large tumour suppressor protein. Proc. Natl. Acad. Sci. U.S.A. *94*, 11612–11616.

Kleylein-Sohn, J., Westendorf, J., Le Clech, M., Habedanck, R., Stierhof, Y.D., and Nigg, E.A. (2007). Plk4-induced centriole biogenesis in human cells. Dev. Cell *13*, 190–202.

Klingelhutz, A.J., Foster, S.A., and McDougall, J.K. (1996). Telomerase activation by the E6 gene product of human papillomavirus type 16. Nature *380*, 79–82.

Korzeniewski, N., Zheng, L., Cuevas, R., Parry, J., Chatterjee, P., Anderton, B., Duensing, A., Munger, K., and Duensing, S. (2009). Cullin 1 functions as a centrosomal suppressor of centriole multiplication by regulating polo-like kinase 4 protein levels. Cancer Res. *69*, 6668–6675.

Kutler, D.I., Auerbach, A.D., Satagopan, J., Giampietro, P.F., Batish, S.D., Huvos, A.G., Goberdhan, A., Shah, J.P., and Singh, B. (2003a). High incidence of head and neck squamous cell carcinoma in patients with Fanconi anemia. Arch. Otolaryngol. Head Neck Surg. *129*, 106–112.

Kutler, D.I., Wreesmann, V.B., Goberdhan, A., Ben-Porat, L., Satagopan, J., Ngai, I., Huvos, A.G., Giampietro, P., Levran, O., Pujara, K., *et al.* (2003b). Human papillomavirus DNA and p53 polymorphisms in squamous cell carcinomas from Fanconi anemia patients. J. Natl. Cancer Inst. *95*, 1718–1721.

Kuzmichev, A., Jenuwein, T., Tempst, P., and Reinberg, D. (2004). Different EZH2–containing complexes target methylation of histone H1 or nucleosomal histone H3. Mol. Cell *14*, 183–193.

Kuzmichev, A., Margueron, R., Vaquero, A., Preissner, T.S., Scher, M., Kirmizis, A., Ouyang, X., Brockdorff, N., Abate-Shen, C., Farnham, P., and Reinberg, D. (2005). Composition and histone substrates of polycomb repressive group complexes change during cellular differentiation. Proc. Natl. Acad. Sci. U.S.A. *102*, 1859–1864.

de Lange, T. (2005). Shelterin: the protein complex that shapes and safeguards human telomeres. Genes Dev. *19*, 2100–2110.Liu, X., Han, S., Baluda, M.A., and Park, N.H. (1997). HPV16 oncogenes E6 and E7 are mutagenic in normal human oral keratinocytes. Oncogene *14*, 2347–2353.

Lanni, J.S., and Jacks, T. (1998). Characterization of the p53-dependent postmitotic checkpoint following spindle disruption. Mol. Cell Biol. *18*, 1055–1064.

Laurson, J., Khan, S., Chung, R., Cross, K., and Raj, K. (2010). Epigenetic repression of E-cadherin by human papillomavirus 16 E7 protein. Carcinogenesis *31*, 918–926.

Liu, X., Yuan, H., Fu, B., Disbrow, G.L., Apolinario, T., Tomaic, V., Kelley, M.L., Baker, C.C., Huibregtse, J., and Schlegel, R. (2005). The E6AP ubiquitin ligase is required for transactivation of the hTERT promoter by the human papillomavirus E6 oncoprotein. J. Biol. Chem. *280*, 10807–10816.

Longworth, M.S., and Laimins, L.A. (2004a). The binding of histone deacetylases and the integrity of zinc finger-like motifs of the E7 protein are essential for the life cycle of human papillomavirus type 31. J. Virol. *78*, 3533–3541.

Longworth, M.S., and Laimins, L.A. (2004b). Pathogenesis of human papillomaviruses in differentiating epithelia. Microbiol. Mol. Biol. Rev. *68*, 362–372.

Martin, L.G., Demers, G.W., and Galloway, D.A. (1998). Disruption of the G1/S transition in human papillomavirus type 16 E7- expressing human cells is associated with altered regulation of cyclin E. J. Virol. *72*, 975–985.

McLaughlin-Drubin, M.E., and Munger, K. (2009a). The human papillomavirus E7 oncoprotein. Virology *384*, 335–344.

McLaughlin-Drubin, M.E., and Munger, K. (2009b). Oncogenic activities of human papillomaviruses. Virus Res. *143*, 195–208.

McLaughlin-Drubin, M.E., Huh, K.W., and Munger, K. (2008). Human papillomavirus type 16 E7 oncoprotein associates with E2F6. J. Virol. *82*, 8695–8705.

McLaughlin-Drubin, M.E., Crum, C.P., and Munger, K. (2011). Human papillomavirus E7 oncoprotein induces KDM6A and KDM6B histone demethylase expression and causes epigenetic reprogramming. Proc. Natl. Acad. Sci. U.S.A. *108*, 2130–2135.

Moody, C.A., and Laimins, L.A. (2009). Human papillomaviruses activate the ATM DNA damage pathway for viral genome amplification upon differentiation. PLoS Pathog. *5*, e1000605.

Moody, C.A., and Laimins, L.A. (2010). Human papillomavirus oncoproteins: pathways to transformation. Nat. Rev. Cancer *10*, 550–560.

Munoz, N., Bosch, F.X., de Sanjose, S., Herrero, R., Castellsague, X., Shah, K.V., Snijders, P.J.F., and Meijer, C.J. L. M. (2003). Epidemiologic classification of human papillomavirus types associated with cervical cancer. N. Engl. J. Med. *348*, 518–527.

Nakagawa, S., and Huibregtse, J.M. (2000). Human scribble (Vartul) is targeted for ubiquitin-mediated degradation by the high-risk papillomavirus E6 proteins and the E6AP ubiquitin-protein ligase. Mol. Cell Biol. *20*, 8244–8253.

Narayan, G., Arias-Pulido, H., Nandula, S.V., Basso, K., Sugirtharaj, D.D., Vargas, H., Mansukhani, M., Villella, J., Meyer, L., Schneider, A., *et al.* (2004). Promoter hypermethylation of FANCF: disruption of Fanconi Anemia-BRCA pathway in cervical cancer. Cancer Res. *64*, 2994–2997.

Nguyen, C.L., and Munger, K. (2008). Direct association of the HPV16 E7 oncoprotein with cyclin A/CDK2 and cyclin E/CDK2 complexes. Virology *380*, 21–25.

Nguyen, C.L., and Munger, K. (2009). Human papillomavirus E7 protein deregulates mitosis via an association with nuclear mitotic apparatus protein 1. J. Virol. *83*, 1700–1707.

Nguyen, C.L., Eichwald, C., Nibert, M.L., and Munger, K. (2007). Human papillomavirus type 16 E7 oncoprotein associates with the centrosomal component gamma-tubulin. J. Virol. *81*, 13533–13543.

Nguyen, C.L., McLaughlin-Drubin, M.E., and Munger, K. (2008). Delocalization of the microtubule motor Dynein from mitotic spindles by the human papillomavirus E7 oncoprotein is not sufficient for induction of multipolar mitoses. Cancer Res. *68*, 8715–8722.

Nguyen, D.X., Westbrook, T.F., and McCance, D.J. (2002). Human papillomavirus type 16 E7 maintains elevated levels of the cdc25A tyrosine phosphatase during deregulation of cell cycle arrest. J. Virol. *76*, 619–632.

Nigg, E.A. (2002). Centrosome aberrations: cause or consequence of cancer progression? Nat. Rev. Cancer *2*, 1–11.

Nigg, E.A. (2007). Centrosome duplication: of rules and licenses. Trends Cell. Biol. *17*, 215–221.

Ohki, R., and Ishikawa, F. (2004). Telomere-bound TRF1 and TRF2 stall the replication fork at telomeric repeats. Nucleic Acids Res. *32*, 1627–1637.

Park, J.W., Pitot, H.C., Strati, K., Spardy, N., Duensing, S., Grompe, M., and Lambert, P.F. (2010). Deficiencies in the Fanconi anemia DNA damage response pathway increase sensitivity to HPV-associated head and neck cancer. Cancer Res. *70*, 9959–9968.

Patel, D., and McCance, D.J. (2010). Compromised spindle assembly checkpoint due to altered expression of Ubch10 and Cdc20 in human papillomavirus type 16 E6- and E7-expressing keratinocytes. J. Virol. *84*, 10956–10964.

Patel, D., Incassati, A., Wang, N., and McCance, D.J. (2004). Human papillomavirus type 16 E6 and E7 cause polyploidy in human keratinocytes and up-regulation of G2-M-phase proteins. Cancer Res. *64*, 1299–1306.

Patel, K.J., and Joenje, H. (2007). Fanconi anemia and DNA replication repair. DNA Repair (Amst) *6*, 885–890.

Peschiaroli, A., Dorrello, N.V., Guardavaccaro, D., Venere, M., Halazonetis, T., Sherman, N.E., and Pagano, M. (2006). SCFbetaTrCP-mediated degradation of Claspin regulates recovery from the DNA replication checkpoint response. Mol. Cell *23*, 319–329.

Plug-Demaggio, A.W., and McDougall, J.K. (2002). The human papillomavirus type 16 E6 oncogene induces premature mitotic chromosome segregation. Oncogene *21*, 7507–7513.

Riley, R.R., Duensing, S., Brake, T., Munger, K., Lambert, P.F., and Arbeit, J.M. (2003). Dissection of human papillomavirus E6 and E7 function in transgenic mouse models of cervical carcinogenesis. Cancer Res. *63*, 4862–4871.

Rincon-Orozco, B., Halec, G., Rosenberger, S., Muschik, D., Nindl, I., Bachmann, A., Ritter, T.M., Dondog, B., Ly, R., Bosch, F.X., *et al.* (2009). Epigenetic silencing of interferon-kappa in human papillomavirus type 16-positive cells. Cancer Res. *69*, 8718–8725.

Scheffner, M., Huibregtse, J.M., Vierstra, R.D., and Howley, P.M. (1993). The HPV16 E6 and E6–AP complex functions as a ubiquitin-protein ligase in the ubiquitination of p53. Cell *75*, 495–505.

Shay, J.W., and Wright, W.E. (2005). Senescence and immortalization: role of telomeres and telomerase. Carcinogenesis *26*, 867–874.

Solinas-Toldo, S., Dürst, M., and Lichter, P. (1997). Specific chromosomal imbalances in human papillomavirus-transfected cells during progression toward immortality. Proc. Natl. Acad. Sci. U.S.A. *94*, 3854–3859.

Song, S., Gulliver, G.A., and Lambert, P.F. (1998). Human papillomavirus type 16 E6 and E7 oncogenes abrogate radiation-induced DNA damage response *in vivo* through p53-dependent and p53-independent mechanisms. Proc. Natl. Acad. Sci. U.S.A. *95*, 2290–2295.

Southern, S.A., Evans, M.F., and Herrington, C.S. (1997). Basal cell tetrasomy in low-grade cervical squamous intraepithelial lesions infected with high-risk human papillomaviruses. Cancer Res. *57*, 4210–4213.

Southern, S.A., Noya, F., Meyers, C., Broker, T.R., Chow, L.T., and Herrington, C.S. (2001). Tetrasomy is induced by human papillomavirus type 18 E7 gene expression in keratinocyte raft cultures. Cancer Res. *61*, 4858–4863.

Spardy, N., Covella, K., Cha, E., Hoskins, E.E., Wells, S.I., Duensing, A., and Duensing, S. (2009). Human papillomavirus 16 E7 oncoprotein attenuates DNA damage checkpoint control by increasing the proteolytic turnover of claspin. Cancer Res. *69*, 7022–7029.

Spardy, N., Duensing, A., Charles, D., Haines, N., Nakahara, T., Lambert, P.F., and Duensing, S. (2007). The human papillomavirus type 16 E7 oncoprotein activates the Fanconi anemia (FA) pathway and causes accelerated chromosomal instability in FA cells. J. Virol. *81*, 13265–13270.

Spardy, N., Duensing, A., Hoskins, E.E., Wells, S.I., and Duensing, S. (2008). HPV16 E7 Reveals a Link between DNA Replication Stress, Fanconi Anemia D2 Protein, and Alternative Lengthening of Telomere-Associated Promyelocytic Leukemia Bodies. Cancer Res. *68*, 9954–9963.

Stoppler, H., Hartmann, D.P., Sherman, L., and Schlegel, R. (1997). The human papillomavirus type 16 E6 and E7 oncoproteins dissociate cellular telomerase activity from the maintenance of telomere length. J. Biol. Chem. *272*, 13332–13337.

Strnad, P., and Gonczy, P. (2008). Mechanisms of procentriole formation. Trends Cell. Biol. *18*, 389–396.

Tanaka, S., and Diffley, J.F. (2002a). Deregulated G1-cyclin expression induces genomic instability by preventing efficient pre-RC formation. Genes Dev. *16*, 2639–2649.

Tanaka, S., and Diffley, J.F. X. (2002b). Deregulated G1-cyclin expression induces genomic instability by preventing efficient pre-RC formation. Genes Dev. *16*, 2639–2649.

Thomas, J.T., and Laimins, L.A. (1998). Human papillomavirus oncoproteins E6 and E7 independently abrogate the mitotic spindle checkpoint. J. Virol. *72*, 1131–1137.

Thomas, M., Laura, R., Hepner, K., Guccione, E., Sawyers, C., Lasky, L., and Banks, L. (2002). Oncogenic human papillomavirus E6 proteins target the MAGI-2 and MAGI-3 proteins for degradation. Oncogene *21*, 5088–5096.

Thomas, M.C., and Chiang, C.M. (2005). E6 oncoprotein represses p53-dependent gene activation via inhibition of protein acetylation independently of inducing p53 degradation. Mol. Cell *17*, 251–264.

Thompson, D.A., Belinsky, G., Chang, T.H., Jones, D.L., Schlegel, R., and Munger, K. (1997). The human papillomavirus-16 E6 oncoprotein decreases the vigilance of mitotic checkpoints. Oncogene *15*, 3025–3035.

Tommasino, M., Adamczewski, J.P., Carlotti, F., Barth, C.F., Manetti, R., Contorni, M., Cavalieri, F., Hunt, T., and Crawford, L. (1993). HPV16 E7 protein associates with the protein kinase p33CDK2 and cyclin A. Oncogene *8*, 195–202.

Tsou, M.F., Wang, W.J., George, K.A., Uryu, K., Stearns, T., and Jallepalli, P.V. (2009). Polo kinase and separase regulate the mitotic licensing of centriole duplication in human cells. Dev. Cell *17*, 344–354.

van Vugt, M.A., and Medema, R.H. (2004). Checkpoint adaptation and recovery: back with polo after the break. Cell Cycle *3*, 1383–1386.

Wang, S.S., Bratti, M.C., Rodriguez, A.C., Herrero, R., Burk, R.D., Porras, C., Gonzalez, P., Sherman, M.E., Wacholder, S., Lan, Z.E., *et al.* (2009). Common variants in immune and DNA repair genes and risk for human papillomavirus persistence and progression to cervical cancer. J. Infect. Dis. *199*, 20–30.

Wang, X., and D'Andrea, A.D. (2004). The interplay of Fanconi anemia proteins in the DNA damage response. DNA Repair (Amst) *3*, 1063–1069.

Wang, X., Andreassen, P.R., and D'Andrea, A.D. (2004). Functional interaction of monoubiquitinated FANCD2 and BRCA2/FANCD1 in chromatin. Mol. Cell Biol. *24*, 5850–5862.

Wei, L., Gravitt, P.E., Song, H., Maldonado, A.M., and Ozbun, M.A. (2009). Nitric oxide induces early viral transcription coincident with increased DNA damage and mutation rates in human papillomavirus-infected cells. Cancer Res. *69*, 4878–4884.

White, A.E., Livanos, E.M., and Tlsty, T.D. (1994). Differential disruption of genomic integrity and cell cycle regulation in normal human fibroblasts by the HPV oncoproteins. Genes Dev *8*, 666–677.

van Zeeburg, H.J., Snijders, P.J., Wu, T., Gluckman, E., Soulier, J., Surralles, J., Castella, M., van der Wal, J.E., Wennerberg, J., Califano, J., *et al.* (2008). Clinical and molecular characteristics of squamous cell carcinomas from Fanconi anemia patients. J. Natl. Cancer Inst. *100*, 1649–1653.

Zerfass-Thome, K., Zwerschke, W., Mannhardt, B., Tindle, R., Botz, J.W., and Jansen-Durr, P. (1996). Inactivation of the cdk inhibitor p27KIP1 by the human papillomavirus type 16 E7 oncoprotein. Oncogene *13*, 2323–2330.

Zhang, A., Wang, J., Zheng, B., Fang, X., Angstrom, T., Liu, C., Li, X., Erlandsson, F., Bjorkholm, M., Nordenskjord, M., *et al.* (2004). Telomere attrition predominantly occurs in precursor lesions during *in vivo* carcinogenic process of the uterine cervix. Oncogene *23*, 7441–7447.

Targeting of Promyelocytic Leukaemia Proteins and Promyelocytic Leukaemia Nuclear Bodies by DNA Tumour Viruses

13

Keith N. Leppard and Jordan Wright

Abstract

Promyelocytic leukaemia (PML) protein is the principal and much studied component of subnuclear structures known as PML nuclear bodies (PML NBs). These structures and/or their components have been implicated in a very wide range of cellular processes, without the precise mechanism of their involvement being determined. One of the key areas of PML study has been the interactions that many viruses make with PML NBs during infection, leading to the suggestion that PML NBs play a role in antiviral responses. This chapter reviews what is known about the interactions between various DNA tumour viruses and other viruses with PML NBs, and places this information in the wider context of the functions ascribed to PML NBs in the uninfected cell to consider what is the underlying purpose of this class of virus:host interactions.

Introduction

One of the most fruitful areas of virology research over the past 20 years has been the exploration of virus–host interaction at the intracellular level. Typically beginning with a chance observation, perhaps of an association between a viral and a cellular protein in the infected cell that was unanticipated or unexplained, this research has gone on to establish much about the normal workings of host biology. Such interactions are best rationalized in the context of the evolutionary arms race between a virus and its host. A virus entering a host cell requires many things of that host in order to replicate successfully, and commissioning these to the cause of virus replication will demand specific molecular interactions. Equally, we now understand that the host cell has many intrinsic defences that it will deploy to slow down or block infection long before an adaptive immune response comes into play. These too will generate molecular interactions between virus and host, though this time to benefit the host. And finally the virus, being capable of more rapid evolution than its host, has typically developed countermeasures against these host defences, even to the point of turning them to productive use; this too demands interactions between virus and host components.

The focus of this chapter is the interaction between viruses and promyelocytic leukaemia (PML) protein, and the structures that this protein forms in the nucleus of the cell (PML bodies). A large amount of work on PML has revealed huge complexity, including links with some of the other classic targets of virus–host interactions such as p53, DNA damage responses and innate immunity, but as yet no unifying view of PML function has emerged. These topics have been extensively reviewed in recent years (Bernardi and Pandolfi, 2007; Salomoni *et al.*, 2008; Nichol *et al.*, 2009; Geoffroy and Chelbi-Alix, 2011). Here we explore what has been learned from studying virus interactions with PML, with particular emphasis on the small DNA viruses but expanding to include other viruses too, and attempt to address the question of whether there is a common overarching reason for the apparent commonality of virus–PML interaction between such diverse virus types.

PML protein and PML bodies

Discovery and characterization of PML and PML NBs

The human *pml* gene is located on chromosome 15 and is approximately 35 kb in length (Goddard *et al.*, 1991). It was originally identified by its involvement in a chromosomal translocation with the retinoic receptor α (RARα) gene that is associated with a high proportion of acute promyelocytic leukaemia (APL) cases (de The *et al.*, 1990; Goddard *et al.*, 1991). Normal PML protein is expressed in almost all tissues and, by immunofluorescent staining, is found in a distinctive pattern of spherical nuclear dots, each up to 1 μm in diameter (Daniel *et al.*, 1993; Dellaire and Bazett-Jones, 2004); these are variously known as ND10, PML oncogenic domains (PODS), Kremer bodies or PML nuclear bodies (PML NBs). Examination by electron microscopy had originally suggested PML NBs were ring-like structures. However, it is now clear that, during interphase, PML NBs are hollow shells composed of both PML and another defining component, Sp100 (Boisvert *et al.*, 2000; Lang *et al.*, 2010). In APL, expression of a functional fusion protein, PML-RARα, causes dispersal of PML NBs; since specific therapy that is associated with regression of the disease causes PML-NB structures to reform, the proper organization of PML protein in PML-NBs is thought to be important for normal cell function (Salomoni *et al.*, 2008).

Both the size and number of NBs is dependent on numerous factors, reflecting their highly dynamic nature. The average number of NBs per nucleus varies between cell types, and the average size and number change over the cell cycle. This is thought to be mainly due to modifications in chromatin structure during S phase that cause fission of associated NBs, resulting in an increased number of PML NBs in G2 phase (Dellaire *et al.*, 2006b). Once a cell enters mitosis, PML becomes dispersed throughout the cytoplasm, leading to the formation of PML aggregates termed mitotic accumulations of PML protein (MAPPs) (Dellaire *et al.*, 2006c). These are distinct from PML NBs as they lack other NB components.

The number of PML NBs in a cell may vary further when cells are subjected to certain stresses. For example, DNA damage by UV irradiation or chemical agents leads to increases in both the size and number of PML NBs (Dellaire and Bazett-Jones, 2004; Dellaire *et al.*, 2006a). This reflects the involvement of these bodies in DNA damage responses, as a number of DNA damage proteins co-localize with PML NBs (Krieghoff-Henning and Hofmann, 2008). Under certain circumstances PML NBs have also been shown to be formed *de novo*. Incoming exogenous DNA, whether it be a viral genome or plasmid, is often associated with the formation of new PML NBs, an observation thought to be a manifestation of a general cellular defence mechanism that initiates gene repression (Bishop *et al.*, 2006; Tavalai and Stamminger, 2008).

The complexity of PML protein isoforms and structural elements

The *pml* gene consists of nine exons, which are alternatively spliced during gene expression to generate numerous mRNA and hence protein isoforms (Fig. 13.1). This fact significantly complicates attempts to understand the function(s) of PML and PML NBs. There are currently seven described principal isoforms of PML, termed PML I-VII, that are defined by their unique carboxyl-terminal sequences. Only PMLs I-VI contain a nuclear localization signal (NLS), encoded in exon 6, and are known to be incorporated into PML NBs (Jensen *et al.*, 2001). Thus PML species lacking this exon, such as PML VII, exhibit a cytoplasmic localization. However, recently a secondary NLS was demonstrated in the PML II C-terminus (Matic *et al.*, 2010). PML I uniquely carries a nuclear export signal (NES) in its C-terminal domain, meaning it has the potential to shuttle between the nucleus and cytoplasm (Condemine *et al.*, 2006).

With the exception of PML VII, the principal PML isoforms include all of exons 1–6, though they differ in their content deriving from the remaining exons 7–9. Further complexity derives from the potential for alternative internal splicing among PML exons 4–6 to generate, in principle, three further variants (Jensen *et al.*, 2001). Exon 5 may be excluded, maintaining the reading frame, to create a presumptively nuclear form of each of PML I-VI that lacks 48 internal residues.

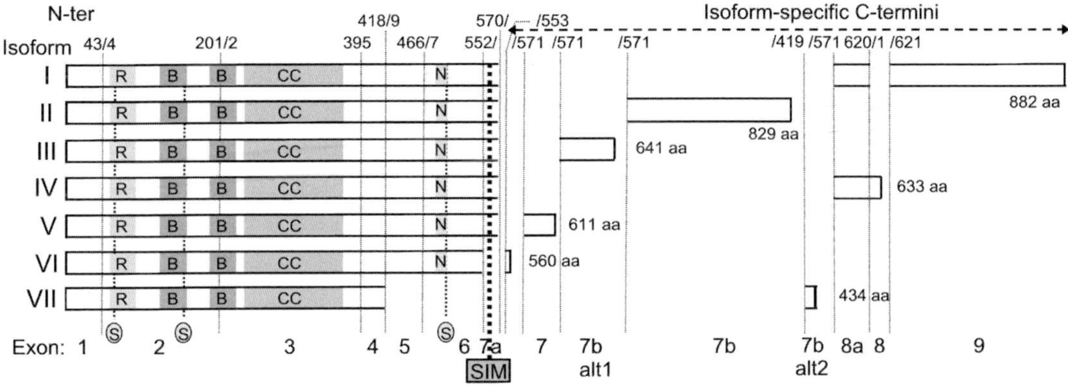

Figure 13.1 The structure of PML isoforms I–VII. PML I–VI share a common N-terminal domain encoded by exons 1–6, but differ in C-terminal sequence owing to variable use of exons 7a–9. PML VI is formed by an alternative splice from exon 6 to exon 7a that alters the reading frame as compared with other isoforms. PML VII is formed by an alternative splice from exon 4 to a site in the distal part of exon 7b, accessing a unique C-terminal sequence. R, RING finger; B, B-Box; CC, Coiled coil; N, nuclear localization signal. The position of the three sumoylatable lysine residues (K65, K160, K490) are indicated (S) and the SUMO interaction motif in exon 7a is indicated (SIM). The amino acid positions of splice junctions are indicated above the diagram and the lengths of each isoform are indicated on the right hand side.

Alternatively, exons 4, 5 and 6 may be excluded together, again maintaining the reading frame into the PML I-V unique C-termini, or exons 5 and 6 may be excluded together which alters the reading frame to give a C-terminus equivalent to that of PML VI. Whilst some examples of these forms have been identified as cDNA, it is unclear whether all of the distinct C-terminal isoforms exist in all variant forms.

All PML isoforms so far described share exons 1–3. These encode a distinctive configuration of three sequence elements: a RING finger (a particular form of zinc-binding motif; Borden and Freemont, 1996), two B-boxes (also zinc-binding) and a coiled-coil region. Collectively, these are known as an RBCC or a tripartite motif (TRIM). This motif is characteristic of a large family of TRIM proteins produced by mammalian cells; PML protein is also known as TRIM19 (Ozato *et al.*, 2008). The RBCC domain is thought to play a key role in the biological function of PML protein, and is required for the oncogenic activity of the PML–RARα fusion protein. All three RBCC elements are crucial for the formation of PML NBs (Borden, 2002).

PML protein is subject to covalent modification with the small ubiquitin-like modifier (SUMO) as discussed below. PMLs I-V also possess a SUMO-interacting motif (SIM) that is encoded in exon 7a and which mediates non–covalent associations with SUMO moieties covalently attached to other proteins (Shen *et al.*, 2006). Like the RBCC domain, the SIM is crucial for the formation of PML NBs and has been suggested to mediate PML – PML interactions via covalently attached SUMO. The coiled-coil within the PML TRIM is also thought to allow both homo-oligomerization of PML proteins and their hetero-oligomerization with other TRIM family members, and this activity too may therefore play a role in the assembly of PML NBs (Cao *et al.*, 1997, 1998; Jensen *et al.*, 2001).

Post-translational modification of PML protein

Further complexity arises in the constellation of PML proteins from the fact that the various isoforms can each be reversibly post-translationally modified in a variety of ways (Nichol *et al.*, 2009). Many of the known modifications affect regions of the protein that are shared by multiple isoforms. However it is unclear whether a given modification occurs equally in all of the potential PML isoform targets.

PML may be covalently conjugated with SUMO at K65, K160 and K490 (Duprez *et al.*,

1999; Kamitani *et al.*, 1998a), the last of these sites being absent in cytoplasmic variants of PML. While all studies of the importance of PML sumoylation to date have focused on these three sites, very recently three further sites of sumoylation have been reported (Galisson *et al.*, 2010). Of the four SUMO homologues described in mammalian cells, PML protein may be modified by each of SUMO1–3 (Kamitani *et al.*, 1998b). Mono-sumoylation occurs by SUMO1, which is unable to form polymeric chains. However PML may be poly sumoylated by SUMO2 and/or 3, with SUMO1 forming the last subunit of the chain. Sumoylation requires Ubc9, the E2 SUMO-conjugating enzyme (Duprez *et al.*, 1999), working with an E3 ligase that may be PML itself (Chu and Yang, 2011). The reversal of PML sumoylation is mediated by at least two members of the SUMO-specific protease family, also known as SENPs. SENP2 is likely to be responsible for de-conjugation of SUMO1 and SUMO3 modifications, whereas SENP5 removes all of SUMO1–3 (Gong and Yeh, 2006). Therefore PML sumoylation is a reversible and dynamic process.

Poly sumoylation of PML is critical for the assembly of PML NBs (Fu *et al.*, 2005; Bernardi and Pandolfi, 2007; Nichol *et al.*, 2009). Dispersal of these bodies during mitosis correlates with desumoylation (Everett *et al.*, 1999) while overexpression of SUMO proteases leads to the loss of PML NBs (Best *et al.*, 2002). Sumoylation has also been shown to regulate the dynamics of exchange of components between PML NBs and the nucleoplasm, including PML itself as well as other NB components such as Sp100 (Weidtkamp-Peters *et al.*, 2008). Finally, PML sumoylation regulates PML stability, either promoting or inhibiting its degradation depending on the pattern of modification. An E3 ubiquitin ligase RNF4/SNURF recognizes both mono- and poly sumoylated PML via its own SIMs and promotes the ubiquitination and subsequent degradation of PML (Tatham *et al.*, 2008; Percherancier *et al.*, 2009).

PML function can also be modified by phosphorylation. During mitosis, appearance of a phosphorylated form of PML coincides with PML desumoylation and dispersal of PML NBs (Everett *et al.*, 1999). Hyperphosphorylation induced by phosphatase inhibition also correlates with reduced PML sumoylation (Muller *et al.*, 1998). In contrast, PML phosphorylation by the MAP kinase, ERK, which targets several residues in the N-terminus as well as S527 and S530, or by the DNA-damage induced kinase HIPK2, which targets some of the same N-terminal residues, promotes sumoylation (Hayakawa and Privalsky, 2004; Gresko *et al.*, 2009). PML is also phosphorylated by the DNA damage activated kinase ATR (Bernardi *et al.*, 2004), and the checkpoint kinase Chk2; following ionizing irradiation, Chk2 phosphorylates PML S117 leading to release of Chk2 from PML-NBs (Yang *et al.*, 2002). Phosphorylation can also regulate PML stability. Modification of S565 (S517 in the Δexon5 PML used in the study) by casein kinase 2 results in the ubiquitination of PML and its subsequent degradation by the proteasome (Scaglioni *et al.*, 2006); S565 is encoded within exon 7a so this mechanism can only affect PML I-V. Phosphorylation has also been associated with PML degradation dependent on the prolyl isomerize Pin1. Pin1 binds S527-phospho-PML IV and initiates its degradation; this activity is opposed by PML sumoylation (Reineke *et al.*, 2008). Since S527 is encoded within exon 6, this mechanism may apply to all nuclear PML isoforms.

Finally, PML is subject to acetylation by the histone acetyltransferase p300 on Lys487 and Lys515, two residues encoded by exon 6, in response to trichostatin A (TSA), an inducer of apoptosis (Hayakawa *et al.*, 2008); this acetylation promotes PML sumoylation. Since exon 6 is excluded from cytoplasmic PMLs, acetylation can only occur on nuclear PML proteins. Although Lys487 is actually located within the NLS, its acetylation does not appear to affect the subcellular localization of PML.

In summary, it is clear that PML is subject to multiple post-translational modifications which can 'cross-talk', adding a further layer of complexity to PML structure and to its functional regulation. Some of these modifications are due to well-established cell stress signalling pathways. As discussed later, PML also has effects on the downstream response to such signalling. PML can therefore be placed firmly within the network of responses to such stresses.

Composition and diversity of PML NBs

Changes in the relative levels of expression of different PML isoforms in a cell are an obvious potential source of functional variation, and specific isoform imbalances have been shown to alter the size and number of PML NBs (Beech *et al.*, 2005). All the major PML isoforms are expressed in both primary cells and cell lines, but each cell type displays a distinct and consistent pattern of isoform expression (Condemine *et al.*, 2006). PML I and PML II are the most abundant isoforms in normal cells, whereas PMLs III-V are present at much lower levels; PML VI appears to be the least abundant isoform. However, in tumour cell lines, the relative abundance of PML I and PML II is reduced relative to PML III-VI (Condemine *et al.*, 2006).

As described above, sumoylated forms of PML I-VI are key components of PML NBs. The mechanism of PML NB assembly is thought to incorporate homo- and hetero-oligomerization of PML isoforms via the RBCC motif as well as interactions between conjugated SUMO and the PML SIM (Shen *et al.*, 2006). On this basis, the nuclear PML isoforms might be expected to act equivalently in these assembly events. However, experiments by Weidtkamp-Peters *et al.* have shown that among these isoforms, PML V has a uniquely stable association with PML NBs, suggesting that it is likely to be a particularly important scaffold component (Weidtkamp-Peters *et al.*, 2008). Interestingly, this study also suggested that the sumoylation status of PML was an important regulator of its exchange dynamics between NBs and the nucleoplasm. Thus changes in PML isoform balance and modification could alter PML NB organization and hence function.

A large number of proteins unrelated to PML have been found in PML NBs. A few of these, such as Sp100 and Daxx, form more stable associations with NBs but most associate only transiently and/or under specific circumstances (Bernardi and Pandolfi, 2007; Van Damme *et al.*, 2010). Such proteins are now thought to pass relatively freely through the PML/Sp100 shell into the hollow interior, which is enriched with polymeric SUMO 2/3 chains (Lang *et al.*, 2010). These chains provide possible binding targets for any proteins possessing a SIM, and therefore facilitate the association of proteins with PML NBs (Shen *et al.*, 2006). Moreover, Sp100 and Daxx, as well as the PML NB-associated fraction of many other transiently resident proteins, are themselves sumoylated, which is also thought to enhance their association with these bodies (Van Damme *et al.*, 2010).

In the light of the functions proposed for PML NBs (summarized below), it is important to consider whether these structures contain RNA or DNA. Except for a distinct subset of PML NBs in those cell lines that lack telomerase (Yeager *et al.*, 1999), PML NBs do not contain any detectable levels of nucleic acid within their structure (Boisvert *et al.*, 2000). They do though closely associate with transcriptionally active chromatin (Wang *et al.*, 2004) and with nascent RNA (Kiesslich *et al.*, 2002).

PML NBs are described as such, despite their complex composition, because it is widely held that other NB components do not form such structures in the absence of PML protein (Ishov *et al.*, 1999). However, Sp100 is reported, when overexpressed, to form structures that appear similar to PML NBs and which can recruit at least some other PML NB components (Borden, 2008). Perhaps more significantly, the nuclear fraction of the translation factor eIF4E is clearly associated with at least a subset of PML NBs and, in cells where PML is absent, the distribution of eIF4E is unaltered (Cohen *et al.*, 2001). This protein, which is far more conserved in evolution than PML, has been argued to be the more ancient founding component of the PML NB functional compartment (Borden, 2008).

Involvement of PML and PML NBs in many areas of cell biology

The main focus of this chapter is on the interaction of viruses with PML and PML NBs, as will be discussed in detail later. However, rationalizing the various virus effects on PML and PML NBs during infection needs to be done in the context of information about the functions of these components in uninfected cells. Although there is a very large literature on this topic, it has not yet produced simple clear conclusions about the role(s) of PML in the cell. Many of the functions ascribed to NBs are often equated with functions of PML protein, though this is clearly an unjustified assumption given the complex composition of PML NBs. Equally, demonstrated functions of PML protein need not necessarily be functions of PML-NBs since PML protein also exists outside these structures in both the nucleoplasm and cytoplasm. Confusion also arises because many early studies used overexpression of single cloned PML cDNAs and assumed that the results of these experiments applied equally to all PML isoforms.

Throughout the extensive research into PML, two distinct scenarios have been proposed for how PML NBs might function. Either these structures could be active centres in which the assembled proteins perform function(s) not otherwise available in the cell or else they might represent stores from which proteins could be released to perform their functions under appropriate stimuli. These scenarios are of course not mutually exclusive, and they may apply differently in respect of each area of PML function. The following sections consider briefly the areas of cell biology where PML and/or PML NBs have been implicated.

Roles in the DNA damage and other stress responses

PML NBs have been implicated in both the sensing of DNA damage and some aspects of the subsequent repair process (Dellaire and Bazett-Jones, 2004). Rapidly after DNA double-strand break induction, PML NBs disperse or increase in number, dependent on the activity of signalling components such as ATR and Chk2 (Dellaire *et al.*, 2006a; Varadaraj *et al.*, 2007). Several proteins involved in DNA damage signalling and repair become mobilized from PML NBs to sites of DNA damage (Mirzoeva and Petrini, 2001; Barr *et al.*, 2003; Krieghoff-Henning and Hofmann, 2008; Salomoni *et al.*, 2008). Only with some delay does PML itself associate with any persisting sites of DNA damage (Dellaire *et al.*, 2006a). PML NBs also become dispersed following UV light exposure (Seker *et al.*, 2003) but PML itself rapidly associates with ssDNA sites following such UV-induced DNA damage (Boe *et al.*, 2006). Thus the precise nature of PML NB changes during DNA damage depends on the type of damage insult.

The response of PML NBs to DNA double-strand breaks has similarities with the response to a variety of other stresses. Heat shock and treatment with heavy metals also cause PML NB fragmentation (Maul *et al.*, 1995; Eskiw *et al.*, 2003), though the composition of the resulting structures can vary between different stresses. Conversely, specific amino acid starvation causes PML NB aggregation (Kamei, 1997). For sublethal stress, once the source of stress is removed, cell recovery is accompanied by reformation of apparently normal PML NBs.

Role in apoptosis and senescence

PML protein is a tumour suppressor. As well as its expression being frequently lost in human tumours (Gurrieri *et al.*, 2004), its overexpression in mouse skin inhibits chemically induced carcinogenesis (Virador *et al.*, 2009). PML protein plays an integral role in numerous apoptotic mechanisms and is required for caspase-1 and caspase-3-dependent apoptosis (Wang *et al.*, 1998b). PML-null cells are less sensitive than wild-type to multiple apoptotic stimuli, such as ionizing radiation, interferon treatment and ceramide treatment, and PML-null animals are more susceptible to tumours (Wang *et al.*, 1998b). Caspase-2 and caspase-6 have also been physically and/or functionally associated with PML NBs (Tang *et al.*, 2005; Tan *et al.*, 2008).

The tumour suppressor p53 is one very well-documented example of a transient PML NB protein. Upon DNA damage, p53 is recruited to PML NBs via an interaction with PML IV where it is subject to post-translational modifications by acetyl transferases such as CBP and kinases such as HIPK2 that activate its pro-apoptotic functions (Fogal *et al.*, 2000; Pearson *et al.*, 2000). Consequently, PML relocation to the cytoplasm is thought to inhibit p53 activation (Bellodi *et al.*, 2006). Many factors involved in the stability of p53 are also regulated by PML. For instance, casein kinase 1 activity is enhanced by PML (Alsheich-Bartok *et al.*, 2008). Also, the E3 ubiquitin ligase MDM2, a negative regulator of p53, is itself inhibited by direct binding to PML, and thus PML prevents p53 degradation (Louria-Hayon *et al.*, 2003). Upon DNA damage, PML protein binds MDM2 and sequesters it in the nucleolus, so stabilizing p53 (Bernardi *et al.*, 2004).

Possibly connected with its role in p53 regulation, PML IV is implicated in the development of premature senescence in response to oncogene stress (Bischof *et al.*, 2002). Roles for other PML isoforms in this process are likely, however, as PML IV cannot induce senescence in PML-null cells (Bischof *et al.*, 2002). PML IV also binds the oncoprotein Myc and induces its degradation, a function that is dependent on the PML RING domain (Buschbeck *et al.*, 2007). This too would be expected to oppose cell growth and oncogenesis. There may also be a role for cytoplasmic PML in regulating growth and apoptosis in response to TGF-β. A cytoplasmic PML variant lacking exons 5 and 6 is required for efficient TGF-β signalling in mouse embryonic fibroblasts, and nuclear sequestration of cPML results in a decrease in TGF-β signalling (Salomoni and Bellodi, 2007).

Role in telomere maintenance

Possibly related to their roles in regulating DNA damage responses, in some tumour cells a subset of PML NBs plays a role in telomere maintenance by a telomerase-independent mechanism known as alternative lengthening of telomeres (ALT). Termed ALT-associated PML NBs (APBs), these bodies contain telomeric DNA, telomeric repeat binding factors (TRFs) 1 and 2 as well as factors required for DNA repair (Jiang *et al.*, 2007). PML III directly binds to TRF1 and is required for the formation of APBs, but not PML NBs (Yu *et al.*, 2010). In contrast, PML IV interacts directly with TERT and inhibits it, apparently opposing telomere maintenance (Oh *et al.*, 2009). This action of PML IV could also be related to its role in senescence.

Role in cell division

Studies on PML III have revealed a possible role in the maintenance of the centrosome (Xu *et al.*, 2005). PML III was shown to associate with the centrosome and to inhibit its duplication by inhibiting the kinase, Aurora A, by binding to it and regulating its post-translational modification (Xu *et al.*, 2005). However, in subsequent studies by another group, PML III was not observed to localize to the centrosome (Condemine *et al.*, 2006), so the status of this potential PML function is uncertain.

Role in innate or intrinsic immunity

PML and PML NBs have been clearly linked to innate and/or intrinsic immunity, not least by many experiments studying their effects on specific virus infections (Geoffroy and Chelbi-Alix, 2011); these are discussed in detail in the following sections. However, the most obvious link between PML and antiviral activity is that PML expression is stimulated by interferon (IFN); many other IFN-stimulated genes (ISG) are known to encode mediators of the antiviral state. The first intron of the *pml* gene contains both ISRE and GAS elements, responsive to type I (α and β) and type II (γ) IFN respectively. Upon IFN treatment, PML transcription is up-regulated, resulting in an increase in both the size and number of PML NBs (Lavau *et al.*, 1995). Other key components of PML NBs such as Sp100 (Guldner *et al.*, 1992) and Daxx (Shimoda *et al.*, 2002), are also up-regulated by IFN providing further evidence of an antiviral role for these structures.

As already noted, PML is a member of the TRIM family of proteins (Nisole *et al.*, 2005). Although TRIM family members have a variety of roles within the cell, there is increasing evidence to suggest that a number of them perform antiviral duties (Nisole *et al.*, 2005; Ozato *et al.*, 2008). For example, TRIM5α acts as a restriction factor against retrovirus replication (Stremlau *et al.*, 2004; Towers, 2007). Indeed, a significant number of TRIM proteins are induced by IFN. Out of 72 TRIM proteins tested in two cell types, 24 were directly up-regulated by either type I or type II IFN, including PML and TRIM5 (Carthagena *et al.*, 2009). TRIM21 has recently been shown to have antiviral activity against adenovirus, binding with high affinity to Ig:virus particle complexes that enter the cell by endocytosis and targeting them for degradation by the proteasome (Mallery *et al.*, 2010). Since the virus itself does not become ubiquitylated in this process, this effect would be predicted to apply to any non-enveloped virus.

A further link between PML NBs and innate immunity comes from a study of the PLZF protein, which is expressed from a gene that is disrupted by translocation in the small minority of promyelocytic leukaemia cases not due to a translocation affecting the *pml* gene. PLZF, a transcription factor, is recruited to PML NBs by IFN treatment and this is necessary for the induction of a specific subset of ISGs (Xu *et al.*, 2009). The protection of mice from lethal Semliki Forest virus infection that is provided by prior administration of IFNα in wild-type mice is lost in PLZF-null animals, demonstrating the importance of PLZF for the production of an effective antiviral response (Xu *et al.*, 2009).

PML and protein aggregation

Another area of cell biology in which PML has been implicated is the handling of protein aggregates in the nucleus. In situations where such aggregates form, such as poly glutamine (polyQ) expansion diseases, PML reorganizes to form a shell around the aggregates (Skinner *et al.*, 1997; Yamada *et al.*, 2001). This association, which in one experimental system involved specifically PML IV (Janer *et al.*, 2006), may be linked with the cell seeking to resolubilize or to degrade the aggregates. Sumoylation of a polyQ protein (which would probably be promoted by association with PML NBs) has been shown to promote its solubility (Janer *et al.*, 2010) while induction of PML by IFN has been linked with recruitment of proteasome components to PML NBs (Fabunmi *et al.*, 2001; Lafarga *et al.*, 2002) and with the removal of aggregates from cells (Janer *et al.*, 2006).

Role in controlling chromatin/gene expression

The juxtaposition of PML NBs and transcriptionally active chromatin suggests a role in controlling gene expression. Such an activity might underpin one or more of the functional areas already discussed. One gene locus that has been studied in detail is the major histocompatibility complex (MHC). PML NBs associate with this locus, where they are thought to organize the higher-order structure of chromatin by associating with matrix attachment regions (MARs) to form chromatin loops (Shiels *et al.*, 2001; Kumar *et al.*, 2007). These associations are functionally significant since silencing PML expression both disrupts the associations and causes specific defects in the expression of genes within the locus, such as *HLA-A*; some of these effects are PML isoform-specific (Kumar *et al.*, 2007). More recently, the MHC class II gene *DRA* has been shown to move into association with PML NBs upon induction of its transcription by IFNγ. Most interestingly, this association is maintained for several cell generations thereafter, keeping the *DRA* gene poised for greater and more rapid induction upon re-exposure to IFNγ via maintenance of an accessible histone H3 modification state in the DRA promoter (Gialitakis *et al.*, 2010).

The eIF4E theory of PML NB function

Another activity of PML that may underpin one or more of its functions is its association with eIF4E. Although best understood as a translation initiation factor, there is a nuclear fraction of eIF4E that appears to be involved in regulating the export, via its association with mRNA cap structures, of a select subset of mRNAs (Culjkovic *et al.*, 2006). The translation products of these mRNAs typically function within discrete pathways of cell growth regulation such as the Akt pathway (Culjkovic *et al.*, 2008). By interacting with eIF4E, PML inhibits both aspects of eIF4E activity. It has been argued that many of the diverse effects of PML/PML NBs are aspects of this regulation of eIF4E (Borden and Culjkovic, 2009).

PML proteins as E3 SUMO ligases

PML proteins have long been hypothesized to possess activity similar to that of an E3 ubiquitin-ligase, since other RING-domain proteins have such activity (Freemont, 2000; Meroni and Diez-Roux, 2005). The finding that overexpression of PML in yeast led to a general increase in sumoylation and to PML self-sumoylation suggested its actual substrate might be SUMO rather than ubiquitin (Quimby *et al.*, 2006). This was also supported by the fact that the PML RING can bind the SUMO E2 conjugating enzyme Ubc9 (Duprez *et al.*, 1999), a component which can also be found at PML NBs (Nacerddine *et al.*, 2005). In direct confirmation of these ideas, PML IV, as well as several other TRIM proteins, has been shown recently to have E3 SUMO ligase activity in mammalian cells with specificity for p53 among other targets (Chu and Yang, 2011).

E3 SUMO ligases bring together an E2 conjugating enzyme carrying covalently attached activated SUMO and a sumoylation target by providing binding sites for each protein; the binding site for the E2 enzyme is provided by the RING domain while the substrate may be bound via many other parts of the protein. The diverse C-terminal domains of PML provide the opportunity for binding of multiple sumoylation target proteins and would be expected to impart unique E3 ligase substrate specificities to each PML isoform. This therefore offers a further overarching PML function that could explain the involvement of these proteins in so many different areas of cell activity.

The PML-null phenotype in mice

One problem with any description of a key function for PML in the life of a cell is that mice can survive quite well without this protein; PML-null animals have only subtle phenotypic differences from normal mice. They have a somewhat elevated rate of tumorigenesis and increased sensitivity to carcinogens (Wang *et al.*, 1998a) and their haematopoietic cells are largely resistant to the induction of apoptosis by various stimuli (Wang *et al.*, 1998b). They have reduced levels of some types of immune cell and an elevated susceptibility to botryomycotic bacterial lesions (Wang *et al.*, 1998a). Studies with PML-null animals have also revealed roles for the protein in restricting proliferation of haematopoietic stem cells (Ito *et al.*, 2008) and neural progenitor cells (Regad *et al.*, 2009); loss of PML also disturbs mammary gland development (Li *et al.*, 2009). Whilst these deficiencies do not impact severely on the animals within the context of stress- and pathogen-free animal husbandry environments, they might conceivably have a more severe phenotype in real-world situations. Alternatively, it may be argued that there is redundancy in the system, meaning the absence of PML is largely compensated for by other proteins in the animals.

Small DNA tumour virus interactions with PML protein and PML NBs

Studies of the effects of viruses on PML and PML NBs, and conversely of the effects of these cell components on virus infection, have been significant strands of PML research from its earliest stages. In the course of this work, a diverse collection of viruses has been found to interact with or have effects on PML and/or PML NBs. These studies provide perhaps the most compelling line of evidence supporting an anti-viral role of PML/PML NBs (Leppard and Dimmock, 2006; Everett and Chelbi-Alix, 2007). Given the focus of this book on the small DNA tumour viruses, this section considers first what has been learnt about virus interactions with PML and PML NBs in the context of the adenovirus, papillomavirus and polyomavirus infections before summarizing findings from studies of a broader range of viruses.

Adenoviruses

PML NBs become disrupted in a unique fashion during the early phase of infection with human species C adenovirus type 5 (Ad5; Fig. 13.2). Puvion-Dutilleul *et al.* (1995) first observed that Ad5 infection redistributed PML NBs from their native spherical structures into elongated 'tracks'. Through the use of viral mutants, this redistribution of PML NBs was found to be dependent on the early gene product E4 Orf3, which co-located in tracks with PML; indeed E4 Orf3 expressed from a plasmid was sufficient for PML NB reorganization (Carvalho *et al.*, 1995; Doucas *et al.*, 1996). Doucas *et al.* also found reorganized PML NB tracks in a small proportion of cells infected at standard multiplicities of infection (MOI) with an Ad5 mutant unable to make E4 Orf3, but which could still produce a second protein E4 Orf6 that has other overlapping functions with E4 Orf3 (Leppard, 1997); even mutants lacking both proteins could cause this reorganization to some extent when used at very high MOI. One explanation for this apparent E4 Orf3-independent effect might be that monkey CV1 cells rather than human cells were used; we have not observed track formation during Ad5 infection in the absence of Orf3 in human cell types (K. Leppard, unpublished data). It may also be that PML NB reorganization similar to true track formation can result from cell stress due to Ad5 infection in a few individual cells of some types, independent of E4 Orf3 expression. However, it is now generally held that E4 Orf3 is necessary for PML NB reorganization into tracks in the context of Ad5 infection.

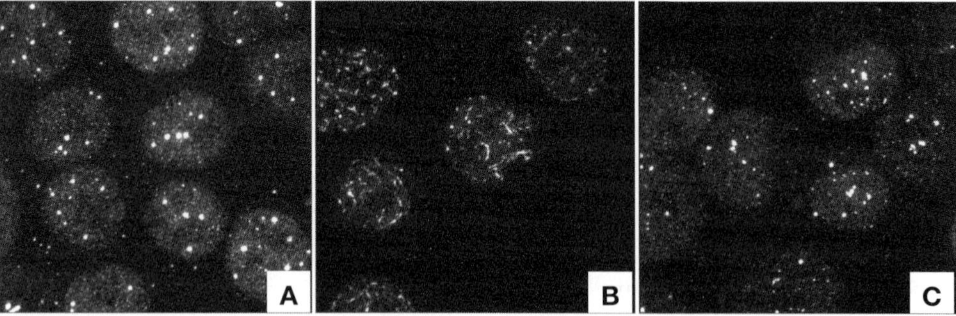

Figure 13.2 PML nuclear bodies in HEp2 cells. Human cancer cell line HEp2 was either mock-infected (A), or infected with wild-type Ad5 (B) or a mutant unable to make E4 Orf3 protein, inOrf3 (C). Sixteen hours post-infection, cells were fixed and stained for PML. Staining patterns were detected by confocal microscopy. Images shown are maximum projections of z-stacks in each case. A is reproduced from Beech *et al.* (2005) Exp. Cell Res. *307*, 109–117, Fig. 2D.

This function is conserved across E4 Orf3 proteins from representative viruses of other human Ad species (subgroups) (Hosel *et al.*, 2001; Evans and Hearing, 2003).

A second important feature of the interaction of Ad5 with PML NBs concerns the sites of virus colonization in the nucleus upon infection. Mutant Ad5 genomes, deficient in E4 Orf3 expression and hence unable to reorganize PML NBs, locate close to PML NBs prior to replication. The same association can be seen to develop over the first few hours of infection with wild-type Ad5, before E4 Orf3-mediated PML NB rearrangement occurs (Ishov and Maul, 1996). Intriguingly, not all PML NB components behave in the same way during this rearrangement. Initially, PML and Sp100 are both reorganized to virtually identical tracks in the early phase of infection. However, once viral replication centres are established in the cell the Sp100 and other components tested, apart from PML itself, further relocate into these centres while PML remains in the tracks (Doucas *et al.*, 1996; Ishov and Maul, 1996).

The molecular basis of E4 Orf3-mediated reorganization of PML NBs is now understood in some detail. E4 Orf3 binds directly to one specific PML isoform, PML II, and across a panel of E4 Orf3 mutant proteins, there is an absolute correlation between ability to bind PML II and ability to reorganize PML NBs (Hoppe *et al.*, 2006). Furthermore, neoformed PML NBs created from transiently expressed PML in PML-null cells are only rearranged by Orf3 if they are formed of PML II (Hoppe *et al.*, 2006). Thus it is this interaction that causes PML NBs to become reorganized. It could be inferred from the fact that only PML II was directly bound by E4 Orf3 that its binding site would lie in the PML II unique C-terminal region. This was confirmed in a study which identified a 40 amino acid sequence unique to PML II that could confer E4 Orf3 interaction on an irrelevant protein (Leppard *et al.*, 2009).

Importantly, despite E4 Orf3 being a small protein (~11 kDa) and PML II about ten times larger, it is the E4 Orf3 protein that forms the basis of nuclear tracks with which PML NB components then associate. This conclusion is based on the observation that E4 Orf3 expressed alone in PML-null cells still forms nuclear tracks (Hoppe *et al.*, 2006). E4 Orf3 protein is known to be highly insoluble (Sarnow *et al.*, 1982) and, since its formation of nuclear tracks and its interaction with PML II both apparently require it to form self-interactions (Hoppe *et al.*, 2006), the underlying architecture of a nuclear track may be an Orf3 oligomer or polymer.

As well as its effects on PML NB organization, E4 Orf3 also exerts two apparently linked effects on the PML protein population. In the later phases of infection with wild-type Ad5, higher molecular weight sumoylated PMLs are lost, and this coincides with the appearance of a novel, infection-specific, species of PML, which becomes a significant PML component (Leppard and Everett, 1999). Both effects on PML require the expression of E4 Orf3 from the virus. Intriguingly, this infection-specific PML is very similar to a mitosis-specific, phosphorylated form of PML observed in another study (Everett *et al.*, 1999).

There is evidence to suggest that the purpose of E4 Orf3 interaction with PML NBs is to mitigate an innate antiviral response. In cells pretreated with IFNα or IFNγ, E4 Orf3-mutant Ad5 replication is significantly impaired compared with untreated cells, in contrast to the minimal effect of IFN pretreatment on the replication of wild-type virus (Ullman *et al.*, 2007). The ability of E4 Orf3 to overcome the antiviral state induced by IFN correlates with its ability to rearrange PML NBs, and is thus conserved among human Ads (Ullman *et al.*, 2007). These results clearly link PML NBs with the creation of an effective IFN-induced antiviral state. Indeed, depletion of either PML itself or the PML NB component Daxx, but not depletion of Sp100, completely relieves the inhibition of viral DNA replication imposed on an E4 Orf3-deficient virus by prior IFN treatment (Ullman and Hearing, 2008). The inference that the crucial effect of E4 Orf3-mediated PML NB reorganization in this context is to inactivate a PML NB function is further supported by the fact that expression of HSV1 ICP0 and HCMV IE72, two other viral proteins known to disrupt PML NBs (see next section), restores the growth of E4 Orf3 mutant virus in IFN-treated cells (Ullman and Hearing, 2008). Therefore there is strong evidence that E4 Orf3-mediated disruption of PML NBs is a mechanism employed by Ad5 to overcome cellular antiviral activities that are performed by PML NBs or their components.

In addition to redistributing the components of PML NBs, the action of E4 Orf3 also leads to the recruitment of other host proteins to PML tracks. TIF1α (TRIM24) is normally diffusely distributed in the nucleus but some is drawn into association with PML tracks via a direct association with E4 Orf3 (Yondola and Hearing, 2007). This E4 Orf3 activity is separable from its PML reorganizing activity but is still conserved across several human Ad species. Another Ad5 E4 Orf3 function is its ability to reorganize and disrupt the action of the Mre11-Rad50-Nbs1 (MRN) complex within DNA damage signalling and repair pathways (Stracker *et al.*, 2002, 2005; Carson *et al.*, 2009). In the absence of the E1B 55K/E4 Orf6 complex (which also targets MRN, but in a different way), E4 Orf3 causes the MRN complex to associate with PML tracks early after infection (Evans and Hearing, 2005); MRN is later exported to cytoplasmic aggresomes (Araujo *et al.*, 2005; Liu *et al.*, 2005). This E4 Orf3 activity is apparently restricted to Ad5 (representing human Ad species C) (Stracker *et al.*, 2005).

Ad5 proteins other than E4 Orf3 can also associate with PML. A small proportion of the transcriptional regulator E1A protein is found in PML tracks during infection, and with normal PML NBs after E1A plasmid transfection (Carvalho *et al.*, 1995). This association depends on E1A conserved region 2A, but does not correlate with the well-recognized ability of this E1A region to interact with any of the Rb family members. E1A overexpression has also been shown to partially disassemble PML NBs into microstructures in a manner very similar to the effect of other cell stresses such as heat shock or heavy metal treatment (Eskiw *et al.*, 2003). This disassembly may relate to an effect on SUMO conjugation of E1A binding to Ubc9 (Yousef *et al.*, 2010).

The E1B 55K protein, which together with E4 Orf6 and host proteins forms an E3 ubiquitin ligase that targets p53 and other host proteins for degradation (Querido *et al.*, 2001), also associates with PML tracks or PML NBs early during infection (Doucas *et al.*, 1996). However, its partner protein E4 Orf6 does not make the same association. Rather, once E4 Orf6 is expressed, E1B 55K moves with it to viral replication centres (Ornelles and Shenk, 1991); thus the E1B 55K association with rearranged PML tracks is normally only transient. However, in the absence of E4 Orf6, a proportion of E1B 55K remains tightly associated with PML tracks throughout infection (Konig *et al.*, 1999; Leppard and Everett, 1999; Lethbridge *et al.*, 2003). It was suggested that the normally transient retention of E1B 55K in the tracks prevents premature inhibition of p53 by E1B 55K during infection (Konig *et al.*, 1999). However, recent demonstrations that p53-responsive genes remain repressed during infection even by an E1B 55K mutant virus (Miller *et al.*, 2009) and that E4

Orf3 itself induces silencing heterochromatin at p53-responsive promoters to inhibit this activation (Soria *et al.*, 2010) calls this interpretation into question.

The E1B 55K protein is subject to sumoylation on a small fraction of the protein (Endter *et al.*, 2001). Since the amount of this SUMO-E1B 55K is greatest relatively early in infection and is enhanced if the E4 Orf6 protein is absent (Lethbridge *et al.*, 2003) or if its nuclear export signal is mutated (Kindsmuller *et al.*, 2007), this modification can be linked with the phase of infection when E1B 55K is associating with E4 Orf3, PML NBs and with the nuclear matrix (Leppard and Everett, 1999; Lethbridge *et al.*, 2003). Sumoylation of E1B 55K has now been shown to be needed for it to associate with PML NBs (Pennella *et al.*, 2010).

Interestingly, E1B 55K is now known to bind directly to PML. In a human cell transfection-overexpression system, it binds specifically to PML IV and PML V (Wimmer *et al.*, 2010). However, these interactions are affected by other Ad5 proteins since, in infected cells overexpressing each of the nuclear PML isoforms, additional isoforms PML I and VI are strongly bound while PML IV binding is lost (Wimmer *et al.*, 2010). Binding this latter isoform depends on E1B 55K sumoylation which, as just discussed, is substantially inhibited by E4 Orf6 during virus infection. It should be noted that E1B 55K does not bind PML II, which is the direct binding target of E4 Orf3. Thus, E1B 55K has the potential to inhibit or modify the function of PML NBs and specific PML isoforms in a manner distinct from the effects of E4 Orf3.

The link between E1B 55K, sumoylation and PML NBs has been given a new and exciting twist by the discovery that E1B 55K can act as an E3 SUMO ligase; this is of course in addition to its role as part of a ubiquitin ligase. E1B 55K, without its partner protein E4 Orf6, induces p53 sumoylation both in a transfection-overexpression context (Muller and Dobner, 2008) and during Ad5 infection (Pennella *et al.*, 2010) and the direct action of E1B 55K as an E3 SUMO ligase has been shown *in vitro* with purified components (Pennella *et al.*, 2010). Importantly, this action of E1B 55K drives the association of p53 with PML NBs by forming large p53-E1B 55K aggregates; the effect is to increase substantially the dwell-time of p53 in these structures and hence inhibit its activity. One final observation that is that E1B 55K can, like E4 Orf3 (Ullman and Hearing, 2008), overcome the antiviral effect of the PML NB component Daxx and it does so by driving its proteasomal degradation (Schreiner *et al.*, 2010). Since this activity is independent of E4 Orf6, it may reflect either a dual specificity of the intrinsic E3 ligase activity of E1B 55K for both SUMO and ubiquitin, or else the degradation of Daxx might result indirectly from sumoylation of it or another target.

As well as these complex interactions of early proteins with PML and PML NBs, during the late stage of infection the minor structural protein IX is reported to associate with PML. Protein IX localizes to PML NBs to form amorphous inclusion bodies, and it has been proposed that it may act to maintain the disruption of PML NBs, caused initially by E4 Orf3, throughout the later phases of infection when Orf3 expression wanes (Rosa-Calatrava *et al.*, 2003). Furthermore, impairing the ability of protein IX to form these inclusions results in reduced viral yields, suggesting this mechanism is important for the viral life cycle (Rosa-Calatrava, 2003).

Papillomaviruses

Like Ad5 genomes, bovine papillomavirus (BPV) DNA is deposited close to PML NB structures upon infection in cell culture (Day *et al.*, 2004). However, in contrast to other virus examples, using PML-null cells reconstituted or not with a PML cDNA (probably PML IV), the presence of PML increased rather than decreased viral gene expression. Either papillomaviruses are different from other viruses in their response to PML or possibly this PML isoform is simply not the appropriate one to inhibit their infection. Along with human papillomavirus (HPV) DNA, its two replication proteins E1 and E2 also localize to foci adjacent to PML NBs (Swindle *et al.*, 1999), suggesting that PV replication centres are founded in the vicinity of PML NBs.

Upon overexpression, the HPV L2 capsid protein associates with PML NBs through an interaction with Daxx (Day *et al.*, 1998; Florin *et al.*, 2002). L2 also deposits at PML NBs with incoming genomes from infectious BPV or HPV particles (Day *et al.*, 2004; Karanam *et al.*, 2010). L2 expression causes disruption to PML NBs by causing a decrease in levels of Sp100 although their appearance by PML immunofluorescence is not much altered (Florin *et al.*, 2002). Depletion of Sp100 coincides with the recruitment of another capsid protein, L1, and also the transcription/replication factor E2 to PML NBs. Although it has been proposed that L2 association with PML NBs facilitates virus assembly, L1/L2 nuclear focus formation and assembly occur in PML-null cells (Becker *et al.*, 2004) and L2 only seldom associates with PML NBs when artificially expressed at lower levels (Kieback and Muller, 2006). Moreover, Nakahara and Lambert showed, in organotypic raft cultures, that while cells in the basal layers that contained HPV DNA had increased numbers of PML NBs, these were not detectable in the granular layers where virus structural protein expression and assembly would occur (Nakahara and Lambert, 2007). Thus, the association of L2 with PML NBs is more likely to reflect the early events of genome entry into the nucleus than later events in the life cycle.

Two early proteins encoded by HPV have also been associated with PML NBs. The E6 proteins of HPV11 and HPV18 interact directly with PML I, II and IV, and also co-localize with neoformed NBs created from these isoforms in PML-null cells, but they do not bind or co-locate with PML III, V or VI (Guccione *et al.*, 2004). The E6 protein of HPV18, but not HPV11, also induced the proteasome-mediated degradation of PML IV, thereby inhibiting its ability to induce senescence (Guccione *et al.*, 2004). More recently, E6-AP, the host protein that forms an E3 ubiquitin ligase with E6 to target proteins for degradation, has been shown to negatively regulate PML protein levels even in the absence of E6 (Louria-Hayon *et al.*, 2009). Meanwhile, the other HPV16 transforming protein, E7, can overcome PML-IV induced senescence by inhibiting its ability to activate p53 (Bischof *et al.*, 2005). E7 also induces ALT-associated PML NBs in primary human keratinocytes, thus allowing telomere maintenance (Spardy *et al.*, 2008). These findings may be relevant to the specific cancer-causing potential of HPV16 and HPV18.

Polyomaviruses

Various polyomaviruses have been studied for their interactions with PML NBs. In the course of their study of Ad5 effects on PML NBs, Carvalho *et al.* showed that large T antigen from the simian polyomavirus SV40 could localize to PML NBs when expressed from a plasmid (Carvalho *et al.*, 1995). Also, like Ad5 and the HPVs, SV40 genomes locate close to PML NBs so that transcription and replication occur in proximity to these structures (Ishov and Maul, 1996; Jul-Larsen *et al.*, 2004).

Work with the human polyomavirus JC concerning association of the viral late proteins with PML NBs also suggests parallels with the papillomaviruses. Major and minor capsid proteins VP1, 2 and 3 localize with PML NBs in experimental JCV late gene plasmid transfection and in brain tissue obtained at autopsy from a progressive multifocal leukoencephalopathy patient (Shishido-Hara *et al.*, 2004). A further study indicated that both JCV replication and assembly occurred in association with PML NBs in such patient tissues (Shishido-Hara *et al.*, 2008). However, the suggested interpretation of these findings, that PML NBs exert a positive influence on JCV replication, is contradicted by another study in a cell culture system based on human astrocytes where nuclear PML was found to be inhibitory to JCV (Gasparovic *et al.*, 2009). Although JCV large T formed microdomains closely associated with PML NBs, PML knock-down at a single cell level did not inhibit either the formation of these domains or JCV capsid protein expression. Moreover, global PML reduction by arsenite treatment actually enhanced virus replication while, conversely, increasing the number of PML NBs in advance of infection by IFNβ treatment inhibited replication; this IFN effect was completely lost if PML was depleted from the cells (Gasparovic *et al.*, 2009). These results are more consistent with a model in which the association of virus components with PML NBs is driven by a host response designed to limit the infection.

BK virus is another human polyomavirus. Like SV40 and JCV, its replicating genomes are associated with PML-NBs (Jul-Larsen *et al.*, 2004). A recent study in a physiologically relevant cell type has shown that, like its relatives, BKV large

T antigen associates with PML NBs in the early stage of infection. Later, dependent on viral DNA replication, PML NBs are disrupted with PML forming larger aggregates while other NB components (Sp100 and Daxx) become dissociated from these sites of PML accumulation (Jiang *et al.*, 2011). This reorganization represents functional inactivation of an antiviral function of PML NBs since BK virus infection could complement the defect of a herpes simplex virus mutant that is unable to destroy PML protein (see below). These data are very reminiscent of what happens during Ad5 infection.

Other viruses and PML/PML NBs

To fully interpret and understand the interactions between the small DNA tumour viruses and PML NBs, it is important also to consider briefly what has been learned from the study of other viruses, not least because one of these viruses – herpes simplex virus type 1(HSV1) – has been the most widely studied in terms of its interaction with PML NBs. Some key findings from such studies are summarized in this section.

Herpesviruses

All the human herpesviruses examined make immediate early proteins that interfere with PML NBs, though the mechanisms by which they do so vary. One of the immediate-early proteins encoded by HSV1 in the first phase of its lytic cycle gene expression, ICP0, causes complete disruption of PML NBs by acting as an E3 ubiquitin ligase that targets PML protein, and also Sp100, for degradation (Everett *et al.*, 1998; Chelbi-Alix and de The, 1999; Boutell *et al.*, 2002). When studied under conditions where this rapid loss of PML NBs is prevented, HSV1 genomes and the transcriptional activator ICP4 co-locate in proximity to PML NBs and later develop into virus replication centres (Maul *et al.*, 1996; Everett *et al.*, 2003), results similar to those seen for the smaller DNA viruses. However, an important additional finding in the HSV1 system is that, when the earliest stages of the infection process are studied in live cells, pre-existing PML NBs can be seen to deconstruct and reform in those locations in the nucleus where incoming viral genomes are arriving (Everett and Murray, 2005). Thus, rather than being specific target sites in the nucleus that viruses have to reach when establishing an infection, PML NBs are in fact dynamic structures mobilized by the host in response to infection.

There is compelling evidence that ICP0-mediated disruption of PML NBs is closely linked to overcoming cellular antiviral responses. Unlike wild-type virus, ICP0-deficient HSV1 is sensitive to IFN *in vitro* and grows to lower titres in mice (Leib *et al.*, 1999; Mossman *et al.*, 2000; Everett *et al.*, 2006). This growth defect is partially corrected in transgenic mice defective in IFN α/β signalling (Leib *et al.*, 1999). This function of ICP0 clearly relates to its effects on PML NBs since, in both PML-null mouse embryo fibroblasts (Chee *et al.*, 2003) and PML-knockdown human fibroblasts (Everett *et al.*, 2008b), the absence of PML substantially reduces the inhibitory effect of IFN on the growth of ICP0-defective virus. This suggests that the ability of IFN to inhibit HSV1 replication is dependent on the PML protein and is overcome by ICP0 expression in the early phase of infection. However PML is not the only component of PML NBs with activity against HSV1. In the absence of PML, former NB components Sp100 and Daxx still move to form foci in association with incoming viral genomes (Everett *et al.*, 2006) and co-depletion of PML and Sp100 additively restores the growth of an ICP0 mutant to near normal levels (Everett *et al.*, 2008a).

The mechanism whereby ICP0 overcomes the IFN response is still unclear, as there are contrasting reports whether it can mitigate IFN signalling pathways and inhibit the expression of ISGs. Initial reports found that ICP0 inhibited the activity of both IRF3 and IRF7, leading to a decrease in the expression of ISGs (Lin *et al.*, 2004). However, Everett and colleagues subsequently found that the growth of ICP0-deficient mutants was not improved in IRF3-deficient HFs (Everett *et al.*, 2008b), and that ICP0 did not interfere with the IRF3-dependent induction of ISGs (Everett and Orr, 2009).

Despite the clear positive effect on HSV1 replication of removing PML, overexpressing PML isoforms III, IV or VI had no effect on virus replication (Tavalai *et al.*, 2008). However, in a recent study using a cell system in which endogenous PML is removed, so permitting growth of an HSV1 mutant unable to degrade PML, reconstitution with either PML I or PML II but not with any of PML III-VI partially restored inhibition of that mutant (Cuchet *et al.*, 2011). Thus PML I and II are each inhibitory to HSV1 replication while other isoforms are not. This result serves to emphasize the importance of making a full comparison of PML isoforms.

One further area where studies of HSV have been informative concerns changes in the balance between PML isoforms. HSV1 infection increase the proportion of cytoplasmic PML protein within the PML population (McNally *et al.*, 2008). A novel cytoplasmic variant, termed PML Ib, which lack exons 5 and 6 and possesses a stop codon in exon 7a, was demonstrated to exert a strong antiviral activity against HSV1, suggesting this change in the population of PML proteins might also be part of a host antiviral response. Herpes simplex type 2 (HSV2) infection also alters the relative ratio of PML isoforms, in this case increasing the ratio of PML V to PML II transcripts (Nojima *et al.*, 2009). The viral protein ICP27 inhibits splicing from the 3′ splice site of exon 7a to cause retention of the following intron, so converting a PML II mRNA into a PML V mRNA. The reason why the virus promotes this isoform switching remains unclear, as the authors found that PML II, when overexpressed, actually enhanced viral replication (Nojima *et al.*, 2009). The authors suggested that this may be a mechanism employed by the virus to favour a persistent infection. However, in the light of the more recent finding that PML II is inhibitory to HSV1 infection (Cuchet *et al.*, 2011), this may need to be reconsidered; HSV1 and HSV2 are closely related and are thought to replicate and to establish latency in very similar ways.

Whilst both PML and Sp100 contribute to antiviral activity against HSV1, the picture for varicella-zoster virus (VZV), another human alpha-herpesvirus, is different. Knock-down of PML or Daxx, but not Sp100, shortens the time course of replication and enhances spread in cell monolayers (Kyratsous and Silverstein, 2009). Interestingly, in the cell type studied (human melanoma cell line, not the physiologically relevant cell type) the VZV protein analogous to ICP0 does not alter PML but does cause some loss of Sp100. Either VZV has not evolved a PML targeting function despite this protein having some inhibitory effect on the virus, or else the ability to counter PML may be cell type-restricted. Very recently, a mechanism has been demonstrated for PML activity against VZV. In infected neurons and skin cells (two natural targets of VZV), PML shells arise in the nucleus around newly generated VZV capsids, formed in particular of PML IV; this limits the formation of infectious particles (Reichelt *et al.*, 2011). These PML IV cages could co-accumulate polyQ protein aggregates, demonstrating a relationship between this PML response to virus infection and the response to aggregated nuclear proteins.

In the case of human cytomegalovirus (HCMV), the repressive effect of PML NBs on the virus begins at the earliest stage of gene expression – immediate early gene expression. This is overcome by pp71, a tegument protein found in the incoming particle, which migrates to the nucleus and localizes to PML NBs via a direct interaction with the NB component Daxx, which is then degraded (Cantrell and Bresnahan, 2006; Saffert and Kalejta, 2006). PML NBs themselves are then disrupted through the action of a viral immediate early protein IE72 (Xu *et al.*, 2001). As expected from these observations, PML knockdown can mitigate the growth defect of HCMV deleted for this immediate-early function (Tavalai *et al.*, 2006).

Epstein–Barr virus (EBV) operates a little differently but with a similar result. Firstly, during the lytic cycle its principal transcriptional activator BZLF 1 becomes sumoylated and is thought to out-compete PML for SUMO binding (Adamson and Kenney, 2001); unsumoylated PML therefore accumulates and, because PML NB structure depends on PML sumoylation, the NBs disintegrate into the nucleoplasm. This activity of BZLF-1 correlates with its ability to activate viral gene transcription, thus further suggesting that PML NBs act to repress viral gene expression (Adamson and Kenney, 2001). Secondly, the EBNA-1 protein, which is expressed in EBV latently infected cells, independently disrupts PML nuclear bodies and causes a decrease in

PML protein levels, dependent on the cellular ubiquitin-specific protease USP-7 (Sivachandran *et al.*, 2008); as the basis for this activity, EBNA-1 was found to preferentially bind a particular isoform of PML, in this case PML IV.

RNA viruses

The range of viruses involved with PML NBs extends beyond the nucleus-replicating DNA viruses to include several RNA viruses that replicate in the cytoplasm. The Z proteins of two arenaviruses, lymphocytic choriomeningitis virus (LCMV) and Lassa fever virus, both target PML NBs through a direct interaction with PML and subsequently cause PML re-localization to the cytoplasm (Borden *et al.*, 1998; Kentsis *et al.*, 2001). Once there, PML directly interacts with the translation initiation factor eIF4E and reduces its affinity for the 5′ cap of cellular mRNA (Kentsis *et al.*, 2001), thus contributing to translational shutoff during these viral infections. Redistribution of PML and Sp100 from PML NBs to the cytoplasm is also observed during infection with respiratory syncytial virus, a paramyxovirus (Brasier *et al.*, 2004).

The M1, NS1 and NS2 proteins of the orthomyxovirus influenza A virus all migrate to the nucleus and associate with PML NBs during infection, though the reason for this remains unclear (Sato *et al.*, 2003). Influenza A virus replication is significantly repressed by overexpression of PML III, PML IV or PML VI, and correspondingly, depletion of total PML content leads to an enhancement in viral propagation (Chelbi-Alix *et al.*, 1998; Iki *et al.*, 2005). This suggests that PML protein exerts an antiviral activity against influenza A. However, there is evidence to suggest that the effectiveness of this antiviral activity varies depending on the subtype of influenza A virus (Li *et al.*, 2009).

The overexpression of PML III has also been shown to exert an inhibitory effect on some other RNA virus infections, though often a comparison with a full range of PML isoforms was not done. Overexpression of PML III decreases viral gene expression during infection by the retrovirus human (now primate) foamy virus (HFV) by binding and sequestering the viral transactivator Tas (Regad *et al.*, 2001). Also, like several other viruses already discussed, IFN-mediated inhibition of HFV is PML-dependent (Regad *et al.*, 2001), though by this stage in its replication cycle, HFV should be considered as a nucleus-replicating DNA virus rather than an RNA virus. The rhabdovirus, vesicular stomatitis virus (VSV), also shows markedly reduced viral titres in cells overexpressing PML III (Chelbi-Alix *et al.*, 1998). However, PML III overexpression has no effect on the replication of rabies virus, another rhabdovirus, or the picornavirus encephalomyocarditis virus (EMCV) (Chelbi-Alix *et al.*, 1998, 2006). The distinction between rabies virus and VSV may be that the rabies virus P protein, which is responsible for overcoming the IFN response during rabies virus infection, interacts with PML NBs and may therefore negate the effect of exogenous PML III on infection (Chelbi-Alix *et al.*, 2006).

PML protein is implicated in the host antiviral response to poliovirus (PV), another picornavirus (Pampin *et al.*, 2006). Upon PV infection, additional PML recruitment to NBs, then PML phosphorylation by ERK and subsequent sumoylation leads rapidly to p53 recruitment to NBs and to its activation; this opposes infection. PV counters this by initiating the degradation of p53 in a manner that is partly dependent on MDM2 (Pampin *et al.*, 2006). Given the lack of effect of PML III on EMCV infection, how can this role of PML during PV infection be rationalized? Perhaps there are genuine differences between the host responses to these two picornaviruses meaning that PML is relevant only to PV infection. Alternatively it may be that PML III is simply not the correct isoform to inhibit EMCV. Although EMCV is pathogenic in a range of mammals, its natural host is believed to mice or rats; genome sequence alignments suggest that, unlike other mammals, PML III cannot be encoded by these species (KNL, unpublished data).

Relating PML/PML NB functions to virus infection: conclusions and perspectives

At the outset of this review, we set out three broad categories of virus–host interaction: those set up by the virus to harness host systems for virus replication; those set up by the host to inhibit virus replication; and those set up by the virus to overcome such inhibitory host responses. Based on the evidence to date, is it possible to say into which of these categories virus interactions with PML and PML NBs fall?

Fig. 13.3 makes an attempt to summarize the data discussed in this chapter concerning viruses and PML. When considering these data, one of the most striking things is the great diversity of viruses that have been shown to make such interactions. Notably, this includes not just the nucleus-replicating DNA viruses but also a variety of RNA viruses with cytoplasmic replication cycles. Among this collection of viruses, there are very different mechanisms of replication and, given the great differences in their genome sizes, considerable differences in the extent of demands on host factors to carry out essential replication processes. Thus it would seem inherently implausible that all these viruses would interact with PML NBs to liberate from them a common activity that they each needed from the host. Such a proposition would also predict that the absence of PML NBs, or of selected components, would be inhibitory to virus replication. In fact, there is considerable evidence that the loss of such components, far from inhibiting infection, actually enhances virus replication; this has been demonstrated with varying degrees of rigour for many of the viruses that have been examined in this context. However, this does not exclude the possibility that some individual PML NB components yet to be tested in a knock-down scenario – perhaps specific to a type of virus – might nonetheless make a positive contribution to the outcome of infection.

Among the nucleus-replicating DNA viruses that have been tested, all appear to deposit their genomes in foci close to PML NBs at the start of infection and to establish replication centres

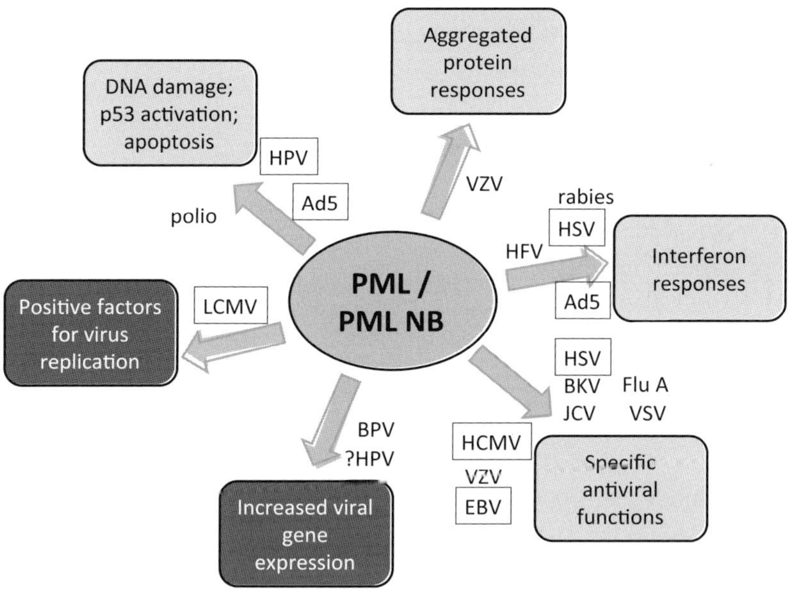

Figure 13.3 Overview of the activities of PML/PML NBs during virus infection. The activity areas of PML and/or PML NBs that have been demonstrated are represented in the rectangles; light grey, virus inhibitory activities; dark grey, virus enhancing activities. Virus acronyms associated with each activity area are those where evidence for the involvement of that activity with that virus has been discussed in the text. Absence of an acronym in any area does not indicate that this area is irrelevant to the virus concerned, simply that it has not been tested. Virus acronyms that are boxed are those where one or more viral functions have been described that reorganize, disrupt or degrade PML and/or PML NB. The distinction between 'Interferon Responses' and 'Specific Antiviral Functions' is not well delineated; some of the activities currently defined as specific may, with further investigation, be linked to the interferon response. For virus acronyms, see text.

there. Whilst there have been suggestions that this location favours virus replication, the observation that in HSV1 infection PML-NBs deconstruct and then reform in the part of the nucleus suggests that any such benefit is secondary to an initial mobilization of PML NBs by the host in response to infection. Certainly, the failure of global PML knock-down or transgenic PML removal to inhibit replication by any virus tested indicates that the juxtaposition of viral replication centres with PML NBs is not essential for productive infection. But viruses can evolve more rapidly than their hosts. What may have begun as a host response to infection may nonetheless have come to be used by the virus to its benefit. The fact that active genes, as has been particularly shown for genes activated by IFN, move into close proximity to PML NBs suggests these locations will accumulate components of the gene expression pathway that might make them favourable sites for viral gene expression during infection.

If PML NBs are mobilized by the host in response to infection, it might be predicted that these structures or their components have antiviral activity. This has now been demonstrated for several viruses. Importantly, even though only a few specific components have been tested, the relevant factors are not the same for each virus. For example, while PML removal allows the replication of both HSV1 and Ad5 mutants that lack PML NB disrupting functions, Sp100 apparently only inhibits HSV1 infection while Daxx affects Ad5 and HCMV growth. Thus, there is no single unified mechanism by which PML NBs or their components operate to combat virus infection. That said, the evidence also suggests that the antiviral effect associated with PML/PML NBs can have multiple additive elements so it remains possible that an overarching common mechanism of antiviral activity is complemented by one or more virus-specific mechanisms to give the overall effect on virus growth that is observed.

There is a great diversity in PML isoforms and, so far, the full range of these has often not been tested for specific effects on individual viruses. However, even from the incomplete data available it is possible to say that there is no one PML isoform that is consistently antiviral. The fact that different isoform(s) can have antiviral activity against specific viruses suggests that PML does more than simply form the architecture of a structure (PML NB) that is antiviral through the activity of some other component. In this context, the demonstration that PML has SUMO ligase activity immediately opens up the possibility that selective antiviral effects might emerge from each PML isoform having a distinct sumoylation substrate specificity. The notion that sumoylation of specific targets might form part of the innate host response to infection is supported by the recent observation that a virulence determinant of the bacterial pathogen *Listeria monocytogenes* acts to reduce global sumoylation by targeting the E2 conjugating enzyme Ubc9 for degradation (Ribet *et al.*, 2010).

The best studied elements of the initial cellular response to infection are the stimulation of IFN production and its downstream effects, and there is considerable evidence to link PML NBs with these events during virus infection. Both the induction of an IFN response and the resulting antiviral state have been found to depend, to an extent, on the integrity of PML or PML NBs. In several cases, viral proteins that are known to target PML NBs are also implicated in the ability of that virus to overcome an IFN response. IFN responses are triggered by the cell's recognition of specific molecular patterns (PAMPs) by cytoplasmic or membrane-associated sensors in the very early stages of infection. Whilst the particular elements recognized differ between classes of virus, all viruses are likely to trigger such responses unless viral functions are deployed to counter this induction; the extent to which a virus can counter the IFN response in a particular host species is often a determinant of virulence (Randall and Goodbourn, 2008). Thus, an involvement of PML/PML NBs in the IFN response would represent a unifying reason why so many different types of virus have PML targeting functions.

Another potential antiviral activity associated with PML NBs is stress or DNA damage-induced apoptosis. Many viruses induce such responses during the course of infection and it is generally thought that these are deleterious to the outcome of infection, meaning that there has been selective pressure for viruses to evolve functions that inhibit the induction of apoptosis. One key mediator of apoptosis induction is p53, the activation of which as a transcription factor has been associated with its recruitment to PML NBs. For adenoviruses, polyomaviruses and papillomaviruses, there is a striking conjunction between the viral proteins that are known to disrupt p53 function and interaction with PML/PML NBs. For Ad5 E1B 55K, it has been shown that PML is involved in the blocking of p53 activity by these viral functions. So perhaps the involvement of PML NBs in the induction of apoptosis is a further reason why many viruses target these structures.

One final feature of the PML/PML NB response to virus infection is a link with the response of these cell components to protein aggregates. The demonstration that VZV capsids can be found wrapped in a shell of PML protein, similar to the interaction of PML with polyQ aggregates, and that this inhibits production of infectious progeny suggests this might be a more general mechanism of PML antiviral activity, which some viruses may counter more successfully than others. Viral replication centres and newly assembling progeny might very plausibly stimulate an aggregated protein response in the cell. Possibly even an incoming virus particle, or a partially uncoated particle or virus core, might be enough to stimulate such a response. The signal that starts the mobilization of PML NBs during HSV-1 infection has not yet been defined but the process clearly precedes all viral gene expression so must be triggered by a feature of the incoming particle.

In conclusion, there is now sufficient evidence to say beyond doubt that PML NBs and/or some of their components have antiviral activity, and that many if not all virus interactions with these components have, at their heart, the purpose of counteracting this activity. Although elements of this antiviral activity may be related to and mediated by the IFN response, at least in some cases it appears that this does not fully account for the activity observed. There remains no firm conclusion as to what triggers the initial response of the cell that mobilizes these antiviral activities. Possibly, all aspects of the antiviral activity are initiated in some way through the detection of PAMPs that are known to activate the IFN response. Alternatively, any non-IFN component of PML/PML NB antiviral activity might be separately triggered, perhaps by a nuclear aggregated protein response or an as yet unidentified pathway. Despite the enormous amount of data accumulated on PML and PML NB function, there is much still to learn about their activities and roles.

Acknowledgements

Parts of this work have been modified from the introduction to JW's PhD thesis (University of Warwick, 2010). JW was supported during his studies by a Doctoral Training Grant from the UK Biotechnology and Biological Sciences Research Council.

References

Adamson, A.L., and Kenney, S. (2001). Epstein–Barr virus immediate-early protein BZLF1 is SUMO-1 modified and disrupts promyelocytic leukemia bodies. J. Virol. *75*, 2388–2399.

Alsheich-Bartok, O., Haupt, S., Alkalay-Snir, I., Saito, S., Appella, E., and Haupt, Y. (2008). PML enhances the regulation of p53 by CK1 in response to DNA damage. Oncogene *27*, 3653–3661.

Araujo, F.D., Stracker, T.H., Carson, C.T., Lee, D.V., and Weitzman, M.D. (2005). Adenovirus type 5 E4orf3 protein targets the Mre11 complex to cytoplasmic aggresomes. J. Virol. *79*, 11382–11391.

Barr, S.M., Leung, C.G., Chang, E.E., and Cimprich, K.A. (2003). ATR kinase activity regulates the intranuclear translocation of ATR and RPA following ionizing radiation. Curr. Biol. *13*, 1047–1051.

Becker, K.A., Florin, L., Sapp, C., Maul, G.G., and Sapp, M. (2004). Nuclear localization but not PML protein is required for incorporation of the papillomavirus minor capsid protein L2 into virus-like particles. J. Virol. *78*, 1121–1128.

Beech, S.J., Lethbridge, K.J., Killick, N., McGlincy, N., and Leppard, K.N. (2005). Isoforms of the promyelocytic leukemia protein differ in their effects on ND10 organization. Exp. Cell Res. *307*, 109–117.

Bellodi, C., Kindle, K., Bernassola, F., Cossarizza, A., Dinsdale, D., Melino, G., Heery, D., and Salomoni, P. (2006). A cytoplasmic PML mutant inhibits p53 function. Cell Cycle *5*, 2688–2692.

Bernardi, R., and Pandolfi, P.P. (2007). Structure, dynamics and functions of promyelocytic leukaemia nuclear bodies. Nat. Rev. Mol. Cell Biol. *8*, 1006–1016.

Bernardi, R., Scaglioni, P.P., Bergmann, S., Horn, H.F., Vousden, K.H., and Pandolfi, P.P. (2004). PML regulates p53 stability by sequestering Mdm2 to the nucleolus. Nat. Cell Biol. *6*, 665–672.

Best, J.L., Ganiatsas, S., Agarwal, S., Changou, A., Salomoni, P., Shirihai, O., Meluh, P.B., Pandolfi, P.P., and Zon, L.I. (2002). SUMO-1 protease-1 regulates gene transcription through PML. Mol. Cell *10*, 843–855.

Bischof, O., Kirsh, O., Pearson, M., Itahana, K., Pelicci, P.G., and Dejean, A. (2002). Deconstructing PML-induced premature senescence. EMBO J. *21*, 3358–3369.

Bischof, O., Nacerddine, K., and Dejean, A. (2005). Human papillomavirus oncoprotein E7 targets the promyelocytic leukemia protein and circumvents cellular senescence via the Rb and p53 tumor suppressor pathways. Mol. Cell. Biol. *25*, 1013–1024.

Bishop, C.L., Ramalho, M., Nadkarni, N., Kong, W.M., Higgins, C.F., and Krauzewicz, N. (2006). Role for centromeric heterochromatin and PML nuclear bodies in the cellular response to foreign DNA. Mol. Cell. Biol. *26*, 2583–2594.

Boe, S.O., Haave, M., Jul-Larsen, A., Grudic, A., Bjerkvig, R., and Lonning, P.E. (2006). Promyelocytic leukemia nuclear bodies are predetermined processing sites for damaged DNA. J. Cell Sci. *119*, 3284–3295.

Boisvert, F.M., Hendzel, M.J., and Bazett-Jones, D.P. (2000). Promyelocytic leukemia (PML) nuclear bodies are protein structures that do not accumulate RNA. J. Cell Biol. *148*, 283–292.

Borden, K.L.B. (2002). Pondering the promyelocytic leukemia protein (PML) puzzle: Possible functions for PML nuclear bodies. Mol. Cell. Biol. *22*, 5259–5269.

Borden, K.L.B. (2008). Pondering the puzzle of PML (promyelocytic leukemia) nuclear bodies: Can we fit the pieces together using an RNA regulon? Biochimica et Biophysica Acta-Mol. Cell Research *1783*, 2145–2154.

Borden, K.L.B., and Culjkovic, B. (2009). Perspectives in PML: a unifying framework for PML function. Front. Biosci. *14*, 497–509.

Borden, K.L.B., and Freemont, P.S. (1996). The RING finger domain: a recent example of a sequence-structure family. Curr. Opin. Struct. Biol. *6*, 395–401.

Borden, K.L.B., Dwyer, E.J.C., and Salvato, M.S. (1998). An arenavirus RING (zinc-binding) protein binds the oncoprotein promyelocyte leukemia protein (PML) and relocates PML nuclear bodies to the cytoplasm. J. Virol. *72*, 758–766.

Boutell, C., Sadis, S., and Everett, R.D. (2002). Herpes simplex virus type 1 immediate-early protein ICP0 and its isolated RING finger domain act as ubiquitin E3 ligases *in vitro*. J. Virol. *76*, 841–850.

Brasier, A.R., Spratt, H., Wu, Z., Boldogh, I., Zhang, Y., Garofalo, R.P., Casola, A., Pashmi, J., Haag, A., Luxon, B., *et al.* (2004). Nuclear heat shock response and novel nuclear domain 10 reorganization in respiratory syncytial virus-infected A549 cells identified by high resolution two-dimensional gel electrophoresis. J. Virol. *78*, 11461–11476.

Buschbeck, M., Uribesalgo, I., Ledl, A., Gutierrez, A., Minucci, S., Muller, S., and Di Croce, L. (2007). PML4 induces differentiation by Myc destabilization. Oncogene *26*, 3415–3422.

Cantrell, S.R., and Bresnahan, W.A. (2006). Human cytomegalovirus (HCMV) UL82 gene product (pp71) relieves hDaxx-mediated repression of HCMV replication. J. Virol. *80*, 6188–6191.

Cao, T.Y., Borden, K.L.B., Freemont, P.S., and Etkin, L.D. (1997). Involvement of the rfp tripartite motif in protein–protein interactions and subcellular distribution. J. Cell Sci. *110*, 1563–1571.

Cao, T.Y., Duprez, E., Borden, K.L.B., Freemont, P.S., and Etkin, L.D. (1998). Ret finger protein is a normal component of PML nuclear bodies and interacts directly with PML. J. Cell Sci. *111*, 1319–1329.

Carson, C.T., Orazio, N.I., Lee, D.V., Suh, J.H., Bekker-Jensen, S., Araujo, F.D., Lakdawala, S.S., Lilley, C.E., Bartek, J., Lukas, J., *et al.* (2009). Mislocalization of the MRN complex prevents ATR signalling during adenovirus infection. EMBO J. *28*, 652–662.

Carthagena, L., Bergamaschi, A., Luna, J.M., David, A., Uchil, P.D., Margottin-Goguet, F., Mothes, W., Hazan, U., Transy, C., Pancino, G., *et al.* (2009). Human TRIM gene expression in response to interferons. PLoS One *4*, e4894.

Carvalho, T., Seeler, J.S., Ohman, K., Jordan, P., Pettersson, U., Akusjarvi, G., Carmo-Fonseca, M., and Dejean, A. (1995). Targeting of adenovirus E1A and E4-ORF3 proteins to nuclear matrix-associated PML bodies. J. Cell Biol. *131*, 45–56.

Chee, A.V., Lopez, P., Pandolfi, P.P., and Roizman, B. (2003). Promyelocytic leukemia protein mediates interferon-based anti- herpes simplex virus 1 effects. J. Virol. *77*, 7101–7105.

Chelbi-Alix, M.K., and de The, H. (1999). Herpes virus induced proteasome-dependent degradation of the nuclear bodies-associated PML and Sp100 proteins. Oncogene *18*, 935–941.

Chelbi-Alix, M.K., Quignon, F., Pelicano, L., Koken, M.H.M., and De The, H. (1998). Resistance to virus infection conferred by the interferon-induced promyelocytic leukemia protein. J. Virol. *72*, 1043–1051.

Chelbi-Alix, M.K., Vidy, A., El Bougrini, J., and Blondel, D. (2006). Rabies viral mechanisms to escape the IFN system: The viral protein P interferes with IRF-3, Stat1, and PML nuclear bodies. J. Interferon Cytokine Res. *26*, 271–280.

Chu, Y., and Yang, X. (2011). SUMO E3 ligase activity of TRIM proteins. Oncogene *30*, 1108–1116.

Cohen, N., Sharma, M., Kentsis, A., Perez, J.M., Strudwick, S., and Borden, K.L. B. (2001). PML RING suppresses oncogenic transformation by reducing the affinity of eIF4E for mRNA. EMBO J. *20*, 4547–4559.

Condemine, W., Takahashi, Y., Zhu, J., Puvion-Dutilleul, F., Guegan, S., Janin, A., and de The, H. (2006). Characterization of endogenous human promyelocytic leukemia isoforms. Cancer Res. *66*, 6192–6198.

Cuchet, D., Sykes, A., Nicolas, A., Orr, A., Murray, J., Sirma, H., Heeren, J., Bartelt, A., and Everett, R.D. (2011). PML isoforms I and II participate in PML-dependent restriction of HSV-1 replication. J. Cell Sci. *124*, 280–291.

Culjkovic, B., Topisirovic, I., Skrabanek, L., Ruiz-Gutierrez, M., and Borden, K.L. B. (2006). eIF4E is a central node of an RNA regulon that governs cellular proliferation. J. Cell Biol. *175*, 415–426.

Culjkovic, B., Tan, K., Orolicki, S., Amri, A., Meloche, S., and Borden, K.L. B. (2008). The eIF4E RNA regulon promotes the Akt signalling pathway. J. Cell Biol. *181*, 51–63.

Daniel, M.T., Koken, M., Romagne, O., Barbey, S., Bazarbachi, A., Stadler, M., Guillemin, M.C., Degos, L., Chomienne, C., and de The, H. (1993). PML protein expression in hematopoietic and acute promyelocytic leukemia cells. Blood *82*, 1858–1867.

Day, P.M., Roden, R.B.S., Lowy, D.R., and Schiller, J.T. (1998). The papillomavirus minor capsid protein, L2, induces localization of the major capsid protein, L1, and the viral transcription/replication protein, E2, to PML oncogenic domains. J. Virol. *72*, 142–150.

Day, P.M., Baker, C.C., Lowy, D.R., and Schiller, J.T. (2004). Establishment of papillomavirus infection is enhanced by promyelocytic leukemia protein (PML) expression. Proc. Natl. Acad. Sci. U.S.A. *101*, 14252–14257.

Dellaire, G., and Bazett-Jones, D.P. (2004). PML nuclear bodies: dynamic sensors of DNA damage and cellular stress. BioEssays *26*, 963–977.

Dellaire, G., Ching, R.W., Ahmed, K., Jalali, F., Tse, K.C.K., Bristow, R.G., and Bazett-Jones, D.P. (2006a). Promyelocytic leukemia nuclear bodies behave as DNA damage sensors whose response to DNA double-strand breaks is regulated by NBS1 and the kinases ATM, Chk2, and ATR. J. Cell Biol. *175*, 55–66.

Dellaire, G., Ching, R.W., Dehghani, H., Ren, Y., and Bazett-Jones, D.P. (2006b). The number of PML nuclear bodies increases in early S phase by a fission mechanism. J. Cell Sci. *119*, 1026–1033.

Dellaire, G., Eskiw, C.H., Dehghani, H., Ching, R.W., and Bazett-Jones, D.P. (2006c). Mitotic accumulations of PML protein contribute to the re-establishment of PML nuclear bodies in G1. J. Cell Sci. *119*, 1034–1042.

Doucas, V., Ishov, A.M., Romo, A., Juguilon, H., Weitzman, M.D., Evans, R.M., and Maul, G.G. (1996). Adenovirus replication is coupled with the dynamic properties of the PML nuclear structure. Genes Dev. *10*, 196–207.

Duprez, E., Saurin, A.J., Desterro, J.M., Lallemand-Breitenbach, V., Howe, K., Boddy, M.N., Solomon, E., de The, H., Hay, R.T., and Freemont, P.S. (1999). SUMO-1 modification of the acute promyelocytic leukaemia protein PML: implications for nuclear localisation. J. Cell Sci. *112*, 381–393.

Endter, C., Kzhyshkowska, J., Stauber, R., and Dobner, T. (2001). SUMO-1 modification required for transformation by adenovirus type 5 early region 1B 55-kDa oncoprotein. Proc. Natl. Acad. Sci. U.S.A. *98*, 11312–11317.

Eskiw, C.H., Dellaire, G., Mymryk, J.S., and Bazett-Jones, D.R. (2003). Size, position and dynamic behavior of PML nuclear bodies following cell stress as a paradigm for supramolecular trafficking and assembly. J. Cell Sci. *116*, 4455–4466.

Evans, J.D., and Hearing, P. (2003). Distinct roles of the adenovirus E4 ORF3 protein in viral DNA replication and inhibition of genome concatenation. J. Virol. *77*, 5295–5304.

Evans, J.D., and Hearing, P. (2005). Relocalization of the Mre11–Rad50-Nbs1 complex by the adenovirus E4 ORF3 protein is required for viral replication. J. Virol. *79*, 6207–6215.

Everett, R., Sourvinos, G., and Orr, A. (2003). Recruitment of herpes simplex virus type 1 transcriptional regulatory protein ICP4 into foci juxtaposed to ND10 in live, infected cells. J. Virol. *77*, 3680–3689.

Everett, R.D., and Chelbi-Alix, M.K. (2007). PML and PML nuclear bodies: Implications in antiviral defence. Biochimie *89*, 819–830.

Everett, R.D., and Murray, J. (2005). ND10 components relocate to sites associated with herpes simplex virus type 1 nucleoprotein complexes during virus infection. J. Virol. *79*, 5078–5089.

Everett, R.D., and Orr, A. (2009). Herpes simplex virus type 1 regulatory protein ICP0 aids infection in cells with a preinduced interferon response but does not impede interferon-induced gene induction. J. Virol. *83*, 4978–4983.

Everett, R.D., Freemont, P., Saitoh, H., Dasso, M., Orr, A., Kathoria, M., and Parkinson, J. (1998). The disruption of ND10 during herpes simplex virus infection correlates with the Vmw110- and proteasome-dependent loss of several PML isoforms. J. Virol. *72*, 6581–6591.

Everett, R.D., Lamonte, P., Sternsdorf, T., van Driel, R., and Orr, A. (1999). Cell cycle regulation of PML modification and ND10 composition. J. Cell Sci. *112*, 4581–4588.

Everett, R.D., Rechter, S., Papior, P., Tavalai, N., Stamminger, T., and Orr, A. (2006). PML contributes to a cellular mechanism of repression of herpes simplex virus type 1 infection that is inactivated by ICP0. J. Virol. *80*, 7995–8005.

Everett, R.D., Parada, C., Gripon, P., Sirma, H., and Orr, A. (2008a). Replication of ICP0-Null mutant herpes simplex virus type 1 is restricted by both PML and Sp100. J. Virol. *82*, 2661–2672.

Everett, R.D., Young, D.F., Randall, R.E., and Orr, A. (2008b). STAT-1- and IRF-3-dependent pathways are not essential for repression of ICP0-null mutant herpes simplex virus type 1 in human fibroblasts. J. Virol. *82*, 8871–8881.

Fabunmi, R.P., Wigley, W.C., Thomas, P.J., and DeMartino, G.N. (2001). Interferon gamma regulates accumulation of the proteasome activator PA28 and immunoproteasomes at nuclear PML bodies. J. Cell Sci. *114*, 29–36.

Florin, L., Schafer, F., Sotlar, K., Streeck, R.E., and Sapp, M. (2002). Reorganization of nuclear domain 10 induced by papillomavirus capsid protein L2. Virology *295*, 97–107.

Fogal, V., Gostissa, M., Sandy, P., Zacchi, P., Sternsdorf, T., Jensen, K., Pandolfi, P.P., Will, H., Schneider, C., and Del Sal, G. (2000). Regulation of p53 activity in nuclear bodies by a specific PML isoform. EMBO J. *19*, 6185–6195.

Freemont, P.S. (2000). Ubiquitination: RING for destruction? Curr. Biol. *10*, R84-R87.

Fu, C.H., Ahmed, K., Ding, H.S., Ding, X., Lan, J.P., Yang, Z.H., Miao, Y., Zhu, Y.Y., Shi, Y.Y., Zhu, J.D., *et al.* (2005). Stabilization of PML nuclear localization by conjugation and oligomerization of SUMO-3. Oncogene *24*, 5401–5413.

Galisson, F., Mahrouche, L., Courcelles, M., Bonneil, E., Meloche, S., Chelbi-Alix, M.K., and Thibault, P. (2010). A novel proteomics approach to identify SUMOylated proteins and their modification sites in human cells. Mol. Cell Proteomics *10*, mcp. M110.004796.

Gasparovic, M.L., Maginnis, M.S., O'Hara, B.A., Dugan, A.S., and Atwood, W.J. (2009). Modulation of PML protein expression regulates JCV infection. Virology *390*, 279–288.

Geoffroy, M.C., and Chelbi-Alix, M.K. (2011). Role of Promyelocytic Leukemia Protein in Host Antiviral Defense. J. Interferon. Cytokine Res. *31*, 145–158.

Gialitakis, M., Arampatzi, P., Makatounakis, T., and Papamatheakis, J. (2010). Gamma Interferon-Dependent Transcriptional Memory via Relocalization of a Gene Locus to PML Nuclear Bodies. Mol. Cell. Biol. *30*, 2046–2056.

Goddard, A.D., Borrow, J., Freemont, P.S., and Solomon, E. (1991). Characterization of a zinc finger gene disrupted by the t (15, 17) in acute promyelocytic leukemia. Science *254*, 1371–1374.

Gong, L., and Yeh, E.T. H. (2006). Characterization of a family of nucleolar SUMO-specific proteases with preference for SUMO-2 or SUMO-3. J. Biol. Chem. *281*, 15869–15877.

Gresko, E., Ritterhoff, S., Sevilla-Perez, J., Roscic, A., Frobius, K., Kotevic, I., Vichalkovski, A., Hess, D., Hemmings, B.A., and Schmitz, M.L. (2009). PML tumor suppressor is regulated by HIPK2-mediated phosphorylation in response to DNA damage. Oncogene *28*, 698–708.

Guccione, E., Lethbridge, K.J., Killick, N., Leppard, K.N., and Banks, L. (2004). HPV E6 proteins interact with specific PML isoforms and allow distinctions to be made between different POD structures. Oncogene *23*, 4662–4672.

Guldner, H.H., Szostecki, C., Grotzinger, T., and Will, H. (1992). IFN enhances expression of Sp100, an autoantigen in primary biliary cirrhosis. J. Immunol. *149*, 4067–4073.

Gurrieri, C., Capodieci, P., Bernardi, R., Scaglioni, P.P., Nafa, K., Rush, L.J., Verbel, D.A., Cordon-Cardo, C., and Pandolfi, P.P. (2004). Loss of the tumour suppressor PML in human cancers of multiple histologic origins. J. Natl. Cancer Inst. *96*, 269–279.

Hayakawa, F., and Privalsky, M.L. (2004). Phosphorylation of PML by mitogen-activated protein kinases plays a key role in arsenic trioxide-mediated apoptosis. Cancer Cell *5*, 389–401.

Hayakawa, F., Abe, A., Kitabayashi, I., Pandolfi, P.P., and Naoe, T. (2008). Acetylation of PML is involved in histone deacetylase inhibitor-mediated apoptosis. J. Biol. Chem. *283*, 24420–24425.

Hoppe, A., Beech, S.J., Dimmock, J., and Leppard, K.N. (2006). Interaction of the adenovirus type 5 E4 Orf3 protein with promyelocytic leukemia protein isoform II is required for ND10 disruption. J. Virol. *80*, 3042–3049.

Hosel, M., Schroer, J., Webb, D., Jaroshevskaja, E., and Doerfler, W. (2001). Cellular and early viral factors in the interaction of adenovirus type 12 with hamster cells: the abortive response. Virus Res. *81*, 1–16.

Iki, S., Yokota, S., Okabayashi, T., Yokosawa, N., Nagata, K., and Fujii, N. (2005). Serum-dependent expression of promyelocytic leukemia protein suppresses propagation of influenza virus. Virology *343*, 106–115.

Ishov, A.M., and Maul, G.G. (1996). The periphery of nuclear domain 10 (ND10) as site of DNA virus deposition. J. Cell Biol. *134*, 815–826.

Ishov, A.M., Sotnikov, A.G., Negorev, D., Vladimirova, O.V., Neff, N., Kamitani, T., Yeh, E.T.H., Strauss, J.F., and Maul, G.G. (1999). PML is critical for ND10 formation and recruits the PML- interacting protein Daxx to this nuclear structure when modified by SUMO-1. J. Cell Biol. *147*, 221–233.

Ito, K., Bernardi, R., Morotti, A., Matsuoka, S., Saglio, G., Ikeda, Y., Rosenblatt, J., Avigan, D.E., Teruya-Feldstein, J., and Pandolfi, P.P. (2008). PML targeting eradicates quiescent leukaemia-initiating cells. Nature *453*, 1072–1078.

Janer, A., Martin, E., Muriel, M., Latouche, M., Fujigasaki, H., Ruberg, M., Brice, A.T., Rottier, Y., and Sittler, A. (2006). PML clastosomes prevent nuclear accumulation of mutant ataxin-7 and other polyglutamine proteins. J. Cell Biol. *174*, 65–76.

Janer, A., Werner, A., Takahashi-Fujigasaki, J., Daret, A., Fujigasaki, H., Takada, K., Duyckaerts, C., Brice, A., Dejean, A., and Sittler, A. (2010). SUMOylation attenuates the aggregation propensity and cellular toxicity of the polyglutamine expanded ataxin-7. Hum. Mol. Genet. *19*, 181–195.

Jensen, K., Shiels, C., and Freemont, P.S. (2001). PML protein isoforms and the RBCC/TRIM motif. Oncogene *20*, 7223–7233.

Jiang, M., Entezami, P., Gamez, M., Stamminger, T., and Imperiale, M.J. (2011). Functional reorganization of promyelocytic leukemia nuclear bodies during BK virus infection. mBio *2*, e00281.

Jiang, W.Q., Zhong, Z.H., Henson, J.D., and Reddel, R.R. (2007). Identification of candidate alternative lengthening of telomeres genes by methionine restriction and RNA interference. Oncogene *26*, 4635–4647.

Jul-Larsen, A., Visted, T., Karlsen, B.O., Rinaldo, C.H., Bjerkvig, R., Lonning, P.E., and Boe, S.O. (2004). PML-nuclear bodies accumulate DNA in response to polyomavirus BK and simian virus 40 replication. Exp. Cell Res. *298*, 58–73.

Kamei, H. (1997). Cystine starvation induces reversible large-body formation from nuclear bodies in T24 cells. Exp. Cell Res. *237*, 207–216.

Kamitani, T., Kito, K., Nguyen, H.P., Wada, H., Fukuda-Kamitani, T., and Yeh, E.T. H. (1998a). Identification of three major sentrinization sites in PML. J. Biol. Chem. *273*, 26675–26682.

Kamitani, T., Nguyen, H.P., Kito, K., FukudaKamitani, T., and Yeh, E.T. H. (1998b). Covalent modification of PML by the sentrin family of ubiquitin-like proteins. J. Biol. Chem. *273*, 3117–3120.

Karanam, B., Peng, S.W., Li, T., Buck, C., Day, P.M., and Roden, R.B. S. (2010). Papillomavirus Infection Requires gamma Secretase. J. Virol. *84*, 10661–10670.

Kentsis, A., Dwyer, E.C., Perez, J.M., Sharma, M., Chen, A., Pan, Z.Q., and Borden, K.L. B. (2001). The RING domains of the promyelocytic leukemia protein PML and the arenaviral protein Z repress translation by directly inhibiting translation initiation factor eIF4E. J. Mol. Biol. *312*, 609–623.

Kieback, E., and Muller, M. (2006). Factors influencing subcellular localization of the human papillomavirus L2 minor structural protein. Virology *345*, 199–208.

Kiesslich, A., von Mikecz, A., and Hemmerich, P. (2002). Cell cycle-dependent association of PML bodies with sites of active transcription in nuclei of mammalian cells. J. Struct. Biol. *140*, 167–179.

Kindsmuller, K., Groitl, P., Hartl, B., Blanchette, P., Hauber, J., and Dobner, T. (2007). Intranuclear targeting and nuclear export of the adenovirus E1B-55K protein are regulated by SUMO1 conjugation. Proc. Natl. Acad. Sci. U.S.A. *104*, 6684–6689.

Konig, C., Roth, J., and Dobbelstein, M. (1999). Adenovirus type 5 E4orf3 protein relieves p53 inhibition by E1B-55- kilodalton protein. J. Virol. *73*, 2253–2262.

Krieghoff-Henning, E., and Hofmann, T.G. (2008). Role of nuclear bodies in apoptosis signalling. Biochim. Biophys. Acta Mol. Cell Res. *1783*, 2185–2194.

Kumar P, P., Bischof, O., Purbey, P.K., Notani, D., Urlaub, H., Dejean, A., and Galande, S. (2007). Functional interaction between PML and SATB1 regulates chromatin-loop architecture and transcription of the MHC class I locus. Nat. Cell Biol. *9*, 45-U57.

Kyratsous, C.A., and Silverstein, S.J. (2009). Components of nuclear domain 10 bodies regulate Varicella-Zoster Virus replication. J. Virol. *83*, 4262–4274.

Lafarga, M., Berciano, M.T., Pena, E., Mayo, I., Castano, J.G., Bohmann, D., Rodrigues, J.P., Tavanez, J.P., and Carmo-Fonseca, M. (2002). Clastosome: A subtype of nuclear body enriched in 19S and 20S proteasomes, ubiquitin, and protein substrates of proteasome. Mol. Biol. Cell *13*, 2771–2782.

Lang, M., Jegou, T., Chung, I., Richter, K., Munch, S., Udvarhelyi, A., Cremer, C., Hemmerich, P., Engelhardt, J., Hell, S.W., *et al.* (2010). Three-dimensional organization of promyelocytic leukemia nuclear bodies. J. Cell Sci. *123*, 392–400.

Lavau, C., Marchio, A., Fagioli, M., Jansen, J., Falini, B., Lebon, P., Grosveld, F., Pandolfi, P.P., Pelicci, P.G., and Dejean, A. (1995). The acute promyelocytic leukemia-associated PML gene is induced by interferon. Oncogene *11*, 871–878.

Leib, D.A., Harrison, T.E., Laslo, K.M., Machalek, M.A., Moorman, N.J., and Virgin, H.W. (1999). Interferons regulate the phenotype of wild-type and mutant herpes simplex viruses *in vivo*. J. Exp. Med. *189*, 663–672.

Leppard, K.N. (1997). E4 gene function in adenovirus, adenovirus vector and adeno-associated virus infections. J. Gen. Virol. *78*, 2131–2138.

Leppard, K.N., and Dimmock, J. (2006). Virus interactions with PML nuclear bodies. In Viruses and the nucleus, J. Hiscox, ed. (J. Wiley, Chichester, UK), pp. 213–245.

Leppard, K.N., and Everett, R.D. (1999). The adenovirus type 5 E1b 55K and E4 Orf3 proteins associate in infected cells and affect ND10 components. J. Gen. Virol. *80*, 997–1008.

Leppard, K.N., Emmott, E., Cortese, M.S., and Rich, T. (2009). Adenovirus type 5 E4 Orf3 protein targets promyelocytic leukaemia (PML) protein nuclear domains for disruption via a sequence in PML isoform II that is predicted as a protein interaction site by bioinformatic analysis. J. Gen. Virol. *90*, 95–104.

Lethbridge, K.J., Scott, G.E., and Leppard, K.N. (2003). Nuclear matrix localization and SUMO-1 modification of adenovirus type 5 E1b 55K protein are controlled by E4 Orf6 protein. J. Gen. Virol. *84*, 259–268.

Li, W., Ferguson, B.J., Khaled, W.T., Tevendale, M., Stingl, J., Poli, V., Rich, T., Salomoni, P., and Watson, C.J. (2009). PML depletion disrupts normal mammary gland development and skews the composition of the mammary luminal cell progenitor pool. Proc. Natl. Acad. Sci. U.S.A. *106*, 4725–4730.

Lin, R., Noyce, R.S., Collins, S.E., Everett, R.D., and Mossman, K.L. (2004). The herpes simplex virus ICP0 RING finger domain inhibits IRF3- and IRF7-mediated activation of interferon-responsive genes. J. Virol. 78, 1675–1684.

Liu, Y., Shevchenko, A., and Berk, A.J. (2005). Adenovirus exploits the cellular aggresome response to accelerate inactivation of the MRN complex. J. Virol. *79*, 14004–14016.

Louria-Hayon, I., Alsheich-Bartok, O., Levav-Cohen, Y., Silberman, I., Berger, M., Grossman, T., Matentzoglu, K., Jiang, Y.H., Muller, S., Scheffner, M., *et al.* (2009). E6AP promotes the degradation of the PML tumour suppressor. Cell Death Diff. *16*, 1156–1166.

Louria-Hayon, I., Grossman, S.R., Sionov, R.V., Alsheich, O., Pandolfi, P.P., and Haupt, Y. (2003). The promyelocytic leukemia protein protects p53 from Mdm2-mediated inhibition and degradation. J. Biol. Chem. *278*, 33134–33141.

McNally, B.A., Trgovcich, J., Maul, G.G., Liu, Y., and Zheng, P. (2008). A Role for Cytoplasmic PML in Cellular Resistance to Viral Infection. Plos One 3, e2277.

Mallery, D.L., McEwan, W.A., Bidgood, S.R., Towers, G.J., Johnson, C.M., and James, L.C. (2010). Antibodies mediate intracellular immunity through tripartite motif-containing 21 (TRIM21). Proc. Natl. Acad. Sci. U.S.A. *107*, 19985–19990.

Matic, I., Schimmel, J., Hendriks, I.A., van Santen, M.A., van de Rijke, F., van Dam, H., Gnad, F., Mann, M., and Vertegaal, A.C.O. (2010). Site-Specific Identification of SUMO-2 Targets in Cells Reveals an Inverted SUMOylation Motif and a Hydrophobic Cluster SUMOylation Motif. Mol. Cell *39*, 641–652.

Maul, G.G., Yu, E., Ishov, A.M., and Epstein, A.L. (1995). Nuclear domain 10 (ND10) associated proteins are also present in nuclear bodies and redistribute to hundreds of nuclear sites after stress. J. Cell. Biochem. *59*, 498–513.

Maul, G.G., Ishov, A.M., and Everett, R.D. (1996). Nuclear domain 10 as preexisting potential replication start sites of herpes simplex virus type-1. Virology *217*, 67–75.

Meroni, G., and Diez-Roux, G. (2005). TRIM/RBCC, a novel class of 'single protein RING finger' E3 ubiquitin ligases. BioEssays *27*, 1147–1157.

Miller, D.L., Rickards, B., Mashiba, M., Huang, W.Y., and Flint, S.J. (2009). The adenoviral E1B 55-kilodalton protein controls expression of immune response genes but not p53-dependent transcription. J. Virol. *83*, 3591–3603.

Mirzoeva, O.K., and Petrini, J.H. J. (2001). DNA damage-dependent nuclear dynamics of the Mre11 complex. Mol. Cell. Biol. *21*, 281–288.

Mossman, K.L., Saffran, H.A., and Smiley, J.R. (2000). Herpes simplex virus ICP0 mutants are hypersensitive to interferon. J. Virol. *74*, 2052–2056.

Muller, S., and Dobner, T. (2008). The adenovirus E1B-55K oncoprotein induces SUMO modification of p53. Cell Cycle *7*, 754–758.

Muller, S., Matunis, M.J., and DeJean, A. (1998). Conjugation of the ubiquitin-related modifier SUMO-1 regulates the partitioning of PML within the nucleus. EMBO J. *17*, 61–70.

Nacerddine, K., Lehembre, F., Bhaumik, M., Artus, J., Cohen-Tannoudji, M., Babinet, C., Pandolfi, P.P., and Dejean, A. (2005). The SUMO pathway is essential for nuclear integrity and chromosome segregation in mice. Dev. Cell *9*, 769–779.

Nakahara, T., and Lambert, P.F. (2007). Induction of promyelocytic leukemia (PML) oncogenic domains (PODs) by papillomavirus. Virology *366*, 316–329.

Nichol, J.N., Petruccelli, L.A., and Miller, W.H. (2009). Expanding PML's functional repertoire through post-translational mechanisms. Front. Biosci. *14*, 2293–2306.

Nisole, S., Stoye, J.P., and Saib, A. (2005). TRIM family proteins: Retroviral restriction and antiviral defence. Nat. Rev. Microbiol. 3, 799–808.

Nojima, T., Oshiro-Ideue, T., Nakanoya, H., Kawamura, H., Morimoto, T., Kawaguchi, Y., Kataoka, N., and Hagiwara, M. (2009). Herpesvirus protein ICP27 switches PML isoform by altering mRNA splicing. Nucleic Acids Res. *37*, 6515–6527.

Oh, W., Ghim, J., Lee, E.W., Yang, M.R., Kim, E.T., Ahn, J.H., and Song, J. (2009). PML-IV functions as a negative regulator of telomerase by interacting with TERT. J. Cell Sci. *122*, 2613–2622.

Ornelles, D.A., and Shenk, T. (1991). Localization of the adenovirus early region 1b 55-kilodalton protein during lytic infection: association with nuclear viral inclusions requires the early region 4 34-kilodalton protein. J. Virol. *65*, 424–439.

Ozato, K., Shin, D.M., Chang, T.H., and Morse, H.C. (2008). TRIM family proteins and their emerging roles in innate immunity. Nat. Revs. Immunol. *8*, 849–860.

Pampin, M., Simonin, Y., Blondel, B., Percherancier, Y., and Chelbi-Alix, M.K. (2006). Cross talk between PML and p53 during poliovirus infection: Implications for antiviral defense. J. Virol. *80*, 8582–8592.

Pearson, M., Carbone, R., Sebastiani, C., Cioce, M., Fagioli, M., Saito, S., Higashimoto, Y., Appella, E., Minucci, S., Pandolfi, P.P., *et al.* (2000). PML regulates p53 acetylation and premature senescence induced by oncogenic Ras. Nature *406*, 207–210.

Pennella, M.A., Liu, Y., Woo, J.L., Kim, C.A., and Berk, A.J. (2010). Adenovirus E1B 55-kilodalton protein is a p53-SUMO1 E3 ligase that represses p53 and stimulates its nuclear export through interactions with promyelocytic leukemia nuclear bodies. J. Virol. *84*, 12210–12225.

Percherancier, Y., Germain-Desprez, D., Galisson, F., Mascle, X.H., Dianoux, L., Estephan, P., Chelbi-Alix, M.K., and Aubry, M. (2009). Role of SUMO in RNF4-mediated promyelocytic leukemia protein (PML) degradation. J. Biol. Chem. *284*, 16595–16608.

Puvion-Dutilleul, F., Chelbi-Alix, M.K., Koken, M., Quignon, F., Puvion, E., and de The, H. (1995). Adenovirus infection induces rearrangements in the intranuclear distribution of the nuclear body-associated PML protein. Exp. Cell Res. *218*, 9–16.

Querido, E., Blanchette, P., Yan, Q., Kamura, T., Morrison, M., Boivin, D., Kaelin, W.G., Conaway, R.C., Conaway, J.W., and Branton, P.E. (2001). Degradation of p53 by adenovirus E4orf6 and E1B55K proteins occurs via a novel mechanism, involving a Cullin-containing complex. Genes Dev. *15*, 3104–3117.

Quimby, B.B., Yong-Gonzalez, V., Anan, T., Strunnikov, A.V., and Dasso, M. (2006). The promyelocytic leukemia protein stimulates SUMO conjugation in yeast. Oncogene *25*, 2999–3005.

Randall, R.E., and Goodbourn, S. (2008). Interferons and viruses: an interplay between induction, signalling, antiviral responses and virus countermeasures. J. Gen. Virol. *89*, 1–47.

Regad, T., Saib, A., Lallemand-Breitenbach, V., Pandolfi, P.P., de The, H., and Chelbi-Alix, M.K. (2001). PML mediates the interferon-induced antiviral state against a complex retrovirus via its association with the viral transactivator. EMBO J. *20*, 3495–3505.

Regad, T., Bellodi, C., Nicotera, P., and Salomoni, P. (2009). The tumour suppressor Pml regulates cell fate in the developing neocortex. Nat. Neurosci. *12*, 132–140.

Reichelt, M., Wang, L., Sommer, M., Perrino, J., Nour, A.M., Sen, N., Baiker, A., Zerboni, L., and Arvin, A.M. (2011). Entrapment of viral capsids in nuclear PML cages is an intrinsic antiviral host defense against Varicella-Zoster Virus. Plos Pathog. *7*, e1001266.

Reineke, E.L., Lam, M., Liu, O., Liu, Y., Stanya, K.J., Chang, K.S., Means, A.R., and Kao, H.Y. (2008). Degradation of the tumour suppressor PML by Pin1 contributes to the cancer phenotype of breast cancer MDA-MB-231 cells. Mol. Cell. Biol. *28*, 997–1006.

Ribet, D., Hamon, M., Gouin, E., Nahori, M.A., Impens, F., Neyret-Kahn, H., Gevaert, K., Vandekerckhove, J., Dejean, A., and Cossart, P. (2010). Listeria monocytogenes impairs SUMOylation for efficient infection. Nature *464*, 1192–1195.

Rosa-Calatrava, M., Puvion-Dutilleul, F., Lutz, P., Dreyer, D., de The, H., Chatton, B., and Kedinger, C. (2003). Adenovirus protein IX sequesters host-cell promyelocytic leukaemia protein and contributes to efficient viral proliferation. EMBO Rep. *4*, 969–975.

Saffert, R.T., and Kalejta, R.F. (2006). Inactivating a cellular intrinsic immune defense mediated by Daxx is the mechanism through which the human cytomegalovirus pp71 protein stimulates viral immeiate-early gene expression. J. Virol. *80*, 3863–3871.

Salomoni, P., and Bellodi, C. (2007). New insights into the cytoplasmic function of PML. Histol. Histopathol. *22*, 937–946.

Salomoni, P., Ferguson, B.J., Wyllie, A.H., and Rich, T. (2008). New insights into the role of PML in tumour suppression. Cell Res. *18*, 620–638.

Sarnow, P., Hearing, P., Anderson, C.W., Reich, N., and Levine, A.J. (1982). Identification and characterization of an immunologically conserved adenovirus early region 11, 000 Mr protein and its association with the nuclear matrix. J. Mol. Biol. *162*, 565–583.

Sato, Y., Yoshioka, K., Suzuki, C., Awashima, S., Hosaka, Y., Yewdell, J., and Kuroda, K. (2003). Localization of influenza virus proteins to nuclear dot 10 structures in influenza virus-infected cells. Virology *310*, 29–40.

Scaglioni, P.P., Yung, T.M., Cai, L.F., Erdjument-Bromage, H., Kaufman, A.J., Singh, B., Teruya-Feldstein, J., Tempst, P., and Pandolfi, P.P. (2006). A CK2-dependent mechanism for degradation of the PML tumor suppressor. Cell *126*, 269–283.

Schreiner, S., Wimmer, P., Sirma, H., Everett, R.D., Blanchette, P., Groitl, P., and Dobner, T. (2010). Proteasome-dependent degradation of Daxx by the viral E1B-55K protein in human adenovirus-infected cells. J. Virol. *84*, 7029–7038.

Seker, H., Rubbi, C., Linke, S.P., Bowman, E.D., Garfield, S., Hansen, L., Borden, K.L.B., Milner, J., and Harris, C.C. (2003). UV-C induced DNA damage leads to p53-dependent trafficking of PML. Oncogene *22*, 1620–1628.

Shen, T.H., Lin, H. -K., Scaglioni, P.P., Yung, T.M., and Pandolfi, P.P. (2006). The mechanisms of PML-nuclear body formation. Mol. Cell *24*, 331–339.

Shiels, C., Islam, S.A., Vatcheva, R., Sasieni, P., Sternberg, M.J.E., Freemont, P.S., and Sheer, D. (2001). PML bodies associate specifically with the MHC gene cluster in interphase nuclei. J. Cell Sci. *114*, 3705–3716.

Shimoda, K., Kamesaki, K., Numata, A., Aoki, K., Matsuda, T., Oritani, K., Tamiya, S., Kato, K., Takase, K., Imamura, R., *et al.* (2002). Cutting edge: Tyk2 is required for the induction and nuclear translocation of Daxx which regulates IFN-alpha-induced suppression of B lymphocyte formation. J. Immunol. *169*, 4707–4711.

Shishido-Hara, Y., Higuchi, K., Ohara, S., Duyckaerts, C., Hauw, J.J., and Uchihara, T. (2008). Promyelocytic leukemia nuclear bodies provide a scaffold for human polyomavirus JC replication and are disrupted after development of viral inclusions in progressive multifocal leukoencephalopathy. J. Neuropathol. Exp. Neurol. *67*, 299–308.

Shishido-Hara, Y., Ichinose, S., Higuchi, K., Hara, Y., and Yasui, K. (2004). Major and minor capsid proteins of human polyomavirus JC cooperatively accumulate to nuclear domain 10 for assembly into virions. J. Virol. *78*, 9890–9903.

Sivachandran, N., Sarkari, F., and Frappier, L. (2008). Epstein–Barr nuclear antigen 1 contributes to nasopharyngeal carcinoma through disruption of PML nuclear bodies. Plos Pathog. *4*, e1000170.

Skinner, P.J., Koshy, B.T., Cummings, C.J., Klement, I.A., Helin, K., Servadio, A., Zoghbi, H.Y., and Orr, H.T. (1997). Ataxin-1 with an expanded glutamine tract alters nuclear matrix-associated structures. Nature *389*, 971–974.

Soria, C., Estermann, F.E., Espantman, K.C., and O'Shea, C.C. (2010). Heterochromatin silencing of p53 target genes by a small viral protein. Nature *466*, 1076–1081.

Spardy, N., Duensing, A., Hoskins, E.E., Wells, S.I., and Duensing, S. (2008). HPV16 E7 Reveals a link between DNA Replication stress, Fanconi anemia D2 protein, and alternative lengthening of telomere-associated promyelocytic leukemia bodies. Cancer Res. *68*, 9954–9963.

Stracker, T.H., Carson, C.T., and Weitzman, M.D. (2002). Adenovirus oncoproteins inactivate the Mre11-Rad50-NBS1 DNA repair complex. Nature *418*, 348–352.

Stracker, T.H., Lee, D.V., Carson, C.T., Araujo, F.D., Ornelles, D.A., and Weitzman, M.D. (2005). Serotype-specific reorganization of the Mre11 complex by adenoviral E4orf3 proteins. J. Virol. *79*, 6664–6673.

Stremlau, M., Owens, C.M., Perron, M.J., Kiessling, M., Autissier, P., and Sodroski, J. (2004). The cytoplasmic body component TRIM5 alpha restricts HIV-1 infection in Old World monkeys. Nature *427*, 848–853.

Swindle, C.S., Zou, N.X., Van Tine, B.A., Shaw, G.M., Engler, J.A., and Chow, L.T. (1999). Human papillomavirus DNA replication compartments in a transient DNA replication system. J. Virol. *73*, 1001–1009.

Tan, J.A.T., Sun, Y.J., Song, J., Chen, Y., Krontiris, T.G., and Durrin, L.K. (2008). SUMO conjugation to the matrix attachment region-binding protein, special AT-rich sequence-binding protein-1 (SATB1), targets SATB1 to promyelocytic nuclear bodies where it undergoes caspase cleavage. J. Biol. Chem. *283*, 18124–18134.

Tang, J., Xie, W.S., and Yang, X.L. (2005). Association of caspase-2 with the promyelocytic leukemia protein nuclear bodies. Cancer Biol. Ther. *4*, 645–649.

Tatham, M.H., Geoffroy, M.C., Shen, L., Plechanovova, A., Hattersley, N., Jaffray, E.G., Palvimo, J.J., and Hay, R.T. (2008). RNF4 is a poly SUMO-specific E3 ubiquitin ligase required for arsenic-induced PML degradation. Nat. Cell Biol. *10*, 538–546.

Tavalai, N., Papior, P., Rechter, S., and Stamminger, T. (2008). Nuclear domain 10 components promyelocytic leukemia protein and hDaxx independently contribute to an intrinsic antiviral defense against human cytomegalovirus infection. J. Virol. *82*, 126–137.

Tavalai, N., and Stamminger, T. (2008). New insights into the role of the subnuclear structure ND10 for viral infection. Biochim. Biophys. Acta Mol. Cell Res. *1783*, 2207–2221.

de The, H., Chomienne, C., Lanotte, M., Degos, L., and Dejean, A. (1990). The t (15;17) translocation of promyelocytic leukaemia fuses the retinoic acid receptor a gene to a novel transcribed locus. Nature *347*, 558–561.

Towers, G.J. (2007). The control of viral infection by tripartite motif proteins and cyclophilin A. Retrovirology *4*, 40.

Ullman, A.J., and Hearing, P. (2008). Cellular proteins PML and Daxx mediate an innate antiviral defense antagonized by the adenovirus E4 ORF3 protein. J. Virol. *82*, 7325–7335.

Ullman, A.J., Reich, N.C., and Hearing, P. (2007). Adenovirus E4 ORF3 protein inhibits the interferon-mediated antiviral response. J. Virol. *81*, 4744–4752.

Van Damme, E., Laukens, K., Dang, T.H., and Van Ostade, X. (2010). A manually curated network of the PML nuclear body interactome reveals an important role for PML-NBs in SUMOylation dynamics. Int. J. Biol. Sci. *6*, 51–67.

Varadaraj, A., Dovey, C.L., Laredj, L., Ferguson, B., Alexander, C.E., Lubben, N., Wyllie, A.H., and Rich, T. (2007). Evidence for the receipt of DNA damage stimuli by PML nuclear domains. J. Pathol. *211*, 471–480.

Virador, V.M., Flores-Obando, R.E., Berry, A., Patel, R., Zakhari, J., Lo, Y.C., Strain, K., Anders, J., Cataisson, C., Hansen, L.A., *et al.* (2009). The human promyelocytic leukemia protein is a tumor suppressor for murine skin carcinogenesis. Mol. Carcinog. *48*, 599–609.

Wang, J., Shiels, C., Sasieni, P., Wu, P.J., Islam, S.A., Freemont, P.S., and Sheer, D. (2004). Promyelocytic leukemia nuclear bodies associate with transcriptionally active genomic regions. J. Cell Biol. *164*, 515–526.

Wang, Z.G., Delva, L., Gaboli, M., Rivi, R., Giorgio, M., Cordon-Cardo, C., Grosveld, F., and Pandolfi, P.P. (1998a). Role of PML in cell growth and the retinoic acid pathway. Science *279*, 1547–1551.

Wang, Z.G., Ruggero, D., Ronchetti, S., Zhong, S., Gaboli, M., Rivi, R., and Pandolfi, P.P. (1998b). PML is essential for multiple apoptotic pathways. Nat. Genet. *20*, 266–272.

Weidtkamp-Peters, S., Lenser, T., Negorev, D., Gerstner, N., Hofmann, T.G., Schwanitz, G., Hoischen, C., Maul, G., Dittrich, P., and Hemmerich, P. (2008). Dynamics of component exchange at PML nuclear bodies. J. Cell Sci. *121*, 2731–2743.

Wimmer, P., Schreiner, S., Everett, R.D., Sirma, H., Groitl, P., and Dobner, T. (2010). SUMO modification of E1B-55K oncoprotein regulates isoform-specific binding to the tumour suppressor protein PML. Oncogene *29*, 5511–5522.

Xu, D.K., Holko, M., Sadler, A.J., Scott, B., Higashiyama, S., Berkofsky-Fessler, W., McConnell, M.J., Pandolfi, P.P., Licht, J.D., and Williams, B.R.G. (2009). Promyelocytic leukemia zinc finger protein regulates interferon-mediated innate immunity. Immunity *30*, 802–816.

Xu, Y.X., Ahn, J.H., Cheng, M.F., Ap Rhys, C.M., Chiou, C.J., Zong, J.H., Matunis, M.J., and Hayward, G.S. (2001). Proteasome-independent disruption of PML oncogenic domains (PODs), but not covalent modification by SUMO-1, is required for human cytomegalovirus immediate-early protein IE1 to inhibit PML-mediated transcriptional repression. J. Virol. *75*, 10683–10695.

Xu, Z.X., Zou, W.X., Lin, P., and Chang, K.S. (2005). A role for PML3 in centrosome duplication and genome stability. Mol. Cell *17*, 721–732.

Yamada, M., Sato, T., Shimohata, T., Hayashi, S., Igarashi, S., Tsuji, S., and Takahashi, H. (2001). Interaction between neuronal intranuclear inclusions and promyelocytic leukemia protein nuclear and coiled bodies in CAG repeat diseases. Am. J. Pathol. *159*, 1785–1795.

Yang, S.T., Kuo, C., Bisi, J.E., and Kim, M.K. (2002). PML-dependent apoptosis after DNA damage is regulated by the checkpoint kinase hCds1/Chk2. Nat. Cell Biol. *4*, 865–870.

Yeager, T.R., Neumann, A.A., Englezou, A., Huschtscha, L.I., Noble, J.R., and Reddel, R.R. (1999). Telomerase-negative immortalized human cells contain a novel type of promyelocytic leukemia (PML) body. Cancer Res. *59*, 4175–4179.

Yondola, M.A., and Hearing, P. (2007). The adenovirus E4 ORF3 protein binds and reorganizes the TRIM family member transcriptional intermediary factor 1 alpha. J. Virol. *81*, 4264–4271.

Yousef, A.F., Fonseca, G.J., Pelka, P., Ablack, J.N.G., Walsh, C., Dick, F.A., Bazett-Jones, D.P., Shaw, G.S., and Mymryk, J.S. (2010). Identification of a molecular recognition feature in the E1A oncoprotein that binds the SUMO conjugase UBC9 and likely interferes with polySUMOylation. Oncogene *29*, 4693–4704.

Yu, J., Lan, J.P., Wang, C., Wu, Q., Zhu, Y.Y., Lai, X.Y., Sun, J., Jin, C.J., and Huang, H. (2010). PML3 interacts with TRF1 and is essential for ALT-associated PML bodies assembly in U2OS cells. Cancer Lett. *291*, 177–186.

Adenoviruses and Gene Therapy: The Role of the Immune System

14

Laura White and G. Eric Blair

Abstract

Adenovirus (Ad)-based vectors have been frequently used as gene therapy vectors owing to their ability to infect a wide range of dividing and non-dividing cells, their efficient growth to high titres in complementing cell lines and ease of genome manipulation. However, the transition of Ad vectors from *in vitro* studies to clinical application has been limited by suboptimal efficacy and robust inflammatory responses elicited upon administration. In recent years it has become clear that multiple innate and adaptive immune responses limit the efficacy and safety of Ad vectors. In this review, we focus on the current understanding of the immune response to Ads with a particular focus on the innate immune response, and how this information can be used to design safer and more efficacious vectors.

Introduction

Adenoviruses (Ads) are non-enveloped, double-stranded DNA viruses believed to hold promise as gene therapy vectors. There has been a considerable expansion in the use of Ad vectors in clinical studies over the last few years and Ad vectors currently account for the highest proportion of virus vectors in clinical trials to date. Ads possess many characteristics ideal for a virus vector: their genome can be easily manipulated, they are able to infect a wide range of dividing and non-dividing cells and they can be easily and inexpensively grown to high titres (reviewed in Shen and Post, 2007). In addition, they are believed to hold promise for many applications, including vaccination, gene-addition for inherited diseases and as oncolytics for cancer therapy. However, host immune responses to Ad vectors have hampered therapeutic progress. Most gene therapy and oncolytic vectors have been based on Ads 2 or 5, however the high prevalence of these serotypes within the general population means most humans have been exposed to these Ads (Ersching *et al.*, 2010). Therefore, upon administration in immune hosts, there is rapid clearance of these vectors by pre-existing antibodies (reviewed in Bessis *et al.*, 2004). Moreover, upon intravenous (IV) administration, Ad vectors are sequestered in the liver and rapidly cleared by liver macrophages (Kupffer cells) (Lieber *et al.*, 1997; Tao *et al.*, 2001). In addition to rapid clearance by the adaptive response, Ad vectors induce overt innate inflammatory responses upon administration that can be dangerous to the patient (Raper *et al.*, 2003). Although the adaptive immune response to Ads has been well documented, it is only in recent years that an insight has been gained into the innate immune response to Ads. The innate response appears to involve activation of multiple pattern recognition receptors (PRRs), including Toll-like Receptors (TLRs) (Cerullo *et al.*, 2007; Iacobelli-Martinez and Nemerow, 2007; Appledorn *et al.*, 2008c) and the NALP3 inflammasome (Muruve *et al.*, 2008; Hornung *et al.*, 2009). Activation of these receptors leads to production of inflammatory cytokines and type-1 interferons (IFNs) resulting in an acute inflammatory response, and also facilitates stimulation of the subsequent adaptive response by activating dendritic cells (reviewed in Pasare and Medzhitov, 2004).

The expansion of knowledge of the immune response to Ads in recent years has allowed investigators to address the lack of efficacy and heightened immune responses to Ad vectors. The strategies currently being tried to manipulate the immune response generally involve either modulation of the host's immune system or modification of the vector. In this chapter, an introduction to Ads and Ad vectors will be presented, followed by the innate immune response to Ads. Finally, the methods currently being tested to manipulate the immune response to Ad vectors will be described.

Adenoviruses

Adenoviruses (Ads) are non-enveloped icosahedral viruses with double-stranded DNA (dsDNA) genomes of approximately 36 kb. Ads are known to cause acute respiratory, gastrointestinal and ocular infections which are generally self-limiting but can cause more serious infection in the immunocompromised host (Hale *et al.*, 1999). Classification of Ads is based on multiple parameters including neutralization by specific antisera, haemagglutination properties and genome sequence homology (reviewed in Berk, 2007). There are currently over 50 different serotypes of human Ads divided into six species (A to F) and species B is further subdivided into Groups B1 and B2 (Wadell *et al.*, 1986; Segerman *et al.*, 2003).

The structure of Ads comprises an icosahedral protein capsid with each of the 20 triangular faces composed of 240 hexon capsomeres, each of which is a homotrimer (Silvestry *et al.*, 2009) (Fig. 14.1). The 12 vertices comprise penton capsomeres which consist of a non-covalent complex of homopentameric penton base and homotrimeric fibre protein (reviewed in Russell, 2009; Hall *et al.*, 2010). The other minor capsid components are proteins IIIa, VI, VIII and IX. The remaining structural polypeptides V, VII, Mu and terminal protein are associated with the dsDNA genome, comprising the core of the virus (Russell, 2009; Hall *et al.*, 2010).

The Ad genome encodes five early transcription units (E1A, E1B, E2A, E2B, E3 and E4), two delayed early and one major late transcription unit (Berk, 2007) (Fig. 14.2). The early gene regions are transcribed prior to virus DNA replication and act to promote cell cycle progression,

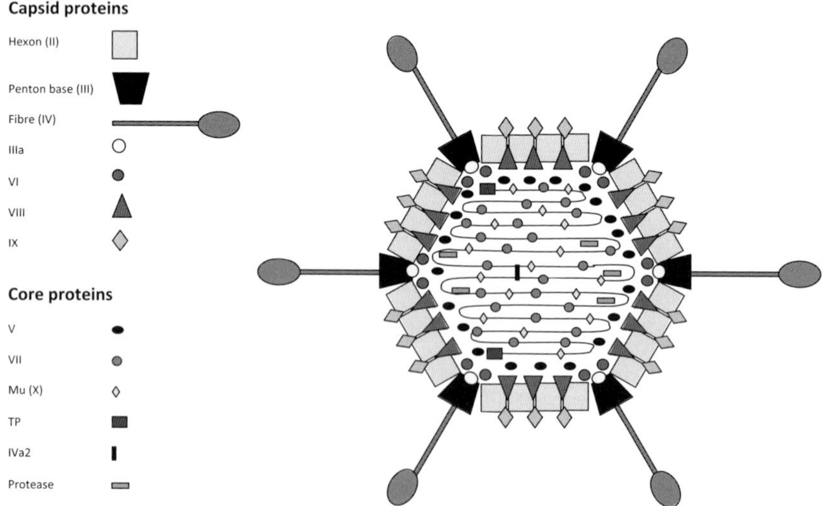

Figure 14.1 Adenovirus structure. A schematic representation of the capsid and core proteins comprising the Ad capsid derived from X-ray crystallography and cryo-electron microscopy. The major capsid proteins are hexon, penton base and fibre whilst the minor capsid proteins are IIIa, VI, VIII and IX. The core proteins V, VII, Mu, terminal protein (TP), IVa2 and protease are found associated with the double-stranded DNA genome. Adapted from Hall *et al.* (2010).

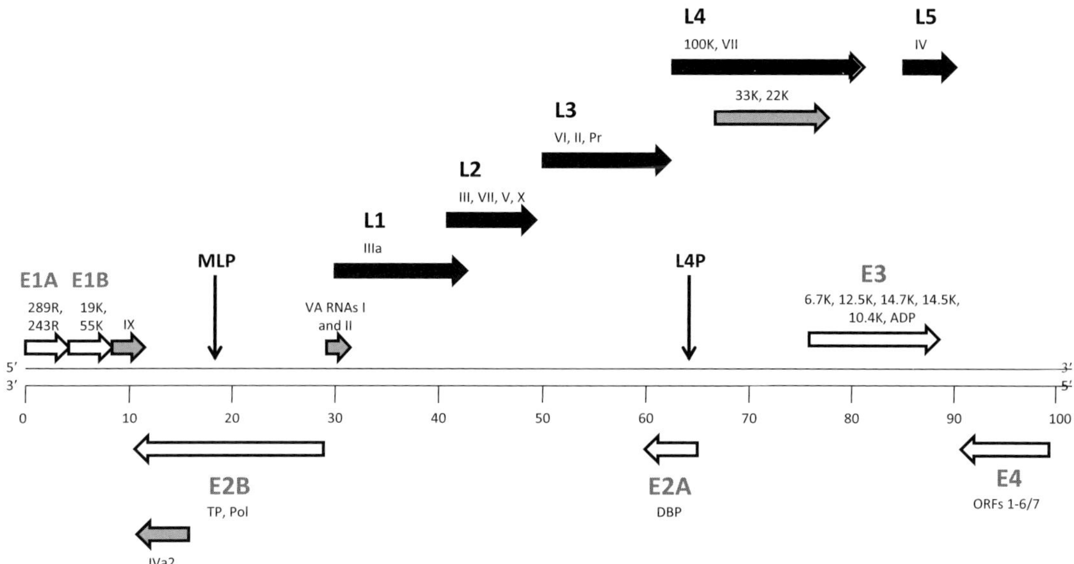

Figure 14.2 Genome organization of species C adenovirus. The genome is conventionally divided into 100 map units (MU). Arrows represent the direction of transcription. Early gene regions are denoted by the white arrows, intermediate gene regions by grey arrows and late gene regions by the black arrows. MLP, major late promoter; L4P, late region 4 (L4) promoter; VA RNAs, virus-associated RNAs; Pol, Adenovirus DNA polymerase; Pr, protease; DBP, DNA-binding protein; ADP, Adenovirus death protein; ORFs 1–6/7, open reading frames 1–6/7.

transcription of virus genes and suppression of host cell anti-virus mechanisms. The late genes are transcribed after the onset of virus replication and function mainly to co-ordinate virus assembly and the subsequent release of progeny virions from the host cell.

Life cycle of human adenoviruses

Cell entry

The classical pathway of Ad entry has been investigated for species C Ads (i.e. Ads 2 and 5) in great detail. However, there are differences in recognition of host cell-surface molecules by species B and certain species D Ads and this leads to alternate cellular entry pathways for these Ads (Bergelson *et al.*, 1997; Tuve *et al.*, 2006; Wang *et al.*, 2010).

The classical infection pathway involves initial attachment of Ad species A, C, D, E, and F to host cells *in vitro* by fibre binding to the Coxsackie and Adenovirus Receptor (CAR) (Bergelson *et al.*, 1997) (Fig. 14.3). However, the biodistribution of Ad vectors *in vivo* appears to be primarily CAR-independent (Fechner *et al.*, 1999; Alemany and Curiel, 2001; Martin *et al.*, 2003). Certain blood coagulation factors, including Factor X, appear to bind the Ad5 hexon capsomeres, bridging the capsid to heparan sulphate proteoglycans (HSPGs) on the cell surface of target cells such as liver hepatocytes and pancreatic cancer cells, leading to enhanced uptake of Ads into these cells (Hamdan *et al.* 2011; Shayakhmetov *et al.*, 2005; Parker *et al.*, 2006, 2007; Waddington *et al.*, 2008). It should be noted that Ad dissemination in the bloodstream is unlikely during infection; however, intravenous (IV) administration of Ad vectors is a common administration route in gene therapy regimens.

Unlike the other Ad species, the species B Ads do not utilize CAR as the primary cell surface attachment receptor *in vitro* (Roelvink *et al.*, 1998; Segerman *et al.*, 2003). The species B Ads have been classified into two groups according

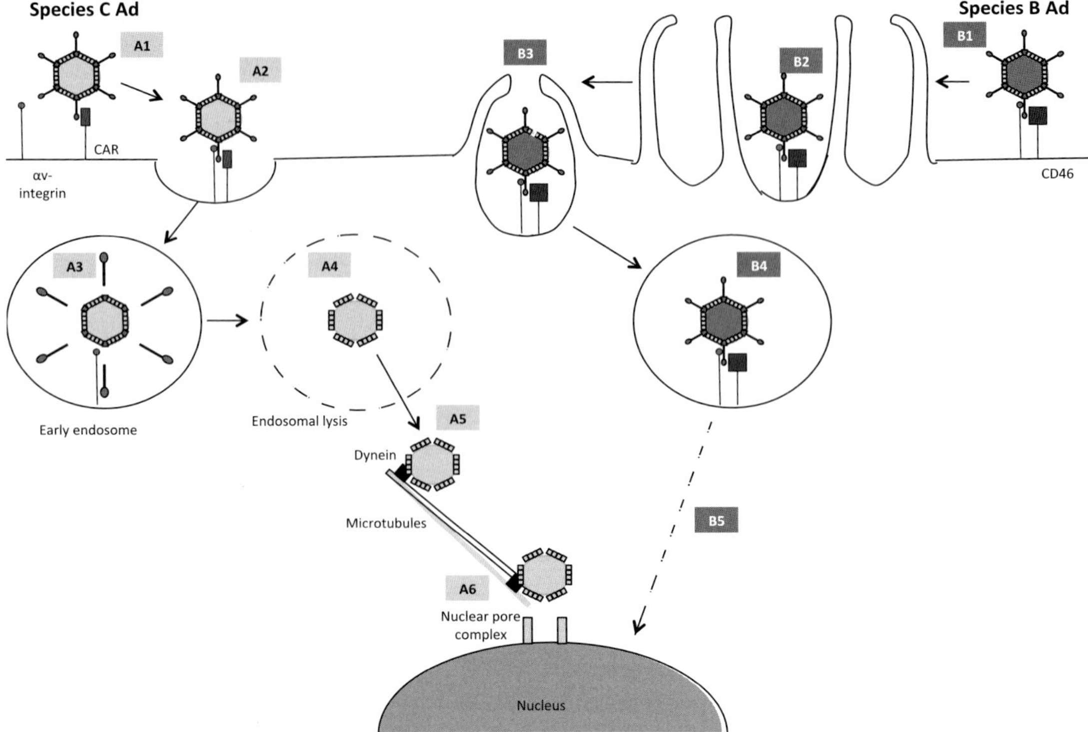

Figure 14.3 Cell entry pathways utilized by adenoviruses. The entry pathways defined for (A) Species C Ads and (B) Species B Ads 3 and 35. The entry pathway for Species C Ads involves (A1) attachment of the fibre protein to cellular CAR. (A2) Interaction of the penton base with α_v integrins. (A3) Virus internalization into endosomes via clathrin-mediated endocytosis. (A4) Dissociation of Ad capsid components in the acidic environment of the early endosome and disruption of the endosomal membrane by protein VI leads to release of the partially disassembled particle into the cytoplasm. (A5) The Ad particle is trafficked along microtubules in a dynein-dependent manner to the nucleus where it associates with the nuclear pore complex to promote entry of the Ad genome into the cell nucleus (A6). In contrast, the entry pathway for species B Ad3 into epithelial cells occurs by the following mechanism. (B1) Attachment of the fibre protein to CD46 (and/or desmoglein-2). (B2) Interaction of penton base protein with α_v integrins stimulates (B3) membrane ruffling leading to (B4) endocytosis of the virus into macropinosomes. The subsequent release of the particle from the macropinosome and transport to the nucleus (B5) is currently undefined. Adapted from Hall *et al.* (2010).

to conserved restriction endonuclease cleavage sites, which also broadly correlates with tissue tropism of the viruses (Segerman *et al.*, 2003). Group B1 comprises Ads 3, 7, 16, 21 and 50 whilst Group B2 includes Ads 11, 14, 34 and 35. Attachment by most species B Ads, apart from Ads 3 and 7, was found to be via CD46 in cultured human cells (Gaggar *et al.*, 2003). Desmoglein 2 (DSG-2) has been recently identified as a primary cell-surface attachment molecule for Ads 3, 7, 11 and 14 (Wang *et al.*, 2010). Ad11 appears to be able to utilize either CD46 or DSG-2 for attachment (Segerman *et al.*, 2003; Wang *et al.*, 2010).

Following interaction of Ads with the primary attachment molecule, there is secondary, low-affinity binding between the arginine-glycine-aspartate (RGD) motif (conserved among all Ads except species F) of the Ad penton base protein and αv integrins on the host cell surface (Wickham *et al.*, 1993, 1994; Mathias *et al.*, 1994). This leads to internalization of species C Ads by endocytosis into clathrin-coated vesicles (Wickham *et al.*, 1994; Li *et al.*, 1998; Stupack *et al.*, 1998; Meier *et al.*, 1999) (Fig. 14.3). In contrast, both Ad3 (from Group B2) and Ad35 (from Group B1) have been reported to utilize macropinocytosis for cell entry, suggesting this may be a common entry pathway for species B Ads (Amstutz *et al.*, 2008; Kalin *et al.*, 2010).

Transit of adenoviruses to the nucleus

Following internalization of species C Ads, the acidic environment of the early endosome induces conformational changes in certain virus capsid components and facilitates capsid disassembly (Greber *et al.*, 1993; Wiethoff *et al.*, 2005). This leads to endosomal lysis allowing the partially uncoated virus particle to be released into the cytoplasm (Wiethoff *et al.*, 2005). A PPxY motif in pVI appears to interact with Nedd4 ubiquitin ligases, leading to ubiquitination of pVI and endosomolysis (Wodrich *et al.*, 2010). The pathway of Ad3 from macropinosomes into the cytoplasm is less well-defined, however it appears to be a much slower process than escape of species C Ads from endosomes (Miyazawa *et al.*, 2001; Shayakhmetov *et al.*, 2003; Amstutz *et al.*, 2008). This indicates that a much lower pH for species B capsid disassembly is required than for species C.

Once species C Ads escape from the endosome, the partially uncoated virus particles associate with microtubules in the cytoplasm (Mabit *et al.*, 2002; Kelkar *et al.*, 2006). The Ad particles are then trafficked along the microtubules in a dynein-dependent manner to the nucleus (Greber *et al.*, 1997; Suomalainen *et al.*, 1999; Leopold *et al.*, 2000; Strunze *et al.*, 2005).

Adenovirus vectors

Species C Ads (Ads 2 and 5) were first shown to serve as efficient vectors for gene delivery to target cells *in vitro* (Hajahmad and Graham, 1986), and remain the most studied for vector application to date. There are currently two major types of Ad vectors: replication-deficient and replication-competent vectors.

Replication-deficient adenovirus vectors

Replication-deficient vectors have deletions in early region genes necessary for virus replication and were first studied as potential gene-replacement vectors. The E1A and E1B regions were deleted in the first-generation Ad vectors, which had a dual function of preventing virus replication and providing coding capacity for insertion of a therapeutic transgene (Yang *et al.*, 1995). Some vectors also combined the E1 deletion with deletions in E3, since this region is dispensable for replication, and this further increased capacity for insertion of transgenes. However these vectors were rapidly cleared by the host immune system (Yang *et al.*, 1995). The E1 and/or E3 deletions were combined with additional deletions in various E2 and E4 genes in the second and third generation Ad vectors (Armentano *et al.*, 1995; Engelhardt *et al.*, 1994). Decreased toxicity was observed in these vectors; however there was still stimulation of the inflammatory response following administration and rapid antibody neutralization upon subsequent inoculation. This lead to the development of helper-dependent ('gutless') vectors, which have all the coding genes deleted (Schiedner *et al.*, 1998). A recent study found that, unlike wild-type Ads or replication-deficient Ad vectors, gutless Ads may be capable of integrating vector DNA into host chromosomes (Stephen *et al.*, 2010). In addition, gutless vectors generally have much improved toxicity and immunogenic profiles owing to lack of immunogenic peptide expression. However, immune responses to the virus capsid of gutless Ads still cause problems with re-administration (reviewed in Brunetti-Pierri and Ng, 2008). An additional problem with gutless viruses is the contamination of gutless virus stocks with the helper virus used to grow them (reviewed in Morsy and Caskey, 1999). Approaches to eliminate helper virus contamination include recombination systems (Chen *et al.*, 1996; Hardy *et al.*, 1997; Ng *et al.*, 2001; Umana *et al.*, 2001; Alba *et al.*, 2007), use of hybrid viruses (Cheshenko *et al.*, 2001; Kubo *et al.*, 2003) and introduction of packaging size or timing constraints (Sargent *et al.*, 2004; Alba *et al.*, 2007). Recombination systems have been most intensely studied: this approach utilizes a helper virus with recombinase enzyme recognition sites integrated into the regions flanking the packaging signal (ψ) and a complementing cell line expressing recombinase enzymes such as Flp or Cre (Chen *et al.*, 1996; Umana *et al.*, 2001). This allows for removal of the packaging signal by the recombinase enzyme and prevention of helper virus propagation. However, the presence of ψ-deleted helper virus has been detected in gutless virus preparations (Sakhuja *et al.*, 2003). Reduced helper virus contamination was detected when a phage DNA sequence was inserted between the Ad packaging signal and inverted terminal repeats (ITRs) (Alba *et al.*, 2007). In order to completely eliminate contamination, combination of two or more strategies may be necessary (Alba *et al.*, 2007; Segura *et al.*, 2008). In addition, methods to accurately quantify gutless and contaminating helper viruses are also being investigated (Crettaz *et al.*, 2008).

Replication-deficient Ads have also been investigated as vaccine carriers, as their ability to induce potent immune responses may be beneficial in this context (reviewed in Lasaro and Ertl, 2009). Ad5-based vectors have been developed to treat numerous infectious and non-infectious human diseases, including the malaria parasite *Plasmodium* species (Reyes-Sandoval *et al.*, 2008), bacteria such as *Mycobacterium tuberculosis* (Magalhaes *et al.*, 2008), viruses including HIV-1 (Buchbinder *et al.*, 2008; McElrath *et al.*, 2008) as well as cancer (Osada *et al.*, 2009) and Alzheimer's disease (Zou *et al.*, 2008). However, the efficacy of these vectors was generally lower than expected, and, worryingly, in a clinical trial for a Ad5 HIV-1 vaccine vector (the STEP trial) the individuals receiving the vector demonstrated increased susceptibility to HIV-1 infection if they carried Ad5-neutralizing antibodies (Buchbinder *et al.*, 2008; McElrath *et al.*, 2008). Although the mechanism for this is still unclear (Koup *et al.*, 2009), focus is now moving away from Ad5-based vectors to other, more novel serotypes (McCoy *et al.*, 2007; Tatsis *et al.*, 2009).

Replication-competent adenovirus vectors

Replication-competent Ad vectors include oncolytic vectors, which are intended to target specifically to tumour tissue. In theory, these vectors should specifically invade and replicate within tumour cells resulting in tumour cell lysis, releasing progeny virions able to infect and kill surrounding tumour cells. The most well-studied oncolytic Ad has been ONYX-015, which contains a mutation in the E1B-55K region of the Ad genome (Bischoff *et al.*, 1996). This mutation abolished the ability of the E1B-55K protein to bind and inhibit p53, thus restricting virus replication to cells in which the p53 pathway is already inhibited, which occurs in a high proportion of human cancers. Although the efficacy of ONYX-015 was found to be inadequate in Phase I and II trials when used as a single agent (Mulvihlll *et al.*, 2001; Nemunaitis *et al.*, 2001), utilizing ONYX-015 in combination with chemotherapy or radiotherapy resulted in complete tumour regression (Khuri *et al.*, 2000; Geoerger *et al.*, 2003). A similar E1B-deleted virus, termed H101, has been certified for use against head and neck cancer in China (Garber, 2006).

Safety and efficacy issues with adenovirus vectors

Despite the advances in Ad vector development, most vectors to date have exhibited a suboptimal capacity to produce beneficial therapeutic effect (efficacy) and often induced potent host immune responses. The high prevalence of species C Ads within the general population means that most humans have been exposed to these Ads (Ersching *et al.*, 2010), although the prevalence of immunity to different serotypes varies between populations (Holterman *et al.*, 2004; Kostense *et al.*, 2004; Metzgar *et al.*, 2005; Abbink *et al.*, 2007). Therefore upon administration in immune hosts, there is rapid clearance of these vectors by pre-existing antibodies (Bessis *et al.*, 2004). Neutralizing antibodies are generally serotype-specific and block transduction of Ad vectors into target tissue, thus decreasing efficacy. Even in naive hosts, adaptive immune responses are rapidly generated to Ads, resulting in rapid clearance of the vector upon re-administration (Bessis *et al.*, 2004). T-cell responses are also generated against Ads, with $CD4^+$ and $CD8^+$ T-cell responses apparently cross-reactive between Ad serotypes (Olive *et al.*, 2002; Onion *et al.*, 2007; Leen *et al.*, 2008).

In contrast to gene-replacement vectors, vaccine vectors rely on induction of a potent immune response in order to elicit immunity to the transgene antigen. However, the efficacy of these vectors has generally been lower than expected, indicating that the anti-antigen immune responses being generated by the vectors are not sufficiently robust (Buchbinder *et al.*, 2008; McElrath *et al.*, 2008). In addition, the response to Ad vaccine vectors in prime-boost regimens has also been less robust than expected (Pinto *et al.*, 2003; Liu *et al.*, 2008; Tatsis *et al.*, 2009).

With respect to oncolytic vectors, the adaptive immune response limits virus spread in the majority of tumours, as it rapidly eliminates the virus within two weeks of administration in immunocompetent hamster models (Thomas *et al.*, 2008). However, tumour cells are generally in an immunosuppressed environment, and there is evidence that activation of the immune response is beneficial in enhancing the anti-tumour response by the host (Todo *et al.*, 1999; Parato *et al.*, 2005; Ino *et al.*, 2006; Li *et al.*, 2007).

Whereas the adaptive response is mainly responsible for limiting efficacy, the innate response limits the safety of Ad vectors. The expression of virus proteins by Ad vectors and the presence of Ad DNA in host cells stimulates innate inflammatory responses which can lead to toxicity (Everett *et al.*, 2003; Gabitzsch *et al.*, 2009). In addition, IV-administered vectors become sequestered by the liver, resulting in low levels of vector transduction at the target site and local toxicity in the liver (Engelhardt *et al.*, 1994; Lieber *et al.*, 1997; Everett *et al.*, 2003). High doses of Ad vectors can lead to exaggerated responses, which have been shown to be lethal in mouse studies (Varnavski *et al.*, 2005) and also one case in a human clinical trial (Raper *et al.*, 2003). With respect to oncolytic Ads, vector clearance by Kupffer cells results in cell death (Manickan *et al.*, 2006), thus the initial systemic delivery of oncolytic vectors results in depletion of Kupffer cells. A second dose has been shown to be more efficacious as a result (Tao *et al.*, 2001; Schiedner *et al.*, 2003; Manickan *et al.*, 2006). However, there is increased hepatocyte transduction once Kupffer cells are depleted, which can lead to hepatotoxicity (Tao *et al.*, 2001; Shashkova *et al.*, 2008a; Koski *et al.*, 2009).

The immune response is crucial in not only reducing efficacy by clearance of vectors before therapeutic levels have been reached at the target site, but also in the potential toxicity of the inflammatory response induced upon administration. As well as being responsible for the immediate inflammation, stimulation of the innate response ultimately shapes the adaptive response. Therefore a detailed understanding of the innate response to Ads will be crucial for the development of safer and more efficacious Ad vectors. In the following section, the current understanding of the innate immune response to Ads is presented, followed by the strategies currently being investigated to manipulate the immune response to Ad vectors.

The innate immune response to adenoviruses

The responses elicited upon administration of Ads and Ad vectors include activation of complement (Kiang *et al.*, 2006a; Appledorn *et al.*, 2008a,b; Hartman *et al.*, 2008a), release of cytokines and chemokines (Muruve *et al.*, 1999), activation of macrophages, dendritic cells and endothelial cells (Lieber *et al.*, 1997; Zhang *et al.*, 2001; Schiedner *et al.*, 2003; Muruve *et al.*, 2008), dysregulation of transcription in host tissues (Hartman *et al.*, 2007b, 2008b), and thrombocytopenia (Wolins *et al.*, 2003). Ads and Ad vectors are first detected by complement and the pattern recognition receptors (PRRs) of the innate immune system, as discussed below.

Pattern recognition receptors

Pathogens are discriminated from self by a set of germline-encoded pattern recognition receptors (PRRs), which recognize a limited repertoire of pathogen antigens [known as pathogen-associated molecular patterns (PAMPs)] (reviewed in Mogensen, 2009). The PRRs involved in detection of virus infection are the Toll-like receptors (TLRs), the nucleotide-binding oligomerization domain (NOD)-like receptors (NLRs) and the retinoic acid-inducible gene 1 (RIG-1)-like receptors (RLRs). TLRs detect PAMPs located in the extracellular and endosomal compartments, whereas NLRs and RLRs detect intracellular PAMPs (Mogensen, 2009). Detection of PAMPs by PRRs (with the exception of certain NLRs) leads to activation of signalling pathways resulting in production of inflammatory cytokines and type-I IFNs. Inflammatory cytokines include tumour necrosis factor (TNF) which stimulates apoptosis in infected cells and also induces maturation of dendritic cells (DCs) (Caux *et al.*, 1992; Szabolcs *et al.*, 1995; Benihoud *et al.*, 2007). Type-I IFNs stimulate transcription of IFN-stimulated genes (ISGs) such as protein kinase R (PKR) and 2′,5′-oligoadenylate synthetase (OAS) (Meurs *et al.*, 1990; Hovanessian, 1991; Marie and Hovanessian, 1992). These ISGs then bring about the anti-virus effects, such as PKR-induced shutdown of protein synthesis in the infected cell and surrounding cells, and inhibition of virus proliferation by OAS stimulation of RNAse L (reviewed in Sen, 2001). Activation of natural killer (NK) cells and DCs then leads to initiation of an adaptive immune response, orchestrated by T and B cells.

Toll-like receptors

The TLRs are the best characterized members of the PRR family and are characterized by the presence of N-terminal leucine-rich repeats (LRRs), a transmembrane domain and a cytoplasmic C-terminal Toll/IL-R (TIR) homology domain (Rock *et al.*, 1998). TLRs are expressed on non-immune cells such as epithelial cells and fibroblasts and also on immune cells such as B cells, certain T cells, macrophages and DCs (Krug *et al.*, 2001b; Diebold *et al.*, 2003; Mogensen, 2009). TLRs 2 and 4 are located at the cell surface whilst TLRs 3, 7, 8 and 9 are located to intracellular compartments such as endosomes, where they are able to sense nucleic acids revealed after virus uncoating (Ahmad-Nejad *et al.*, 2002; Latz *et al.*, 2004). The virus antigens recognized by TLRs include surface glycoproteins (TLRs 2 and 4) and nucleic acids (TLRs 3, 7/8 and 9) (Fig. 14.4).

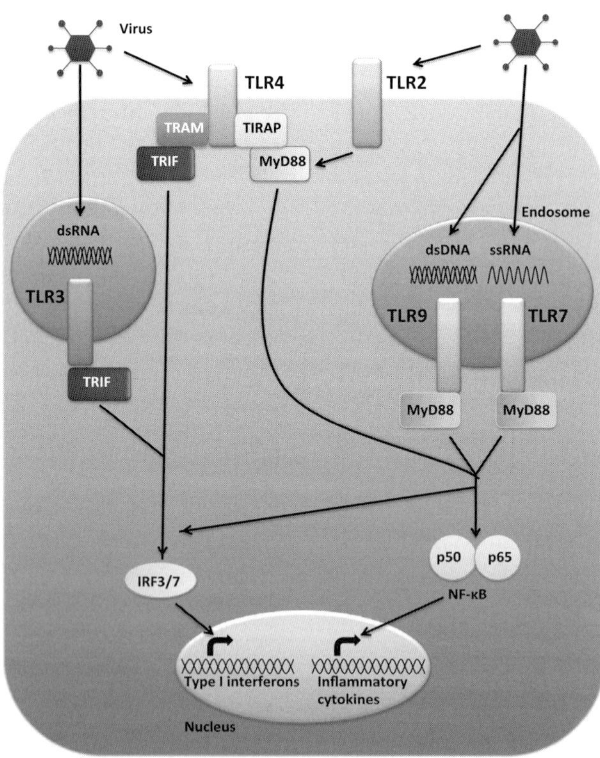

Figure 14.4 The role of the Toll-like receptors during virus infection. Detection of virus infection can occur by recognition of virus glycoproteins by cell surface Toll-like receptors (TLRs) such as TLR4 and 2, and/or recognition of virus DNA/RNA in the endosomal compartment by TLRs 3, 7/8 or 9. TLR signalling is conducted by the MyD88-dependent pathway (all TLRs apart from TLR3) and/or the TRIF-dependent pathway (TLRs 3 and 4). This leads to production of pro-inflammatory cytokines and type I interferons (IFNs). dsDNA, double-stranded DNA; dsRNA, double-stranded RNA; ssRNA, single-stranded RNA. Adapted from Kanneganti (2010) and Kumar *et al.* (2011).

Recognition of virus glycoproteins

TLR2 recognizes a multitude of lipoprotein antigens, including bacterial peptidoglycan and fungal zymosan (Takeuchi *et al.*, 1999; Underhill *et al.*, 1999). The mechanism for recognizing such a wide variety of microbial components resides in the ability of TLR2 to form heterodimers with TLRs 1 and 6 (Ozinsky *et al.*, 2000). TLR2 is also important for detection of Ad infection. TLR2 is activated by an as yet unidentified Ad capsid component (Appledorn *et al.*, 2008c) and is thought to be responsible for sustained activation of ERK and the transcription factor NF-κB during Ad infection (Appledorn *et al.*, 2008c).

TLR4 in conjunction with myeloid differentiation factor 2 (MD2) recognizes the Gram-negative bacterial ligand lipopolysaccharide (LPS) on the cell surface (Nagai *et al.*, 2002). TLR4 alongside TLR3 appears to negatively regulate the innate response to Ads during infection (Appledorn *et al.*, 2009). TLR4 may also be critical in the subsequent stimulation of an adaptive response to Ads (Biragyn *et al.*, 2002; Eisenbarth *et al.*, 2002; Rhee *et al.*, 2010).

Recognition of virus nucleic acids

The TLRs responsible for detection of nucleic acids are the endosomally located TLRs 3, 7/8 and 9 (Fig. 14.4). TLR3 is responsible for detection of double-stranded RNA (dsRNA) (Alexopoulou *et al.*, 2001) whilst TLRs 7/8 recognize single-stranded RNA (ssRNA) from various virus families (Heil *et al.*, 2004; Lund *et al.*, 2004). TLR9 has been identified as a receptor for recognition of bacterial and virus CpG DNA motifs (Hemmi *et al.*, 2000; Krug *et al.*, 2001a) and appears to be partially responsible for induction of antivirus responses following Ad infection. Infection of plasmacytoid DCs (pDCs) by CD46-utilizing Ads was shown to be detected by TLR9, but this was not seen for CAR-utilizing Ads (Iacobelli-Martinez and Nemerow, 2007) indicating that differing virus entry pathways may impact on TLR9 exposure to Ad DNA. As certain species B Ads (CD46-utilizing) appear to remain in the endosome for longer than species C Ads (CAR-utilizing) (Miyazawa *et al.*, 2001; Shayakhmetov *et al.*, 2003), it is possible that the prolonged exposure of species B Ads in the endosome may facilitate detection of these Ads by TLR9.

Downstream signalling pathways of TLRs

Binding of an antigenic ligand to a TLR triggers receptor dimerization, facilitating recruitment of adaptor proteins containing a TIR domain to the cytoplasmic TIR domain of the TLR. There are four adaptor molecules; myeloid differentiation factor 88 (MyD88), TIR-associated protein (TIRAP), TIR-domain-containing adaptor protein-inducing IFN-β (TRIF) and TRIF-related adaptor molecule (TRAM) (Oshiumi *et al.*, 2003). Activation of a TIR-containing adaptor protein leads to stimulation of signalling cascades resulting in changes in host gene expression, with the main responses being up-regulation of inflammatory cytokines and IFNs.

Pro-inflammatory cytokine production

MyD88-dependent pathway MyD88 associates with the TIR domain of TLRs 7/8 and 9 (Xu *et al.*, 2000) (Fig. 14.4). For TLRs 2 and 4, TIRAP is required for transduction of the signal from the TLR to MyD88 (Horng *et al.*, 2002; Yamamoto *et al.*, 2002). MyD88 activation leads to propagation of two distinct signalling pathways, facilitating activation of the AP-1 and NF-κB families of transcription factors and resulting in transcription of inflammatory cytokines (Schnare *et al.*, 2000; Honda *et al.*, 2004, 2005).

MyD88 appears to be critical in the innate response to Ads. Entry of Ads into host cells was shown to induce host gene dysregulation and up-regulation of certain cytokines by a MyD88-dependent mechanism (Hartman *et al.*, 2007a). Immune responses to the capsid and TLR9 recognition of Ad DNA also appear to be MyD88-dependent (Cerullo *et al.*, 2007; Hartman *et al.*, 2007a).

TRIF-dependent pathway The MyD88-independent pathway is seen for TLRs 3 and 4, and is TRIF-dependent (Hoebe *et al.*, 2003; Oshiumi *et al.*, 2003) (Fig. 14.4). For TLR4, TRAM is required for transduction of the signal from the TLR to TRIF (Yamamoto *et al.*, 2003b). Induction of inflammatory cytokines by TLR4 appears to require both the MyD88-dependent and the MyD88-independent TLR4 pathways (Kawai *et al.*, 1999; Yamamoto *et al.*, 2003a).

Type-I IFN production

MyD88-dependent pathway Induction of type-I IFNs via TLRs 7/8 and 9 is MyD88-mediated. Activation of MyD88 by TLRs 7 or 9 in pDCs results in activation of interferon regulatory factor (IRF) 7, leading to induction of IFN-β and ISGs (Honda *et al.*, 2004). The protein kinase IRAK-1 appears to have a specific role in the MyD88-mediated IFN pathway, whereas MyD88 and IRAK-4 are crucial to both MyD88-dependent inflammatory and IFN responses (Uematsu *et al.*, 2005).

TRIF-dependent pathway Activation of TRIF by TLR3 or 4 leads to activation of NF-κB (Meylan *et al.*, 2004) and also the protein kinase TBK1 (Fitzgerald *et al.*, 2003). Both TBK1 and IKKi/IKKε appear to play roles in the MyD88-independent induction of IRF3 by TLRs 3 and 4 (Kawai *et al.*, 2001; Fitzgerald *et al.*, 2003; Sharma *et al.*, 2003; Hemmi *et al.*, 2004; Perry *et al.*, 2004). IRFs appear to be an important dysregulation target for Ads; binding of Ad E1A to the transcriptional co-activator CBP/p300 prevents interaction of IRF3 with CBP/p300 (Yoneyama *et al.*, 1998) and E1A has also been shown to target IRF9 (Leonard and Sen, 1997), preventing up-regulation of IFN-γ. This indicates that bypassing the IFN response is crucial for the Ad life cycle.

Cytosolic PRRs

The cytosolic PRRs identified to date are RNA-induced protein kinase (PKR), retinoic acid-inducible gene 1 (RIG-1), melanoma differentiation-associated gene 5 (MDA-5), DNA-dependent activator of interferon (DAI), and NLRs (Fig. 14.5).

RNA sensors

PKR

PKR is a cytosolic RNA sensor capable of detecting both dsRNA and 5′ triphosphate ssRNA with short stem loops (Garcia *et al.*, 2006). Activation of PKR upon ligand binding leads to phosphorylation of the protein synthesis initiation factor 2α (eIF2a) (Garcia *et al.*, 2006). This results in inhibition of general translation therefore preventing virus replication in the cell. PKR can also elicit an IFN response in response to dsDNA binding (Diebold *et al.*, 2003). The Ad virus-associated (VA) RNAs are able to bind and inhibit PKR activation (Mathews and Shenk, 1991) indicating PKR is critical for detection of Ad infection.

RIG-1-like receptors

This subgroup of cytosolic nucleic acid sensors consists of RIG-1, MDA-5 and LGP2 (Yoneyama *et al.*, 2005). RIG-1 is able to stimulate an IFN response upon detection of dsRNA (Yoneyama *et al.*, 2004) and has been implicated in the induction of the interferon response to Ad infection by recognition of Ad VA RNAs (Minamitani *et al.*, 2011). MDA-5 is also a dsRNA sensor and appears to recognize virus dsRNA ligands that are different to those recognized by RIG-1 (Kato *et al.*, 2006). Although LGP2 binds dsRNA, it appears to have a role in regulation of RIG-1 and MDA-5 signalling rather than RNA detection. However, it is currently unclear whether the regulation by LGP2 is negative (Yoneyama *et al.*, 2005) or positive (Satoh *et al.*, 2010).

Recruitment of RIG-1/MDA-5 and IKK kinases to the adaptor protein mitochondrial antivirus signalling (MAVS) facilitates activation of IRF3 and NF-κB (Meylan *et al.*, 2005). The cellular proteins FADD, RIP1 and STING have also been reported to interact with MAVS and are required for type-I IFN production in response to cytosolic dsRNA detection (Balachandran *et al.*, 2004; Ishikawa and Barber, 2008).

DNA sensors

DAI

DAI was observed to act as a cytoplasmic DNA sensor in L929 cells, and was not activated on exposure to RNA (Takaoka *et al.*, 2007). Signalling appears to be independent of RIG-1, yet dependent on MAVS and TBK1/IKKi (Ishii *et al.*, 2006). Binding to DNA allows association of DAI with TBK1, NFκB, IRF3 and possibly IRF7 (Takaoka *et al.*, 2007).

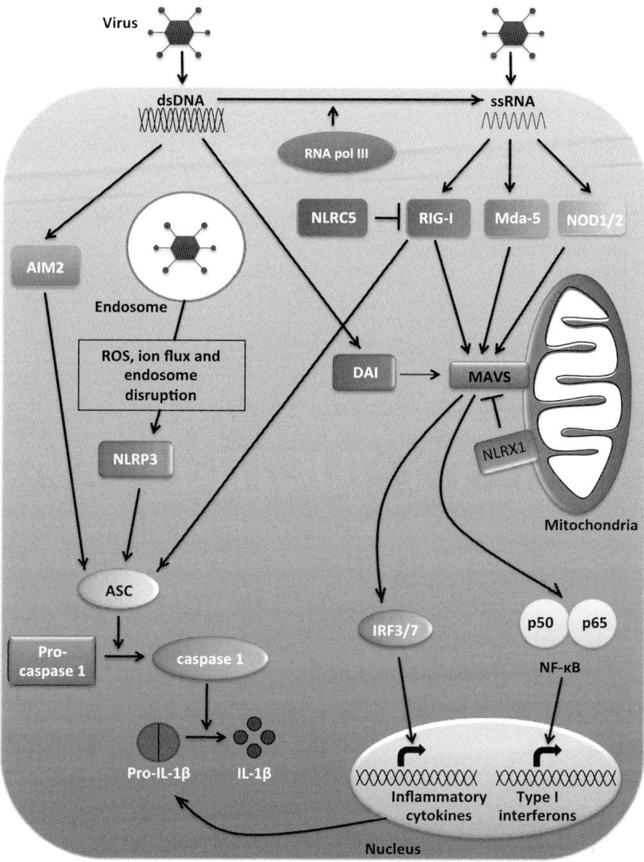

Figure 14.5 Detection of virus infection by the cytosolic pattern-recognition receptors. Virus infection activates NOD-like receptors (NLRs), RIG-1-like receptors (RLRs) and AIM2-like receptors (ALRs) resulting in induction of pro-inflammatory cytokines and interferons (IFNs). Virus pathogen-associated molecular patterns (PAMPs) stimulate absent in melanoma 2 (AIM2) to recruit ASC (apoptosis-associated speck-like protein containing a CARD [caspase activation and recruitment domain]) forming the AIM2 inflammasome complex. ASC then cleaves pro-caspase 1 to mature caspase 1, resulting in production of IL-1β. Activation of RIG-1 by single-stranded RNA (ssRNA) triggers a signalling pathway involving the mitochondrial antivirus signalling protein (MAVS), leading to phosphorylation and activation of the transcription factors interferon regulatory factor (IRF) 3 and 7, leading to induction of IFNs. RIG-1 can also form an inflammasome complex with ASC to induce production of IL-1β and IL-18. The NLR NOD2 also interacts with MAVS to induce inflammatory cytokine and IFN production, and NALP3 can associate with ASC to form the NALP3 inflammasome. Certain NLRs play roles in negative regulation of immune signalling; NLRX1 negatively regulates MAVS activity at the mitochondrial membrane, whilst NLRC5 negatively regulates interaction of RIG-1 with MAVS. ROS, reactive oxygen species. Adapted from Kanneganti (2010) and Kumar *et al.* (2011).

RNA polymerase III

Another novel mechanism of DNA detection is by transcription of cytosolic poly (dA:dT) DNA into 5′ triphosphate RNA by RNA polymerase III (Ablasser *et al.*, 2009; Chiu *et al.*, 2009). This allows recognition of the 5′ triphosphate RNA by RIG-1, facilitating activation of a type-I IFN response (Ablasser *et al.*, 2009; Chiu *et al.*, 2009).

AIM2-like receptors

The ability of certain exogenous DNAs, in particular non-poly (dA:dT) DNAs, to elicit IFN-mediated responses in host cells independently of either DAI or RNA polymerase III indicates further intracellular DNA sensors remain to be identified (reviewed in Hornung and Latz, 2010). Recently, a member of the PYHIN protein family known as IFI16 was found to directly bind exogenous DNA and elicit an IFN-β response (Unterholzner *et al.*, 2010). The mechanism was found to be independent of TLRs, DAI or RNA polymerase III and dependent on STING, TBKI and IRF3 (Unterholzner *et al.*, 2010). In addition, another PYHIN protein known as AIM2 has also been identified as a DNA sensor for stimulating release of interleukin-1β (IL-1β) (Burckstummer *et al.*, 2009; Fernandes-Alnemri *et al.*, 2009; Hornung *et al.*, 2009; Roberts *et al.*, 2009). The mechanism for this involves recruitment of AIM2 into a large molecular weight complex known as an inflammasome (Burckstummer *et al.*, 2009; Fernandes-Alnemri *et al.*, 2009; Hornung *et al.*, 2009; Roberts *et al.*, 2009). Activation of inflammasomes leads to activation of inflammatory caspase 1, resulting in cleavage of pro-IL-1β and pro-IL-18 into their mature forms which are then secreted from the cell (reviewed in Martinon and Tschopp, 2007). IL-1β functions to generate fever, activate lymphocytes and promote leucocyte infiltration to infected sites (reviewed in Dinarello, 1996) whilst IL-18 stimulates the release of IFN-γ from activated T cells and NK cells, leading to production of a T helper 1 (T_H1) response (Okamura, 2005).

Nod-like receptors

Nod-like receptors (NLRs) are a class of intracellular pathogen sensors characterized by the presence of a conserved nucleotide binding and oligomerization domain (NOD) motif (Inohara *et al.*, 2001; Harton *et al.*, 2002). Detection of PAMPs by NLRs results in conformational rearrangement and oligomerization, facilitating interaction with effector molecules and propagation of immune signalling (reviewed in Kanneganti, 2010; Fig. 14.5).

NLRs NOD1/2 interact with the protein kinase RIP2K leading to activation of NF-κB and MAPK pathways (Girardin *et al.*, 2003; Hidajat *et al.*, 2009; Xiang *et al.*, 2003) and are also involved in triggering autophagy (Travassos *et al.*, 2010). Following virus infection, NOD2 also associates with MAVS facilitating induction of type-1 IFNs (Sabbah *et al.*, 2009). Some NLRs possess immune regulatory functions; NLRX1 and NLRC5 have both been shown to negatively regulate MAVS and RIG-1 respectively, preventing induction of type-1 IFNs and inflammatory cytokines (Moore *et al.*, 2008; Arnoult *et al.*, 2009).

Other NLR family members, such as NLRP1, NLRP3 and NLRC4 assemble into inflammasome complexes leading to production of IL-1β and IL-18 (Martinon *et al.*, 2002). Activation of the NLRP3 inflammasome appears to occur in response to a wide range of stimuli including microbial PAMPs, endogenous danger-associated molecular patterns (DAMPs) and exogenous non-microbial stimuli such as crystalline particles (Dostert *et al.*, 2008; Allen *et al.*, 2009; Maitra *et al.*, 2009; Zhou *et al.*, 2010). This activation appears to occur via a two step process, involving a 'primer' stimulus and an 'activator' stimulus (reviewed in Bauernfeind *et al.*, 2011). Primer stimuli include TLR, RLR and NLR ligands that lead to increased transcription of NLRP3 and pro-IL1β via NF-κB activation (Bauernfeind *et al.*, 2009). NLRP3 activators, such as potassium (K^+) efflux, lysosomal degradation or production of reactive oxygen species (ROS), then stimulate caspase 1 activation via the inflammasome (Cruz *et al.*, 2007; Petrilli *et al.*, 2007; Halle *et al.*, 2008; Hornung *et al.*, 2008; Duewell *et al.*, 2010). NLRP3 activation during Ad infection has been shown to require both TLR9 recognition of Ad DNA (Muruve *et al.*, 2008) and ROS production owing to Ad endosome penetration (Lasaro *et al.*, 2008).

The complement system

Complement activation can occur via three distinct pathways; the classical (CP) pathway, the alternative (AP) pathway and/or the mannose-binding lectin (MBL) pathway (reviewed in Walport, 2001a,b). CP is activated when complement protein C1q binds to antibody–pathogen complexes, AP by the continuous non-specific hydrolysis of complement component protein C3 and the MBL pathway by the binding of microbial proteins to MBL or ficolins (Guo and Xin, 2006). These pathways converge at the level of C3 and C5 convertase formation resulting in the generation of opsonins, anaphylotoxins and the membrane attack complex (MAC) which facilitate opsonization, acute inflammation and bacterial cell lysis respectively (Walport, 2001a,b). Activation of complement also leads to induction of adaptive immune responses (reviewed in Morgan *et al.*, 2005; Kemper and Atkinson, 2007).

Following systemic administration, Ads activate the CP by direct binding of C1q to Ad-antibody complexes (Cichon *et al.*, 2001). The AP also appears to play a role during Ad infection as both Ad4 and Ad5 appear to bind C3 directly (Hartman *et al.*, 2008a) and are able to activate complement when both CP and MBL pathway are blocked (Jiang *et al.*, 2004; Kiang *et al.*, 2006a). Interestingly, the complement response differs between Ad serotypes. Human Ad3 and simian Ad23 (SAd23) appear to activate the alternative complement pathway to a greater extent than Ad5 (Jiang *et al.*, 2004; Kiang *et al.*, 2006a; Appledorn *et al.*, 2008a) and, whereas Ad5 complement responses were found to be primarily dependent on functional C3, Ad3 complement responses were primarily dependent on Factor B (FB) (Kiang *et al.*, 2006a; Appledorn *et al.*, 2008a).

Certain responses to IV-administered Ads appear to be partially complement-dependent (Kiang *et al.*, 2006a; Appledorn *et al.*, 2008a). Activation of NF-κB and release of cytokines and chemokines appeared to be partially dependent on C3 (Appledorn *et al.*, 2008a,b). The alternative pathway is implicated in thrombocytopenia and the complement-dependent transcriptome response following Ad transduction of the liver, as both these responses were shown to be dependent upon functional FB and C3 (Appledorn *et al.*, 2008a,b). The subsequent induction of an adaptive response to Ads also appears to be complement-dependent (Appledorn *et al.*, 2008a,b).

Approaches to overcome immune responses to adenovirus vectors

Manipulation of the immune system is critical for improving both the safety and the efficacy of Ad vectors. The two broad strategies for manipulating immune responses to Ad vectors are modification of the vector or pre-emptive modification of the host immune system.

Modification of the host immune system

Suppression of innate inflammatory responses

General immunosuppression

This method uses chemotherapy to transiently suppress the host immune system upon administration of Ad vectors. Intraperitoneal injection (IP) of the glucocorticoid dexamethasone (DEX) in mice one day prior to IV administration of a β-galactosidase (β-gal)-expressing Ad vector significantly reduced release of pro-inflammatory cytokines such as IL-6 and TNF-α, reduced inflammatory infiltrate in the lung and increased β-gal expression in the lung (Otake *et al.*, 1998; Seregin *et al.*, 2009b). These studies have also been extended to non-human primates, with DEX treatment in African green monkeys reducing the inflammatory response upon Ad vector administration (Lawrence *et al.*, 1999). Several studies have also highlighted that suppressing the immune system prior to treatment with oncolytic Ads increases virus efficacy (Friedman *et al.*, 2006; Fulci *et al.*, 2006; Lamfers *et al.*, 2006; Thomas *et al.*, 2008).

Selective immunosuppression

As TLR9 and ERK are both important in innate immune signalling during Ad infection (Cerullo *et al.*, 2007; Iacobelli-Martinez and Nemerow, 2007; Appledorn *et al.*, 2008c), TLR9 antagonists and ERK inhibitors have been investigated as potential inflammatory suppressors. Treatment of mice with the TLR9 antagonist oligonucleotide ODN-2088 or the ERK inhibitor U0126 prior to and post-administration with an Ad vector both resulted in significantly reduced innate immune responses upon Ad administration, including decreased production of inflammatory cytokines (Tibbles *et al.*, 2002; Cerullo *et al.*, 2007). However, ERK also appears critical in the uptake of Ads into the host cell therefore ERK inhibition may impede Ad transduction (Tibbles *et al.*, 2002). In addition, ERK may also be required for Ad propagation as reduced propagation of intratumorally administered oncolytic vectors has been reported (Schumann and Dobbelstein, 2006).

Targeting pro-inflammatory cytokines has also been attempted. Administration of anti-TNF-α antibodies (Wilderman *et al.*, 2006) or antagonism of the TNF-α receptor (TNFα-R1) (Benihoud *et al.*, 2007) significantly reduced inflammatory responses to Ad vectors. The latter study also detected reduced infiltration of immune cells into the transduced liver, allowing prolonged transgene expression in the liver. There was also reduced induction of anti-Ad antibodies, indicating that the innate response is also important in inducing cellular immunity to Ads. However, TNF-α may play a role in Ad propagation, as intratumorally administered oncolytic Ads were shown to have reduced propagation when anti-TNF-α antibodies were used (Wilderman *et al.*, 2006).

Taken together, this indicates that selective immunomodulation strategies may hold promise for Ad gene therapy. However, they may reduce the efficiency of Ad replication, therefore alternative immunomodulatory strategies may be required for oncolytic vectors.

Macrophage depletion

The majority of Ad vectors administered systemically are taken up and sequestered by the liver (Worgall *et al.*, 1997a,b). Here Kupffer cells are critical in the efficacy of Ad vectors as they take up and degrade up to 90% of IV-administered Ad vectors (Worgall *et al.*, 1997a,b). This tropism decreases the therapeutic efficacy of Ad vectors, as an insufficient proportion of the vector reaches the target site. Therefore depletion of Kupffer cells has been attempted in order to improve efficacy and subvert induction of immune responses to Ad vectors. The macrophage-depletion agents dichloromethylene bisphosphate (Cl_2MBP) and gadolinium chloride ($GdCl_3$) appear to increase the persistence of Ad DNA in the liver and lungs of mice (Lieber *et al.*, 1997; Worgall *et al.*, 1997a,b) and increased hepatic gene transfer by Ad vectors was also seen upon macrophage depletion with liposome-encapsulated clodronate (Kuzmin *et al.*, 1997). Reduced antibody responses to the transgene (Kuzmin *et al.*, 1997) and reduced production of inflammatory cytokines and chemokines (Hortobagyi *et al.*, 2001) were also evident. In addition, ERK activation upon Ad5 vector administration was shown to be abrogated in macrophage-depleted livers (Appledorn *et al.*, 2008c), indicating that Kupffer cells are the major source of ERK signalling in the liver.

Combining macrophage-depleting agents with other drugs has also been investigated. Combining macrophage depletion with the anticoagulant drug warfarin was shown to significantly reduce liver toxicity of an oncolytic vector (Shashkova *et al.*, 2008a). The anticoagulant effects of warfarin decrease blood factors necessary for hepatocyte transduction, thus reducing toxicity.

Modification of the vector

Genome modification

Inherent immune-evasive properties of Ads

Multiple Ad genes encode proteins with immunomodulatory activities. E1A is the first gene transcribed upon entry of Ads to the host cell nucleus, and is responsible for the subsequent transcription of the remaining early genes (E1B, E2A, E2B, E3 and E4) (Nevins *et al.*, 1979; Leff *et al.*, 1984; Lillie and Green, 1989; Bridge *et al.*, 1991; Liu and Green, 1994). One of the major functions of E1A during infection is binding and inactivation of pRb, leading to activation of the transcription factor E2F and subsequent progression of the cell cycle to S phase (Whyte *et al.*, 1988; Egan *et al.*, 1989). E1A is also involved in immune subversion; Ad5 E1A abrogates IFN-γ-mediated gene activation by inhibition of STAT1 (Look *et al.*, 1998) and IRF9 (Leonard and Sen, 1997), factors involved in the IFN signalling pathways. In addition, Ad12 E1A, but not Ad5 E1A, has also been shown to block MHC class 1 expression at the cell surface (Ackrill and Blair, 1988).

The gene products of the E1B region are the E1B-55K and E1B-19K proteins. Both proteins act to suppress p53-induced cell cycle arrest and apoptosis triggered by the E1A-mediated abolition of pRb function (Rao *et al.*, 1992; Debbas and White, 1993; Yew *et al.*, 1994; Harada *et al.*, 2002; Cuconati *et al.*, 2003). The E3 region encodes a multitude of proteins with immunomodulatory functions. E3–19K binds to MHC class 1 molecules, preventing trafficking to the cell surface (Burgert and Kvist, 1985; Bennett *et al.*, 1999) thus reducing exposure of MHC class I exhibiting virus peptides to circulating cytotoxic T-cells. E3–10.4K and 14.5K also affect cell surface receptors by forming the Receptor Internalization and Degradation (RID) complex (Gooding *et al.*, 1991). The RID complex acts to down-regulate cell surface expression of the Fas and TRAIL ligands, preventing circulating NK and T-cells from killing the infected cell (Shisler *et al.*, 1997). E3 14.7K also prevents host cell death by blocking TNF-α and TRAIL-induced apoptosis (Gooding *et al.*, 1991; Tollefson *et al.*, 2001).

Attenuated Ads

In replication-deficient Ad vectors the E1 region is deleted to prevent virus replication and most vectors are also E3-deleted to increase cloning capacity. Removal of virus genes also reduces vector immunogenicity by removal of immunogenic virus proteins (Everett *et al.*, 2003; Kothari *et al.*, 2008; Gabitzsch *et al.*, 2009; Yamaguchi *et al.*, 2010). However, removal of E1 and/or E3 gene regions often results in increased inflammatory responses to Ad vectors compared with infection by wildtype virus (Amalfitano and Parks, 2002; Alba *et al.*, 2005; Hartman *et al.*, 2008b; Seregin and Amalfitano, 2009; Toth and Wold, 2010; Vetrini and Ng, 2010), indicating that it may be beneficial to retain the immune subversive properties of the proteins encoded in these regions. Indeed, re-introduction of the E3 region in some studies has shown reduced humoral and $CD8^+$ T-cell responses and prolonged transgene expression (Ilan *et al.*, 1997). Conversely, with regard to vaccine vectors, removal of genes responsible for immune subversion may be beneficial in eliciting potent immune responses required for inducing immunity to the transgene. Indeed, an Ad5-based vector deleted for E1, E2B and E3 regions displayed potent immunogenicity compared with an E1-deleted vector in naïve mice (Osada *et al.*, 2009).

In the case of oncolytic vectors, the E1 region is mostly retained to facilitate virus replication in target cells, for example ONYX-015 contains a single mutation in E1B-55K (Heise *et al.*, 1997). However, individual virus proteins often have multiple functions during infection, and many virus proteins have overlapping functions that can compensate for each other when one is deleted. The mutant E1B-55K in ONYX-015 is unable to bind and inhibit p53 (Bischoff *et al.*, 1996) but this did not confer selective replication or selective killing by ONYX-015 to p53-negative tumour cells (Goodrum and Ornelles, 1998; Rothmann *et al.*, 1998; Turnell *et al.*, 1999). Indeed, p53 activity was found to be attenuated in cells infected with E1B-55K-deleted Ads (Hobom and Dobbelstein, 2004; O'Shea *et al.*, 2004). It was later shown that it was in fact late virus RNA export that was dictating the tumour selectivity of ONYX-015 rather than p53 expression. In addition to abolition of p53 activity, E1B-55K is also required for export of virus RNAs, and this function was complemented in many cancer cells by up-regulation of heat shock proteins (HSPs) (O'Shea *et al.*, 2004, 2005). Moreover, it was recently shown that both E1A and E4 ORF3 are also able to subvert p53 activity (Soria *et al.*, 2010; Savelyeva and Dobbelstein, 2011). This highlights the complexity of modifying genes encoding multifunctional proteins, and the potential of genes that are retained in the vector being able to compensate functions of an absent gene.

Non-attenuated Ads

In order to produce oncolytic vectors with cancer cell-specific replication, the Ad vector genome is modified, either deleting or mutating virus genes. However, genome modification often results in a vector that replicates more slowly in the tumour tissue than wild-type virus. To increase oncolytic vector efficacy, the Ad5-based vector VRX-007 was constructed, which possesses no attenuated modifications (Doronin *et al.*, 2003). Instead it has the immunomodulatory E3 region deleted, allowing the immune system to detect the virus. It also overexpresses the Adenovirus Death Protein (ADP) resulting in rapid release of progeny from host cells, increasing the speed of spread from cell to cell. When compared with the parental tumour-attenuated vector KD3, survival rates in subcutaneous tumour xenograft models were significantly increased following IV administration but not following intratumoral injection (Doronin *et al.*, 2003; Toth *et al.*, 2003, 2010). VRX-007 also increased survival rates to a greater extent than ONYX-015 in an orthotopic lung tumour model (Toth *et al.*, 2010). In addition, the immune status was shown to significantly impact the efficacy of VRX-007. Hamsters that had been pre-immunized with Ad5 displayed decreased toxicity upon intratumoral administration of VRX-007 compared with naïve hamsters, which may be due to pre-existing antibodies neutralizing the vector that escapes into the bloodstream (Dhar *et al.*, 2009a,b). Naïve immunocompetent hamsters rapidly developed immunity to the vector, resulting in decreased efficacy and tumour regrowth within a month of vector administration. However, vector efficacy was retained in naïve immunosuppressed hamsters (Thomas *et al.*, 2008). This indicates that naïve hosts develop an anti-vector immune response which reduces vector efficacy. In addition, immunosuppressed hamsters that had been previously immunized with Ad5 displayed decreased vector efficacy against subcutaneous tumours compared with immunocompetent pre-immunized hamsters (Dhar *et al.*, 2009a,b). This indicates that pre-existing neutralizing antibodies reduce efficacy but immunosuppression is necessary to reveal this (Dhar *et al.*, 2009a,b).

The spread of intratumorally administered VRX-007 from the tumour site was also investigated. It was found that pre-existing neutralizing antibodies were beneficial in reducing 'spill-over' of vector to non-target organs such as the liver (Dhar *et al.*, 2009a,b). Pre-existing antibodies may also be beneficial in preventing lethal IV titres of Ads in both immunocompetent and immunosuppressed animals (Dhar *et al.*, 2009a,b).

Intrinsic immunogenicity of the transgene

Adaptive immune responses can reduce the expression of encoded transgenes, although this varies greatly depending on the intrinsic immunogenicity of the transgene itself (Ding *et al.*, 2002). For treatment of genetic disorders, there has been some success with long-term expression of homologous genes in immunocompetent hosts (Kiang *et al.*, 2006b; Tripathy *et al.*, 1994). However, Ad vectors carrying heterologous genes have been less successful as these genes are more immunogenic, resulting in transient expression due to induction of cellular immune responses (Lemarchand *et al.*, 1992; Smith *et al.*, 1993; Sarukhan *et al.*, 2001). Indeed, it was recently shown that strong foreign promoters such as CMV contribute greatly to the increased inflammation seen with Ad vectors, and the degree of inflammation appears to correlate with the strength of the promoter (Schaack *et al.*, 2011).

Specific targeting

The following methods are intended to increase tropism for a specific tissue type whilst decreasing tropism for other tissue types, thus increasing efficacy and reducing potential activation of host immune responses. Methods for specific targeting of Ad vectors include direct administration to the target site, placing certain Ad gene regions under the control of tissue-specific promoters (TSPs) or incorporation of targeting ligands onto the Ad capsid surface.

Route of administration

Administration routes differ depending on the target tissue. IV, intratumoral (IT) or intraperitoneal (IP) administration of ONYX-015 have been fully assessed in both murine models and in clinical trials (Alemany, 2007; Crompton and Kirn, 2007). In nude mice, the maximum IV dose before toxicity was apparent was 1.7×10^9 PFU of ONYX-015 split into 5 daily doses (Heise *et al.*, 1999b), whereas a single IT injection of 10^9 PFU did not appear to evoke adverse toxicity (Heise *et al.*, 1999a,b). In addition, no dose-limiting toxicity was observed in clinical trials following administration of the ONYX-015 vector by five IT administrations (Habib *et al.*, 2002).

Efficient delivery of oncolytic Ads to target cells is also influenced by the tumour microenvironment and the initial tumour distribution (Heise *et al.*, 1997). Even when injected IT, virus-infected cells occupy only a small percentage of the whole tumour mass, and are usually located close to the site of administration and once administered, virus titres decrease rapidly (Sauthoff *et al.*, 2003). To prevent impedance of virus spread within a tumour, multiple injections within a tumour mass has been investigated (Barton *et al.*, 2004), and also injection into the major artery supplying the tumour (Reid *et al.*, 2002).

Delivery of oncolytic Ads to tumours using mesenchymal stem cells (MSCs) has also been investigated. MSCs home to tumours upon systemic delivery (reviewed in Brunt *et al.*, 2007), and can be infected with Ads (Ishihara *et al.*, 2006), therefore infected MSCs deliver oncolytic Ads to tumour cells. Increased efficacy of the Ad-infected MSCs compared with Ad alone has been shown in murine orthotopic lung and breast cancer models (Komarova *et al.*, 2006; Hakkarainen *et al.*, 2007).

Differential administration routes are also being investigated in Ad vaccine vector regimens in order to bypass pre-existing immunity (Hidajat *et al.*, 2009). Indeed, in a newborn mouse model, orally administered replication-competent Ad vectors expressing HIV-1 proteins appeared to bypass maternally acquired antibodies (Xiang *et al.*, 2003).

Modification of the capsid

Both hexon and fibre proteins have been shown to be highly immunogenic (Gall *et al.*, 1996, 1998; Roy *et al.*, 1998; Molinier-Frenkel *et al.*, 2002; Wu *et al.*, 2002b), therefore covalent modification of these epitopes has been attempted in order to reduce induction of an immune response to Ads. Attachment of synthetic polymers of polyethylene glycol (PEG) (O'Riordan *et al.*, 1999), polyactic glycolic acid (PLGA) (Matthews *et al.*, 1999) or polycationic liposomes (Lee *et al.*, 2000) to the Ad capsid have all been utilized in an attempt to shield immunogenic capsid proteins.

PEGylation of Ad capsids has been intensively studied. PEGylated Ads have significantly reduced innate immune responses compared with unmodified vectors, with reduced plasma cytokine and chemokine release and reduced toxic side effects such as thrombocytopenia (Croyle *et al.*, 2002, 2005; De Geest *et al.*, 2005). In addition, administration of high doses of PEGylated Ad vectors resulted in prolonged transgene expression in both murine liver and lung (O'Riordan *et al.*, 1999; Croyle *et al.*, 2002). The adaptive immune response profile was also improved, with reduced $CD8^+$ and $CD4^+$ T-cell responses and reduced neutralizing antibody titres compared with non-PEGylated vectors (Croyle *et al.*, 2002, 2005; De Geest *et al.*, 2005). Combination therapy of PEGylated Ad and treatment with the glucocorticoid methylprednisolone resulted in a dramatic reduction of inflammatory cytokine expression in the liver, resulting in reduced infiltration of lymphocytes (De Geest *et al.*, 2005). Studies in mice have also been extended to non-human primates, with a reduction in inflammatory cytokine production reported (reviewed in Wonganan and Croyle, 2010). However, PEGylation may reduce the efficiency of Ad transduction *in vivo* (Ogawara *et al.*, 2004; Mok *et al.*, 2005).

Chemical modification has also been investigated for oncolytic Ads (reviewed in Kreppel and Kochanek, 2008). Shielding the Ad from circulating neutralizing antibodies permits repeated administration of the vector. Although PEGylation reduces capsid interaction with CAR *in vitro* (Fisher *et al.*, 2001), it does not appear to prevent CAR-independent liver transduction *in vivo* (Kalyuzhniy *et al.*, 2008) unless high molecular weight PEG is used that indirectly prevents liver transduction by preventing access to hepatocytes through the liver endothelium (Doronin *et al.*, 2009). To target PEGylated Ads in a more specific manner, tumour ligands such as VEGF have been incorporated onto the surface of PEGylated virions (Eto *et al.*, 2008; Park *et al.*, 2008).

Genome modification of the Ad vector either involves removing select portions of the genome and replacing it with the corresponding genes from an alternative Ad serotype, or replacing the entire genome with an alternative Ad serotype (Wu *et al.*, 2002a; Youil *et al.*, 2002; Noureddini and Curiel, 2005; Roberts *et al.*, 2006; McCoy *et al.*, 2007; Liu *et al.*, 2009b).

Incorporation of targeting ligands

Modification of epitopes critical for Ad entry have been attempted in order to alter the tropism of species C Ad vectors (Dmitriev *et al.*, 1998; Waddington *et al.*, 2008; Coughlan *et al.*, 2009b; Tang *et al.*, 2009). The capsid proteins targeted for genetic manipulation include hexon, fibre, penton base and pIX (Crompton *et al.*, 1994; Wickham *et al.*, 1997; Dmitriev *et al.*, 1998; Einfeld *et al.*, 1999; Vellinga *et al.*, 2004). Vectors with modified capsid proteins have been produced with the intention of either reducing tropism from certain tissue types (Xia *et al.*, 2000) or increasing tropism for a selected tissue (Su *et al.*, 2001). Targeting to $\alpha_v\beta_{3/5}$ integrins (Nokisalmi *et al.*, 2010), HS-GAGs (Wickham *et al.*, 1996) and CD40 (Belousova *et al.*, 2003) have all been attempted. Although most studies have not translated from *in vitro* findings to *in vivo*, targeting to $\alpha_v\beta_{3/5}$ integrins by an oncolytic Ad with an RGD motif inserted into the fibre protein resulted in increased tumour uptake and decreased liver sequestration (Coughlan *et al.*, 2009a). However, it appears that incorporation of some targeting ligands into the capsid proteins does not necessarily direct them to the intended target cells (van Geer *et al.*, 2009).

Bridging molecules have also been utilized to deliver Ad vectors to target cells efficiently. In order to target the folate receptor, which is overexpressed on the surface of multiple cancer cell lines (Coney *et al.*, 1991; Weitman *et al.*, 1992; Ross *et al.*, 1994), folate was conjugated to the fragment antigen binding (Fab) region of a neutralizing monoclonal anti-fibre antibody and mixed with the virus prior to administration in mice (Douglas *et al.*, 1996). The conjugate was shown to redirect the tropism of the vector to cells expressing the folate receptor with high efficiency and vectors carrying the HSV-TK gene were shown to kill target cells efficiently (Douglas *et al.*, 1996). Antibody fragments have also been directly conjugated to Ad capsid components. One such vector displayed pIX conjugated to single-domain (consisting of the heavy chain only) antibodies to the AFA1 protein, which is a ligand of CD66c [carcinoembryonic antigen-related cell adhesion molecule family member 6 (CEACAM6)] (Poulin *et al.*, 2010). This pIX-conjugated vector was shown to increase transduction of CD66c-expressing target cells. It was also evident from this study that the targeting ligand in question can drastically impact upon the stability of the recombinant pIX, as conjugation of single-chain variable fragment antibodies (scFv) to pIX resulted in low level incorporation of the conjugate into the capsid (Poulin *et al.*, 2010).

Chimaeric vectors

Chimaeric vectors can be utilized to change the tissue tropism of Ad5 vectors, as CAR is found on many tissue types and CAR-independent liver transduction is responsible for most of the toxicity detected with IV-administered Ad vectors. As CAR expression is generally low or down-regulated on cancer cells (Li *et al.*, 1999; Okegawa *et al.*, 2000), Ad5 vectors whose fibre protein has been replaced with that of a species B Ad have been designed to increase the host range of the vector (Shayakhmetov *et al.*, 2000). One Ad5-based oncolytic vector was engineered to replace the Ad5 fibre with that of Ad3, providing CAR-independent cell attachment (Nokisalmi *et al.*, 2010; Pesonen *et al.*, 2010). Chimaeric Ads expressing structurally homologous attachment proteins from other viruses have also been attempted. These include the mucosal-targeting sigma1 protein of reovirus type 3 Dearing (T3D), which targets the Ad vector to cells expressing Junctional adhesion molecule 1 (JAM1) (Mercier *et al.*, 2004).

Although these retargeting methods successfully extend the tissue tropism of the vector, they do not address the inherent liver tropism of Ad vectors, which is CAR-independent via plasma proteins such as FX which bridges the Ad5 hexon protein to HSPGs (Kalyuzhniy *et al.*, 2008; Waddington *et al.*, 2008). As FX demonstrates the greatest impact upon Ad liver transduction and the FX binding site on hexon has recently been defined (Alba *et al.*, 2009), an Ad vector ablated for FX binding has been constructed. Although the vector still initially localized to the liver and spleen of mice following IV administration, the transduction of the liver was shown to be reduced (Alba *et al.*, 2010). However, the inflammatory response to these FX-binding ablated vectors was not significantly different from the parental vector, and induction of certain inflammatory cytokines was even increased.

Capsid display of antigenic epitopes

A novel approach to improve the efficacy of vaccine vectors involves modification of capsid proteins to express antigenic epitopes (Crompton *et al.*, 1994; Worgall *et al.*, 2005). A vector displaying a HIV epitope incorporated into the hypervariable region 2 (HVR2) of hexon protein was shown to elicit a HIV antigen-specific cellular immune response in mice (Matthews *et al.*, 2010). However, it seems that the conformation of the HIV epitope in hexon is different to that in the wild-type protein, which may impact on immune responses to the epitope. In addition, pre-existing Ad5 antibodies were still able to recognize the capsid-modified vector (Matthews *et al.*, 2010). A recent study found that the choice of capsid component for the ligand conjugate can also impact on the anti-antigen immunogenicity. A comparison of four Ad vectors displaying a haemagglutination (HA) epitope of influenza incorporated into either hexon, penton base, fibre or pIX found that the greatest cellular and humoral responses were generated against the vector with the epitope incorporated into the fibre protein (Krause *et al.*, 2006). This indicates that epitope display on the fibre protein may be necessary to achieve optimal induction of anti-antigen immunity to vaccine vectors.

Tissue-specific promoters

Tissue-specific promoters (TSPs) tend to reduce immune responses to heterologous transgenes and facilitate long-term expression of the transgene in the target tissue (Schneider *et al.*, 1998a; Alemany *et al.*, 2000; Ding *et al.*, 2002; De Geest *et al.*, 2003). Therefore it seems that expression of heterologous transgenes in non-target tissues, such as antigen-presenting cells (APCs) (De Geest *et al.*, 2003), may be the primary cause of immune response induction to heterologous transgenes in Ad gene-replacement therapy.

TSPs have also been utilized in oncolytic Ad therapy, with transgenes placed under the control of prostate-specific antigen (PSA)-driven promoters, resulting in targeting of the vectors to prostate cancer cells (Rodriguez *et al.*, 1997). No dose-limiting toxicities were observed, and, at high virus titres, neither IP nor IV injection of vector resulted in toxic side-effects (DeWeese *et al.*, 2001; Small *et al.*, 2006). However, only a modest reduction in PSA levels was observed and this was found only in a subset of patients (Small *et al.*, 2006).

Immune modulation

Display of immunoevasive proteins on the Ad capsid

This has been attempted in order to circumvent immune responses to capsid proteins. As complement plays a key role in detection of Ad vectors upon administration (Tian *et al.*, 2009), Ad vectors displaying complement inhibitory proteins (COMPinh) or decay accelerating factor (DAF) on the capsid surface have been engineered (Rein *et al.*, 2006; Seregin *et al.*, 2010a,b). Both recombinant vectors showed promise in *in vitro* studies, with the COMPinh-bearing vector showing a significantly decreased complement activation (Seregin *et al.*, 2010b) and the DAF-displaying vector resulting in reduced release of inflammatory cytokines and chemokines, as well as reduced toxicity effects such as thrombocytopenia (Seregin *et al.*, 2010a). Importantly, the transduction of Ads with modified capsids did not appear to be hampered. These results indicate that subversion of complement activation may be a promising method of dampening the immune response to Ad vectors.

Use of alternative Ad serotypes

A major drawback in Ad vector gene therapy is the high prevalence of pre-existing immunity to Ad5 within the population. As DSG-2 (the attachment molecule for species B Ads 3, 7, 11 and 14) appears to be up-regulated in a number of cancers (Harada *et al.*, 1996; Biedermann *et al.*, 2005; Trojan *et al.*, 2005; Schmitt *et al.*, 2007; Wang *et al.*, 2010), this favours a role for species B Ads in cancer therapy. Indeed, pre-treatment of breast cancer cell cultures with UV-inactivated Ad3 prior to treatment with the Her2 receptor-targeting anti-cancer agent herceptin resulted in a significant increase in the cytotoxic effect of herceptin (Wang *et al.*, 2010). This is believed to occur owing to reorganization of cell junctions during infection by species B2 Ads, resulting in the relocalization of cell junction-sequestered proteins such as Her2/neu (Wang *et al.*, 2010).

Use of alternative serotypes has also been investigated in Ad vaccine vector regimens for delivery into patients with immunity against Ad5 (Barouch *et al.*, 2004; Liu *et al.*, 2008). Although the underlying mechanism for the increased susceptibility to HIV-1 detected in the STEP trial is still unclear, research efforts are now moving away from Ad5-based vectors towards novel human serotypes (Liu *et al.*, 2009a) and primate serotypes (Tatsis *et al.*, 2009). Simian Ad26 has shown promise in HIV vaccine regimens (de Souza *et al.*, 2007). Combination of Ad vectors of different serotypes or combination with other virus platforms appears to induce greater prime-boost responses (Schneider *et al.*, 1998b; Pinto *et al.*, 2003; McCoy *et al.*, 2007; Barefoot *et al.*, 2008; Liu *et al.*, 2008; Yu *et al.*, 2008). The use of sequential administration of different Ad serotype vectors has been shown to induce potent immune responses to simian immunodeficiency virus (SIV) in macaques (Liu *et al.*, 2009a).

However, the immune responses induced by alternative serotype vectors are largely unknown. As other serotypes such as species B and certain species D Ads utilize alternative entry pathways to Ad5, it would be reasonable to consider that they may induce different immune responses and may exhibit different biodistribution. Indeed, comparative studies of representative wild-type human Ad serotypes from each species A-D and F alongside simian Ad23 revealed differing biodistribution following systemic delivery into immunocompetent naive mice (Paielli *et al.*, 2000; Everett *et al.*, 2003; Appledorn *et al.*, 2008a). Induction of inflammatory cytokines also differs between serotypes, with simian Ad23 and human Ad3 inducing higher levels of IL-6 than Ad5, and Ad41 inducing higher levels of IL-12 than Ad5 (Kiang *et al.*, 2006a; Hartman *et al.*, 2007a,b; Appledorn *et al.*, 2008a). Crucially, side effects such as thrombocytopenia and macrophage depletion were also shown to vary between serotypes, with human Ad3, 37 and 41 inducing significant thrombocytopenia at 24 hours post-infection (h p.i.) which was not observed upon administration of Ad5, 41 or simian Ad23. The extent of Kupffer cell depletion was also found to be serotype-dependent, with Ad3, 5 and simian Ad23 being the most cytotoxic (Appledorn *et al.*, 2008a). Owing to its much lower immunostimulatory profile than Ad5, the species A virus Ad31 has been suggested for investigation as a gene therapy vector, although biodistribution studies are required (Appledorn *et al.*, 2008a).

In addition, non-human serotypes, such as chimpanzee Ads, have also been investigated for Ad vaccine regimens (McCoy *et al.*, 2007). However, care must be taken when utilizing alternative serotype Ads from different species as the immune response to these vectors is almost entirely unknown and could potentially be exaggerated compared with the response to human Ads owing to lack of exposure during co-evolution. Indeed, when compared with human Ads, immune responses to non-human Ads have been shown to be much more robust in mouse models, and toxicity has also been reported (Hensley *et al.*, 2007; Appledorn *et al.*, 2008a; Hartman *et al.*, 2008b; Seregin *et al.*, 2009a). The biodistribution of the vectors may also be quite different to that of human Ads.

With regard to vaccine vectors, prime and boost regimens utilizing Ad vectors in combination with alternative viruses appear to increase anti-antigen immune responses. In addition to HIV-1, there has been much work on the generation of Ad vaccine vectors against antigens of the causative agent of malaria, the protozoan *Plasmodium* species. An Ad5 vector expressing a 42 kDa peptide of the blood stage merozoite surface protein 1 (MSP-1) of *Plasmodium yoelii*-induced high-level neutralizing antibody titres to the peptide antigen in mice when boosted with a modified vaccinia virus Ankara (MVA) vector expressing the same peptide (Draper *et al.*, 2008). In addition, the time between administration of the priming dose and subsequent boost injection also appeared critical (Draper *et al.*, 2008); a prime-boost regime with an eight week interval protected significantly more mice against infection with *P. yoelii* than mice that received the boost 2 weeks after the priming dose (Draper *et al.*, 2008). It has been suggested that a more sufficient population of memory T cells are formed during the 8-week interval compared with the 2-week interval, resulting in a more robust response to the MVA boost (Draper *et al.*, 2008).

Chimaeric vectors

As the fibre protein and hexon are the main capsid components targeted by neutralizing antibodies (Gall *et al.*, 1996, 1998; Roy *et al.*, 1998; Molinier-Frenkel *et al.*, 2002; Wu *et al.*, 2002a), these have been investigated for modification. Not all chimaeric Ads appear to be viable, however several studies have highlighted promising reductions in evasion of pre-existing immunity to Ads (reviewed in Seregin and Amalfitano, 2009), indicating that this method may hold promise for future development.

Increasing efficacy of oncolytic vectors

Selection of highly oncolytic Ads

The hypothesis behind this method is that mutations beneficial for oncolysis will result in increased growth of the mutant virus compared with the wild-type. This may therefore increase vector efficacy allowing lysis of all target cells before immune clearance occurs. The method typically involves propagation of a mixed population of Ad serotypes under conditions that promote virus recombination, thus increasing diversity in the pool. The pool of serotypes is then propagated in a cancer cell line for which there is initially poor virus replication. This method was utilized to generate ColAd1, an Ad3/11p chimaera with potent oncolytic activity in colon cancer cell lines (Kuhn *et al.*, 2008).

A similar method involves random mutagenesis of Ad vectors and subsequent propagation of fast-growing isolates in cancer cell lines or human tumour xenografts (Subramanian *et al.*, 2006; Gros *et al.*, 2008). The mutant Ad5 vector AdT1 was produced by this method and exhibited potent oncolytic activity (Gros *et al.*, 2008). This enhanced oncolysis was due to a mutation in E3–19K which caused the relocation of E3–19K to the plasma membrane, resulting in enhanced release of progeny virions from the host cell. Importantly, this mutation did not appear to impact on the other functions of E3–19K, such as MHC class 1 down-regulation or induction of apoptosis (Gros *et al.*, 2008).

Expression of chemotherapeutic drugs

To enhance the killing of tumour cells by oncolytic Ads, expression of therapeutic anti-cancer drugs by the vector has been investigated. The drug-metabolizing enzyme is often inserted into the E3 region; owing to the differing promoters and splicing in this region the vector can be engineered to express the drug at either early or late time points of infection (Jin *et al.*, 2005) or can be induced by constitutive or foreign promoters. The infected cell will inevitably be destroyed by the oncolytic Ad, whilst activation of the pro-drug produces the bystander effect, killing cells in close proximity to the infected cell. The types of drug molecules used include molecules increasing virus spread, pro-drug converting enzymes, immunomodulatory molecules, anti-angiogenic molecules or anti-tumour molecules.

Molecules increasing virus spread

To increase the speed of virus spread across a tumour, the ADP over-expressing vector VRX-007 was developed (Doronin *et al.*, 2000, 2003; Toth *et al.*, 2010), which lead to rapid lysis and release of progeny virions. Vectors encoding fusogenic proteins from measles virus or gibbon ape leukaemia virus (GALV) have also been utilized (Galanis *et al.*, 2001). The infected cell membrane fuses with surrounding cells, forming a large syncytium, thus increasing spread of the virus.

In addition to being limited by replication kinetics, the efficacy of oncolytic Ads is also limited by the tumour architecture. Tumours are often composed of separate tissue compartments, which impedes dissemination of the virus to all compartments of the tumour (Sauthoff *et al.*, 2003). Pre-treatment with detergents allowed removal of glycosaminoglycans from the bladder surface, and this was shown to increase infection and efficacy of an oncolytic Ad in bladder cancer (Ramesh *et al.*, 2004). Ads have also been engineered to express the peptide hormone relaxin which facilitates remodelling of the tumour microenvironment by inducing production of certain matrix metalloproteinases whilst decreasing collagen (Kim *et al.*, 2006; Ganesh *et al.*, 2007). This was shown to augment the spread of the vector significantly within the tumour, thus increasing efficacy.

Pro-drug converters

Another way to improve the efficacy of oncolytic vectors is by modification of ONYX-015 to express fusion proteins of either cytosine deaminase (CD) or herpes simplex virus thymidine kinase (TK). Administration of the vector is followed by administration of the pro-drug. Once expressed in target tissues during virus replication, these enzymes are capable of cleaving the pro-drug, resulting in toxicity at the tumour site. Phase I clinical trials for prostate cancer have indicated low toxicity for intra-prostatic administration of these vectors, which appear to have high efficacy (Freytag *et al.*, 2002, 2003, 2007a,b).

Anti-angiogenic molecules

The growth of tumours leads to hypoxic conditions in regions away from the blood supply and this stimulates formation of blood vessels (angiogenesis) within the tumour. Thus, the vascular architecture of the tumour is an important target for preventing the supply of vital nutrients, oxygen and growth factors to tumour cells (reviewed by Hanna *et al.*, 2009). Ads are physiologically targeted to newly formed blood vessels as they express high levels of $\alpha_v\beta_3$ integrins (Hood and Cheresh, 2002). Vectors encoding anti-angiogenic molecules such as endostatin (Li *et al.*, 2005) VEGF antagonists (Yoo *et al.*, 2007) and IFNs (Shashkova *et al.*, 2007) have all been tested. An E1B-55K-deleted Ad5 vector expressing endostatin exhibited a greater capacity than ONYX-015 to suppress growth of tumour xenografts in a mouse model (Li *et al.*, 2005). In addition, an Ad5 vector expressing a short hairpin RNA (shRNA) against VEGF was shown to have greater anti-tumour activity against a glioma xenograft model compared with the parental vector (Yoo *et al.*, 2007).

Expression of apoptosis-inducing factors

Oncolytic Ads have also been engineered to express inducers of apoptosis such as the Fas and TRAIL ligands (Kirkman *et al.*, 2001; Sova *et al.*, 2004) and TNF-α (Shashkova *et al.*, 2008b). However, inefficient transduction of the tumours has been observed which may be due to the tumour microenvironment and tissue architecture impeding viral spread (Kirkman *et al.*, 2001; Sova *et al.*, 2004).

Immunomodulatory molecules

Tumour cells develop numerous mechanisms of evading the immune response, including down-regulation of MHC class 1 molecules and expression of Fas ligand to kill circulating CD8$^+$ T-cells (reviewed in de Souza and Bonorino, 2009). Therefore even when a tumour expresses tumour antigens, the immune system may be prevented from detecting it (de Souza and Bonorino, 2009). However, infection with Ads induces a strong inflammatory response, which changes the tumour microenvironment and may alleviate the immune suppression. Although enhancing the immune response to the tumour will also enhance immune responses to the vector, this may also result in enhanced oncolysis if the tumour is sufficiently infected. To this end, enhancement of the immune response has been attempted by constructing vectors encoding immunomodulatory molecules. In nude mice, GM-CSF stimulated production of APCs, thus facilitating an adaptive response to the tumour cells (Ramesh *et al.*, 2004; Lei *et al.*, 2009). To overcome defects in antigen presentation, an oncolytic Ad expressing HSP70 has also been developed (Haviv *et al.*, 2001). Antigenic peptides are bound by HSP70 in the infected cell and, following cell lysis, are released and taken up by DCs which then present the peptides to CD4$^+$ T-cells. In addition, expression of type-I and -II IFNs in murine models have been shown to increase the inflammatory response and direct tumour cell killing by infiltrating lymphocytes (Su *et al.*, 2006; Shashkova *et al.*, 2007;He *et al.*, 2008). An IFN-α expressing vector was also shown to have increased efficacy and reduced toxicity against kidney tumours in the Syrian hamster model (Shashkova *et al.*, 2008a).

Expression of immunomodulatory modules has also been attempted with Ad vaccine vectors in order to boost anti-antigen immune responses. Adjuvants such as TLR4 ligands (Rhee *et al.*, 2010), cytokines (Hoffmann *et al.*, 2007) and inhibitors of immunosuppressive signalling pathways (Lasaro *et al.*, 2008) have all been investigated. Expression of the complement protein adjuvant C4 binding protein (C4BP) by a *P. yoelii* vaccine vector was found to enhance both antibody and T-cell responses to the antigen in mice (Draper *et al.*, 2008).

Concluding remarks

As Ad vectors generally induce potent immune responses upon administration (Amalfitano and Parks, 2002; Alba *et al.*, 2005; Hartman *et al.*, 2008b; Seregin and Amalfitano, 2009; Toth and Wold, 2010; Vetrini and Ng, 2010), the use of Ads in vaccine regimens appears promising as this requires potent stimulation of the immune response in order to elicit immunity to the transgene antigen. Although responses to these vectors have generally been suboptimal to date, the use of alternative serotype Ads or alternative viruses for the booster vaccine along with expression of adjuvants may hold promise for enhancing the immune response to transgene antigens (Draper *et al.*, 2008). In addition, altering the time interval between the priming dose and subsequent booster vaccination also appears critical for optimization of vector efficacy (Draper *et al.*, 2008). However, the immune response to alternative serotype Ads needs to be defined as it is likely to differ to the immune response to Ad5 (Kiang *et al.*, 2006a; Hartman *et al.*, 2007a,b; Appledorn *et al.*, 2008a). In addition, the increased susceptibility to HIV infection following the STEP trial (Buchbinder *et al.*, 2008; McElrath *et al.*, 2008) is a matter of great concern and the mechanism behind this needs to be elucidated.

With regard to oncolytic Ads, although stimulation of the immune response by the vector may increase the anti-tumour effect, this stimulation will also facilitate clearance of the vector by the immune system. This, combined with the complexity of tumour architecture (Sauthoff *et al.*, 2003) and the decreased potency of most oncolytic Ads owing to vector attenuation, has often resulted in suboptimal vector efficacy *in vivo*. It has also become apparent from studies with ONYX-015 that the efficacy of attenuated oncolytic Ads can also be limited owing to the multifunctional nature of many Ad proteins, as removal of certain genes may affect numerous functions (O'Shea *et al.*, 2004, 2005). Moreover, Ad genes remaining in the vector may be able to compensate for the functions of deleted genes (Soria *et al.*, 2010; Savelyeva and Dobbelstein, 2011). However, methods to increase infiltration of the virus into the tumour such as increasing the speed of vector replication (Doronin *et al.*, 2000, 2003; Toth *et al.*, 2010) along with expression of chemotherapeutic prodrugs (Freytag *et al.*, 2002, 2003, 2007a,b) appear promising in pre-clinical studies. Moreover, combination therapy of oncolytic Ads with existing regimens appears to be beneficial (Reid *et al.*, 2001, 2002, 2005; Sze *et al.*, 2003) indicating that a multimodality of treatments may increase efficacy.

References

Abbink, P., Lemckert, A.A.C., Ewald, B.A., Lynch, D.M., Denholtz, M., Smits, S., Holterman, L., Damen, I., Vogels, R., Thorner, A.R., *et al.* (2007). Comparative seroprevalence and immunogenicity of six rare serotype recombinant adenovirus vaccine vectors from subgroups B and D. J. Virol. *81*, 4654–4663.

Ablasser, A., Bauernfeind, F., Hartmann, G., Latz, E., Fitzgerald, K.A., and Hornung, V. (2009). RIG-I-dependent sensing of poly (dA:dT) through the induction of an RNA polymerase III-transcribed RNA intermediate. Nat. Immunol. *10*, 1065-U1040.

Ackrill, A.M., and Blair, G.E. (1988). Regulation of the major histocompatibility class-I gene-expression at the level of transcription in highly oncogenic adenovirus transformed rat cells. Oncogene *3*, 483–487.

Ahmad-Nejad, P., Hacker, H., Rutz, M., Bauer, S., Vabulas, R.M., and Wagner, H. (2002). Bacterial CpG-DNA and lipopolysaccharides activate Toll-like receptors at distinct cellular compartments. Eur. J. Immunol. *32*, 1958–1968.

Alba, R., Bosch, A., and Chillon, M. (2005). Gutless adenovirus: last-generation adenovirus for gene therapy. Gene Ther. *12*, S18-S27.

Alba, R., Hearing, P., Bosch, A., and Chillon, M. (2007). Differential amplification of adenovirus vectors by flanking the packaging signal with attB/attP-Phi C31 sequences: Implications for helper-dependent adenovirus production. Virology *367*, 51–58.

Alba, R., Bradshaw, A.C., Parker, A.L., Bhella, D., Waddington, S.N., Nicklin, S.A., van Rooijen, N., Custers, J., Goudsmit, J., Barouch, D.H., *et al.* (2009). Identification of coagulation factor (F) X binding sites on the adenovirus serotype 5 hexon: effect of mutagenesis on FX interactions and gene transfer. Blood *114*, 965–971.

Alba, R., Bradshaw, A.C., Coughlan, L., Denby, L., McDonald, R.A., Waddington, S.N., Buckley, S.M.K., Greig, J.A., Parker, A.L., Miller, A.M., *et al.* (2010). Biodistribution and retargeting of FX-binding ablated adenovirus serotype 5 vectors. Blood *116*, 2656–2664.

Alemany, R. (2007). Cancer selective adenoviruses. Mol. Aspects Med. *28*, 42–58.

Alemany, R., and Curiel, D.T. (2001). CAR-binding ablation does not change biodistribution and toxicity of adenoviral vectors. Gene Ther. *8*, 1347–1353.

Alemany, R., Balague, C., and Curiel, D.T. (2000). Replicative adenoviruses for cancer therapy. Nat. Biotechnol. *18*, 723–727.

Alexopoulou, L., Holt, A.C., Medzhitov, R., and Flavell, R.A. (2001). Recognition of double-stranded RNA and activation of NF-kappa B by Toll-like receptor 3. Nature *413*, 732–738.

Allen, I.C., Scull, M.A., Moore, C.B., Holl, E.K., McElvania-TeKippe, E., Taxman, D.J., Guthrie, E.H., Pickles, R.J., and Ting, J.P. Y. (2009). The NLRP3 inflammasome mediates *in vivo* innate immunity to influenza A virus through recognition of viral RNA. Immunity *30*, 556–565.

Amalfitano, A., and Parks, R.J. (2002). Separating fact from fiction: Assessing the potential of modified adenovirus vectors for use in human gene therapy. Curr. Gene Ther. 2, 111–133.

Amstutz, B., Gastaldelli, M., Kalin, S., Imelli, N., Boucke, K., Wandeler, E., Mercer, J., Hemmi, S., and Greber, U.F. (2008). Subversion of CtBP1-controlled macropinocytosis by human adenovirus serotype 3. EMBO J. 27, 956–969.

Appledorn, D.M., Kiang, A., McBride, A., Jiang, H., Seregin, S., Scott, J.M., Stringer, R., Kousa, Y., Hoban, M., Frank, M.M., *et al.* (2008a). Wild-type adenoviruses from groups A-F evoke unique innate immune responses, of which HAd3 and SAd23 are partially complement dependent. Gene Ther. *15*, 885–901.

Appledorn, D.M., McBride, A., Seregin, S., Scott, J.M., Schuldt, N., Kiang, A., Godbehere, S., and Amalfitano, A. (2008b). Complex interactions with several arms of the complement system dictate innate and humoral immunity to adenoviral vectors. Gene Ther. *15*, 1606–1617.

Appledorn, D.M., Patial, S., McBride, A., Godbehere, S., Van Rooijen, N., Parameswaran, N., and Amalfitano, A. (2008c). Adenovirus vector-induced innate inflammatory mediators, MAPK signalling, as well as adaptive immune responses are dependent upon both TLR2 and TLR9 *in vivo*. J. Immunol. *181*, 2134–2144.

Appledorn, D.M., Patial, S., Godbehere, S., Parameswaran, N., and Amalfitano, A. (2009). TRIF, and TRIF-interacting TLRs differentially modulate several adenovirus vector-induced immune responses. J. Innate Immun. *1*, 376–388.

Armentano, D., Sookdeo, C.C., Hehir, K.M., Gregory, R.J., Stgeorge, J.A., Prince, G.A., Wadsworth, S.C., and Smith, A.E. (1995). Characterization of an adenovirus gene-transfer vector containing an E4 deletion. Hum. Gene The. *6*, 1343–1353.

Arnoult, D., Soares, F., Tattoli, I., Castanier, C., Philpott, D.J., and Girardin, S.E. (2009). An N-terminal addressing sequence targets NLRX1 to the mitochondrial matrix. J. Cell Sci. *122*, 3161–3168.

Balachandran, S., Thomas, E., and Barber, G.N. (2004). A FADD-dependent innate immune mechanism in mammalian cells. Nature *432*, 401–405.

Barefoot, B., Thornburg, N.J., Barouch, D.H., Yu, J.S., Sample, C., Johnston, R.E., Liao, H.X., Kepler, T.B., Haynes, B.F., and Ramsburg, E. (2008). Comparison of multiple vaccine vectors in a single heterologous prime-boost trial. Vaccine *26*, 6108–6118.

Barouch, D.H., Pau, M.G., Custers, J., Koudstaal, W., Kostense, S., Havenga, M.J.E., Truitt, D.M., Sumida, S.M., Kishko, M.G., Arthur, J.C., *et al.* (2004). Immunogenicity of recombinant adenovirus serotype 35 vaccine in the presence of pre-existing anti-Ad5 immunity. J. Immunol. *172*, 6290–6297.

Barton, K.N., Xia, X.Q., Yan, H., Stricker, H., Heisey, G., Yin, F.F., Nagaraja, T.N., Zhu, G.P., Kolozsvary, A., Fenstermacher, J.D., *et al.* (2004). A quantitative method for measuring gene expression magnitude and volume delivered by gene therapy vectors. Mol. Ther. *9*, 625–631.

Bauernfeind, F.G., Horvath, G., Stutz, A., Alnemri, E.S., MacDonald, K., Speert, D., Fernandes-Alnemri, T., Wu, J.H., Monks, B.G., Fitzgerald, K.A., *et al.* (2009). Cutting edge: NF-kappa B activating pattern recognition and cytokine receptors license NLRP3 inflammasome activation by regulating NLRP3 expression. J. Immunol. *183*, 787–791.

Bauernfeind, F., Ablasser, A., Bartok, E., Kim, S., Schmid-Burgk, J., Cavlar, T., and Hornung, V. (2011). Inflammasomes: current understanding and open questions. Cell Mol. Life Sci. *68*, 765–783.

Belousova, N., Korokhov, N., Krendelshchikova, V., Simonenko, V., Mikheeva, G., Triozzi, P.L., Aldrich, W.A., Banerjee, P.T., Gillies, S.D., Curiel, D.T., *et al.* (2003). Genetically targeted adenovirus vector directed to CD40-expressing cells. J. Virol. *77*, 11367–11377.

Benihoud, K., Esselin, S., Descamps, D., Jullienne, B., Salone, B., Bobe, P., Bonardelle, D., Connault, E., Opolon, P., Saggio, I., *et al.* (2007). Respective roles of TNF-alpha and IL-6 in the immune response-elicited by adenovirus-mediated gene transfer in mice. Gene Ther. *14*, 533–544.

Bennett, E.M., Bennink, J.R., Yewdell, J.W., and Brodsky, F.M. (1999). Cutting edge: Adenovirus E19 has two mechanisms for affecting class I MHC expression. J. Immunol. *162*, 5049–5052.

Bergelson, J.M., Cunningham, J.A., Droguett, G., KurtJones, E.A., Krithivas, A., Hong, J.S., Horwitz, M.S., Crowell, R.L., and Finberg, R.W. (1997). Isolation of a common receptor for coxsackie B viruses and adenoviruses 2 and 5. Science *275*, 1320–1323.

Berk, A.J. (2007). Adenoviridae: the viruses and their replication. In Fields Virology, Knipe, D.M., Howley, P.M., Griffin, D.E., Lamb, R.A., and Martin, M.A., eds. (Philadelphia: Lippincott, Williams and Wilkins), pp. 2355–2394.

Bessis, N., GarciaCozar, F.J., and Boissier, M.C. (2004). Immune responses to gene therapy vectors: influence on vector function and effector mechanisms. Gene Ther. *11*, S10-S17.

Biedermann, K., Vogelsang, H., Becker, I., Plaschke, S., Siewert, J.R., Hofler, H., and Keller, G. (2005). Desmoglein 2 is expressed abnormally rather than mutated in familial and sporadic gastric cancer. J. Pathol. *207*, 199–206.

Biragyn, A., Ruffini, P.A., Leifer, C.A., Klyushnenkova, E., Shakhov, A., Chertov, O., Shirakawa, A.K., Farber, J.M., Segal, D.M., Oppenheim, J.J., *et al.* (2002). Toll-like receptor 4-dependent activation of dendritic cells by beta-defensin 2. Science *298*, 1025–1029.

Bischoff, J.R., Kim, D.H., Williams, A., Heise, C., Horn, S., Muna, M., Ng, L., Nye, J.A., SampsonJohannes, A., Fattaey, A., *et al.* (1996). An adenovirus mutant that replicates selectively in p53-deficient human tumor cells. Science *274*, 373–376.

Bridge, E., Hemstrom, C., and Pettersson, U. (1991). Differential regulation of adenovirus late transcriptional units by the products of early region. Virology *183*, 260–266.

Brunetti-Pierri, N., and Ng, P. (2008). Progress and prospects: gene therapy for genetic diseases with helper-dependent adenoviral vectors. Gene Ther. *15*, 553–560.

Brunt, K.R., Hall, S.R.R., Ward, C.A., and Melo, L.G. (2007). Endothelial progenitor cell and mesenchymal stem cell isolation, characterization, viral transduction. Methods Mol. Med. *139*, 197–210.

Buchbinder, S.P., Mehrotra, D.V., Duerr, A., Fitzgerald, D.W., Mogg, R., Li, D., Gilbert, P.B., Lama, J.R., Marmor, M., del Rio, C., *et al.* (2008). Efficacy assessment of a cell-mediated immunity HIV-1 vaccine (the Step Study): a double-blind, randomised, placebo-controlled, test-of-concept trial. Lancet *372*, 1881–1893.

Burckstummer, T., Baumann, C., Bluml, S., Dixit, E., Durnberger, G., Jahn, H., Planyavsky, M., Bilban, M., Colinge, J., Bennett, K.L., *et al.* (2009). An orthogonal proteomic-genomic screen identifies AIM2 as a cytoplasmic DNA sensor for the inflammasome. Nat. Immunol. *10*, 266–272.

Burgert, H.G., and Kvist, S. (1985). An adenovirus type-2 glycoprotein blocks cell-surface expression of human histocompatibility class-I antigens. Cell *41*, 987–997.

Caux, C., Dezutterdambuyant, C., Schmitt, D., and Banchereau, J. (1992). GM-CSF and TNF-alpha cooperate in the generation of dendritic langerhans cells. Nature *360*, 258–261.

Cerullo, V., Seiler, M.P., Mane, V., Brunetti-Pierri, N., Clarke, C., Bertin, T.K., Rodgers, J.R., and Lee, B. (2007). Toll-like receptor 9 triggers an innate immune response to helper-dependent adenoviral vectors. Mol. Ther. *15*, 378–385.

Chen, L.N., Anton, M., and Graham, F.L. (1996). Production and characterization of human 293 cell lines expressing the site-specific recombinase Cre. Somatic Cell Mol. Gen. *22*, 477–488.

Cheshenko, N., Krougliak, N., Eisensmith, R.C., and Krougliak, V.A. (2001). A novel system for the production of fully deleted adenovirus vectors that does not require helper adenovirus. Gene Ther. *8*, 846–854.

Chiu, Y.H., MacMillan, J.B., and Chen, Z.J. J. (2009). RNA polymerase III detects cytosolic DNA and induces Type I interferons through the RIG-I pathway. Cell *138*, 576–591.

Cichon, G., Boeckh-Herwig, S., Schmidt, H.H., Wehnes, E., Muller, T., Pring-Akerblom, P., and Burger, R. (2001). Complement activation by recombinant adenoviruses. Gene Ther. *8*, 1794–1800.

Coney, L.R., Tomassetti, A., Carayannopoulos, L., Frasca, V., Kamen, B.A., Colnaghi, M.I., and Zurawski, V.R. (1991). Cloning of a tumour-associated antigen – MOV18 and MOV19 antibodies recognize a folate-binding protein. Cancer Res. *51*, 6125–6132.

Coughlan, L., Vallath, S., Saha, A., Flak, M., McNeish, I.A., Vassaux, G., Marshall, J.F., Hart, I.R., and Thomas, G.J. (2009a). *In vivo* retargeting of adenovirus type 5 to alpha v beta 6 integrin results in reduced hepatotoxicity and improved tumor uptake following systemic delivery. J. Virol. *83*, 6416–6428.

Coughlan, L., Vallath, S., Saha, A., McNeish, I.A., Vassaux, G., Marshall, J.F., Hart, I.R., and Thomas, G.J. (2009b). An Ad5 Vector Retargeted to alpha v beta 6 Displays Improved Tumour Transduction Combined with Reduced Liver Uptake Following Systemic Delivery. Molecular Therapy *17*, S323-S323.

Crettaz, J., Olague, C., Vales, A., Aurrekoetxea, I., Berraondo, P., Otano, I., Kochanek, S., Prieto, J., and Gonzalez-Aseguinolaza, G. (2008). Characterization of high-capacity adenovirus production by the quantitative real-time polymerase chain reaction: a comparative study of different titration methods. J. Gene Med. *10*, 1092–1101.

Crompton, A.M., and Kirn, D.H. (2007). From ONYX-015 to armed vaccinia viruses: The education and evolution of oncolytic virus development. Curr. Cancer Drug Targets *7*, 133–139.

Crompton, J., Toogood, C.I.A., Wallis, N., and Hay, R.T. (1994). Expression of a foreign epitope on the surface of the adenovirus hexon. J. Gen. Virol. *75*, 133–139.

Croyle, M.A., Chirmule, N., Zhang, Y., and Wilson, J.M. (2002). PEGylation of E1-deleted adenovirus vectors allows significant gene expression on readministration to liver. Hum. Gene Ther. *13*, 1887–1900.

Croyle, M.A., Le, H.T., Linse, K.D., Cerullo, V., Toietta, G., Beaudet, A., and Pastore, L. (2005). PEGylated helper-dependent adenoviral vectors: highly efficient vectors with an enhanced safety profile. Gene Ther. *12*, 579–587.

Cruz, C.M., Rinna, A., Forman, H.J., Ventura, A.L.M., Persechini, P.M., and Ojcius, D.M. (2007). ATP activates a reactive oxygen species-dependent oxidative stress response and secretion of proinflammatory cytokines in macrophages. J. Biol. Chem. *282*, 2871–2879.

Cuconati, A., Mukherjee, C., Perez, D., and White, E. (2003). DNA damage response and MCL-1 destruction initiate apoptosis in adenovirus-infected cells. Genes Dev. *17*, 2922–2932.

De Geest, B.R., Van Linthout, S.A., and Collen, D. (2003). Humoral immune response in mice against a circulating antigen induced by adenoviral transfer is strictly dependent on expression in antigen-presenting cells. Blood *101*, 2551–2556.

De Geest, B., Snoeys, J., Van Linthout, S., Lievens, J., and Collen, D. (2005). Elimination of innate immune responses and liver inflammation by PEGylation of adenoviral vectors and methylprednisolone. Hum. Gene Ther. *16*, 1439–1451.

Debbas, M., and White, E. (1993). Wild-type p53 mediates apoptosis by E1A, which is inhibited by E1B. Genes Dev. *7*, 546–554.

DeWeese, T.L., van der Poel, H., Li, S.D., Mikhak, B., Drew, R., Goemann, M., Hamper, U., DeJong, R., Detorie, N., Rodriguez, R., *et al.* (2001). A phase I trial of CV706, a replication-competent, PSA selective oncolytic adenovirus, for the treatment of locally recurrent prostate cancer following radiation therapy. Cancer Res. *61*, 7464–7472.

Dhar, D., Spencer, J.F., Toth, K., and Wold, W.S. M. (2009a). Effect of preexisting immunity on oncolytic adenovirus vector INGN 007 anti-tumour efficacy in immunocompetent and immunosuppressed Syrian hamsters. J. Virol. *83*, 2130–2139.

Dhar, D., Spencer, J.F., Toth, K., and Wold, W.S. M. (2009b). Pre-existing immunity and passive immunity to adenovirus 5 prevents toxicity caused by an oncolytic adenovirus vector in the Syrian hamster model. Mol. Ther. *17*, 1724–1732.

Diebold, S.S., Montoya, M., Unger, H., Alexopoulou, L., Roy, P., Haswell, L.E., Al-Shamkhani, A., Flavell, R., Borrow, P., and Sousa, C.R. E. (2003). Viral infection switches non-plasmacytoid dendritic cells into high interferon producers. Nature *424*, 324–328.

Dinarello, C.A. (1996). Biologic basis for interleukin-1 in disease. Blood *87*, 2095–2147.

Ding, E.Y., Hu, H.M., Hodges, B.L., Migone, F., Serra, D., Xu, F., Chen, Y.T., and Amalfitano, A. (2002). Efficacy of gene therapy for a prototypical lysosomal storage disease (GSD-II) is critically dependent on vector dose, transgene promoter, and the tissues targeted for vector transduction. Mol. Ther. *5*, 436–446.

Dmitriev, I., Krasnykh, V., Miller, C.R., Wang, M.H., Kashentseva, E., Mikheeva, G., Belousova, N., and Curiel, D.T. (1998). An adenovirus vector with genetically modified fibers demonstrates expanded tropism via utilization of a coxsackievirus and adenovirus receptor-independent cell entry mechanism. J. Virol. 72, 9706–9713.

Doronin, K., Shashkova, E.V., May, S.M., Hofherr, S.E., and Barry, M.A. (2009). Chemical modification with high molecular weight polyethylene glycol reduces transduction of hepatocytes and increases efficacy of intravenously delivered oncolytic adenovirus. Hum. Gene Ther. *20*, 975–988.

Doronin, K., Toth, K., Kuppuswamy, M., Ward, P., Tollefson, A.E., and Wold, W.S. M. (2000). Tumor-specific, replication-competent adenovirus vectors overexpressing the adenovirus death protein. J. Virol. *74*, 6147–6155.

Doronin, K., Toth, K., Kuppuswamy, M., Krajcsi, P., Tollefson, A.E., and Wold, W.S. M. (2003). Overexpression of the ADP (E3–11.6K) protein increases cell lysis and spread of adenovirus. Virology *305*, 378–387.

Dostert, C., Petrilli, V., Van Bruggen, R., Steele, C., Mossman, B.T., and Tschopp, J. (2008). Innate immune activation through Nalp3 inflammasome sensing of asbestos and silica. Science *320*, 674–677.

Douglas, J.T., Rogers, B.E., Rosenfeld, M.E., Michael, S.I., Feng, M.Z., and Curiel, D.T. (1996). Targeted gene delivery by tropism-modified adenoviral vectors. Nat. Biotechnol. *14*, 1574–1578.

Draper, S.J., Moore, A.C., Goodman, A.L., Long, C.A., Holder, A.A., Gilbert, S.C., Hill, F., and Hill, A.V. S. (2008). Effective induction of high-titer antibodies by viral vector vaccines. Nat. Med. *14*, 819–821.

Duewell, P., Kono, H., Rayner, K.J., Sirois, C.M., Vladimer, G., Bauernfeind, F.G., Abela, G.S., Franchi, L., Nunez, G., Schnurr, M., *et al.* (2010). NLRP3 inflammasomes are required for atherogenesis and activated by cholesterol crystals. Nature *464*, 1357-U1357.

Egan, C., Bayley, S.T., and Branton, P.E. (1989). Binding of the Rb1 protein to E1A products is required for adenovirus transformation. Oncogene *4*, 383–388.

Einfeld, D.A., Brough, D.E., Roelvink, P.W., Kovesdi, I., and Wickham, T.J. (1999). Construction of a pseudoreceptor that mediates transduction by adenoviruses expressing a ligand in fiber or penton base. J. Virol. *73*, 9130–9136.

Eisenbarth, S.C., Piggott, D.A., Huleatt, J.W., Visintin, I., Herrick, C.A., and Bottomly, K. (2002). Lipopolysaccharide-enhanced, toll-like receptor 4-dependent T helper cell type 2 responses to inhaled antigen. J. Exp. Med. *196*, 1645–1651.

Engelhardt, J.F., Ye, X.H., Doranz, B., and Wilson, J.M. (1994). Ablation of E2A in recombinant adenoviruses improves transgene persistance and decreases inflammatory response in mouse liver. Proc. Natl. Acad. Sci. U.S.A. *91*, 6196–6200.

Ersching, J., Hernandez, M.I.M., Cezarotto, F.S., Ferreira, J.D.S., Martins, A.B., Switzer, W.M., Xiang, Z.Q., Ertl, H.C.J., Zanetti, C.R., and Pinto, A.R. (2010). Neutralizing antibodies to human and simian adenoviruses in humans and New-World monkeys. Virology *407*, 1–6.

Eto, Y., Yoshioka, Y., Mukai, Y., Okada, N., and Nakagawa, S. (2008). Development of PEGylated adenovirus vector with targeting ligand. Int. J. Pharma. *354*, 3–8.

Everett, R.S., Hodges, B.L., Ding, E.Y., Xu, F., Serra, D., and Amalfitano, A. (2003). Liver toxicities typically induced by first-generation adenoviral vectors can be reduced by use of E1, E2b-deleted adenoviral vectors. Hum. Gene Ther. *14*, 1715–1726.

Fechner, H., Haack, A., Wang, H., Wang, X., Eizema, K., Pauschinger, M., Schoemaker, R.G., van Veghel, R., Houtsmuller, A.B., Schultheiss, H.P., *et al.* (1999). Expression of Coxsackie adenovirus receptor and alpha (v)-integrin does not correlate with adenovector targeting *in vivo* indicating anatomical vector barriers. Gene Ther. *6*, 1520–1535.

Fernandes-Alnemri, T., Yu, J.W., Datta, P., Wu, J.H., and Alnemri, E.S. (2009). AIM2 activates the inflammasome and cell death in response to cytoplasmic DNA. Nature *458*, 509-U505.

Fisher, K.D., Stallwood, Y., Green, N.K., Ulbrich, K., Mautner, V., and Seymour, L.W. (2001). Polymer-coated adenovirus permits efficient retargeting and evades neutralising antibodies. Gene Ther. *8*, 341–348.

Fitzgerald, K.A., McWhirter, S.M., Faia, K.L., Rowe, D.C., Latz, E., Golenbock, D.T., Coyle, A.J., Liao, S.M., and Maniatis, T. (2003). IKK epsilon and TBK1 are essential components of the IRF3 signalling pathway. Nat. Immunol. *4*, 491–496.

Freytag, S.O., Khil, M., Stricker, H., Peabody, J., Menon, M., DePeralta-Venturina, M., Nafziger, D., Pegg, J., Paielli, D., Brown, S., *et al.* (2002). Phase I study of replication-competent adenovirus-mediated double suicide gene therapy for the treatment of locally recurrent prostate cancer. Cancer Res. *62*, 4968–4976.

Freytag, S.O., Stricker, H., Pegg, J., Paielli, D., Pradhan, D.G., Peabody, J., DePeralta-Venturina, M., Xia, X.Q., Brown, S., Lu, M., *et al.* (2003). Phase I study of replication-competent adenovirus-mediated double-suicide gene therapy in combination with conventional-dose three-dimensional conformal radiation therapy for the treatment of newly diagnosed, intermediate- to high-risk prostate cancer. Cancer Res. *63*, 7497–7506.

Freytag, S.O., Movsas, B., Aref, I., Stricker, H., Peabody, J., Pegg, J., Zhang, Y.S., Barton, K.N., Brown, S.L., Lu, M., *et al.* (2007a). Phase I trial of replication-competent adenovirus-mediated suicide gene therapy combined with IMRT for prostate cancer. Mol. Ther. *15*, 1016–1023.

Freytag, S.O., Stricker, H., Peabody, J., Pegg, J., Paielli, D., Movsas, B., Barton, K.N., Brown, S.L., Lu, M., and Kim, J.H. (2007b). Five-year follow-up of trial of replication-competent adenovirus-mediated suicide gene therapy for treatment of prostate cancer. Mol. Ther. *15*, 636–642.

Friedman, A., Tian, J.J.P., Fulci, G., Chiocca, E.A., and Wang, J. (2006). Glioma virotherapy: Effects of innate immune suppression and increased viral replication capacity. Cancer Res. *66*, 2314–2319.

Fulci, G., Breymann, L., Gianni, D., Kurozomi, K., Rhee, S.S., Yu, J.H., Kaur, B., Louis, D.N., Weissleder, R., Caligiuri, M.A., *et al.* (2006). Cyclophosphamide enhances glioma virotherapy by inhibiting innate immune responses. Proc. Natl. Acad. Sci. U.S.A. *103*, 12873–12878.

Gabitzsch, E.S., Xu, Y.N., Yoshida, L.H., Balint, J., Gayle, R.B., Amalfitano, A., and Jones, F.R. (2009). A preliminary and comparative evaluation of a novel Ad5 E1-, E2b- recombinant-based vaccine used to induce cell mediated immune responses. Immunol. Lett. *122*, 44–51.

Gaggar, A., Shayakhmetov, D.M., and Lieber, A. (2003). CD46 is a cellular receptor for group B adenoviruses. Nat. Med. *9*, 1408–1412.

Galanis, E., Bateman, A., Johnson, K., Diaz, R.M., James, C.D., Vile, R., and Russell, S.J. (2001). Use of viral fusogenic membrane glycoproteins as novel therapeutic transgenes in gliomas. Hum. Gene Ther. *12*, 811–821.

Gall, J., KassEisler, A., Leinwand, L., and FalckPedersen, E. (1996). Adenovirus type 5 and 7 capsid chimera: Fiber replacement alters receptor tropism without affecting primary immune neutralization epitopes. J. Virol. *70*, 2116–2123.

Gall, J.G.D., Crystal, R.G., and Falck-Pedersen, E. (1998). Construction and characterization of hexon-chimeric adenoviruses: Specification of adenovirus serotype. J. Virol. *72*, 10260–10264.

Ganesh, S., Edick, M.G., Idamakanti, N., Abramova, M., VanRoey, M., Robinson, M., Yun, C.O., and Jooss, K. (2007). Relaxin-expressing, fiber chimeric oncolytic adenovirus prolongs survival of tumor-bearing mice. Cancer Res. *67*, 4399–4407.

Garber, K. (2006). China approves world's first oncolytic virus therapy for cancer treatment. J. Nat. Cancer Institute *98*, 298–300.

Garcia, M.A., Gil, J., Ventoso, I., Guerra, S., Domingo, E., Rivas, C., and Esteban, M. (2006). Impact of protein kinase PKR in cell biology: from antiviral to antiproliferative action. Microbiol. Mol. Biol. Rev. *70*, 1032-+.

van Geer, M.A., Bakker, C.T., Koizumi, N., Mizuguchi, H., Wesseling, J.G., Elferink, R., and Bosma, P.J. (2009). Ephrin A2 receptor targeting does not increase adenoviral pancreatic cancer transduction *in vivo*. World J. Gastroenterol. *15*, 2754–2762.

Geoerger, B., Grill, J., Opolon, P., Morizet, J., Aubert, G., Lecluse, Y., van Beusechem, V.W., Gerritsen, W.R., Kirn, D.H., and Vassal, G. (2003). Potentiation of radiation therapy by the oncolytic adenovirus dl 1520 (ONYX-015) in human malignant glioma xenografts. Br. J. Cancer *89*, 577–584.

Girardin, S.E., Boneca, I.G., Viala, J., Chamaillard, M., Labigne, A., Thomas, G., Philpott, D.J., and Sansonetti, P.J. (2003). Nod2 is a general sensor of peptidoglycan through muramyl dipeptide (MDP) detection. J. Biol. Chem. *278*, 8869–8872.

Gooding, L.R., Ranheim, T.S., Tollefson, A.E., Aquino, L., Duerksenhughes, P., Horton, T.M., and Wold, W.S. M. (1991). The 10.400-dalton and 14, 500-dalton proteins encoded by region E3 of adenovirus function together to protect many but not all mouse cell-lines against lysis by tumor-necrosis-factor. J. Virol. *65*, 4114–4123.

Goodrum, F.D., and Ornelles, D.A. (1998). p53 status does not determine outcome of E1B 55-kilodalton mutant adenovirus lytic infection. J. Virol. *72*, 9479–9490.

Greber, U.F., Suomalainen, M., Stidwill, R.P., Boucke, K., Ebersold, M.W., and Helenius, A. (1997). The role of the nuclear pore complex in adenovirus DNA entry. Embo J. *16*, 5998–6007.

Greber, U.F., Willetts, M., Webster, P., and Helenius, A. (1993). Stepwise dismantling of adenovirus-2 during entry into cells. Cell *75*, 477–486.

Gros, A., Martinez-Quintanilla, J., Puig, C., Guedan, S., Mollevi, D.G., Alemany, R., and Cascallo, M. (2008). Bioselection of a gain of function mutation that enhances adenovirus 5 release and improves its anti-tumoral potency. Cancer Res. *68*, 8928–8937.

Guo, J., and Xin, H. (2006). Chinese gene therapy – Splicing out the west? Science *314*, 1232–1235.

Habib, N., Salama, H., Abu Median, A., Anis, II, Al Aziz, R.A.A., Sarraf, C., Mitry, R., Havlık, R., Seth, P., Hartwigsen, J., *et al.* (2002). Clinical trial of E1B-deleted adenovirus (dl1520) gene therapy for hepatocellular carcinoma. Cancer Gene Ther. *9*, 254–259.

Hajahmad, Y., and Graham, F.L. (1986). Development of a helper-independent human adenovirus vector and its use in the transfer of the herpes-simplex virus thymidine kinase gene. J. Virol. *57*, 267–274.

Hakkarainen, T., Sarkioja, M., Lehenkari, P., Miettinen, S., Ylikomi, T., Suuronen, R., Desmond, R.A., Kanerva, A., and Hemminki, A. (2007). Human mesenchymal stem cells lack tumor tropism but enhance the anti-tumor activity of oncolytic adenoviruses in orthotopic lung and breast tumors. Hum. Gene Ther. *18*, 627–641.

Hale, G.A., Heslop, H.E., Krance, R.A., Brenner, M.A., Jayawardene, D., Srivastava, D.K., and Patrick, C.C. (1999). Adenovirus infection after pediatric bone marrow transplantation. Bone Marrow Transplant. *23*, 277–282.

Hall, K., Zajdel, M.E.B., and Blair, G.E. (2010). Unity and diversity in the human adenoviruses: exploiting alternative entry pathways for gene therapy. Biochem. J. *431*, 321–336.

Halle, A., Hornung, V., Petzold, G.C., Stewart, C.R., Monks, B.G., Reinheckel, T., Fitzgerald, K.A., Latz, E., Moore, K.J., and Golenbock, D.T. (2008). The NALP3 inflammasome is involved in the innate immune response to amyloid-beta. Nat. Immunol. *9*, 857–865.

Hamdan, S., Verbeke, C.S., Fox, N., Booth, J., Bottley, G., Pandha, H.S., and Blair, G.E. (2011). The roles of cell surface attachment molecules and coagulation Factor X in adenovirus 5-mediated gene transfer in pancreatic cancer cells. Cancer Gene Therapy *18*, 478–488.

Hanna, E., Quick, J., and Libutti, S.K. (2009). The tumour microenvironment: a novel target for cancer therapy. Oral Dis. *15*, 8–17.

Harada, H., Iwatsuki, K., Ohtsuka, M., Han, G.W., and Kaneko, F. (1996). Abnormal desmoglein expression by squamous cell carcinoma cells. Acta Dermato-Venereolog. *76*, 417–420.

Harada, J.N., Shevchenko, A., Pallas, D.C., and Berk, A.J. (2002). Analysis of the adenovirus E1B-55K-anchored proteome reveals its link to ubiquitination machinery. J. Virol. *76*, 9194–9206.

Hardy, S., Kitamura, M., HarrisStansil, T., Dai, Y.M., and Phipps, M.L. (1997). Construction of adenovirus vectors through Cre-lox recombination. J. Virol. *71*, 1842–1849.

Hartman, Z.C., Black, E.P., and Amalfitano, A. (2007a). Adenoviral infection induces a multifaceted innate cellular immune response that is mediated by the toll-like receptor pathway in A549 cells. Virology *358*, 357–372.

Hartman, Z.C., Kiang, A., Everett, R.S., Serra, D., Yang, X.Y., Clay, T.M., and Amalfitano, A. (2007b). Adenovirus infection triggers a rapid, MyD88-regulated transcriptome response critical to acute-phase and adaptive immune responses *in vivo*. J. Virol. *81*, 1796–1812.

Hartman, Z.C., Appledom, D.M., Serra, D., Glass, O., Mendelson, T.B., Clay, T.M., and Amalfitano, A. (2008a). Replication-attenuated Human Adenoviral Type 4 vectors elicit capsid dependent enhanced innate immune responses that are partially dependent upon interactions with the complement system. Virology *374*, 453–467.

Hartman, Z.C., Appledorn, D.M., and Amalfitano, A. (2008b). Adenovirus vector induced innate immune responses: Impact upon efficacy and toxicity in gene therapy and vaccine applications. Virus Res. *132*, 1–14.

Harton, J.A., Linhoff, M.W., Zhang, J.H., and Ting, J.P. Y. (2002). Cutting edge: CATERPILLER: A large family of mammalian genes containing CARD, pyrin, nucleotide-binding, and leucine-rich repeat domains. J. Immunol. *169*, 4088–4093.

Haviv, Y.S., Blackwell, J.L., Li, H., Wang, M.H., Lei, X.S., and Curiel, D.T. (2001). Heat shock and heat shock protein 70i enhance the oncolytic effect of replicative adenovirus. Cancer Res. *61*, 8361–8365.

He, L.F., Gu, J.F., Tang, W.H., Fan, J.K., Wei, N., Zou, W.G., Zhang, Y.H., Zhao, L.L., and Liu, X.Y. (2008). Significant anti-tumor activity of oncolytic adenovirus expressing human interferon-beta for hepatocellular carcinoma. J. Gene Med. *10*, 983–992.

Heil, F., Hemmi, H., Hochrein, H., Ampenberger, F., Kirschning, C., Akira, S., Lipford, G., Wagner, H., and Bauer, S. (2004). Species-specific recognition of single-stranded RNA via toll-like receptor 7 and 8. Science *303*, 1526–1529.

Heise, C., SampsonJohannes, A., Williams, A., McCormick, F., VonHoff, D.D., and Kirn, D.H. (1997). ONYX-015, an E1B gene-attenuated adenovirus, causes tumor-specific cytolysis and anti-tumoral efficacy that can be augmented by standard chemotherapeutic agents. Nature Medicine 3, 639–645.

Heise, C.C., Williams, A., Olesch, J., and Kirn, D.H. (1999a). Efficacy of a replication-competent adenovirus (ONYX-015) following intratumoral injection: Intratumoral spread and distribution effects. Cancer Gene Ther. *6*, 499–504.

Heise, C.C., Williams, A.M., Xue, S., Propst, M., and Kirn, D.H. (1999b). Intravenous administration of ONYX-015, a selectively replicating adenovirus, induces anti-tumoral efficacy. Cancer Res. *59*, 2623–2628.

Hemmi, H., Takeuchi, O., Kawai, T., Kaisho, T., Sato, S., Sanjo, H., Matsumoto, M., Hoshino, K., Wagner, H., Takeda, K., *et al.* (2000). A Toll-like receptor recognizes bacterial DNA. Nature *408*, 740–745.

Hemmi, H., Takeuchi, O., Sato, S., Yamamoto, M., Kaisho, T., Sanjo, H., Kawai, T., Hoshino, K., Takeda, K., and Akira, S. (2004). The roles of two I kappa B kinase-related kinases in lipopolysaccharide and double stranded RNA signalling and viral infection. J. Exp. Med. *199*, 1641–1650.

Hensley, S.E., Cun, A.S., Giles-Davis, W., Li, Y., Xiang, Z.Q., Lasaro, M.O., Williams, B.R.G., Silverman, R.H., and Ertl, H.C. J. (2007). Type I interferon inhibits antibody responses induced by a chimpanzee adenovirus vector. Mol. Ther. *15*, 393–403.

Hidajat, R., Xiao, P., Zhou, Q., Venzon, D., Summers, L.E., Kalyanaraman, V.S., Montefiori, D.C., and Robert-Guroff, M. (2009). Correlation of Vaccine-Elicited Systemic and Mucosal Nonneutralizing Antibody Activities with Reduced Acute Viremia following Intrarectal Simian Immunodeficiency Virus SIVmac251 Challenge of Rhesus Macaques. J. Virol. *83*, 791–801.

Hobom, U., and Dobbelstein, M. (2004). E1B-55-kilodalton protein is not required to block p53-induced transcription during adenovirus infection. J. Virol. *78*, 7685–7697.

Hoebe, K., Du, X., Georgel, P., Janssen, E., Tabeta, K., Kim, S.O., Goode, J., Lin, P., Mann, N., Mudd, S., *et al.* (2003). Identification of Lps2 as a key transducer of MyD88-independent TIR signalling. Nature *424*, 743–748.

Hoffmann, D., Bayer, W., and Wildner, O. (2007). *In situ* tumour vaccination with adenovirus vectors encoding measles virus fusogenic membrane proteins and cytokines. World J. Gastroenterol. *13*, 3063–3070.

Holterman, L., Vogels, R., van der Vlugt, R., Sieuwerts, M., Grimbergen, J., Kaspers, J., Geelen, E., van der Helm, E., Lemckert, A., Gillissen, G., *et al.* (2004). Novel replication-incompetent vector derived from adenovirus type 11 (Ad11) for vaccination and gene therapy: Low seroprevalence and non-cross-reactivity with Ad5. J. Virol. *78*, 13207–13215.

Honda, K., Yanai, H., Mizutani, T., Negishi, H., Shimada, N., Suzuki, N., Ohba, Y., Takaoka, A., Yeh, W.C., and Taniguchi, T. (2004). Role of a transductional-transcriptional processor complex involving MyD88 and IRF-7 in Toll-like receptor signalling. Proc. Natl. Acad. Sci. U.S.A. *101*, 15416–15421.

Honda, K., Ohba, Y., Yanai, H., Negishi, H., Mizutani, T., Takaoka, A., Taya, C., and Taniguchi, T. (2005). Spatiotemporal regulation of MyD88-IRF-7 signalling for robust type-I interferon induction. Nature *434*, 1035–1040.

Hood, J.D., and Cheresh, D.A. (2002). Role of integrins in cell invasion and migration. Nat. Rev. Cancer *2*, 91–100.

Horng, T., Barton, G.M., Flavell, R.A., and Medzhitov, R. (2002). The adaptor molecule TIRAP provides signalling specificity for Toll-like receptors. Nature *420*, 329–333.

Hornung, V., and Latz, E. (2010). Intracellular DNA recognition. Nat. Rev. Immunol. *10*, 123–130.

Hornung, V., Bauernfeind, F., Halle, A., Samstad, E.O., Kono, H., Rock, K.L., Fitzgerald, K.A., and Latz, E. (2008). Silica crystals and aluminum salts activate the NALP3 inflammasome through phagosomal destabilization. Nat. Immunol. *9*, 847–856.

Hornung, V., Ablasser, A., Charrel-Dennis, M., Bauernfeind, F., Horvath, G., Caffrey, D.R., Latz, E., and Fitzgerald, K.A. (2009). AIM2 recognizes cytosolic dsDNA and forms a caspase-1-activating inflammasome with ASC. Nature *458*, 514–518.

Hortobagyi, G.N., Ueno, N.T., Xia, W.Y., Zhang, S., Wolf, J.K., Putnam, J.B., Weiden, P.L., Willey, J.S., Carey, M., Branham, D.L., *et al.* (2001). Cationic liposome-mediated E1A gene transfer to human breast and ovarian cancer cells and its biologic effects: A phase I clinical trial. J. Clin. Oncol. *19*, 3422–3433.

Hovanessian, A.G. (1991). Interferon-induced and double-stranded RNA-activated enzymes – a specific protein-kinase and 2′, 5′-oligoadenylate synthetases. J. Interferon Res. *11*, 199–205.

Iacobelli-Martinez, M., and Nemerow, G.R. (2007). Preferential activation of toll-like receptor nine by CD46-utilizing adenoviruses. J. Virol. *81*, 1305–1312.

Ilan, Y., Droguett, G., Chowdhury, N.R., Li, Y.A., Sengupta, K., Thummala, N.R., Davidson, A., Chowdhury, J.R., and Horwitz, M.S. (1997). Insertion of the adenoviral E3 region into a recombinant viral vector prevents antiviral humoral and cellular immune responses and permits long-term gene expression. Proc. Natl. Acad. Sci. U.S.A. *94*, 2587–2592.

Ino, Y., Saeki, Y., Fukuhara, H., and Todo, T. (2006). Triple combination of oncolytic herpes simplex virus-1 vectors armed with interleukin-12, interleukin-18, or soluble B7–1 results in enhanced anti-tumour efficacy. Clin. Cancer Res. *12*, 643–652.

Inohara, N., Ogura, Y., Chen, F.F., Muto, A., and Nunez, G. (2001). Human Nod1 confers responsiveness to bacterial lipopolysaccharides. J. Biol. Chem. *276*, 2551–2554.

Ishihara, A., Zachos, T.A., Bartlett, J.S., and Bertone, A.L. (2006). Evaluation of permissiveness and cytotoxic effects in equine chondrocytes, synovial cells, and stem cells in response to infection with adenovirus 5 vectors for gene delivery. Am. J. Vet. Res. *67*, 1145–1155.

Ishii, K.J., Coban, C., Kato, H., Takahashi, K., Torii, Y., Takeshita, F., Ludwig, H., Sutter, G., Suzuki, K., Hemmi, H., *et al.* (2006). A Toll-like receptor-independent antiviral response induced by double-stranded B-form DNA. Nat. Immunol. *7*, 40–48.

Ishikawa, H., and Barber, G.N. (2008). STING is an endoplasmic reticulum adaptor that facilitates innate immune signalling. Nature *455*, 674-U674.

Jiang, H., Wang, Z., Serra, D., Frank, M.M., and Amalfitano, A. (2004). Recombinant adenovirus vectors activate the alternative complement pathway, leading to the binding of human complement protein C3 independent of anti-Ad antibodies. Mol. Ther. *10*, 1140–1142.

Jin, F., Kretschmer, P.J., and Hermiston, T.W. (2005). Identification of novel insertion sites in the Ad5 genome that utilize the Ad splicing machinery for therapeutic gene expression. Mol. Ther. *12*, 1052–1063.

Kalin, S., Amstutz, B., Gastaldelli, M., Wolfrum, N., Boucke, K., Havenga, M., DiGennaro, F., Liska, N., Hemmi, S., and Greber, U.F. (2010). Macropinocytotic uptake and infection of human epithelial cells with species B2 adenovirus type 35. J. Virol. *84*, 5336–5350.

Kalyuzhniy, O., Di Paolo, N.C., Silvestry, M., Hofherr, S.E., Barry, M.A., Stewart, P.L., and Shayakhmetov, D.M. (2008). Adenovirus serotype 5 hexon is critical for virus infection of heptocytes *in vivo*. Proc. Natl. Acad. Sci. U.S.A. *105*, 5483–5488.

Kanneganti, T.D. (2010). Central roles of NLRs and inflammasomes in viral infection. Nat. Rev. Immunol. *10*, 688–698.

Kato, H., Takeuchi, O., Sato, S., Yoneyama, M., Yamamoto, M., Matsui, K., Uematsu, S., Jung, A., Kawai, T., Ishii, K.J., *et al.* (2006). Differential roles of MDA5 and RIG-I helicases in the recognition of RNA viruses. Nature *441*, 101–105.

Kawai, T., Adachi, O., Ogawa, T., Takeda, K., and Akira, S. (1999). Unresponsiveness of MyD88-deficient mice to endotoxin. Immunity *11*, 115–122.

Kawai, T., Takeuchi, O., Fujita, T., Inoue, J., Muhlradt, P.F., Sato, S., Hoshino, K., and Akira, S. (2001). Lipopolysaccharide stimulates the MyD88-independent pathway and results in activation of IFN-regulatory factor 3 and the expression of a subset of lipopolysaccharide-inducible genes. J. Immunol. *167*, 5887–5894.

Kelkar, S., De, B.P., Gao, G.P., Wilson, J.M., Crystal, R.G., and Leopold, P.L. (2006). A common mechanism for cytoplasmic dynein-dependent microtubule binding shared among adeno-associated virus and adenovirus serotypes. J. Virol. *80*, 7781–7785.

Kemper, C., and Atkinson, J.P. (2007). T-cell regulation: with complements from innate immunity. Nat. Rev. Immunol. *7*, 9–18.

Khuri, F.R., Nemunaitis, J., Ganly, I., Arseneau, J., Tannock, I.F., Romel, L., Gore, M., Ironside, J., MacDougall, R.H., Heise, C., *et al.* (2000). A controlled trial of intratumoral ONYX-015, a selectively replicating adenovirus, in combination with cisplatin and 5-fluorouracil in patients with recurrent head and neck cancer. Nat. Med. *6*, 879–885.

Kiang, A., Hartman, Z.C., Everett, R.S., Serra, D., Jiang, H.X., Frank, M.M., and Amalfitano, A. (2006a). Multiple innate inflammatory responses induced after systemic adenovirus vector delivery depend on a functional complement system. Mol. Ther. *14*, 588–598.

Kiang, A., Hartman, Z.C., Liao, S.X., Xu, F., Serra, D., Palmer, D.J., Ng, P., and Amalfitano, A. (2006b). Fully deleted adenovirus persistently expressing GAA accomplishes long-term skeletal muscle glycogen correction in tolerant and nontolerant GSD-II mice. Mol. Ther. *13*, 127–134.

Kim, J.H., Lee, Y.S., Kim, H., Huang, J.H., Yoon, A.R., and Yun, C.O. (2006). Relaxin expression from tumor-targeting adenoviruses and its intratumoral spread, apoptosis induction, and efficacy. J. Natl. Cancer Institute *98*, 1482–1493.

Kirkman, W., Chen, P., Schroeder, R., Feneley, M.R., Rodriguez, R., Wickham, T.J., King, C.R., and Bruder, J.T. (2001). Transduction and apoptosis induction in the rat prostate, using adenovirus vectors. Hum. Gene Ther. *12*, 1499–1512.

Komarova, S., Kawakami, Y., Stoff-Khalili, M.A., Curiel, D.T., and Pereboeva, L. (2006). Mesenchymal progenitor cells as cellular vehicles for delivery of oncolytic adenoviruses. Mol. Cancer Ther. *5*, 755–766.

Koski, A., Rajecki, M., Guse, K., Kanerva, A., Ristimaki, A., Pesonen, S., Escutenaire, S., and Hemminki, A. (2009). Systemic adenoviral gene delivery to orthotopic murine breast tumours with ablation of coagulation factors, thrombocytes and Kupffer cells. J. Gene Med. *11*, 966–977.

Kostense, S., Koudstaal, N., Sprangers, M., Weverling, G.J., Penders, G., Helmus, N., Vogels, R., Bakker, M., Berkhout, B., Havenga, M., *et al.* (2004). Adenovirus types 5 and 35 seroprevalence in AIDS risk groups supports type 35 as a vaccine vector. Aids *18*, 1213–1216.

Kothari, V., Joshi, G., and Mulherkar, R. (2008). HDAC inhibitor valproic acid enhances tumour cell kill in adenovirus-HSVtk mediated suicide gene therapy for head and neck squamous cell carcinoma. Hum. Gene Ther. *19*, 1107–1107.

Koup, R.A., Lamoreaux, L., Zarkowsky, D., Bailer, R.T., King, C.R., Gall, J.G.D., Brough, D.E., Graham, B.S., and Roederer, M. (2009). Replication-defective adenovirus vectors with multiple deletions do not induce measurable vector-specific T cells in human trials. J. Virol. *83*, 6318–6322.

Krause, A., Joh, J.H., Hackett, N.R., Roelvink, P.W., Bruder, J.T., Wickham, T.J., Kovesdi, I., Crystal, R.G., and Worgall, S. (2006). Epitopes expressed in different adenovirus capsid proteins induce different levels of epitope-specific immunity. J. Virol. *80*, 5523–5530.

Kreppel, F., and Kochanek, S. (2008). Modification of adenovirus gene transfer vectors with synthetic polymers: A scientific review and technical guide. Mol. Ther. *16*, 16–29.

Krug, A., Rothenfusser, S., Hornung, V., Jahrsdorfer, B., Blackwell, S., Ballas, Z.K., Endres, S., Krieg, A.M., and Hartmann, G. (2001a). Identification of CpG oligonucleotide sequences with high induction of IFN-alpha/beta in plasmacytoid dendritic cells. European J. Immunol. *31*, 2154–2163.

Krug, A., Towarowski, A., Britsch, S., Rothenfusser, S., Hornung, V., Bals, R., Giese, T., Engelmann, H., Endres, S., Krieg, A.M., *et al.* (2001b). Toll-like receptor expression reveals CpG DNA as a unique microbial stimulus for plasmacytoid dendritic cells which synergizes with CD40 ligand to induce high amounts of IL-12. Eur. J. Immunol. *31*, 3026–3037.

Kubo, S.J., Saeki, Y., Chiocca, E.A., and Mitani, K. (2003). An HSV amplicon-based helper system for helper-dependent adenoviral vectors. Biochem. Biophys. Res. Comm. *307*, 826–830.

Kuhn, I., Harden, P., Bauzon, M., Chartier, C., Nye, J., Thorne, S., Reid, T., Ni, S., Lieber, A., Fisher, K., *et al.* (2008). Directed evolution generates a novel oncolytic virus for the treatment of colon cancer. Plos One *3*, e2409.

Kuzmin, A.I., Finegold, M.J., and Eisensmith, R.C. (1997). Macrophage depletion increases the safety, efficacy and persistence of adenovirus-mediated gene transfer *in vivo*. Gene Ther. *4*, 309–316.

Lamfers, M.L.M., Fulci, G., Gianni, D., Tang, Y., Kurozumi, K., Kaur, B., Moeniralm, S., Saeki, Y., Carette, J.E., Weissleder, R., *et al.* (2006). Cyclophosphamide increases transgene expression mediated by an oncolytic adenovirus in glioma-bearing mice monitored by bioluminescence imaging. Mol. Ther. *14*, 779–788.

Lasaro, M.O., and Ertl, H.C. J. (2009). New insights on adenovirus as vaccine vectors. Mol. Ther. *17*, 1333–1339.

Lasaro, M.O., Tatsis, N., Hensley, S.E., Whitbeck, J.C., Lin, S.W., Rux, J.J., Wherry, E.J., Cohen, G.H., Eisenberg, R.J., and Ertl, H.C. (2008). Targeting of antigen to the herpesvirus entry mediator augments primary adaptive immune responses. Nat. Med. *14*, 205–212.

Latz, E., Schoenemeyer, A., Visintin, A., Fitzgerald, K.A., Monks, B.G., Knetter, C.F., Lien, E., Nilsen, N.J., Espevik, T., and Golenbock, D.T. (2004). TLR9 signals after translocating from the ER to CpG DNA in the lysosome. Nat. Immunol. *5*, 190–198.

Lawrence, M.S., Foellmer, H.G., Elsworth, J.D., Kim, J.H., Leranth, C., Kozlowski, D.A., Bothwell, A.L.M., Davidson, B.L., Bohn, M.C., and Redmond, D.E.

(1999). Inflammatory responses and their impact on beta-galactosidase transgene expression following adenovirus vector delivery to the primate caudate nucleus. Gene Ther. *6*, 1368–1379.

Lee, S.G., Yoon, S.J., Kim, C.D., Kim, K., Lim, D.S., Yeom, Y.I., Sung, M.W., Heo, D.S., and Kim, N.K. (2000). Enhancement of adenoviral transduction with polycationic liposomes *in vivo*. Cancer Gene Ther. *7*, 1329–1335.

Leen, A.M., Christin, A., Khalil, M., Weiss, H., Gee, A.P., Brenner, M.K., Heslop, H.E., Rooney, C.M., and Bollard, C.M. (2008). Identification of hexon-specific CD4 and CD8 T-cell epitopes for vaccine and immunotherapy. J. Virol. *82*, 546–554.

Leff, T., Elkaim, R., Goding, C.R., Jalinot, P., Sassonecorsi, P., Perricaudet, M., Kedinger, C., and Chambon, P. (1984). Individual products of the adenovirus 12S and 13S EIA messenger-RNAs stimulate viral EIIA and EIII expression at the transcriptional level. Proc. Natl. Acad. Sci. U.S.A. *81*, 4381–4385.

Lei, N., Shen, F.B., Chang, J.H., Wang, L., Li, H., Yang, C., Li, J., and Yu, D.C. (2009). An oncolytic adenovirus expressing granulocyte macrophage colony-stimulating factor shows improved specificity and efficacy for treating human solid tumors. Cancer Gene Ther. *16*, 33–43.

Lemarchand, P., Jaffe, H.A., Danel, C., Cid, M.C., Kleinman, H.K., Stratfordperricaudet, L.D., Perricaudet, M., Pavirani, A., Lecocq, J.P., and Crystal, R.G. (1992). Adenovirus-mediated transfer of a recombinant human alpha-1-antitrypsin cDNA to human endothelial cells. Proc. Natl. Acad. Sci. U.S.A. *89*, 6482–6486.

Leonard, G.T., and Sen, G.C. (1997). Restoration of interferon responses of adenovirus E1A-expressing HT1080 cell lines by overexpression of p48 protein. J. Virol. *71*, 5095–5101.

Leopold, P.L., Kreitzer, G., Miyazawa, N., Rempel, S., Pfister, K.K., Rodriguez-Boulan, E., and Crystal, R.G. (2000). Dynein- and microtubule-mediated translocation of adenovirus serotype 5 occurs after endosomal lysis. Hum. Gene Ther. *11*, 151–165.

Li, E.G., Stupack, D., Klemke, R., Cheresh, D.A., and Nemerow, G.R. (1998). Adenovirus endocytosis via alpha (v) integrins requires phosphoinositide-3-OH kinase. J. Virol. *72*, 2055–2061.

Li, G.C., Sham, J., Yang, J.M., So, C.Q., Xue, H.B., Chua, D., Sun, L.C., Zhang, Q., Cui, Z.F., Wu, M.C., *et al.* (2005). Potent anti-tumor efficacy of an E1B 55kDa-deficient adenovirus carrying murine endostatin in hepatocellular carcinoma. Int. J. Cancer *113*, 640–648.

Li, H.T., Dutuor, A., Fu, X.P., and Zhang, X.L. (2007). Induction of strong anti-tumor immunity by an HSV-2-based oncolytic virus in a murine mammary tumor model. J. Gene Med. *9*, 161–169.

Li, Y.M., Pong, R.C., Bergelson, J.M., Hall, M.C., Sagalowsky, A.I., Tseng, C.P., Wang, Z., and Hsieh, J.T. (1999). Loss of adenoviral receptor expression in human bladder cancer cells: A potential impact on the efficacy of gene therapy. Cancer Research 59, 325–330.

Lieber, A., He, C.Y., Meuse, L., Schowalter, D., Kirillova, I., Winther, B., and Kay, M.A. (1997). The role of Kupffer cell activation and viral gene expression in early liver toxicity after infusion of recombinant adenovirus vectors. J. Virol. *71*, 8798–8807.

Lillie, J.W., and Green, M.R. (1989). Transcriptional activation by the adenovirus E1A protein. Nature *338*, 39–44.

Liu, F., and Green, M.R. (1994). Promoter targeting by adenovirus-E1A through interaction with different cellular DNA-binding domains. Nature *368*, 520–525.

Liu, J.Y., Ewald, B.A., Lynch, D.M., Denholtz, M., Abbink, P., Lemckert, A.A.C., Carville, A., Mansfield, K.G., Havenga, M.J., Goudsmit, J., *et al.* (2008). Magnitude and phenotype of cellular immune responses elicited by recombinant adenovirus vectors and heterologous prime-boost regimens in rhesus monkeys. J. Virol. *82*, 4844–4852.

Liu, J.Y., O'Brien, K.L., Lynch, D.M., Simmons, N.L., La Porte, A., Riggs, A.M., Abbink, P., Coffey, R.T., Grandpre, L.E., Seaman, M.S., *et al.* (2009a). Immune control of an SIV challenge by a T-cell-based vaccine in rhesus monkeys. Nature *457*, 87–91.

Liu, Y., Wang, H.J., Yumul, R., Gao, W.T., Gambotto, A., Morita, T., Baker, A., Shayakhmetov, D., and Lieber, A. (2009b). Transduction of Liver metastases after intravenous injection of Ad5/35 or Ad35 vectors with and without factor X-binding protein pretreatment. Hum. Gene Ther. *20*, 621–629.

Look, D.C., Roswit, W.T., Frick, A.G., Gris-Alevy, Y., Dickhaus, D.M., Walter, M.J., and Holtzman, M.J. (1998). Direct suppression of Stat1 function during adenoviral infection. Immunity *9*, 871–880.

Lund, J.M., Alexopoulou, L., Sato, A., Karow, M., Adams, N.C., Gale, N.W., Iwasaki, A., and Flavell, R.A. (2004). Recognition of single-stranded RNA viruses by Toll-like receptor 7. Proc. Natl. Acad. Sci. U.S.A. *101*, 5598–5603.

Mabit, H., Nakano, M.Y., Prank, U., Sam, B., Dohner, K., Sodeik, B., and Greber, U.F. (2002). Intact microtubules support adenovirus and herpes simplex virus infections. J. Virol. *76*, 9962–9971.

McCoy, K., Tatsis, N., Korioth-Schmitz, B., Lasaro, M.O., Hensley, S.E., Lin, S.W., Li, Y., Giles-Davis, W., Cun, A., Zhou, D.M., *et al.* (2007). Effect of preexisting immunity to adenovirus human serotype 5 antigens on the immune responses of nonhuman primates to vaccine regimens based on human- or chimpanzee-derived adenovirus vectors. J. Virol. *81*, 6594–6604.

McElrath, M.J., De Rosa, S.C., Moodie, Z., Dubey, S., Kierstead, L., Janes, H., Defawe, O.D., Carter, D.K., Hural, J., Akondy, R., *et al.* (2008). HIV-1 vaccine-induced immunity in the test-of-concept Step Study: a case-cohort analysis. Lancet 372, 1894–1905.

Magalhaes, I., Sizemore, D.R., Ahmed, R.K., Mueller, S., Wehlin, L., Scanga, C., Weichold, F., Schirru, G., Pau, M.G., Goudsmit, J., *et al.* (2008). rBCG induces strong antigen-specific T cell responses in Rhesus macaques in a prime-boost setting with an adenovirus 35 tuberculosis vaccine vector. Plos One 3, e3790.

Maitra, R., Clement, C.C., Scharf, B., Crisi, G.M., Chitta, S., Paget, D., Purdue, P.E., Cobelli, N., and Santambrogio, L. (2009). Endosomal damage and TLR2 mediated inflammasome activation by alkane particles in the generation of aseptic osteolysis. Mol. Immunol. *47*, 175–184.

Manickan, E., Smith, J.S., Tian, J., Eggerman, T.L., Lozier, J.N., Muller, J., and Byrnes, A.P. (2006). Rapid Kupffer cell death after intravenous injection of adenovirus vectors. Mol. Ther. *13*, 108–117.

Marie, I., and Hovanessian, A.G. (1992). The 69-kDa 2–5A synthetase is composed of 2 homologous and adjacent functional domains. J. Biol. Chem. *267*, 9933–9939.

Martin, K., Brie, A., Saulnier, P., Perricaudet, M., Yeh, P., and Vigne, E. (2003). Simultaneous CAR- and alpha (V) integrin-binding ablation fails to reduce Ad5 liver tropism. Mol. Ther. *8*, 485–494.

Martinon, F., and Tschopp, J. (2007). Inflammatory caspases and inflammasomes: master switches of inflammation. Cell Death and Differentiation *14*, 10–22.

Martinon, F., Burns, K., and Tschopp, J. (2002). The inflammasome: A molecular platform triggering activation of inflammatory caspases and processing of proIL-beta. Mol. Cell *10*, 417–426.

Mathews, M.B., and Shenk, T. (1991). Adenovirus virus-associated RNA and translation control. J. Virol. *65*, 5657–5662.

Mathias, P., Wickham, T., Moore, M., and Nemerow, G. (1994). Multiple adenovirus serotypes use alpha-v integrins for infection. J. Virol. *68*, 6811–6814.

Matthews, C.B., Jenkins, G., Hilfinger, J.M., and Davidson, B.L. (1999). Poly L-lysine improves gene transfer with adenovirus formulated in PLGA microspheres. Gene Ther. *6*, 1558–1564.

Matthews, Q.L., Fatima, A., Tang, Y.Z., Perry, B.A., Tsuruta, Y., Komarova, S., Timares, L., Zhao, C.X., Makarova, N., Borovjagin, A.V., *et al.* (2010). HIV antigen incorporation within adenovirus hexon hypervariable 2 for a novel HIV vaccine approach. Plos One *5*, e11815.

Meier, O., Vig, S., Boucke, K., Keller, S., Suomalainen, M., and Stidwill, R.P. (1999). Cell signalling of incoming adenovirus stimulates fluid phase endocytosis. Mol. Biol. Cell *10*, 1275.

Mercier, G.T., Campbell, J.A., Chappell, J.D., Stehle, T., Dermody, T.S., and Barry, M.A. (2004). Chimeric adenovirus vector encoding reovirus attachment protein sigma 1 targets cells expressing junctional adhesion molecule 1. Proc. Natl. Acad. Sci. U.S.A. *101*, 6188–6193.

Metzgar, D., Osuna, M., Yingst, S., Rakha, M., Earhart, K., Elyan, D., Esmat, H., Saad, M.D., Kajon, A., Wu, J.G., *et al.* (2005). PCR analysis of Egyptian respiratory adenovirus isolates, including identification of species, serotypes, and coinfections. J. Clin. Microbiol. *43*, 5743–5752.

Meurs, E., Chong, K., Galabru, J., Thomas, N.S.B., Kerr, I.M., Williams, B.R.G., and Hovanessian, A.G. (1990). Molecular cloning and characterization of the human double-stranded-RNA acitvated protein-kinase induced by interferon. Cell *62*, 379–390.

Meylan, E., Burns, K., Hofmann, K., Blancheteau, V., Martinon, F., Kelliher, M., and Tschopp, J. (2004). RIP1 is an essential mediator of Toll-like receptor 3-induced NF-kappa B activation. Nat. Immunol. *5*, 503–507.

Meylan, E., Curran, J., Hofmann, K., Moradpour, D., Binder, M., Bartenschlager, R., and Tschopp, R. (2005). Cardif is an adaptor protein in the RIG-I antiviral pathway and is targeted by hepatitis C virus. Nature *437*, 1167–1172.

Minamitani, T., Iwakiri, D., and Takada, K. (2011). Adenovirus virus-associated RNAs induce type I interferon expression through a RIG-I-mediated pathway. J. Virol. *85*, 4035–4040.

Miyazawa, N., Crystal, R.G., and Leopold, P.L. (2001). Adenovirus serotype 7 retention in a late endosomal compartment prior to cytosol escape is modulated by fiber protein. J. Virol. *75*, 1387–1400.

Mogensen, T.H. (2009). Pathogen recognition and inflammatory signalling in innate immune defenses. Clin. Microbiol. Rev. *22*, 240–273.

Mok, H., Palmer, D.J., Ng, P., and Barry, M.A. (2005). Evaluation of polyethylene glycol modification of first-generation and helper-dependent adenoviral vectors to reduce innate immune responses. Mol. Ther. *11*, 66–79.

Molinier-Frenkel, V., Lengagne, R., Gaden, F., Hong, S.S., Choppin, J., Gahery-Segard, H., Boulanger, P., and Guillet, J.G. (2002). Adenovirus hexon protein is a potent adjuvant for activation of a cellular immune response. J. Virol. *76*, 127–135.

Moore, C.B., Bergstralh, D.T., Duncan, J.A., Lei, Y., Morrison, T.E., Zimmermann, A.G., Accavitti-Loper, M.A., Madden, V.J., Sun, L.J., Ye, Z.M., *et al.* (2008). NLRX1 is a regulator of mitochondrial antiviral immunity. Nature *451*, 573-U578.

Morgan, B.P., van den Berg, C.W., and Harris, C.L. (2005). 'Homologous restriction' in complement lysis: roles of membrane complement regulators. Xenotransplantation *12*, 258–265.

Morsy, M.A., and Caskey, C.T. (1999). Expanded-capacity adenoviral vectors – the helper-dependent vectors. Mol. Med. Today *5*, 18–24.

Mulvihlll, S., Warren, R., Venook, A., Adler, A., Randlev, B., Heise, C., and Kirn, D. (2001). Safety and feasibility of injection with an E1B-55 kDa gene-deleted, replication-selective adenovirus (ONYX-015) into primary carcinomas of the pancreas: a phase I trial. Gene Ther. *8*, 308–315.

Muruve, D.A., Barnes, M.J., Stillman, I.E., and Libermann, T.A. (1999). Adenoviral gene therapy leads to rapid induction of multiple chemokines and acute neutrophil-dependent hepatic injury *in vivo*. Hum. Gene Ther. *10*, 965–976.

Muruve, D.A., Petrilli, V., Zaiss, A.K., White, L.R., Clark, S.A., Ross, P.J., Parks, R.J., and Tschopp, J. (2008). The inflammasome recognizes cytosolic microbial and host DNA and triggers an innate immune response. Nature *452*, 103-U111.

Nagai, Y., Akashi, S., Nagafuku, M., Ogata, M., Iwakura, Y., Akira, S., Kitamura, T., Kosugi, A., Kimoto, M., and Miyake, K. (2002). Essential role of MD-2 in LPS responsiveness and TLR4 distribution. Nat. Immunol. *3*, 667–672.

Nemunaitis, J., Khuri, F., Ganly, I., Arseneau, J., Posner, M., Vokes, E., Kuhn, J., McCarty, T., Landers, S., Blackburn, A., *et al.* (2001). Phase II trial of intratumoral administration of ONYX-015, a replication-selective adenovirus, in patients with refractory head and neck cancer. J. Clin. Oncol. *19*, 289–298.

Nevins, J.R., Ginsberg, H.S., Blanchard, J.M., Wilson, M.C., and Darnell, J.E. (1979). Regulation of the primary expression of the early adenovirus transcription units. J. Virol. *32*, 727–733.

Ng, P., Beauchamp, C., Evelegh, C., Parks, R., and Graham, F.L. (2001). Development of a FLP/frt system for generating helper-dependent adenoviral vectors. Mol. Ther. *3*, 809–815.

Nokisalmi, P., Pesonen, S., Escutenaire, S., Sarkioja, M., Raki, M., Cerullo, V., Laasonen, L., Alemany, R., Rojas, J., Cascallo, M., *et al.* (2010). Oncolytic Adenovirus ICOVIR-7 in patients with advanced and refractory solid tumours. Clin. Cancer Res. *16*, 3035–3043.

Noureddini, S.C., and Curiel, D.T. (2005). Genetic targeting strategies for adenovirus. Mol. Pharma. *2*, 341–347.

O'Riordan, C.R., Lachapelle, A., Delgado, C., Parkes, V., Wadsworth, S.C., Smith, A.E., and Francis, G.E. (1999). PEGylation of adenovirus with retention of infectivity and protection from neutralizing antibody *in vitro* and *in vivo*. Hum. Gene Ther. *10*, 1349–1358.

O'Shea, C.C., Johnson, L., Bagus, B., Choi, S., Nicholas, C., Shen, A., Boyle, L., Pandey, K., Soria, C., Kunich, J., *et al.* (2004). Late viral RNA export, rather than p53 inactivation, determines ONYX-015 tumour selectivity. Cancer Cell *6*, 611–623.

O'Shea, C.C., Soria, C., Bagus, B., and McCormick, F. (2005). Heat shock phenocopies E1B-55K late functions and selectively sensitizes refractory tumour cells to ONYX-015 oncolytic viral therapy. Cancer Cell *8*, 61–74.

Ogawara, K.I., Rots, M.G., Kok, R.J., Moorlag, H.E., van Loenen, A.M., Meijer, D.K.F., Haisma, H.J., and Molema, G. (2004). A novel strategy to modify adenovirus tropism and enhance transgene delivery to activated vascular endothelial cells *in vitro* and *in vivo*. Hum. Gene Ther. *15*, 433–443.

Okamura, H. (2005). IL-1 family (IL-1alpha/beta, IL-1Ra, IL-18), IL-16, IL-17. Nippon Rinsho *63 (Suppl 4)*, 226–233.

Okegawa, T., Li, Y.M., Pong, R.C., Bergelson, J.M., Zhou, J., and Hsieh, J.T. (2000). The dual impact of coxsackie and adenovirus receptor expression on human prostate cancer gene therapy. Cancer Res. *60*, 5031–5036.

Olive, M., Eisenlohr, L., Flomenberg, N., Hsu, S., and Flomenberg, P. (2002). The adenovirus capsid protein hexon contains a highly conserved human CD4(+) T-cell epitope. Hum. Gene Ther. *13*, 1167–1178.

Onion, D., Crompton, L.J., Milligan, D.W., Moss, P.A.H., Lee, S.P., and Mautner, V. (2007). The CD4(+) T-cell response to adenovirus is focused against conserved residues within the hexon protein. J. Gen. Virol. *88*, 2417–2425.

Osada, T., Yang, X.Y., Hartman, Z.C., Glass, O., Hodges, B.L., Niedzwiecki, D., Morse, M.A., Lyerly, H.K., Amalfitano, A., and Clay, T.M. (2009). Optimization of vaccine responses with an E1, E2b and E3-deleted Ad5 vector circumvents pre-existing anti-vector immunity. Cancer Gene Ther. *16*, 673–682.

Oshiumi, H., Matsumoto, M., Funami, K., Akazawa, T., and Seya, T. (2003). TICAM-1, an adaptor molecule that participates in Toll-like receptor 3-mediated interferon-beta induction. Nat. Immunol. *4*, 161–167.

Otake, K., Ennist, D.L., Harrod, K., and Trapnell, B.C. (1998). Nonspecific inflammation inhibits adenovirus-mediated pulmonary gene transfer and expression independent of specific acquired immune responses. Hum. Gene Ther. *9*, 2207–2222.

Ozinsky, A., Underhill, D.M., Fontenot, J.D., Hajjar, A.M., Smith, K.D., Wilson, C.B., Schroeder, L., and Aderem, A. (2000). The repertoire for pattern recognition of pathogens by the innate immune system is defined by cooperation between Toll-like receptors. Proc. Natl. Acad. Sci. U.S.A. *97*, 13766–13771.

Paielli, D.L., Wing, M.S., Rogulski, K.R., Gilbert, J.D., Kolozsvary, A., Kim, J.H., Hughes, J., Schnell, M., Thompson, T., and Freytag, S.O. (2000). Evaluation of the biodistribution, persistence, toxicity, and potential of germ-line transmission of a replication-competent human adenovirus following intraprostatic administration in the mouse. Mol. Ther. *1*, 263–274.

Parato, K.A., Senger, D., Forsyth, P.A.J., and Bell, J.C. (2005). Recent progress in the battle between oncolytic viruses and tumours. Nat. Rev. Cancer *5*, 965–976.

Park, J.W., Mok, H., and Park, T.G. (2008). Epidermal growth factor (EGF) receptor targeted delivery of PEGylated adenovirus. Biochem. Biophys. Res. Commun. *366*, 769–774.

Parker, A.L., Waddington, S.N., Nicol, C.G., Shayakhmetov, D.M., Buckley, S.M., Denby, L., Kemball-Cook, G., Ni, S.H., Lieber, A., McVey, J.H., *et al.* (2006). Multiple vitamin K-dependent coagulation zymogens promote adenovirus-mediated gene delivery to hepatocytes. Blood *108*, 2554–2561.

Parker, A.L., McVey, J.H., Doctor, J.H., Lopez-Franco, O., Waddington, S.N., Havenga, M.J.E., Nicklin, S.A., and Baker, A.H. (2007). Influence of coagulation factor zymogens on the infectivity of adenoviruses pseudotyped with fibers from subgroup D. J. Virol. *81*, 3627–3631.

Pasare, C., and Medzhitov, R. (2004). Toll-like receptors: linking innate and adaptive immunity. Microbes Infect. *6*, 1382–1387.

Perry, A.K., Chow, E.K., Goodnough, J.B., Yeh, W.C., and Cheng, G. (2004). Differential requirement for TANK-binding kinase-1 in type I interferon responses to toll-like receptor activation and viral infection. J. Exp. Med. *199*, 1651–1658.

Pesonen, S., Nokisalmi, P., Escutenaire, S., Sarkioja, M., Raki, M., Cerullo, V., Kangasniemi, L., Laasonen,

L., Ribacka, C., Guse, K., *et al.* (2010). Prolonged systemic circulation of chimeric oncolytic adenovirus Ad5/3-Cox2L-D24 in patients with metastatic and refractory solid tumors. Gene Ther. *17*, 892–904.

Petrilli, V., Papin, S., Dostert, C., Mayor, A., Martinon, F., and Tschopp, J. (2007). Activation of the NALP3 inflammasome is triggered by low intracellular potassium concentration. Cell Death Diff. *14*, 1583–1589.

Pinto, A.R., Fitzgerald, J.C., Giles-Davis, W., Gao, G.P., Wilson, J.M., and Ertl, H.C. J. (2003). Induction of CD8(+) T cells to an HIV-1 antigen through a prime boost regimen with heterologous E1-deleted adenoviral vaccine carriers. J. Immunol. *171*, 6774–6779.

Poulin, K.L., Lanthier, R.M., Smith, A.C., Christou, C., Quiroz, M.R., Powell, K.L., O'Meara, R.W., Kothary, R., Lorimer, I.A., and Parks, R.J. (2010). Retargeting of adenovirus vectors through genetic fusion of a single-chain or single-domain antibody to capsid protein IX. J. Virol. *84*, 10074–10086.

Ramesh, N., Memarzadeh, B., Ge, Y., Frey, D., VanRoey, M., Rojas, V., and Yu, D.C. (2004). Identification of pretreatment agents to enhance adenovirus infection of bladder epithelium. Mol. Ther. *10*, 697–705.

Rao, L., Debbas, M., Sabbatini, P., Hockenbery, D., Korsmeyer, S., and White, E. (1992). The adenovirus E1A proteins induce apoptosis, which is inhibited by the E1B 19-kDa and BCL-2 proteins. Proc. Natl. Acad. Sci. U.S.A. *89*, 7742–7746.

Raper, S.E., Chirmule, N., Lee, F.S., Wivel, N.A., Bagg, A., Gao, G.P., Wilson, J.M., and Batshaw, M.L. (2003). Fatal systemic inflammatory response syndrome in a ornithine transcarbamylase deficient patient following adenoviral gene transfer. Mol. Genet. Metab. *80*, 148–158.

Reid, T., Galanis, E., Abbruzzese, J., Sze, D., Andrews, J., Romel, L., Hatfield, M., Rubin, J., and Kirn, D. (2001). Intra-arterial administration of a replication-selective adenovirus (dl1520) in patients with colorectal carcinoma metastatic to the liver: a phase I trial. Gene Therapy *8*, 1618–1626.

Reid, T., Galanis, E., Abbruzzese, J., Sze, D., Wein, L.M., Andrews, J., Randlev, B., Heise, C., Uprichard, M., Hatfield, M., *et al.* (2002). Hepatic arterial infusion of a replication-selective oncolytic adenovirus (dl1520): Phase II viral, immunologic, and clinical endpoints. Cancer Res. *62*, 6070–6079.

Reid, T.R., Freeman, S., Post, L., McCormick, F., and Sze, D.Y. (2005). Effects of Onyx-015 among metastatic colorectal cancer patients that have failed prior treatment with 5-FU/leucovorin. Cancer Gene Ther. *12*, 673–681.

Rein, D.T., Breidenbach, M., and Curiel, D.T. (2006). Current developments in adenovirus-based cancer gene therapy. Future Oncol. *2*, 137–143.

Reyes-Sandoval, A., Sridhar, S., Berthoud, T., Moore, A.C., Harty, J.T., Gilbert, S.C., Gao, G.P., Ertl, H.C.J., Wilson, J.C., and Hill, A.V. S. (2008). Single-dose immunogenicity and protective efficacy of simian adenoviral vectors against *Plasmodium berghei*. Eur. J. Immunol. *38*, 732–741.

Rhee, E.G., Kelley, R.P., Agarwal, I., Lynch, D.M., La Porte, A., Simmons, N.L., Clark, S.L., and Barouch, D.H. (2010). TLR4 ligands augment antigen-specific CD8(+) T lymphocyte responses elicited by a viral vaccine vector. J. Virol. *84*, 10413–10419.

Roberts, D.M., Nanda, A., Havenga, M.J.E., Abbink, P., Lynch, D.M., Ewald, B.A., Liu, J., Thorner, A.R., Swanson, P.E., Gorgone, D.A., *et al.* (2006). Hexon-chimaeric adenovirus serotype 5 vectors circumvent pre-existing anti-vector immunity. Nature *441*, 239–243.

Roberts, T.L., Idris, A., Dunn, J.A., Kelly, G.M., Burnton, C.M., Hodgson, S., Hardy, L.L., Garceau, V., Sweet, M.J., Ross, I.L., *et al.* (2009). HIN-200 proteins regulate caspase activation in response to foreign cytoplasmic DNA. Science *323*, 1057–1060.

Rock, F.L., Hardiman, G., Timans, J.C., Kastelein, R.A., and Bazan, J.F. (1998). A family of human receptors structurally related to Drosophila Toll. Proc. Natl. Acad. Sci. U.S.A. *95*, 588–593.

Rodriguez, R., Schuur, E.R., Lim, H.Y., Henderson, G.A., Simons, J.W., and Henderson, D.R. (1997). Prostate attenuated replication competent adenovirus (ARCA) CN706: A selective cytotoxic for prostate-specific antigen-positive prostate cancer cells. Cancer Res. *57*, 2559–2563.

Roelvink, P.W., Lizonova, A., Lee, J.G.M., Li, Y., Bergelson, J.M., Finberg, R.W., Brough, D.E., Kovesdi, I., and Wickham, T.J. (1998). The coxsackievirus-adenovirus receptor protein can function as a cellular attachment protein for adenovirus serotypes from subgroups A, C, D, E, and F. J. Virol. *72*, 7909–7915.

Ross, J.F., Chaudhuri, P.K., and Ratnam, M. (1994). Differential regulation of folate receptor isoforms in normal and malignant tissues *in vivo* and in established cell lines – physiological and clinical implications. Cancer *73*, 2432–2443.

Rothmann, T., Hengstermann, A., Whitaker, N.J., Scheffner, M., and zur Hausen, H. (1998). Replication of ONYX-015, a potential anticancer adenovirus, is independent of p53 status in tumor cells. J. Virol. *72*, 9470–9478.

Roy, S., Shirley, P.S., McClelland, A., and Kaleko, M. (1998). Circumvention of immunity to the adenovirus major coat protein hexon. J. Virol. *72*, 6875–6879.

Russell, W.C. (2009). Adenoviruses: update on structure and function. J. Gen. Virol. *90*, 1–20.

Sabbah, A., Chang, T.H., Harnack, R., Frohlich, V., Tominaga, K., Dube, P.H., Xiang, Y., and Bose, S. (2009). Activation of innate immune antiviral responses by Nod2. Nat. Immunol. *10*, 1073-U1049.

Sakhuja, K., Reddy, P.S., Ganesh, S., Cantaniag, F., Pattison, S., Limbach, P., Kayda, D.B., Kadan, M.J., Kaleko, M., and Connelly, S. (2003). Optimization of the generation and propagation of gutless adenoviral vectors. Hum. Gene Ther. *14*, 243–254.

Sargent, K.L., Ng, P., Evelegh, C., Graham, F.L., and Parks, R.J. (2004). Development of a size-restricted pIX-deleted helper virus for amplification of helper-dependent adenovirus vectors. Gene Ther. *11*, 504–511.

Sarukhan, A., Camugli, S., Gjata, B., von Boehmer, H., Danos, O., and Jooss, K. (2001). Successful interference with cellular immune responses to immunogenic proteins encoded by recombinant viral vectors. J. Virol. *75*, 269–277.

Satoh, T., Kato, H., Kumagai, Y., Yoneyama, M., Sato, S., Matsushita, K., Tsujimura, T., Fujita, T., Akira, S., and Takeuchi, O. (2010). LGP2 is a positive regulator of RIG-I- and MDA5-mediated antiviral responses. Proc. Natl. Acad. Sci. U.S.A. *107*, 1512–1517.

Sauthoff, H., Hu, J., Maca, C., Goldman, M., Heitner, S., Yee, H., Pipiya, T., Rom, W.N., and Hay, J.G. (2003). Intratumoral spread of wild-type adenovirus is limited after local injection of human xenograft tumors: Virus persists and spreads systemically at late time points. Hum. Gene Ther. *14*, 425–433.

Savelyeva, I., and Dobbelstein, M. (2011). Infection with E1B-mutant adenovirus stabilizes p53 but blocks p53 acetylation and activity through E1A. Oncogene *30*, 865–875.

Schaack, J., Bennett, M.L., Shapiro, G.S., DeGregori, J., McManaman, J.L., and Moorhead, J.W. (2011). Strong foreign promoters contribute to innate inflammatory responses induced by adenovirus transducing vectors. Virology *412*, 28–35.

Schiedner, G., Morral, N., Parks, R.J., Wu, Y., Koopmans, S.C., Langston, C., Graham, F.L., Beaudet, A.L., and Kochanek, S. (1998). Genomic DNA transfer with a high-capacity adenovirus vector results in improved *in vivo* gene expression and decreased toxicity. Nat. Genet. *18*, 180–183.

Schiedner, G., Hertel, S., Johnston, M., Dries, V., van Rooijen, N., and Kochanek, S. (2003). Selective depletion or blockade of Kupffer cells leads to enhanced and prolonged hepatic transgene expression using high-capacity Adenoviral vectors. Mol. Ther. *7*, 35–43.

Schmitt, C.J., Franke, W.W., Goerdt, S., Falkowska-Hansen, B., Rickelt, S., and Peitsch, W.K. (2007). Homo- and heterotypic cell contacts in malignant melanoma cells and desmoglein 2 as a novel solitary surface glycoprotein. J. Invest. Dermatol. *127*, 2191–2206.

Schnare, M., Holt, A.C., Takeda, K., Akira, S., and Medzhitov, R. (2000). Recognition of CpG DNA is mediated by signalling pathways dependent on the adaptor protein MyD88. Curr. Biol. *10*, 1139–1142.

Schneider, D.B., Fly, C.A., Dichek, D.A., and Geary, R.L. (1998a). Adenoviral gene transfer in arteries of hypercholesterolemic nonhuman primates. Hum. Gene Ther. *9*, 815–821.

Schneider, J., Gilbert, S.C., Blanchard, T.J., Hanke, T., Robson, K.J., Hannan, C.M., Becker, M., Sinden, R., Smith, G.L., and Hill, A.V. S. (1998b). Enhanced immunogenicity for CD8$^+$ T cell induction and complete protective efficacy of malaria DNA vaccination by boosting with modified vaccinia virus Ankara. Nat. Med. *4*, 397–402.

Schumann, M., and Dobbelstein, M. (2006). Adenovirus-induced extracellular signal-regulated kinase phosphorylation during the late phase of infection enhances viral protein levels and virus progeny. Cancer Res. *66*, 1282–1288.

Segerman, A., Arnberg, N., Erikson, A., Lindman, K., and Wadell, G. (2003). There are two different species B adenovirus receptors: sBAR, common to species B1 and B2 adenoviruses, and sB2AR, exclusively used by species B2 adenoviruses. J. Virol. *77*, 1157–1162.

Segura, M.M., Alba, R., Bosch, A., and Chillon, M. (2008). Advances in helper-dependent adenoviral vector research. Curr. Gene Ther. *8*, 222–235.

Sen, G.C. (2001). Viruses and interferons. Ann. Rev. Microbiol. *55*, 255–281.

Seregin, S.S., and Amalfitano, A. (2009). Overcoming pre-existing adenovirus immunity by genetic engineering of adenovirus-based vectors. Expert Opinion on Biological Therapy 9, 1521–1531.

Seregin, S.S., Aldhamen, Y.A., Appledorn, D.M., Schuldt, N.J., McBride, A.J., Bujold, M., Godbehere, S.S., and Amalfitano, A. (2009a). CR1/2 is an important suppressor of Adenovirus-induced innate immune responses and is required for induction of neutralizing antibodies. Gene Ther. *16*, 1245–1259.

Seregin, S.S., Appledorn, D.M., McBride, A.J., Schuldt, N.J., Aldhamen, Y.A., Voss, T., Junping, W., Bujold, M., Godbehere, S., and Amalfitano, A. (2009b). Transient pre-treatment with glucocorticoid ablates innate toxicity of systemically delivered adenoviral vectors without reducing efficacy. Mol. Ther. *17*, S243-S244.

Seregin, S.S., Aldhamen, Y.A., Appledorn, D.M., Hartman, Z.C., Schuldt, N.J., Scott, J., Godbehere, S., Jiang, H.X., Frank, M.M., and Amalfitano, A. (2010a). Adenovirus capsid-display of the retro-oriented human complement inhibitor DAF reduces Ad vector-triggered immune responses *in vitro* and *in vivo*. Blood *116*, 1669–1677.

Seregin, S.S., Hartman, Z.C., Appledorn, D.M., Godbehere, S., Jiang, H.X., Frank, M.M., and Amalfitano, A. (2010b). Novel Adenovirus Vectors 'Capsid-Displaying' a Human Complement Inhibitor. J. Innate Immun. *2*, 353–359.

Sharma, S., tenOever, B.R., Grandvaux, N., Zhou, G.P., Lin, R.T., and Hiscott, J. (2003). Triggering the interferon antiviral response through an IKK-related pathway. Science *300*, 1148–1151.

Shashkova, E.V., Spencer, J.F., Wold, W.S.M., and Doronin, K. (2007). Targeting interferon-alpha increases anti-tumour efficacy and reduces hepatotoxicity of E1A-mutated spread-enhanced oncolytic adenovirus. Mol. Ther. *15*, 598–607.

Shashkova, E.V., Doronin, K., Senac, J.S., and Barry, M.A. (2008a). Macrophage depletion combined with anticoagulant therapy increases therapeutic window of systemic treatment with oncolytic adenovirus. Cancer Res. *68*, 5896–5904.

Shashkova, E.V., Kuppuswamy, M.N., Wold, W.S.M., and Doronin, K. (2008b). Anticancer activity of oncolytic adenovirus vector armed with IFN-alpha and ADP is enhanced by pharmacologically controlled expression of TRAIL. Cancer Gene Ther. *15*, 61–72.

Shayakhmetov, D.M., Papayannopoulou, T., Stamatoyannopoulos, G., and Lieber, A. (2000). Efficient gene transfer into human CD34(+) cells by a retargeted adenovirus vector. J. Virol. *74*, 2567–2583.

Shayakhmetov, D.M., Li, Z.Y., Ternovoi, V., Gaggar, A., Gharwan, H., and Lieber, A. (2003). The interaction between the fiber knob domain and the cellular attachment receptor determines the intracellular trafficking route of adenoviruses. J. Virol. *77*, 3712–3723.

Shayakhmetov, D.M., Gaggar, A., Ni, S.H., Li, Z.Y., and Lieber, A. (2005). Adenovirus binding to blood factors results in liver cell infection and hepatotoxicity. J. Virol. *79*, 7478–7491.

Shen, Y., and Post, L. (2007). Viral Vectors and Their Applications. In Fields Virology, Knipe, D.M., Howley, P.M., Griffin, D.E., Lamb, R.A., Martin, M.A., Roizman, B., and Straus, S.E., eds. (Philadelphia: Lippincott, Williams and Wilkins), pp. 539–564.

Shisler, J., Yang, C., Walter, B., Ware, C.F., and Gooding, L.R. (1997). The adenovirus E3-10.4K/14.5K complex mediates loss of cell surface Fas (CD95) and resistance to fas-induced apoptosis. J. Virol. *71*, 8299–8306.

Silvestry, M., Lindert, S., Smith, J.G., Maier, O., Wiethoff, C.M., Nemerow, G.R., and Stewart, P.L. (2009). Cryo-electron microscopy structure of adenovirus type 2 temperature-sensitive mutant 1 reveals insight into the cell entry defect. J. Virol. *83*, 7375–7383.

Small, E.J., Carducci, M.A., Burke, J.M., Rodriguez, R., Fong, L., van Ummersen, L., Yu, D.C., Aimi, J., Ando, D., Working, P., *et al.* (2006). A phase I trial of intravenous CG7870, a replication-selective, prostate-specific antigen-targeted oncolytic adenovirus, for the treatment of hormone-refractory, metastatic prostate cancer. Mol. Ther. *14*, 107–117.

Smith, T.A.G., Mehaffey, M.G., Kayda, D.B., Saunders, J.M., Yei, S., Trapnell, B.C., McClelland, A., and Kaleko, M. (1993). Adenovirus-mediated expression of therapeutic plasma-levels of human Factor-IX in mice. Nat. Genet. *5*, 397–402.

Soria, C., Estermann, F.E., Espantman, K.C., and O'Shea, C.C. (2010). Heterochromatin silencing of p53 target genes by a small viral protein. Nature *466*, 1076-U1085.

de Souza, A.P., and Bonorino, C. (2009). Tumor immunosuppressive environment: effects on tumor-specific and nontumor antigen immune responses. Exp. Rev. Anticancer Ther. *9*, 1317–1332.

de Souza, A.P.D., Haut, L.H., Silva, R., Ferreira, S., Zanetti, C.R., Ertl, H.C.J., and Pinto, A.R. (2007). Genital CD8(+) T cell response to HIV-1 gag in mice immunized by mucosal routes with a recombinant simian adenovirus. Vaccine *25*, 109–116.

Sova, P., Ren, X.W., Ni, S.H., Bernt, K.M., Mi, J., Kiviat, N., and Lieber, A. (2004). A tumor-targeted and conditionally replicating oncolytic adenovirus vector expressing TRAIL for treatment of liver metastases. Mol. Ther. *9*, 496–509.

Stephen, S.L., Montini, E., Sivanandam, V.G., Al-Dhalimy, M., Kestler, H.A., Finegold, M., Grompe, M., and Kochanek, S. (2010). Chromosomal integration of adenoviral vector DNA *in vivo*. J. Virol. *84*, 9987–9994.

Strunze, S., Trotman, L.C., Boucke, K., and Greber, U.F. (2005). Nuclear targeting of adenovirus type 2 requires CRM1-mediated nuclear export. Mol. Biol. Cell *16*, 2999–3009.

Stupack, E.L.D., Bokoch, G.M., and Nemerow, G.R. (1998). Adenovirus endocytosis requires actin cytoskeleton reorganization mediated by Rho family GTPases. J. Virol. *72*, 8806–8812.

Su, C.Q., Peng, L.H., Sham, J., Wang, X.H., Zhang, Q., Chua, D., Liu, C., Cui, Z.F., Xue, H.B., Wu, H.P., *et al.* (2006). Immune gene-viral therapy with triplex efficacy mediated by oncolytic adenovirus carrying an interferon-gamma gene yields efficient anti-tumor activity in immunodeficient and immunocompetent mice. Mol. Ther. *13*, 918–927.

Su, E.M.J., Stevenson, S.C., Rollence, M., Marshall-Neff, J., and Liau, G. (2001). A genetically modified adenoviral vector exhibits enhanced gene transfer of human smooth muscle cells. J. Vasc. Res. *38*, 471–478.

Subramanian, T., Vijayalingam, S., and Chinnadurai, G. (2006). Genetic identification of adenovirus type 5 genes that influence viral spread. J. Virol. *80*, 2000–2012.

Suomalainen, M., Nakano, M.Y., Keller, S., Boucke, K., Stidwill, R.P., and Geber, U.F. (1999). Microtubule-dependent plus- and minus end-directed motilities are competing processes for nuclear targeting of adenovirus. J. Cell Biol. *144*, 657–672.

Szabolcs, P., Moore, M.A.S., and Young, J.W. (1995). Expansion of immunostimulatory dendritic cells among the myeloid progeny of human CD34(+) bone marrow precursors cultured with C-KIT ligand, granulocyte-macrophage colony-stimulating factor, and TNF-alpha. J. Immunol. *154*, 5851–5861.

Sze, D.Y., Freeman, S.M., Slonim, S.M., Samuels, S.L., Andrews, J.C., Hicks, M., Ahrar, K., Gupta, S., and Reid, T.R. (2003). Dr. Gary J. Becker Young Investigator Award: Intraarterial adenovirus for metastatic gastrointestinal cancer: Activity, radiographic response, and survival. J. Vasc. Interventional Radiol. *14*, 279–290.

Takaoka, A., Wang, Z., Choi, M.K., Yanai, H., Negishi, H., Ban, T., Lu, Y., Miyagishi, M., Kodama, T., Honda, K., *et al.* (2007). DAI (DLM-1/ZBP1) is a cytosolic DNA sensor and an activator of innate immune response. Nature *448*, 501-U514.

Takeuchi, O., Hoshino, K., Kawai, T., Sanjo, H., Takada, H., Ogawa, T., Takeda, K., and Akira, S. (1999). Differential roles of TLR2 and TLR4 in recognition of Gram-negative and Gram-positive bacterial cell wall components. Immunity *11*, 443–451.

Tang, Y.Z., Wu, H.J., Ugai, H., Matthews, Q.L., and Curiel, D.T. (2009). Derivation of a Triple mosaic adenovirus for cancer gene therapy. Plos One *4*, e8526.

Tao, N.J., Gao, G.P., Parr, M., Johnston, J., Baradet, T., Wilson, J.M., Barsoum, J., and Fawell, S.E. (2001). Sequestration of adenoviral vector by Kupffer cells leads to a nonlinear dose response of transduction in liver. Mol. Ther. *3*, 28–35.

Tatsis, N., Lasaro, M.O., Lin, S.W., Xiang, Z.Q., Zhou, D.M., DiMenna, L., Li, H., Bian, A., Abdulla, S., Li, Y., *et al.* (2009). Adenovirus vector-induced immune responses in nonhuman primates: responses to prime boost regimens. J. Immunol. *182*, 6587–6599.

Thomas, M.A., Spencer, J.F., Toth, K., Sagartz, J.E., Phillips, N.J., and Wold, W.S. M. (2008). Immunosuppression enhances oncolytic adenovirus replication and anti-tumour efficacy in the Syrian hamster model. Mol. Ther. *16*, 1665–1673.

Tian, J., Xu, Z.L., Smith, J.S., Hofherr, S.E., Barry, M.A., and Byrnes, A.P. (2009). Adenovirus activates complement by distinctly different mechanisms *in vitro* and *in vivo*: indirect complement activation by virions *in vivo*. J. Virol. *83*, 5648–5658.

Tibbles, L.A., Spurrell, J.C.L., Bowen, G.P., Liu, Q., Lam, M., Zaiss, A.K., Robbins, S.M., Hollenberg, M.D., Wickham, T.J., and Muruve, D.A. (2002). Activation of p38 and ERK signalling during adenovirus vector cell entry lead to expression of the C-X-C chemokine IP-10. J. Virol. *76*, 1559–1568.

Todo, T., Rabkin, S.D., Chahlavi, A., and Martuza, R.L. (1999). Corticosteroid administration does not affect viral oncolytic activity, but inhibits anti-tumor immunity in replication-competent herpes simplex virus tumor therapy. Hum. Gene Ther. *10*, 2869–2878.

Tollefson, A.E., Toth, K., Doronin, K., Kuppuswamy, M., Doronina, O.A., Lichtenstein, D.L., Hermiston, T.W., Smith, C.A., and Wold, W.S. M. (2001). Inhibition of TRAIL-induced apoptosis and forced internalization of TRAIL receptor 1 by adenovirus proteins. J. Virol. *75*, 8875–8887.

Toth, K., and Wold, W.S. M. (2010). Increasing the efficacy of oncolytic adenovirus vectors. Viruses-Basel *2*, 1844–1866.

Toth, K., Tarakanova, V., Doronin, K., Ward, P., Kuppuswamy, M., Locke, J.E., Dawson, J.E., Kim, H.J., and Wold, W.S. M. (2003). Radiation increases the activity of oncolytic adenovirus cancer gene therapy vectors that overexpress the ADP (E3–11.6K) protein. Cancer Gene Ther. *10*, 193–200.

Toth, K., Kuppuswamy, M., Shashkova, E., Spencer, J.F., and Wold, W.S. M. (2010). A fully replication-competent adenovirus vector with enhanced oncolytic properties. Cancer Gene Ther. *17*, 761–770.

Travassos, L.H., Carneiro, L.A.M., Ramjeet, M., Hussey, S., Kim, Y.G., Magalhaes, J.G., Yuan, L., Soares, F., Chea, E., Le Bourhis, L., *et al.* (2010). Nod1 and Nod2 direct autophagy by recruiting ATG16L1 to the plasma membrane at the site of bacterial entry. Nat. Immunol. *11*, 55-U67.

Tripathy, S.K., Goldwasser, E., Lu, M.M., Barr, E., and Leiden, J.M. (1994). Stable delivery of physiological levels of recombinant erythropoietin to the systemic circulation by intramuscular injection of replication-defective adenovirus. Proc. Natl. Acad. Sci. U.S.A. *91*, 11557–11561.

Trojan, L., Schaaf, A., Steidler, A., Haak, M., Thalmann, G., Knoll, T., Gretz, N., Alken, P., and Michel, M.S. (2005). Identification of metastasis-associated genes in prostate cancer by genetic profiling of human prostate cancer cell lines. Anticancer Res. *25*, 183–191.

Turnell, A.S., Grand, R.J.A., and Gallimore, P.H. (1999). The replicative capacities of large E1B-null group A and group C adenoviruses are independent of host cell p53 status. J. Virol. *73*, 2074–2083.

Tuve, S., Wang, H.J., Ware, C., Liu, Y., Gaggar, A., Bernt, K., Shayakhmetov, D., Li, Z.Y., Strauss, R., Stone, D., *et al.* (2006). A new group B adenovirus receptor is expressed at high levels on human stem and tumor cells. J. Virol. *80*, 12109–12120.

Uematsu, S., Sato, S., Yamamoto, M., Hirotani, T., Kato, H., Takeshita, F., Matsuda, M., Coban, C., Ishii, K.J., Kawai, T., *et al.* (2005). Interleukin-1 receptor-associated kinase-1 plays an essential role for Toll-like receptor (TLR)7- and TLR9-mediated interferon-alpha induction. J. Exp. Med. *201*, 915–923.

Umana, P., Gerdes, C.A., Stone, D., Davis, J.R.E., Ward, D., Castro, M.G., and Lowenstein, P.R. (2001). Efficient FLPe recombinase enables scalable production of helper-dependent adenoviral vectors with negligible helper-virus contamination. Nat. Biotechnol. *19*, 582–585.

Underhill, D.M., Ozinsky, A., Hajjar, A.M., Stevens, A., Wilson, C.B., Bassetti, M., and Aderem, A. (1999). The Toll-like receptor 2 is recruited to macrophage phagosomes and discriminates between pathogens. Nature *401*, 811–815.

Unterholzner, L., Keating, S.E., Baran, M., Horan, K.A., Jensen, S.B., Sharma, S., Sirois, C.M., Jin, T.C., Latz, E., Xiao, T.S., *et al.* (2010). IFI16 is an innate immune sensor for intracellular DNA. Nat. Immunol. *11*, 997-U942.

Varnavski, A.N., Calcedo, R., Bove, M., Gao, G., and Wilson, J.M. (2005). Evaluation of toxicity from high-dose systemic administration of recombinant adenovirus vector in vector-naive and pre-immunized mice. Gene Ther. *12*, 427–436.

Vellinga, J., Rabelink, M., Cramer, S.J., van den Wollenberg, D.J.M., Van der Meulen, H., Leppard, K.N., Fallaux, F.J., and Hoeben, R.C. (2004). Spacers increase the accessibility of peptide ligands linked to the carboxyl terminus of adenovirus minor capsid protein IX. J. Virol. *78*, 3470–3479.

Vetrini, F., and Ng, P. (2010). Gene therapy with helper-dependent adenoviral vectors: current advances and future perspectives. Viruses-Basel 2, 1886–1917.

Waddington, S.N., McVey, J.H., Bhella, D., Parker, A.L., Barker, K., Atoda, H., Pink, R., Buckley, S.M.K., Greig, J.A., Denby, L., *et al.* (2008). Adenovirus serotype 5 hexon mediates liver gene transfer. Cell *132*, 397–409.

Wadell, G., Allard, A., Evander, M., and Li, Q.G. (1986). Genetic-variability and evolution of adenoviruses. Chem. Scripta *26B*, 325–335.

Walport, M.J. (2001a). Advances in immunology: Complement (First of two parts). New Engl. J. Med. *344*, 1058–1066.

Walport, M.J. (2001b). Advances in immunology: Complement (Second of two parts). New Engl. J. Med. *344*, 1140–1144.

Wang, H., Li, Z. -Y., Liu, Y., Persson, J., Beyer, I., Moller, T., Koyuncu, D., Drescher, M.R., Strauss, R., Zhang, X. -B., *et al.* (2010). Desmoglein 2 is a receptor for adenovirus serotypes 3, 7, 11 and 14. Nat. Med. *17*, 96–104.

Weitman, S.D., Lark, R.H., Coney, L.R., Fort, D.W., Frasca, V., Zurawski, V.R., and Kamen, B.A. (1992). Distribution of the folate receptor GP38 in normal and malignant cell lines and tissues. Cancer Res. *52*, 3396–3401.

Whyte, P., Buchkovich, K.J., Horowitz, J.M., Friend, S.H., Raybuck, M., Weinberg, R.A., and Harlow, E. (1988). Association between an oncogene and an anti-oncogene – the adenovirus E1A proteins bind to the retinoblastoma gene-product. Nature *334*, 124–129.

Wickham, T.J., Mathias, P., Cheresh, D.A., and Nemerow, G.R. (1993). Integrin alpha-v-beta-3 and integrin alpha-v-beta-5 promote adenovirus internalization but not virus attachment. Cell 73, 309–319.

Wickham, T.J., Filardo, E.J., Cheresh, D.A., and Nemerow, G.R. (1994). Integrin alpha-v-beta-5 selectively promotes adenovirus-mediated cell-membrane permeabilization. J. Cell Biol. *127*, 257–264.

Wickham, T.J., Roelvink, P.W., Brough, D.E., and Kovesdi, I. (1996). Adenovirus targeted to heparan-containing receptors increases its gene delivery efficiency to multiple cell types. Nat. Biotechnol. *14*, 1570–1573.

Wickham, T.J., Tzeng, E., Shears, L.L., Roelvink, P.W., Li, Y., Lee, G.M., Brough, D.E., Lizonova, A., and Kovesdi, I. (1997). Increased *in vitro* and *in vivo* gene transfer by adenovirus vectors containing chimeric fiber proteins. J. Virol. *71*, 8221–8229.

Wiethoff, C.M., Wodrich, H., Gerace, L., and Nemerow, G.R. (2005). Adenovirus protein VI mediates membrane disruption following capsid disassembly. J. Virol. *79*, 1992–2000.

Wilderman, M.J., Kim, S., Gillespie, C.T., Sun, J., Kapoor, V., Vachani, A., Sterman, D.H., Kaiser, L.R., and Albelda, S.M. (2006). Blockade of TNF-alpha decreases both inflammation and efficacy of intrapulmonary Ad. IFN beta immunotherapy in an orthotopic model of bronchogenic lung cancer. Mol. Ther. *13*, 910–917.

Wodrich, H., Henaff, D., Jammart, B., Segura-Morales, C., Seelmeir, S., Coux, O., Ruzsics, Z., Wiethoff, C.M., and Kremer, E.J. (2010). A capsid-encoded PPxY-motif facilitates adenovirus entry. Plos Pathog. *6*, e1000808.

Wolins, N., Lozier, J., Eggerman, T.L., Jones, E., Aguilar-Cordova, E., and Vostal, J.G. (2003). Intravenous administration of replication-incompetent adenovirus to rhesus monkeys induces thrombocytopenia by increasing *in vivo* platelet clearance. Br. J. Haematol. *123*, 903–905.

Wonganan, P., and Croyle, M.A. (2010). PEGylated adenoviruses: from mice to monkeys. Viruses-Basel 2, 468–502.

Worgall, S., Leopold, P.L., Wolff, G., Ferris, B., VanRoijen, N., and Crystal, R.G. (1997a). Role of alveolar macrophages in rapid elimination of adenovirus vectors administered to the epithelial surface of the respiratory tract. Hum. Gene Ther. *8*, 1675–1684.

Worgall, S., Wolff, G., FalckPedersen, E., and Crystal, R.G. (1997b). Innate immune mechanisms dominate elimination of adenoviral vectors following *in vivo* administration. Hum. Gene Ther. *8*, 37–44.

Worgall, S., Krause, A., Rivara, M., Hee, K.K., Vintayen, E.V., Hackett, N.R., Roelvink, P.W., Bruder, J.T., Wickham, T.J., Kovesdi, I., *et al.* (2005). Protection against P. aeruginosa with an adenovirus vector containing an OprF epitope in the capsid. J. Clin. Invest. *115*, 1281–1289.

Wu, H.J., Dmitriev, I., Kashentseva, E., Seki, T., Wang, M.H., and Curiel, D.T. (2002a). Construction and characterization of adenovirus serotype 5 packaged by serotype 3 hexon. J. Virol. *76*, 12775–12782.

Wu, H.J., Seki, T., Dmitriev, I., Uil, T., Kashentseva, E., Han, T., and Curiel, D.T. (2002b). Double modification of adenovirus fiber with RGD and polylysine motifs improves coxsackievirus-adenovirus receptor-independent gene transfer efficiency. Hum. Gene Ther. *13*, 1647–1653.

Xia, H.B., Anderson, B., Mao, Q.W., and Davidson, B.L. (2000). Recombinant human adenovirus: Targeting to the human transferrin receptor improves gene transfer to brain microcapillary endothelium. J. Virol. *74*, 11359–11366.

Xiang, Z.Q., Gao, G.P., Reyes-Sandoval, A., Li, Y., Wilson, J.M., and Ertl, H.C. J. (2003). Oral vaccination of mice with adenoviral vectors is not impaired by preexisting immunity to the vaccine carrier. J. Virol. *77*, 10780–10789.

Xu, Y.W., Tao, X., Shen, B.H., Horng, T., Medzhitov, R., Manley, J.L., and Tong, L. (2000). Structural basis for signal transduction by the Toll/interleukin-1 receptor domains. Nature *408*, 111–115.

Yamaguchi, T., Kawabata, K., Kouyama, E., Ishii, K.J., Katayama, K., Suzuki, T., Kurachi, S., Sakurai, F., Akira, S., and Mizuguchi, H. (2010). Induction of type I interferon by adenovirus-encoded small RNAs. Proc. Natl. Acad. Sci. U.S.A. *107*, 17286–17291.

Yamamoto, M., Sato, S., Hemmi, H., Sanjo, H., Uematsu, S., Kaisho, T., Hoshino, K., Takeuchi, O., Kobayashi, M., Fujita, T., *et al.* (2002). Essential role for TIRAP in activation of the signalling cascade shared by TLR2 and TLR4. Nature *420*, 324–329.

Yamamoto, M., Sato, S., Hemmi, H., Hoshino, K., Kaisho, T., Sanjo, H., Takeuchi, O., Sugiyama, M., Okabe, M., Takeda, K., *et al.* (2003a). Role of adaptor TRIF in the MyD88-independent toll-like receptor signalling pathway. Science *301*, 640–643.

Yamamoto, M., Sato, S., Hemmi, H., Uematsu, S., Hoshino, K., Kaisho, T., Takeuchi, O., Takeda, K., and Akira, S. (2003b). TRAM is specifically involved in the Toll-like receptor 4-mediated MyD88-independent signalling pathway. Nature Immunology *4*, 1144–1150.

Yang, Y.P., Li, Q., Ertl, H.C.J., and Wilson, J.M. (1995). Cellular and humoral immune-responses to viral-antigens create barriers to lung-directed gene-therapy with recombinant adenoviruses. J. Virol. *69*, 2004–2015.

Yew, P.R., Liu, X., and Berk, A.J. (1994). Adenovirus E1B oncoprotein tethers a transcriptional repression domain to p53. Genes Dev. *8*, 190–202.

Yoneyama, M., Suhara, W., Fukuhara, Y., Fukuda, M., Nishida, E., and Fujita, T. (1998). Direct triggering of the type I interferon system by virus infection: activation of a transcription factor complex containing IRF-3 and CBP/p300. Embo J. *17*, 1087–1095.

Yoneyama, M., Kikuchi, M., Natsukawa, T., Shinobu, N., Imaizumi, T., Miyagishi, M., Taira, K., Akira, S., and Fujita, T. (2004). The RNA helicase RIG-I has an essential function in double-stranded RNA-induced innate antiviral responses. Nat. Immunol. *5*, 730–737.

Yoneyama, M., Kikuchi, M., Matsumoto, K., Imaizumi, T., Miyagishi, M., Taira, K., Foy, E., Loo, Y.M., Gale, M., Akira, S., *et al.* (2005). Shared and unique functions of the DExD/H-box helicases RIG-I, MDA5, and LGP2 in antiviral innate immunity. J. Immunol. *175*, 2851–2858.

Yoo, J.Y., Kim, J.H., Kwon, Y.G., Kim, E.C., Kim, N.K., Choi, H.J., and Yun, C.O. (2007). VEGF-specific short hairpin RNA-expressing oncolytic adenovirus elicits potent inhibition of angiogenesis and tumor growth. Mol. Ther. *15*, 295–302.

Youil, R., Toner, T.J., Su, Q., Chen, M.C., Tang, A.M., Bett, A.J., and Casimiro, D. (2002). Hexon gene switch strategy for the generation of chimeric recombinant adenovirus. Hum. Gene Ther. *13*, 311–320.

Yu, S.Q., Feng, X., Shu, T., Matano, T., Hasegawa, M., Wang, X.L., Ma, H.T., Li, H.X., Li, Z.L., and Zeng, Y. (2008). Potent specific immune responses induced by prime-boost-boost strategies based on DNA, adenovirus, and Sendai virus vectors expressing gag gene of Chinese HIV-1 subtype B. Vaccine *26*, 6124–6131.

Zhang, Y., Chirmule, N., Gao, G.P., Qian, R., Croyle, M., Joshi, B., Tazelaar, J., and Wilson, J.M. (2001). Acute cytokine response to systemic adenoviral vectors in mice is mediated by dendritic cells and macrophages. Mol. Ther. *3*, 697–707.

Zhou, R.B., Tardivel, A., Thorens, B., Choi, I., and Tschopp, J. (2010). Thioredoxin-interacting protein links oxidative stress to inflammasome activation. Nat. Immunol. *11*, 136-U151.

Zou, J.T., Yao, Z.B., Zhang, G., Wang, H.Q., Xu, J., Yew, D.T., and Forster, E.L. (2008). Vaccination of Alzheimer's model mice with adenovirus vector containing quadrivalent foldable A beta (1–15) reduces A beta burden and behavioral impairment without A beta-specific T cell response. J. Neurol. Sci. *272*, 87–98.

Index

A

AAA^{+} (ATPases associated with various cellular activities) 206, 224
Adenovirus 175, 184–187, 263–266, 281–286
- E1A 102–103, Fig. 6.1, 285
- E1B 265–266, 285–286, 294–296
- genome 282–283, Fig. 14.2
- immune response 287, 293
- life cycle 283–285
- non-attenuated 296
- oncolytic 281–297, 301
- replication deficient 285–286
- structure 282, Fig. 14.1
- vectors 285–287
- serotypes 299

ALT *see* Alternative lengthening of telomeres
Alternative lengthening of telomeres (ALT) 245, Fig. 12.4, 261, 267
Amplification *see* HPV amplification
Aneuploidy 184, 187–188, 239, 248
Angelman syndrome 83–84
Angiogenesis 302
Apoptosis 10–12, 43, 63–64, 82, 88, 176, 179, 260–261, 271–273, Fig. 13.3, 302
- activating factor (Apaf1) 10

Arenavirus 270
Ataxia-telangiectasia mutated (ATM) 175–176, 177–188, Fig. 9.2, Fig. 9.3, 227, 230–231, 243
ATM and Rad3-related (ATR) 175–178, Fig. 9.1, Fig. 9.2, 230, 248, 258, 260
ATR *see* ATM and Rad3-related
Aurora kinase 140, 189, 248, 261

B

BK virus (BKV) 211, Fig. 10.14, 267–268, 271, Fig. 13.3
Brd4 43, 87, 128, 133–138
Bub1 136, Table 7.1, 182, Fig. 9.3

C

CAR *see* Coxsackie and adenovirus receptor
Casein kinase 2 (casein kinase II/ CK2/ CKII) 9, 102–103, Fig. 6.1, 103, Fig. 6.2, 107–108, 113, 139
Caspase 10–11, 189, 219, Fig. 11.1, 230, 260, 291–292, Fig. 14.5
CDK *see* Cyclin-dependent kinase
CDKI *see* Cyclin-dependent kinase inhibitor
Cell cycle 7–8, Fig. 1.4
Centriole 188, 241–243, Fig. 12.2
Centrosome 10, 134, 188, 241–243, 261
Cervarix 13
Cervical cancer 1, 6, 12–13, 37, 55, 72, 84, 115–117, 151, 153–156, 159–162
Cervical glandular intraepithelial neoplasia (CGIN) 5–6
Cervical intraepithelial neoplasia (CIN) 5–6, Fig. 1.3, 9, 12, 57, 117, 156
Chaperone 60, Table 4.1, 114, 196
Chk2 230–231, 177–181, Fig. 9.1, Fig. 9.2, 188, 258, 260
ChlR1 43–44, 143–136, Table 7.1
Chromosomal instability *see* Genome instability
Claspin 177–178, 189, 248
Cleavage and polyadenylation specificity factor (CPSF) 41, 47
Cleavage stimulatory factor (CstF) 41
C-myc 153, 246, 261, 263
Complement 287, 292–293, 299
Cottontail rabbit papillomavirus (CRPV) 21, 27–28, 138
Coxsackie and adenovirus receptor (CAR) 283–284, Fig. 14.3
CRPV *see* Cottontail rabbit papillomavirus
Cyclin-dependent kinase (CDK) 7–10, 113, 117–118, 189, 241–245
Cyclin-dependent kinase inhibitor (CDKI / CKI) 7–10, 62
Cyclo-oxygenase-2 (Cox-2) 64
Cytokine 281, 288–294, Fig. 14.4, Fig. 14.5

D

DAI *see* DNA-dependent activator of interferon
Daxx 259, 261, 265–266, 269
Dendritic cells (DCs) 117
Dicer 152–153
DNA
- damage 9–11, 89, 170, 175, 177–181, Fig. 9.1, Fig. 9.2, 181–185, 188–189, 230–232, Fig. 11.5, 243–245, Fig.12.4, 260
- damage response (DDR) 175, 181, 188–189, 229–231, 230

melting 207–210, 205, 221–225, Fig. 11.3
replication 176, 195–198, 211–212, 217–233, 245
unwinding *see* DNA melting
DNA-damage induced kinase (HIPK2) 258–260
DNA polymerase α-primase 183, 195, 225,
DNA-dependent activator of interferon (DAI) 290–292
DNA-dependent protein kinase (DNA-PK) 176–177, Fig. 9.1, 179–185
DNA-PK *see* DNA-dependent protein kinase
Double-strand DNA breaks (DSBs) 176–180, 186, 230
Drosha 152

E

E1 *see* HPV E1 protein
E1^E4 *see* HPV E1^E4 protein
E1A *see* Adenovirus E1A protein
E1B *see* Adenovirus E1B protein
E2 *see* HPV E2 protein
E4 *see* HPV E4 protein
E5 *see* HPV E5 protein
E6 *see* HPV E6 protein
E6* *see* HPV E6* protein
E6-associated protein (E6AP/E6-AP) 9–11, 75–88, Fig. 5.5
E7 *see* HPV E7 protein
E8^E2 *see* HPV E8^E2 protein
EBNA1 *see* Epstein-Barr nuclear antigen 1
eIF4E 259, 262
Epidermal growth factor (EGF) 59–62, 159
Epidermal growth factor receptor (EGFR) 56, Fig. 4.1, 59–62, Table 4.1, Fig 4.2, 159
Epidermodysplasia 3, Table 1.1, 55, 63
Epigenetic 249
Episome 4, 72, 128, 218
Epstein–Barr nuclear antigen 1 (EBNA1) 127, Fig. 7.1, 129–132, Fig. 7.4, 136–139, Fig. 7.5, 269, 270
Epstein–Barr virus (EBV) 126–143, 269, 271

F

Fanconi anaemia (FA) 178, 189, 243–245, Fig. 12.4
Fas-associated death domain (FADD) 10–11, 290

G

Gardasil 13
Genome instability 12, 152, 239–249, 180–184, 187–190

H

HAT *see* Histone acetyltransferase
HDAC *see* Histone deacetylase
Head and neck squamous cell carcinoma (HNSCC) 6, 38, 156, 189
Helicase 2, 38–39, Table 3.1, 127–128, 195–198, 203–210, 217–225, Fig. 11.1
Herpes simplex type 1 (HSV1) 126, 265, 268–272
Herpesviruses 126–141, 268
High-risk HPV *see* HPV
Histone acetyltransferase (HAT) 86–87, 165, 249
Histone deacetylase (HDAC) 89, Fig. 5.6, 100–103, Table 6.1, 111–113, 165, 249
HPV (human papillomavirus)
amplification 5, 39, 58, 125, 188, 229–232
E1 protein 2, 38–39, Fig. 3.1, Table 3.1, 127, 142, 188, 217–232, Fig. 11.1, Fig. 11.3, Fig. 11.4
E1^E4 protein 2, 5, 11, 24–25, 39, Table 3.1, 40, Fig. 3.2, 45–46, 58
E2 protein 2, 11–12, 37–39, Fig. 3.1, Table 3.1, 40–47, Fig. 3.3, Fig. 3.4, Fig. 3.5, 113–114, 188, 217–225, Fig. 11.2
E2C protein 42, 114
E4 protein *see* E1^E4
E5 protein 55–65, Fig. 4.1, 72
E6 protein 8–9, 11–12, 40–42, 188–189, 240–247
and BAK 88
binding to E2 45
binding to E6AP 75–77
binding to p53 81–83
binding to PDZ proteins 79–80
binding to PML 267
evolutionary origin 73–74
p53 binding 81–83
structure 71–90
E6* protein 80–81
E7 protein 9–12, 99–116, 188–189, 240–249
and PML 267
binding to E2 45, 113
binding to HDACs 112–113
binding to pocket proteins 101, Table 6.1, 108–110, 111–112
chaperone holdase activity 114
degradation 110
localisation 114–116
phosphorylation 107–108
soluble oligomers 114
structure 102–107
E8^E2 protein 39–44, 128, 141
genome 2–11, 19–27, 37–41, Fig. 3.1, 73, Fig. 5.2, 125–144
proteins 39, Table 3.1
high risk 2–3, 11–13, 58–64, 74–87, 111, 156, 239–249
L1 protein 5, 13, 20, Table 2.1, 23–30, 48
LCR/URR (long control region/upstream regulatory region) 38, Fig. 3.1, 42–44, 218
low risk 2–3, Table 1.1, 11–12, 79–83, 87, 111–112, 151, 163
Human papillomavirus *see* HPV

I

ICP0 183, 268–269
IFN *see* Interferon
Immune modulation 299
Immunosuppression 293, 296
Integration 11, 41–45, Fig. 3.3, 73, Fig. 5.2, 114, 143, 231–232, 243, 249
Interferon (IFN) 113, 116, 226–127, 261–265, 268, 272–273, 281, 288–290, Fig. 14.4, Fig. 14.5
Integrin 4, 20, 44, 161, 284, Fig. 14.3, 298
Intrinsic disorder 105–106, Fig. 6.3

J

Jagged 88–89, Fig. 5.6
JC polyomavirus (JCV) 184, 211, Fig. 10.14, 267, 271
J-domain 196, 210

K

Kaposi's sarcoma-associated herpesvirus (KSHV) 126, 129–10, 132–133, 135–138, Table 7.1, 142
Koilocytes 5, Fig. 1.3, 62

L

L1 *see* HPV L1 protein
LANA 127, Fig.7.1, 129–133, Fig. 7.4, 135–140, Fig. 7.5, 142
Large T antigen *see* SV40 large T antigen
Lassa fever virus 270
LCR/URR *see* HPV LCR/URR
Low-risk HPV *see* HPV
LxCxE motif 73, 75, 88, 102–103, 109, Fig. 6.5, 111–114, 189
Lymphocytic choriomeningitis virus (LCMV) 270–271, Fig. 13.3

M

Macrophage depletion 294
MAGUK 9
Major histocompatibility complex (MHC) 56, 60–63, 262, 294
Matrix attachment region (MAR) 132, 262
MHC *see* major histocompatibility complex
miRNA (microRNA) 151–166
Mitotic accumulations of PML (MAPPs) 256
Mitotic spindle 101, Table 6.1, 134, 136, Table 7.1, 188, 241, 248
Mouse polyomavirus (MPyV) 183–184
Mre11-Rad50-Nbs1 (MRN) 175, 177–179, Fig. 9.2, 185–186, 265
MRN *see* Mre11-Rad50-Nbs1
mRNA stability 48

N

NHEJ *see* Non-homologous end-joining
NOD-like receptors (NLRs) 287, 291–292, Fig. 14.5
Non-homologous end-joining (NHEJ) 179, 180, 186
Notch 88–90, Fig. 5.6
Nuclear matrix 46, 132, 142

O

ONYX-015 286, 295, 296, 302
Organotypic 19–24, 29
Oropharyngeal squamous cell carcinoma (OPSCC) 156–157

P

p53 tumour suppressor protein (TP53) 7–9, 11–12, 44, 74–76, 78–79, 81–83, 86–87, 177–178, 182–184, 186, 189, 229, 240, 247, 295,
- and PML 260–261

PAMPS *see* Pathogen-associated molecular patterns
Pap (Papanicolaou) smear 6, 62, 116
Papanicolaou *see* Pap smear
Partitioning *see* Viral partitioning
Pathogen-associated molecular patterns (PAMPs) 287, 292
Pattern recognition receptors (PRRs) 281, 287, 290–292
PDZ proteins *see* HPV E6 protein
PML *see* Promyelocytic leukaemia protein
PML nuclear bodies *see* Promyelocytic leukaemia protein
PML oncogenic domains (PODS) *see* Promyelocytic leukaemia protein
Polyomavirus 175, 181–183, 211–212, 267–268
Polyploidy 248–249
pRB *see* Retinoblastoma tumour suppressor protein
Promyelocytic leukaemia protein (PML) 126, 247, 142, 183, 256, 263, 271, Fig. 13.3
- isoforms 256, 257, Fig. 13.1
- nuclear bodies
- PODS

R

Retinoblastoma tumour suppressor protein (pRb) 9–10, 74–75, 100–105, 189, 241
- degradation 108–110
- E7 interaction 111–113
- related proteins 10, 100–101, Table 6.1

RIG-1-like receptors (RLRs) 287, 290

S

Serine-arginine (SR)-rich proteins 47 -49
Single-strand DNA breaks (SSBs) 176, 230
Splicing 40, Fig. 3.2, 45, 47
Squamous cell carcinoma (SCC) 1, 6–7, 38, 156–157, 161–164
Squamous cell carcinoma of the head and neck (SCCHN) 156–158, Table 8.3, Fig. 8.2
SR proteins *see* Serine-arginine (SR)-rich proteins
SV40 large T antigen (T/T-Ag) 81, 102, 103, 114, 181, 184, 195–212
- helicase 195–212
- OBD (origin binding domain) 196 -212
- spiral hexamer 201–203

T

T antigen *see* SV40 large T antigen
Telomerase reverse transcriptase (hTERT) 9, 246, 261
TERT *see* Telomerase reverse transcriptase
Toll-like receptors (TLRs) 287
TopBP1 43, 134–135, 177, 186
Topoisomerase 224 195, 224

U

Ubc9 226, 228, 232, 265
Ubiquitination 82–84, 108, 110, 228, 258
URR *see* HPV long control region

V

Vaccine 12–13, 116–117, 143, 286, 299–300
Vacuolar ATPase 60–61
Varicella zoster virus (VZV) 126, 269–273
Viral partitioning 127–141, Fig. 7.2

Viroporin 57, 65
Virus-like particle (VLP) 13, 20, 28–30
VRX-007 295–296
Vulvar intraepithelial neoplasia (VIN) 5

W

Warts 1 1–2, 19–22, Table 2.2, 29, 37

X

Xenograft 20, 22, 28, 295, 301, 302

Other Books of Interest